Changing Facets *of* Nuclear Structure

Changing Facets *of* Nuclear Structure

Proceedings of the 9th International Spring Seminar on Nuclear Physics

Vico Equense, Italy 20 – 24 May 2007

editor

Aldo Covello

Dipartimento di Scienze Fisiche
Università di Napoli Federico II

World Scientific

NEW JERSEY • LONDON • SINGAPORE • BEIJING • SHANGHAI • HONG KONG • TAIPEI • CHENNAI

Published by

World Scientific Publishing Co. Pte. Ltd.
5 Toh Tuck Link, Singapore 596224
USA office: 27 Warren Street, Suite 401-402, Hackensack, NJ 07601
UK office: 57 Shelton Street, Covent Garden, London WC2H 9HE

British Library Cataloguing-in-Publication Data
A catalogue record for this book is available from the British Library.

CHANGING FACETS OF NUCLEAR STRUCTURE
Proceedings of the 9th International Spring Seminar on Nuclear Physics

ISBN-13 978-981-277-902-1
ISBN-10 981-277-902-7

Printed in Singapore by World Scientific Printers

LOCAL ORGANIZING COMMITTEE

A. Covello, Seminar Chairman

A. Gargano, Scientific Secretary

F. Andreozzi

L. Coraggio

N. Itaco

N. Lo Iudice

D. Pierroutsakou

A. Porrino

INTERNATIONAL ADVISORY COMMITTEE

C. Baktash (*Oak Ridge*)

A. Bracco (*Milano*)

B. A. Brown (*Michigan*)

P. Butler (*Liverpool*)

R. Casten (*Yale*)

J. P. Draayer (*Baton Rouge*)

L. S. Ferreira (*Lisboa*)

S. Gales (*GANIL*)

K. Gelbke (*Michigan*)

F. Iachello (*Yale*)

R. Jolos (*Dubna*)

K. Langanke (*GSI*)

R. Liotta (*Stockholm*)

Jie Meng (*Beijing*)

T. Motobayashi (*RIKEN*)

G. Orlandini (*Trento*)

A. Polls (*Barcelona*)

A. A. Raduta (*Bucharest*)

A. Richter (*Darmstadt*)

C. Signorini (*Padova*)

I. Talmi (*Rehovot*)

J. P. Vary (*Iowa*)

SPONSORS OF THE SEMINAR

Università di Napoli "Federico II"
Istituto Nazionale di Fisica Nucleare

HOST TO THE SEMINAR

Dipartimento di Scienze Fisiche, Università "Federico II"

FOREWORD

The Ninth International Spring Seminar on Nuclear Physics was held in Vico Equense from May 20 to May 24, 2007. This Seminar was the ninth in a series of topical meetings to be held every two or three years in the Naples area. The series began with the Sorrento meeting in 1986 and continued with the Capri meeting in 1988, the Ischia meeting in 1990, the Amalfi meeting in 1992, the Ravello meeting in 1995, the S. Agata meeting in 1998, the Maiori meeting in 2001, and the Paestum meeting in 2004.

For this ninth meeting we came back to the Sorrento area and met in the small town of Vico Equense. What remained invariant is the aim of these meetings, which is to discuss recent advances and new perspectives in nuclear structure experiment and theory in a pleasant and friendly atmosphere.

Nuclear structure studies of exotic nuclei are currently being performed in several laboratories where beams of radioactive nuclei are available. Meanwhile the development of new facilities, which will provide high-intensity beams, is in progress or under discussion in Europe, Asia and North America. At this meeting we had a comprehensive overview of this fascinating field and of future scenarios thanks to the participation of leaders of the most important projects.

Nuclear structure is pushing ahead its frontiers. On the one hand, the new exciting results of spectroscopic studies of nuclei far from stability is giving impetus to theoretical studies of possible changes of nuclear structure. On the other hand, sustained efforts are being made to understand the properties of nuclei in terms of the basic interactions between the constituents. This means that a truly microscopic theory of nuclear structure is on the way.

As usual, the program of the meeting consisted of general talks and of more specialized seminars, the latter including most of the contributions submitted by participants. The speakers covered five main topics: i) Exotic Nuclear Structure; ii) Nuclear Structure and Nuclear Forces; iii) Shell Model and Nuclear Structure; iv) Collective Modes of Nuclear Excitation; v) Special Topics. This volume contains the invited papers and all oral and poster contributions considered relevant to the Seminar according to the judgment of the Advisory Committee. The actual program of the Seminar is also included to give an idea of how it was organized. We are confident that the high quality of both invited and contributed papers collected in these Proceedings will be appreciated by the nuclear physics community.

As was the case for most of the previous Seminars, the Vico Equense Seminar too ended with a Round Table Discussion on the theme "Trends and Perspectives in Nuclear Structure". G. de Angelis, F. Azaiez, C. Baktash, J. P. Draayer, A. Faessler and I. Talmi kindly agreed to be on the panel and their remarks were essential in bringing about the active involvement of the audience.

The Conference had more than 80 participants coming from 20 countries. This is well in line with the tradition of these meetings, as is the fact that about 60% of the present participants attended one or more of the previous Seminars.

I gratefully acknowledge the financial support of the Istituto Nazionale di Fisica Nucleare and the University of Naples Federico II who helped make the Seminar possible. I also acknowledge the support provided in various ways by the Dipartimento di Scienze Fisiche which acted as host to the Seminar.

Aldo Covello

9th INTERNATIONAL SPRING SEMINAR ON NUCLEAR PHYSICS

CHANGING FACETS OF NUCLEAR STRUCTURE

VICO EQUENSE, MAY 20–24, 2007

PROGRAM

Sunday, May 20

9:00 Opening address

L. Merola, Head, Sezione di Napoli of the Istituto Nazionale di Fisica Nucleare

A. Covello, Seminar Chairman

Chairperson: **D. Schwalm** (*Heidelberg*)

9:30 **A. Shotter** (*TRIUMF*): Radioactive Beams at TRIUMF

10:00 **H. Ueno** (*RIKEN*): Status of RI Beam Factory Project at RIKEN

10:30 Coffee break

11:00 **M. Thoennessen** (*Michigan*): Unbound States of Neutron-Rich Oxygen Isotopes

11:30 **C. Baktash** (*Oak Ridge*): Nuclear Structure Studies in the ^{132}Sn Region Using Accelerated Radioactive Ion Beams

12:00 **L. S. Ferreira** (*Lisboa*): Observing the Nuclear Drip Lines

12:20 Session close

Chairperson: **P. Sona** (*Firenze*)

15:00 **J. Äystö** (*Jyväskylä*): Nuclear Structure Studies of N-rich Nuclei Using ISOL Facilities at CERN and Jyväskylä

15:30 **F. Azaiez** (*Orsay*): Shell Structure Evolution Far from Stability: Recent Results from GANIL

16:00 **N. Benczer-Koller** (*Rutgers*): Magnetic Moment Meaurements: Pushing the Limits

16:30 Coffee break

17:00 **A. V. Ramayya** (*Vanderbilt*): Technique for Measuring Angular Correlations and g-Factors of Excited States with Large Multi-Detector Arrays: an Application to Neutron Rich Nuclei Produced by the Spontaneous Fission of ^{252}Cf

17:20 **C. Fahlander** (*Lund*): Isospin Symmetry and Proton Decay; Identification of the 10^+ Isomer in ^{54}Ni

17:40 **G. de Angelis** (*Legnaro*): Exploring the Evolution of the Shell Structure by Means of Deep Inelastic Reactions

18:00 **S. Lunardi** (*Padova*): Spectroscopy of Neutron-Rich Nuclei in the $A \approx 60$ Region

18:20 **D. Q. Fang** (*Shangai*): Studies on the exotic structure of ^{23}Al by measurements of σ_R and $P_{//}$

18:40 **M. Tomaselli** (*Darmstadt*): Extended Cluster Model for Light, Medium and Heavy Nuclei

19:00 Session close

Monday, May 21

Chairperson: **L. Moretto** (*Berkeley*)

9:00 **P. Egelhof** (*GSI*): Nuclear Structure Studies on Exotic Nuclei with Radioactive Beams – Present Status and Future Perspectives at FAIR

9:30 **G. Prete** (*Legnaro*): The SPES Direct Target Project at the Laboratori Nazionali di Legnaro

10:00 **J. Wambach** (*Darmstadt*): Modern Aspects of Nuclear Structure Theory

10:30 **R. Schiavilla** (*Jefferson Lab.*): Correlations in Nuclei: Recent Progress on an Old Problem

11:00 Coffee break

11:30 **I. Sick** (*Basel*): Correlated Nucleons in k- and r-Space

12:00 **T. T. S. Kuo** (*Stony Brook*): Roles of All-Order Core Polarization and Brown-Rho Scaling in Nucleon Effective Interactions

12:30 **A. Schwenk** (*TRIUMF*): Low-Momentum Interactions for Nuclei

13:00 Session Close

Chairperson: **V. Voronov** (*Dubna*)

15:00 **J. P. Vary** (*Iowa*): *Ab-initio* & *ab-Exitu* No-Core Shell Model

15:30 **G. Hagen** (*Oak Ridge*): *Ab-Initio* Coupled Cluster Theory for Nuclear Structure

15:50 **J. P. Draayer** (*Baton Rouge*): Symplectic No-Core Shell Model

16:10 **T. Dytrych** (*Baton Rouge*): Role of Two-Particle-Two-Hole Symplectic Irreducible Representations in *ab Initio* No-Core Shell Model Results for Light Nuclei

16:30 Coffee break

17:00 **A. Faessler** (*Tübingen*): Nuclear Structure, Double Beta Decay and Test of Physics Beyond the Standard Model

17:30 **A. A. Raduta** (*Bucharest*): Double Beta Decay to the First 2^+ State

18:00 **N. Van Giai** (*Orsay*): Pion-Induced Interaction in Relativistic Hartree-Fock

18:20 Session close

Tuesday, May 22

Chairperson: **N. Benczer-Koller** (*Rutgers*)

9:00 **I. Talmi** (*Rehovot*): Structure of $A = 14$ Nuclei: the NCSM and the Shell Model

9:20 **B. A. Brown** (*Michigan*): New Theoretical Results for *sd* and *pf* Shell Nuclei

9:50 **A. Covello** (*Napoli*): Realistic Shell-Model Calculations for Exotic Nuclei around Closed Shells

10:10 **H. Mach** (*Notre Dame/Uppsala*): New Results on the Simple Nuclear Systems just above ^{132}Sn

10:30 Coffee break

11:00 **N. Itaco** (*Napoli*): Shell-Model Calculations
with Low-Momentum Nucleon-Nucleon Interactions
Based upon Chiral Perturbation Theory

11:20 **B. Fornal** (*Kraków*): Shell-Model States in Neutron-Rich
Ca and Ar Nuclei

11:40 **A. Heusler** (*Heidelberg*): Nuclear Structure Information
from ^{208}Pb*(p, p')* via Isobaric Analog Resonances in ^{209}Bi

12:00 **D. L. Balabanski** (*Camerino*): First Results from the
g-RISING Campaign: the g-Factor of the $19/2^+$ Isomer in ^{127}Sn

12:20 Session close

Wednesday, May 24

Chairperson: **L. Ferreira** (*Lisboa*)

9:00 **J. N. Ginocchio** (*Los Alamos*): From the Quark Shell Model
to the Nuclear Shell Model

9:20 **S. S. Dimitrova** (*Sofia*): Density Matrix Renormalization
Group Method for Large-Scale Nuclear Shell Model Problems

9:40 **M. Horoi** (*Michigan*): Shell Model Correlations in Nuclei:
Coexistence of Spherical, Deformed and Superdeformed Bands

10:00 **R. Orlandi** (*Legnaro*): Lifetime Measurements in ^{67}Se and ^{67}As
as Mirror Pair: a Test of Isospin Symmetry Breaking

10:20 **R. Schwengner** (*Dresden*): Dipole-Strength Distributions up to
the Giant Dipole Resonance Deduced from Photon Scattering

10:40 Coffee break

11:10 **N. Pietralla** (*Köln*): Isovector Quadrupole Excitations
in the Valence Shell of Vibrational Nuclei

11:40 **P. von Neuman Cosel** (*Darmstadt*): Soft Electric Dipole Modes:
Some Selected Examples

12:10 **A. Zilges** (*Darmstadt*): The Structure of the
Pygmy Dipole Resonance

12:40 Session close

Chairperson: **J. Wambach** (*Darmstadt*)

15:00 **S. Antalic** (*Bratislava*): Beta-Delayed Fission Activity in the Bi-At Region

15:20 **P. G. Bizzeti** (*Firenze*): Critical Point Behaviour of ^{224}Ra and ^{224}Th

15:40 **I. Hamamoto** (*Lund*): Chiral Bands in Nuclei?

16:00 **S. Leoni** (*Milano*): Nuclear Structure at Extreme Conditions through Gamma Spectroscopy Measurements

16:20 **J. H. Hamilton** (*Vanderbilt*): Even- and Odd-Parity Bands in ^{108,110,112}Ru and Odd-Parity Doublets in ^{106}Mo

16:40 Coffee break

17:10 **K. Sugawara-Tanabe** (*Tokyo*): Selection Rules for the Intra and Interband Electromagnetic Transitions and Quantum Numbers for the Triaxial Rotor in Odd A Nuclei

17:30 **Ch. Stoyanov** (*Sofia*): Nuclear Structure Calculation at Large Domain of Excitation Energy

17:50 **J. Kvasil** (*Prague*): TDDFT with Skyrme Forces: Description of Giant Resonances in Deformed Nuclei

18:10 Session close

Thursday, May 24

Chairperson: **J. H. Hamilton** (*Vanderbilt*)

9:00 **N. Lo Iudice** (*Napoli*): Collective Bands in Superdeformed Nuclei

9:20 **V. V. Voronov** (*Dubna*): Finite Rank Approximation for Skyrme Forces and Effects of the Particle-Particle Channel

9:40 **N. D. Dang** (*RIKEN*): Thermal Pairing in Nuclei

10:00 **F. Palumbo** (*Frascati*): Odd Nuclei in a Number Conserving Approach to Bosonization

10:20 **V. I. Abrosimov** (*Kiev*): Kinetic Equation for Nuclear Response with Pairing

10:40 Coffee break

11:10 **V. Sushkov** (*Dubna*): Low-Lying 0^+ states in Deformed Nuclei

11:30 **F. Andreozzi** (*Napoli*): New Microscopic Approach to Multiphonon Nuclear Spectra

11:50 **K. Tanabe** (*Saitama*): Variational Equation for Quantum Number Projection

12:10 Session close

Chairperson: **A. A. Raduta** (*Bucharest*)

15:00 **L. G. Moretto** (*Berkeley*): Phase Diagram of Uncharged, Symmetric, Infinite Nuclear Matter from Multifragmentation Data

15:20 **J. M. G. Gómez** (*Madrid*): Recent Developments in the Spectral Fluctuations of Nuclei, Hadrons and Other Quantum Systems

15:40 **D. Schwalm** (*Heidelberg*): Bremsstrahlung Accompanied α Decay of ^{210}Po

16:00 **M. Gai** (*Connecticut*): The Coulomb Dissociation of ^{8}B; a New Tool for Nuclear Astrophysics

16:20 Coffee break

Chairperson: **A. Covello** (*Napoli*)

16:50 Round Table Discussion: Trends and Perspectives in Nuclear Structure

G. de Angelis, F. Azaiez, C. Baktash, J. P. Draayer, A. Faessler, I. Talmi

CONTENTS

Organizing Committees v

Foreword vii

Program ix

Section I

EXOTIC NUCLEAR STRUCTURE

Radioactive Beams at TRIUMF 3
A. C. Shotter

Status of RI-Beam Factory Project at RIKEN 13
H. Ueno, for the RIBF Collaboration

Population of Neutron Unbound States via Two-Proton Knockout Reactions 23
N. Frank, T. Baumann, D. Bazin, A. Gade, J.-L. Lecouey, W. A. Peters, H. Scheit, A. Schiller, M. Thoennessen, J. Brown, P. A. DeYoung, J. E. Finck, J. Hinnefeld, R. Howes, and B. Luther

Studies of Neutron-Rich Nuclei Using ISOL Facilities at CERN and Jyväskylä 29
J. Äystö, ISOLDE and IGISOL Collaborations

Shell Structure Evolution Far from Stability: Recent Results from GANIL 39
F. Azaiez

Magnetic Moment Meaurements: Pushing the Limits 49
N. Benczer-Koller

Technique for Measuring Angular Correlations and g-Factors of Excited States with Large Multi-Detector Arrays: An Application to Neutron Rich Nuclei Produced in Spontaneous Fission 57
A. V. Ramayya, C. Goodin, K. Li, J. K. Hwang, J. H. Hamilton, Y. X. Luo, A. V. Daniel, G. M.Ter-Akopian, N. J. Stone, J. R. Stone, J. O. Rasmussen, I. Y. Lee, S. J. Zhu, and M. Stoyer

Isospin Symmetry and Proton Decay: Identification of the 10^+ Isomer in ^{54}Ni 65
C. Fahlander, R. Hoischen, D. Rudolph, M. Hellström, S. Pietri, Zs. Podolyák, P. H. Regan, A. B. Garnsworthy, S. J. Steer, F. Becker, P. Bednarczyk, L. Caceres, P. Doornenbal, J. Gerl, M. Górska, J. Grębosz, I. Kojouharov, N. Kurz, W. Prokopowicz, H. Schaffner H. J. Wollersheim, L.-L. Andersson, L. Atanasova, D. L. Balabanski, M. A. Bentley, A. Blazhev, C. Brandau, J. Brown, E. K. Johansson, and A. Jungclaus

Exploring the Evolution of the Shell Structure by Means of Deep Inelastic Reactions 73
G. de Angelis

Studies on the Exotic Structure of ^{23}Al by Measurements of σ_R and $P_{//}$ 81
D. Q. Fang, Y. G. Ma, W. Guo, C. W. Ma, K. Wang, T. Z. Yan, X. Z. Cai, Z. Z. Ren, Z. Y. Sun, J. G. Chen, W. D. Tian, C. Zhong, W. Q. Shen, M. Hosoi, T. Izumikawa, R. Kanungo, S. Nakajima, T. Ohnishi, T. Ohtsubo, A. Ozawa, T. Suda, K. Sugawara, T. Suzuki, A. Takisawa, K. Tanaka, T. Yamaguchi, and I. Tanihata

Extended Cluster Model for Light and Medium Nuclei 89
M. Tomaselli, T. Kühl, D. Ursescu, and S. Fritzsche

Nuclear Structure Studies on Exotic Nuclei with Radioactive Beams – Present Status and Future Perspectives at FAIR 97
P. Egelhof

The SPES Direct Target Project at the Laboratori Nazionali di Legnaro 111
G. Prete, A. Andrighetto, C. Antonucci, M. Barbui, L. Biasetto, G. Bisoffi, S. Carturan, L. Celona, F. Cervellera, S. Cevolani, F. Chines, M. Cinausero, P. Colombo, M. Comunian, G. Cuttone, A. Dainelli, P. Di Bernardo, E. Fagotti, M. Giacchini, F. Gramegna, M. Lollo, G. Maggioni, M. Manzolaro, G. Meneghetti, G. E. Messina, A. Palmieri, C. Petrovich, A. Pisent, L. Piga, M. Re, V. Rizzi, D. Rizzo, M. Tonezzer, D. Zafiropulos, and P. Zanonato

Section II

NUCLEAR STRUCTURE AND NUCLEAR FORCES

Modern Aspects of Nuclear Structure Theory 123
J. Wambach

Correlations in Nuclei: A Review 135
R. Schiavilla

Correlated Nucleons in k- and r-Space 143
I. Sick

Roles of All-Order Core Polarizations and Brown-Rho Scaling in Nucleon Effective Interactions 153
T. T. S. Kuo, J. W. Holt, G. E. Brown, J. D. Holt, and R. Machleidt

Ab initio and *ab exitu* No Core Shell Model 163
J. P. Vary, P. Navrátil, V. G. Gueorguiev, W. E. Ormand, A. Nogga, P. Maris, and A. Shirokov

Ab-initio Coupled Cluster Theory for Open Quantum Systems 173
G. Hagen, D. J. Dean, T. Papenbrock, and M. Hjorth-Jensen

Symplectic No-Core Shell Model 183
J. P. Draayer, T. Dytrych, K. D. Sviratcheva, C. Bahri, and J. P. Vary

Role of Deformed Symplectic Configurations in *ab initio* No-Core Shell Model Results 191
T. Dytrych, K. D. Sviratcheva, C. Bahri, J. P. Draayer, and J. P. Vary

Nuclear Structure, Double Beta Decay and Neutrino Mass 199
A. Faessler

Double Beta Decay to the First 2^+ State 209
A. A. Raduta and C. M. Raduta

Pion-Induced Interaction and Single-Particle Spectra in Relativistic Hartree-Fock 221
N. Van Giai, W. H. Long, J. Meng, and H. Sagawa

Section III

SHELL MODEL AND NUCLEAR STRUCTURE

Structure of $A = 14$ Nuclei: The NCSM and the Shell Model 231
I. Talmi

CI and EDF Applications in Light Nuclei towards the Drip Line 243
B. A. Brown and W. Richter

Realistic Shell-Model Calculations for Exotic Nuclei around Closed Shells 253
A. Covello, L. Coraggio, A. Gargano, and N. Itaco

Surprising Features of Simple Nuclear Systems just above ^{132}Sn 263
H. Mach, R. Navarro-Pérez. L. M. Fraile, U. Köster, B. A. Brown, A. Covello, A. Gargano, O. Arndt, A. Blazhev, N. Boelaert, M. J. G. Borge, R. Boutami, H. Bradley, N. Braun, Z. Dlouhy, C. Fransen, H. O. U. Fynbo, Ch. Hinke, P. Hoff, A. Joinet, A. Jokinen, J. Jolie, A. Korgul, K.-L. Kratz, T. Kröll, W. Kurcewicz, J. Nyberg, E.-M. Reillo, E. Ruchowska, W. Schwerdtfeger, G. S. Simpson, B. Singh, M. Stanjou, O. Tengblad, P. G. Thirolf, V. Ugryumov, and W. B. Walters

Shell-Model Calculations with Low-Momentum Nucleon-Nucleon Interactions Based upon Chiral Perturbation Theory 273
N. Itaco, L. Coraggio, A. Covello, A. Gargano, D. R. Entem, T. T. S. Kuo, and R. Machleidt

Shell-Model States in Neutron-Rich Ca and Ar Nuclei 283
B. Fornal, R. Broda, W. Królas, T. Pawlat, J. Wrzesiński, R. V. F. Janssens, M. P. Carpenter, T. Lauritsen, D. Seweryniak, S. Zhu, N. Mărginean, L. Corradi, G. de Angelis, F. Della Vedova, E. Fioretto, A. Gadea, B. Guiot, D. R. Napoli, A. M. Stefanini, J. J. Valiente-Dobón, S. Lunardi, S. Beghini, E. Farnea, P. Mason, G. Montagnoli, F. Scarlassara, C. A. Ur, M. Honma, P. F. Mantica, T. Otsuka, G. Pollarolo, S. Szilner, and M. Trotta

Nuclear Structure Information from ^{208}Pb(p, p') via Isobaric Analog Resonances in ^{209}Bi 293
A. Heusler, P. von Brentano, T. Faestermann, G. Graw, R. Hertenberger, J. Jolie, R. Krücken, K. H. Maier, D. Mücher, N. Pietralla, F. Riess, V. Werner, and H.-F. Wirth

From the Quark Shell Model to the Nuclear Shell Model 301
J. N. Ginocchio

Lifetime Measurements in ^{67}Se and ^{67}As Mirror Pair:
A Test of Isospin Symmetry Breaking 307
R. Orlandi, G. de Angelis, F. Della Vedova, A. Gadea, N. Mărginean, D. R. Napoli, F. Recchia, J. J. Valiente-Dobón, F. Brandolini, E. Farnea, S. Lenzi, S. Lunardi, D. Mengoni, C. A. Ur, K. T. Wiedemann, E. Sahin, A. Bracco, S. Leoni, D. Tonev, R. Wadsworth, B. S. Nara Singh, D. G. Sarantites, W. Reviol, C. J. Chiara, O. L. Pechenaya, C. J. Lister, M. Carpenter, J. Greene, D. Seweryniak, and S. Zhu

The ^{120}Sn$(p,t)^{118}$Sn Reaction: Level Structure of ^{118}Sn and Microscopic DWBA Calculations 315
P. Guazzoni, L. Zetta, A. Covello, A. Gargano, B. F. Bayman, G. Graw, R. Hertenberger, T. Faestermann, H.-F. Wirth, and M. Jaskóla

Section IV

COLLECTIVE MODES OF NUCLEAR EXCITATION

Isovector Valence Shell Excitations in Vibrational Nuclei 325
N. Pietralla, T. Ahn, O. Burda, and G. Rainovski

Soft Electric Dipole Modes in Heavy Nuclei: Some Selected Examples 335
P. von Neumann-Cosel

The Structure of the Pygmy Dipole Resonance 345
D. Savran and A. Zilges

Dipole-Strength Distributions up to the Giant Dipole Resonance Deduced from Photon Scattering 355
R. Schwengner, G. Rusev, N. Benouaret, R. Beyer, F. Dönau, M. Erhard, E. Grosse, A. R. Junghans, K. Kosev, J. Klug, C. Nair, N. Nankov, K. D. Schilling, and A. Wagner

Critical Point Behaviour of ^{224}Ra and ^{224}Th 363
P. G. Bizzeti and A. M. Bizzeti-Sona

Study of the ^{119}Sb via the ^{121}Sb$(p,t)^{119}$Sb Reaction 371
P. Guazzoni, L. Zetta, V. Yu. Ponomarev, G. Graw, R. Hertenberger, T. Faestermann, H.-F. Wirth, and M. Jaskóla

Chiral Bands in Nuclei? 379
I. Hamamoto and G. B. Hagemann

Even- and Odd-Parity Bands in ^{108,110,112}Ru and Odd-Parity Doublets in ^{106}Mo 387
J. H. Hamilton, S. J. Zhu, Y. X. Luo, J. O. Rasmussen, A. V. Ramayya, J. K. Hwang, X. L. Che, Z. Jang, C. Goodin, P. M. Gore, E. F. Jones, K. Li, S. Frauendorf, V. Dimitrov, J. Y. Zhang, I. Stefanescu, A. Gelberg, J. Jolie, P. Van Isacker, P. von Brentano, J. L. Wood, M. A. Stoyer, R. Donangelo, J. D. Cole, N. J. Stone, and J. Stone

Identification of Levels in ^{144}Cs 395
E. F. Jones, P. M. Gore, Y. X. Luo, J. H. Hamilton, A. V. Ramayya, J. K. Hwang, H. L. Crowell, K. Li, C. T. Goodin, J. O. Rasmussen, and S. J. Zhu

Nuclear Structure at Extreme Conditions through Gamma Spectroscopy Measurements 403
S. Leoni

Selection Rules for the Intra and Interband Transitions and Quantum Numbers for the Triaxial Rotor in Odd-A Nuclei 411
K. Sugawara-Tanabe and K. Tanabe

Collective Bands in Superdeformed Nuclei 419
J. Kvasil, N. Lo Iudice, F. Andreozzi, A. Porrino, and F. Knapp

Nuclear Structure Calculations in a Large Domain of Excitation Energy 427
Ch. Stoyanov and D. Tarpanov

Time-Dependent Density Functional Theory with Skyrme Forces: Description of Giant Resonances in Deformed Nuclei 437
J. Kvasil, W. Kleinig, V. O. Nesterenko, P.-G. Reinhard, and P. Vesely,

Finite Rank Approximation for Skyrme Forces and Effects of the Particle-Particle Channel 445
A. P. Severyukhin, V. V. Voronov, and N. Van Giai

Low-Lying 0^+ States in Deformed Nuclei 453
A. V. Sushkov, N. Lo Iudice, and N. Yu. Shirikova

New Microscopic Approach to Multiphonon Nuclear Spectra 461
F. Andreozzi, F. Knapp, N. Lo Iudice, A. Porrino, and J. Kvasil

Variational Equation for Quantum Number Projection at Finite Temperature 469
K. Tanabe and H. Nakada

Thermal Pairing in Nuclei 477
N. Dinh Dang

Composite Bosons and Quasiparticles in a Number Conserving Approach 485
F. Palumbo

Kinetic Equation for Nuclear Response with Pairing 495
V. I. Abrosimov, D. M. Brink, A. Dellafiore, and F. Matera

Self-Consistent Quasiparticle RPA for Multi-Level Pairing Model 503
N. Quang Hung and N. Dinh Dang

Section V

SPECIAL TOPICS

The Liquid-Vapor Phase Diagram of Infinite Uncharged Nuclear Matter 513
L. G. Moretto, J. B. Elliott, and L. Phair

Bremsstrahlung Accompanied α Decay of ^{210}Po 523
H. Boie, H. Scheit, U. D. Jentschura, F. Köck, M. Lauer, A. I. Milstein, I. S. Terekhov, and D. Schwalm

The Coulomb Dissociation of ^{8}B in the Coulomb Field and the Validity of the CD Method 531
M. Gai

Nuclear Many-Body Physics where Structure and Reactions Meet 539
N. Ahsan and A. Volya

Recent Developments in the Spectral Fluctuations of Nuclei, Hadrons and Other Quantum Systems 547
J. M. G. Gómez, L. Muñoz, J. Retamosa, R. A. Molina, A. Relaño, and E. Faleiro

List of Participants 559

Author Index 569

SECTION I

EXOTIC NUCLEAR STRUCTURE

RADIOACTIVE BEAMS AT TRIUMF

A. C. SHOTTER
TRIUMF, Vancouver, Canada

Abstract: Some of the technical issues associated with radioactive beam production at TRIUMF are outlined together with a selection of some experimental results.

1. Introduction

TRIUMF is Canada's National Laboratory for particle and nuclear physics. It supports a wide spectrum of indigenous facilities as well as supporting Canadian teams to mount large-scale particle physics experiments at international facilities outside Canada. The laboratory facilities are based on a suite of two cyclotrons and three linear accelerators. These accelerators support a variety of scientific programs in particle physics, nuclear physics, materials, chemistry and life sciences. In addition there are three further cyclotrons dedicated to the production of radioisotopes for medical purposes. MDS Nordion manages the marketing of these isotopes. The main cyclotron is a versatile 500 MeV proton accelerator with the capability of simultaneous extraction of several different beams, and so can support at the same time different experimental programs. Over the last few years one of these beams has been used to irradiate an ISOL target for the production of radioactive ion beams (RIB). This facility together with a post accelerator complex is called ISAC. This paper will address some of the technical issues associated with this facility, and present some experimental highlights.

2. ISOL Technical challenges

The overriding issue for any radioactive beam facility is the intensity of ions that can be produced. For facilities based on the ISOL method, such as ISAC, there are many factors that need to be understood and optimized to obtain maximum yields for particular isotopes. The factors can be grouped into four main categories: production, diffusion, effusion, and ionization; the relationship between them is identified in fig. 1, and each is discussed below.

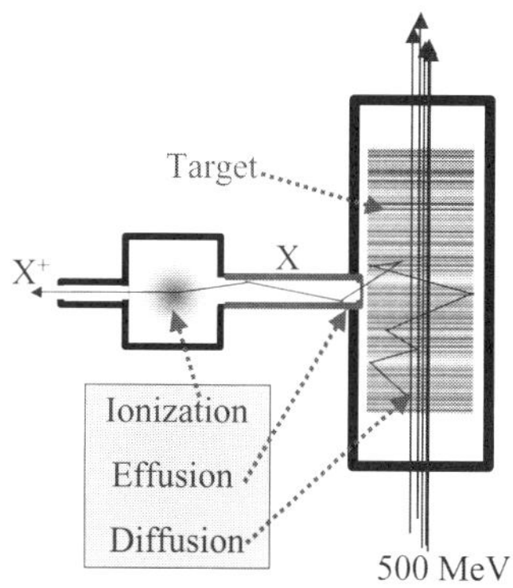

Fig.1 Production of RIBs at ISAC

(a) Production

The radiological license for the ISAC facility limits the proton beam from the cyclotron bombarding the ISOL target to 100 micro amps. Radioactive ions produced by reactions of the incident proton beam with the target material, migrate from the target and its container to a nearby ion source.

The experimenter is generally interested in a range of radioisotopes across the whole nuclear chart. Therefore in order to deliver such a wide range of isotopes different target materials must be selected in order to optimize production. At TRIUMF a range of target materials have been used based on low Z elements such as silicon carbide to higher Z elements such as tantalum. For a given target the quantity of radioisotope production is directly related to the proton bombardment beam intensity. However for any target there is a limit to the power density that can be deposited within the target; above a certain level the target will melt, and as a result the effusion of isotopes from the target will significantly decrease. To optimize the power on the target without melting it, requires efficient removal of heat from the target. At an incident energy of 500 MeV, a 100 micro amps beam will deposit a significant fraction of the 50-kilowatt beam power in a target volume of several cubic centimeters. ISAC targets are typically made from thin discs of the material, about two centimeters in diameter, stacked together. These targets are designed to dissipate the power by relying on thermal conduction to transfer the energy from the target centre, i.e. beam position, to the cylindrical container retaining the target discs. The thermal conductivity of the material therefore is an important issue. Generally metal discs have good thermal conductivity but oxides have very low values. Carbides have intermediate values. However while conductivity is an important physical process to conduct the thermal energy away from the target centre to the container wall, it is essential to transfer this energy from the container wall to the outside environment. Radiation is an efficient transfer process to affect this energy transfer. To maximize this transfer the emissivity of the outside surface of the container must be as high as possible. To aid this radiation transfer process a special target container has been developed with a series of cooling fins on the outside of the target container.

This finned container surface has an emissivity of 0.98 close to the theoretical limit of 1.0. With this configuration targets have been irradiated for several weeks at 100 micro amps proton current at an incident energy of 500 MeV. Beam power dissipation is only one of many factors that have to be optimized to maximize isotope production. Issues concerning diffusion, effusion and ionization need continuous development programs for different radioisotopes, ref.1.

(b) Diffusion

In general, diffusion of isotopes is strongly related to temperature. The temperature at the outer regions of the target will be different from the local situation at the beam position; it is important that outer target regions are at a sufficiently high temperature, while ensuring the temperature at the beam position is below the melting point. It has been known for some time that for material regions that have been exposed to beam radiation, diffusion of isotopes from these regions is enhanced. This radiation-enhanced diffusion shows itself as a non-linear response to the radioisotope production as a function of proton beam intensity. The faster increase of isotope production than for a pure linear function is thought to be associated with the generation and mobility of lattice defects created by the incident proton beam. The question of how well the proton beam should be focused on a particular target must be addressed with consideration of: (a) the maximum allowed temperature to avoid melting, (b) gaining maximum benefit from increased mobility through radiation damage. Rotating a tightly focused beam of ~ mm size on a target of ~ two cm diameter is a method that will be used at ISAC in order to maximize the radiation enhancement while distributing the thermal energy to prevent melting of the target material.

(c) Effusion

Effusion is the physical process that enables isotopes once they are emitted from the material substance of the target to enter the volume of the ionizer. Effusion depends on the physical composition of the target material and the geometrical structure of how the target container is connected to the ionization volume. The sticking coefficients, which are related to the release time of radioisotopes once they are absorbed on surfaces, are very important parameters in designing the physical geometry of the target. Clearly the shorter the lifetime of an isotope the more critical is the role played by the sticking coefficients. A popular method of target construction is to make thin discs of the target material and stack them together in the target cylindrical vessel; the idea is that effusion will occur efficiently between the disc interfaces.

(d) Ionization

The final stage in the ISOL process is to ensure efficient ionization of the isotope of interest; once ionization occurs mass selection can then proceed through an electromagnetic system so that a clean beam of the isotope can be produced, devoid of other unwanted radioisotopes. There are several methods of ionization: surface ionization, ECR, FEBIAD, and resonant ionization. Surface ionization has been used for some years at ISAC and has proved a most efficient way to ionize isotopes of the alkaline elements. It is also the simplest method for an ISOL target; the ionization occurs in a tube that is connected to the side of the target cylinder. As the ions effuse from the target cylinder and bounce down the exit side tube they will be ionized if the inside surface of the tube is coated with a material that has a higher electron affinity then the radioisotope.

Laser ionization is proving to be another very effective method of ionization for ISOL facilities. The basic idea here is to photo-ionize a neutral atom of the isotope of interest by using one laser to resonantly excite the atom from its ground state to an excited state and another laser to photo-ionize the atom from this state to the electron continuum. Since the first laser frequency is set to match the energy of a particular excited state, and this is unique for the particular isotope, there will be a strong selectivity in the ionization process. Even greater selectivity can be obtained if three separate lasers are used; one to excite the neutral atom to a particular excited state, another to excite this state to a higher state and finally a third to photo-ionize this state. This final step can be considerably enhanced if ionization can be induced through a Rydberg state. One of the main advantages of laser ionization, other than its great isotope selectivity, is that the lasers and all the optical elements can be many metres away from the target location. The immediate region around the target is an extremely hostile environment; not only does the target operate at a high temperature of about 2000 degrees centigrade, the nuclear radiation is so severe it can rapidly destroy optical elements. However the laser beams can be generated and controlled many metres away from the target and directed to the target by a series of mirrors through small openings in the target shield. The laser beams are shone through the side exit tube of the target cylinder where the ionization takes place. The aim is to provide enough laser power in the various laser beams to insure that all the atoms of interest are appropriately ionized, i.e. a saturation level is reached. By this technique a range of radioisotopes have been produced at ISAC. Each isotope is a development project, and much work needs to be undertaken to identify the most efficient excitation sequence. Using the latest titanium-sapphire tuned lasers ionization schemes have been developed for a variety of elements, ref. 2.

There are some elements for which surface ionization or laser ionization are inappropriate. For these elements more general ionization methods are needed. ECR and FEBIAD ionizers fall into this category. However one of the main drawbacks of using these sources is that for short-lived isotopes the ionizer must be located near the hostile environment of the target. The use of these ionization sources therefore means that they have to be specially adapted for such an environment. At ISAC a first generation ECR source has been developed but proved to be inefficient to ionize the inert gases. However the lessons learnt from this work are being used to develop the next generation source. A FIBIAD source also has been constructed and will be undergoing tests this summer.

A variety of radioactive beams now can be produced at ISAC (see TRIUMF website). However it should be carefully noted that much work is still needed to fully understand all the technical issues associated with operating ISOL targets at maximum power. The properties of materials that are expected to work at high temperatures and high radiation environments are largely unknown. Each isotope demanded by the experimenter therefore requires extensive development work before the isotope can be used, and obviously this becomes more challenging the higher the beam intensity and purity demanded by the experimenter. The use of actinide targets is also of considerable interest to experimenters since fission of the target nuclei can produce intense beams of neutron-rich fission fragments. However the radiological issues associated with these targets are more complex than for lighter targets. At ISAC a series of controlled tests are being undertaken with various targets with the final aim being to use uranium carbide targets irradiated with 100 micro amps proton beam.

Once ionization occurs, 1+ ions are guided through a mass separator to select the isotope of interest. At the exit of the separator the ions may be used directly by the experimenter. Such experiments could involve studying the decay properties of the radioactive nuclei, or they may involve implantation in various thin film samples for the study of the internal properties of these films.

3. ISAC Accelerator Complex

The ISAC facility complex provides the capability of accelerating RIB ions to higher energy. The first acceleration stage is an RFQ accelerator followed by a conventional DTL linear accelerator. This accelerator combination provides 1+ ion acceleration up to 1.9 MeV/u. The RFQ has a specification of a mass to charge ratio of the ion to be less than or equal to 30. For ions greater than mass 30 a charge state booster is being installed to increase the ion charge state before injection into the RFQ. This RIB accelerator system is called ISAC I.

In the last year an addition to the ISAC accelerator complex has been commissioned; this is a superconducting linear accelerator, called ISAC II. This accelerates ions up to a final energy of 4.3/u MeV. Further extensions of this accelerator are now under construction that will boost the energy still further to 6.7 MeV/u. Figure 2 shows the layout of the ISAC facility. Further details of this accelerator system may be found in ref. 3.

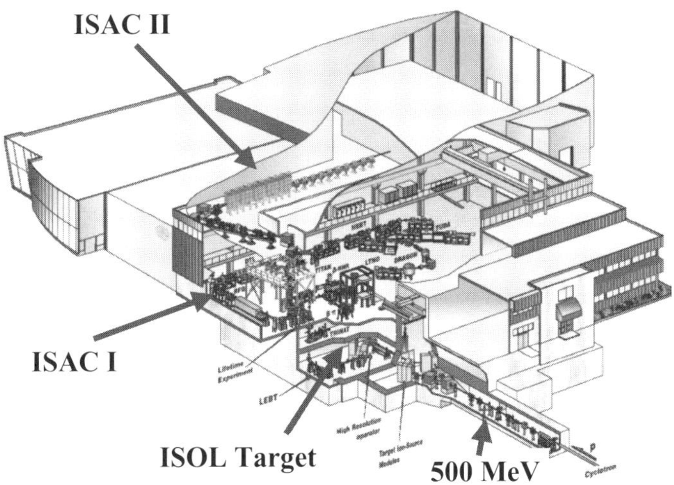

Fig.2 The ISAC facility at TRIUMF

3. Selection of experimental results at ISAC

(a) The Maya Experiment

The production of intense radioactive beams of lithium isotopes has opened up a range of scientific investigations. The lithium isotopes are produced by proton irradiation of tantalum targets followed by surface ionization. In January 2007 a ^{11}Li beam was accelerated through ISAC I and ISAC II accelerators to an energy of 34 MeV. In fact this was the first experiment undertaken with the new ISAC II accelerator. This beam was then injected into the MAYA apparatus that had been brought to TRIUMF from the GANIL Laboratory in France. The MAYA system is an active gaseous target detector, ref. 4. If the detector is filled with a hydrogen based gas then (p,t) or (p,d) reactions can be studied in an essentially 4π solid angle geometry. The system is specially suited to investigate such reactions with the low-intensity beams available at ISOL facilities. The GANIL/ TRIUMF team used this apparatus to study two neutron pick-up reactions for ^{11}Li. The track of one of the first (p,t) events recorded from ISAC II is shown in fig. 3. The aim of the experiment was to gain information about the correlation of the two outer neutrons of ^{11}Li. The experiment worked well with good quality angular distributions being collected for the (p,t) reaction to the ground and first excited states of ^{9}Li. Absolute cross-sections could be measured, so this data has the capability to distinguish between the different theories describing the correlations of the outer neutrons of ^{11}Li.

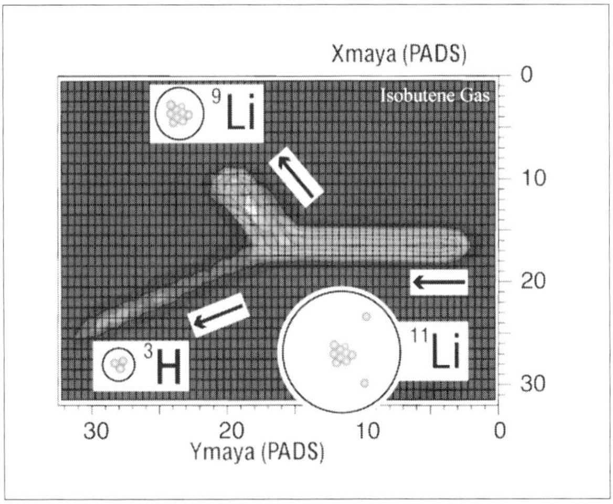

Fig. 3 Track of a ^{11}Li reaction in the MAYA apparatus

(b) Lithium Nuclear Charge Radius

The MAYA experiment for ^{11}Li at TRIUMF has the potential to yield information concerning the outer neutron distributions of this nucleus. Of equal importance for the understanding of this nucleus is its charge distribution. Of course due to its short half-life, normal methods to determine this distribution are not appropriate. For this reason a group at GSI have developed atomic spectroscopy methods to determine the charge radius of lithium isotopes. The

GSI team transported their equipment to TRIUMF to perform the experiment with TRIUMF's high intensity ^{11}Li beam. The idea of the experiment is outlined in fig. 4A, which shows the excitation scheme used. The principle of the experiment is to determine the energy difference between the $2S_{1/2}$ and $3S_{1/2}$ levels for lithium isotopes from ^{6}Li to ^{11}Li. The difference from one isotope to another is mostly determined by the mass difference between isotopes, but there is a small component associated with the slightly differing charge distributions. The excitation from the $2S_{1/2}$ is undertaken by two photon absorption to the $3S_{1/2}$, which then decays to the $2P_{1/2}$, which in turn is ionized by further laser excitation. The measurement involves counting the ions that are produced as a function of the $2S_{1/2}$ -$3S_{1/2}$ laser scan, as shown in fig. 4c. The deduction of the charge radii for the various isotopes is shown in fig. 4b. It is most interesting to note that the charge radius seems to fall for isotopes up to ^{9}Li but shows a significant increase for ^{11}Li. Various theories have been developed to predict these charge radii some of which are shown in fig. 4b, further details can be found in ref. 5.

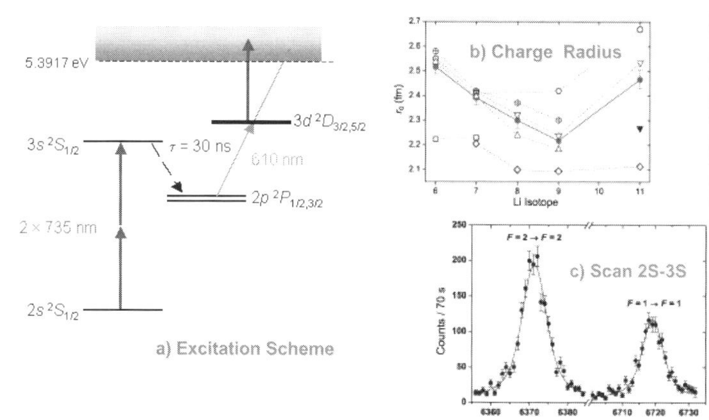

Fig. 4 Determination of the charge radius of lithium isotopes

(c) Sub-barrier Fusion

For neutron rich nuclei, far from stability, the tails of their neutron distributions are probably well extended from the charge distribution. Nuclear reactions for these nuclei could show different reaction characteristics than for stable nuclei. Fusion is one particular reaction that might be affected by such neutron tails. An experiment recently carried out, and led by the Oregon group, concerned the sub-barrier fusion of ^{9}Li on ^{70}Zn.

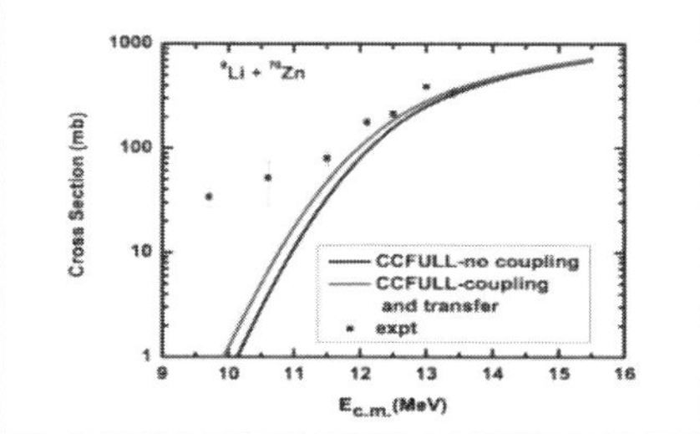

Fig.5 Sub-barrier fusion of ^{9}Li

The cross section was determined by radio-chemical techniques to determine the fusion channel yield of ^{76}As. The fusion cross-sections are shown in fig. 5. Calculations have been undertaken to determine the expected cross-sections taking into account inelastic excitation and transfer processes. The results are shown in fig. 5. Clearly the experimental cross-sections are much higher than the expected values. These results may indicate that a rethink of the fusion process for exotic nuclei may be necessary. Further details are given in ref. 6.

(c) Astrophysics Capture Reactions

For energy production in various stellar sites, proton and alpha particle capture reactions play a very important role. Many of the reactions of interest involve nuclei which are unstable. Therefore for direct measurements of the reactions, one of the goals of radioactive beam facilities like ISAC is to measure these reactions. However such measurements are very challenging. This is because the reaction energy at which these measurements are needed is well below the Coulomb barrier for the reaction. This means the cross-sections are very small, and this coupled with the current limited intensities of radioactive beams and the large backgrounds they generate, presents considerable problems to the experimenter. Special apparatus have to be developed to detect the rare capture events in the midst of a high number of background events. The DRAGON spectrometer, fig. 6a, has been built to measure such reactions.

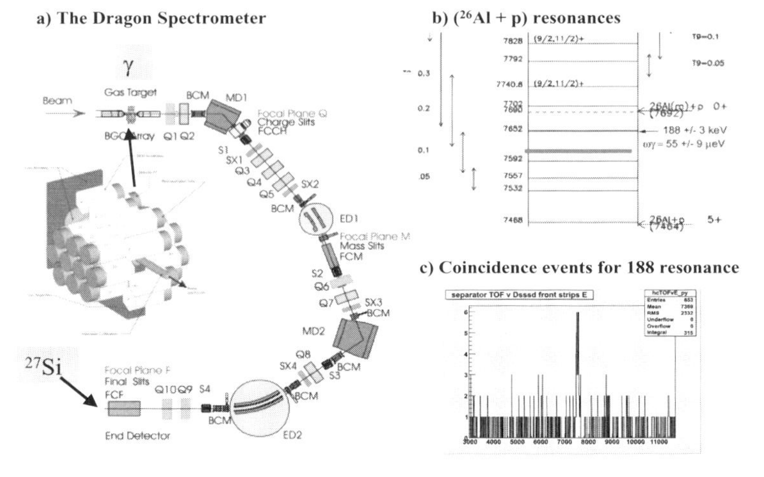

Fig. 6 the Dragon radiative capture Spectrometer

Details of the spectrometer are given in ref. 7. Essentially this spectrometer is a mass separation system to separate the rare capture events from the incident beam particles. One of the best examples of such measurements is ^{26}Al + p capture. This measurement is of interest for the understanding of ^{26}Al distribution in the galaxy. Figure 6b shows the resonant energy of particular astrophysics interest i.e. the 188 keV resonance. Figure 6c shows the results of several weeks of running at an average intensity of 2.5*10^9 pps of ^{26}Al on target. The peak corresponds to the correct time between the gamma capture released at the target and the final ^{27}Si detected at the focal plane for the 188 keV resonance. From the results the resonant strength for this level could be deduced ref. 8. The results also demonstrate the challenge to the experimenter, for if the beam intensity had been $\sim 10^8$ pps, instead of $\sim 10^9$ pps, the experiment would have taken too long to run.

(e) Electron – Neutrino Correlation in Beta-Decay

Radioactive beams provided a good platform to measure some of the parameters of the standard model of particle physics. For example, for pure Fermi beta-decay the standard model identifies the decay mode with the vector boson. This leads to a precise prediction for the angular correlation of the electron-neutrino emission. If there is a component of beta-decay propagated by a scalar boson then this would result in a different electron-neutrino correlation. Measuring the electron- neutrino correlation can therefore determine a possible limit to the presence of a scalar boson. Of course, it is not possible to detect directly neutrinos in such an experiment, but the production of radioactive beams provides a way to infer such correlations. The method is to produce first a radioactive nucleus by the ISOL method, and then trap these atoms in a small volume within a vacuum container, fig. 7a. When nuclei decay within this small volume, the energy of the electron and the recoiling atom are then measured, fig. 7b. Since the decay is a three-body process, measuring just two, i.e. the electron and the atom, will completely determine the reaction kinematics.

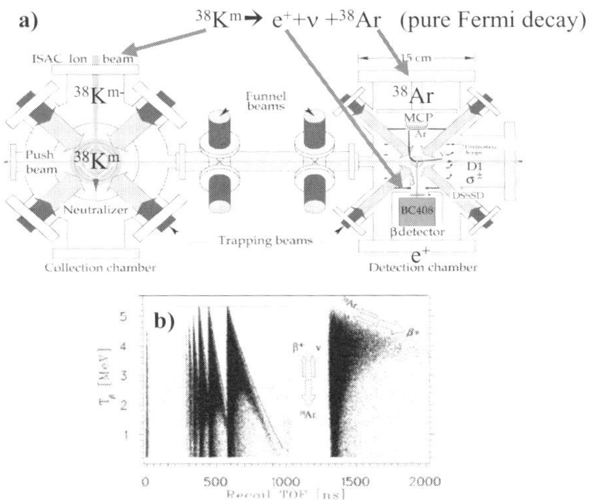

Fig. 7 The neutral atom trap for beta decay investigations.

By this method the team at TRIUMF has been able to establish the world's lowest limit for the presence of a scalar boson in beta-decay, ref. 9. Other similar fundamental questions can be addressed by using such techniques.

4. Outlook

The production of radioactive beams opens up exciting new opportunities in the field of exotic nuclei and nuclear astrophysics. However it is often technically very challenging to produce intense beams of these nuclei. TRIUMF is a laboratory that is pioneering the development of such ISOL beams. Even though some beams have been developed that make certain experiments possible, other experiments need beam intensities that are not yet obtainable; there will be a continuing need for technical development of ISOL beams.

The initiation and successful completion of the first phase of ISAC II accelerator has motivated the development of new apparatus such as the TIGRESS and EMMA spectrometers. These gamma-ray and particle spectrometers have been specially designed to meet the requirements of the experimental program for ISAC II. In addition the TITAN precision mass measurement system will start an exciting new program this summer.

Acknowledgments

Developing intense beams of RIB is technically challenging. I would like to acknowledge the efforts of scientists at TRIUMF who valiantly bear these challenges; mentioning a few, but not to diminishing the efforts of others, Pierre Bricault, Marik Dombsky and Jens Lassen. The successful operation of the ISAC accelerators is largely due to the leadership provided by Paul Schmor, Robert Laxdal, and again I sincerely acknowledge their professionalism.

This work is supported by the Government of Canada through the National Research Council of Canada, and the Natural Sciences and Engineering Research Council of Canada.

References

1) Dombsky M., *et al.* Nucl. Phys. A746 (2004) p. 32
2) Lassen J., *et al.* Hyperfine Interactions 162 (2005) p. 69
3) Laxdal R.E., *et al.* Physica C441 (2006) p.13
4) Demonchy C.E., *et al.* Nucl. Instrum. Method Res. A573 (2007) p. 145
5) Sanchez R., *et al.* Rev. Phy. Lett. 96 (2006) p. 033002
6) Loveland W., *et al.* Phys. Rev. C74 (2006) p. 064609
7) Hutcheon D.A., *et al.* Nucl. Instrum. Method Res. A498 (2003) p. 190
8) Ruiz C., *et al.* Rev. Phy. Lett. 96 (2006) p. 252501
9) Gorelov A., *et al.* Rev. Phy. Lett. 94 (2005) p.142501

STATUS OF RI-BEAM FACTORY PROJECT AT RIKEN

H. UENO*, FOR THE RIBF COLLABORATION

RIKEN Nishina Center,
2-1 Hirosawa, Wako, Saitama 351-0198, Japan
**E-mail: ueno@riken.jp*

The Radioactive-Isotope Beam Factory (RIBF) project is in progress in RIKEN, in which intense primary beams can be provided at the energies $E = 350 - 400$ over the whole range of atomic number in the cascade-cyclotron acceleration including the newly constructed three cyclotrons, fRC, IRC, and SRC. The project proceeds through two phases. In the phase-I program, the superconducting in-flight radioactive-isotope beam separator BigRIPS and the ZeroDegree spectrometer are constructed as well as three new cyclotrons. After the first beam of aluminum from the entire cyclotron system was extracted at $E = 345$ A MeV, a ^{238}U beam was also successfully boosted up to $E = 345$ A MeV. In addition, the first radioactive-isotope beam was produced from ^{86}Kr and ^{238}U and isotope-separated with BigRIPS. These success show that the whole system works as was designed basically.The RIBF project will be fully capitalized in the phase-II program, in which the construction of several experimental key devices has been proposed.

1. Introduction

The Radioactive-Isotope Beam Factory (RIBF) project[1] is in progress in RIKEN. In this project, intense primary beams can be provided at the energies $E = 400A$ and 350 A MeV for lighter- ($A \leq 40$) and heavier-mass ions, respectively, over the whole range of atomic number by using newly constructed three cyclotrons, $K = 570$ fixed-frequency Ring Cyclotron (fRC), $K = 980$ Intermediate-Stage Ring Cyclotron (IRC),[2] and $K = 2500$ Superconducting Ring Cyclotron (SRC).[3] Combining fRC in the cascade acceleration, 1 particle-μA beams will be obtained as a goal for all Z number. In another option, polarized deuteron beams will be given at $E = 440A$ MeV by combining the AVF cyclotron as an injector instead of the linear accelerator RILAC.

The project proceeds through two phases. In the phase-I program, the superconducting in-flight radioactive-isotope beam (RIB) separator BigRIPS[4] and the ZeroDegree spectrometer are constructed downstream of SRC as well as three new cyclotrons fRC, IRC, and SRC. After the cascade acceleration of the cyclotron complex, beams are transported to BigRIPS for the production of large-variety and high-intensity RIBs.

2. Beam Commissioning of the Accelerator Complex

The construction and installation of the new cyclotrons have been finished. The beam commissioning of each cyclotron and cascade acceleration are in progress. After the first beam extraction from IRC at November of 2006, in which Kr ions were used for the acceleration, much effort were paid for the SRC acceleration. The first beam from the entire cyclotron system was successfully extracted at December of 2006, where $q = 10^+$ aluminum ions were extracted at $E = 345A$ MeV.[5] This charge-to-mass ratio q/m was close to that of uranium ions $q/m = 88/238$ in the standard operation so that the aluminum acceleration was a good test for the uranium acceleration. Then, a uranium beam was accelerated in the whole cyclotron-complex system. A uranium beam of $q = 86^+$ was successfully extracted from SRC at the energy of $E = 345A$ MeV recently.[5]

In the present stage, a typical beam current of uranium ions of $I \leq 0.1$ pnA is much lower than the goal value. It is, however, expected that the beam current increases up to $I = 2 \sim 5$ pnA by improving the beam transportation system. The quality of the beam bunching and the extraction efficiencies of each cyclotrons are able to be much improved by the further studies. Also, the beam transmission efficiency around RILAC will be increased by miner changes of the beam-transportation scheme. After these improvements, the renewal of the ion source would be needed in order to realize the goal intensity 1 pμA.[5]

3. BigRIPS Commissioning

3.1. *Superconducting in-flight RIB separator BigRIPS*

BigRIPS is characterized with a two-stage configuration and a large acceptance.[4] In the first stage of BigRIPS, from the production target to the F2 focal plane, RIBs are produced and isotope-separated. For the isotope separation, a wedge-shaped degrader is placed at the momentum-dispersive focal plane F1. A high-power beam dump system and a rotating production target system are equipped inside the gap of the first dipole and a target chamber, respectively, for the use of high intensity primary beams.[4] In the RIBF energy region, medium/heavy-mass RI-ion beams are not fully

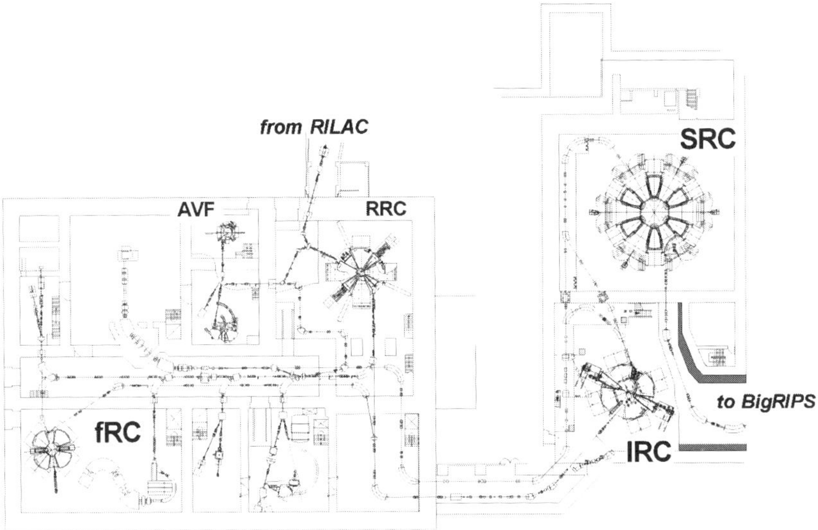

Fig. 1. A layout showing the RIBF cyclotron system. Newly installed $K = 570$ fixed-frequency Ring Cyclotron, $K = 980$ Intermediate-Stage Ring Cyclotron, and $K = 2500$ Superconducting Ring Cyclotron (SRC) are denoted by fRC, IRC, and SRC.

stripped so that they are are produced as a *cocktail* beam. Therefore, the second stage of BigRIPS, from the F3 to F7 focal plane, is designed to give their particle-identification information based on the event-by-event $B\rho$-ΔE-TOF measurement. The total length of BigRIPS is 77 m and the maximum magnetic rigidity is 9 Tm.[4]

In the in-flight uranium fission at $E = 350A$ MeV, with which RIBs far from the stability line can be effectively produced, the momentum and the scattering angle of the fission fragments are widely spread compared with the projectile fragmentation. BigRIPS is designed to accept RIBs produced not only in the projectile-fragmentation reaction but also the in-flight fission of uranium ions. Thus, all quadrupole magnets of BigRIPS are designed to be superconducting and also to have a large aperture. The momentum and angular acceptances of the BigRIPS are designed to be 6 % and 80 mrad$^{\theta}$ $\times$ 100 mrad$^{\phi}$ for this purpose.[4]

3.2. *BigRIPS commissioning*

After the successful beam acceleration, the first RIBs were produced using BigRIPS. A ^{86}Kr beam was first used in the commissioning to bombard a

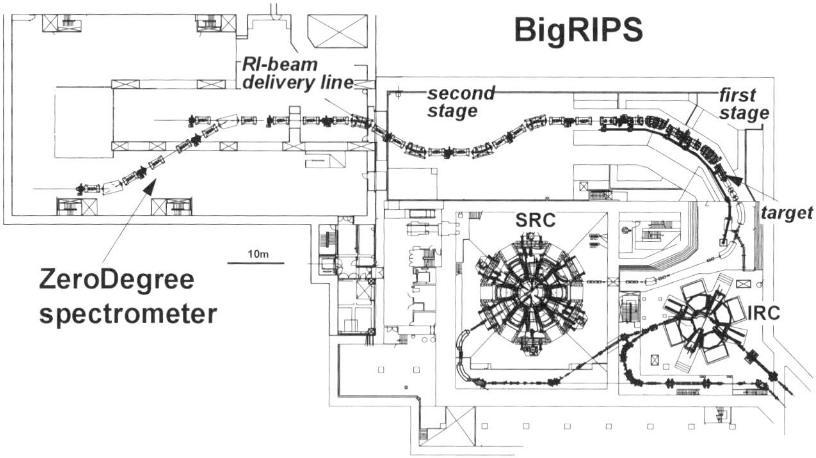

Fig. 2. A layout of the superconducting in-flight radioactive-isotope beam separator BigRIPS. The ZeroDegree spectrometer is also shown downstream of BigRIPS.

2-mm^t Be target at $E = 345A$ MeV.[6] The particle-identification procedure employed with beam-line counters was confirmed with a guide of the isomeric states ^{54m}Fe and ^{43m}Sc. Only their data points were observed in an obtained TOF-ΔE two-dimensional plot in coincidence with the delayed γ-ray emitted from the RIBs stopped after the second stage of BigRIPS.[6] In the next, a ^{238}U beam was used to bombard a 5-mm^t Be target at $E = 345A$ MeV,[6] where RIBs were successfully produced and particle-identified from the uranium beam for the first time. Through these two beam commissioning, it has been shown that whole the BigRIPS system functions as was designed. Details of the BigRIPS commissioning have been reported.[6]

In the phase-I program, the ZeroDegree spectrometer[7] is installed downstream of BigRIPS, which allow us to perform momentum analyses and particle identification of the fragments produced in the secondary reaction of RIBs, so that spectroscopic studies based on inclusive- and/or exclusive measurements can be carried out in combination with BigRIPS. Installation of the the ZeroDegree spectrometer is in progress. It will be fully equipped within 2007.

3.3. *Physics Programs in Phase I*

The first RIBF NP(Nuclear Physics)-PAC meeting was held in February 2007, where nineteen proposals were submitted. A proposed new isotope search project is in progress.[5] Also, the global survey of ground and low-lying excited-state properties for rare isotopes, such as the half lives, Q_β values, reaction cross sections, and 2_1^+-state energies, have been proposed, in which BigRIPS is used and the measurements will be carried out based on the decay spectroscopy and/or the transmission method. In combination with the ZeroDegree spectrometer, studies on the collectivity and matter distribution have been proposed. In these studies, secondary-reaction experiments with RIBs produced by BigRIPS will be employed.

4. RIBF Phase-II Program

The RIBF project will be fully capitalized in the phase-II program, in which the following experimental key devices will be installed.[7]

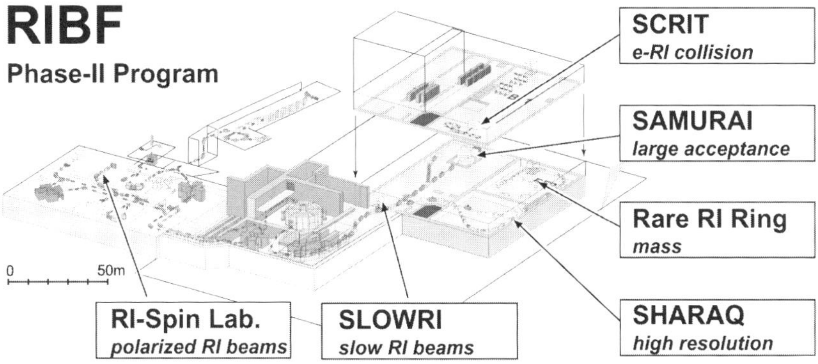

Fig. 3. A schematic plan view of the RIBF phase-II project. Key devices for the phase-II project are shown.

4.1. *Prototypes of SLOWRI and SCRIT*

A slow RI-beam facility (SLOWRI)[7] has been proposed to provide universal slow or trapped RIs with high purity. SLOWRI consists of a gas-catcher system based on the RF ion-guide technique. RIBs produced and transported from BigRIPS are stopped in a gas-catcher cell. They are then drifted and extracted with guided by the *RF carpet*, which is a key device of the SLOWRI system. The SLOWRI system provides a unique opportunity to perform, *e.g.*, precision atomic spectroscopy for a wide variety of RIs, since the production of RIBs in the in-flight fission of uranium and in the projectile-fragmentation reaction are not limited by the chemical process. A prototype of SLOWRI has been developed at the site of the former fragment separator RIPS,[8] with which isotope-shift measurements for ^{7}Be, ^{9}Be, and ^{10}Be have already been carried out.[9]

An innovative Self-Confining Radioactive Ion Target system (SCRIT) has been proposed for the electron scattering experiment from unstable nuclei.[7,10] The SCRIT system is placed in an electron storage ring. RI ions injected from outside the storage ring are not only trapped in the longitudinal direction by the mirror potential produced by a SCRIT electrode but also confined in the transverse potential produced by the circulating electron beam itself. In order to realize the luminosity 10^{26}/s/cm^2, which is needed to employ e-RI collision for the determination of the charge distribution of unstable nuclei, a prototype has been constructed at the electron storage ring KSR, Kyoto University. Using stable isotopes ^{133}Cs, the development of the SCRIT system is in progress now.

4.2. *SHARAQ Spectrograph*

A high-resolution RI-beam spectrograph (SHARAQ)[7] has been proposed for a new type of missing mass spectroscopy. RIBs are used as a probe for the study of, *e.g.*, double Gamow-Teller states, which have been hardly accessible with reactions induced by stable beams. Although the momentum spread of RIBs are very wide, the SHARAQ spectrograph has been designed to provide momentum resolution of 15,000 based on the dispersion-matching technique. The budget for the SHARAQ project has been allocated. It will be installed downstream of BigRIPS in 2008.

4.3. *GARIS and New LINAC*

The recent great success of the discovery of the new super heavy element (SHE), ^{278}113[11] using the RILAC and the GARIS strongly encourages us to further pursue the heavier SHE search and to more extensively study nuclear physical and chemical properties of the SHEs. In the RIBF phase-I scheme, however, BigRIPS experiments and GARIS can not be operated independently, since RILAC is used in both acceleration schemes. In the phase-II program, therefore, the installation of a new LINAC[7] has been proposed, which brings out the operation of BigRIPS experiments and GARIS simultaneously to make it possible to promote further super-heavy element science.

4.4. *Design Study of SAMURAI7 and Rare RI Ring*

Superconducting Analyzer for MUlti particles from RAdioacrive-Isotope beams with 7 Tm (SAMURAI7)[7] has been proposed as a a large-acceptance multi-particle spectrometer to study the particle correlation in unbound states of unstable nuclei. SAMURAI7 is designed to have a large-gap superconducting magnet with bending power of 7 Tm to give the momentum analysis of heavy projectile residues and light-charged particles in the secondary reaction of RIBs at the BigRIPS energies $\sim 250A$ MeV. Its large angular acceptance enables to measure breakup neutrons in coincidence with high efficiency. SAMURAI7 gives us unique opportunities for studies of three-body forces and dynamical properties of isospin-asymmetry nuclear matters by means of, *e.g.*, the invariant mass spectroscopy.

A new mass measurement system (Rare RI ring)[7] has been proposed, which consists of a fast response and feedback kicker-magnet system for the individual injection of RIBs of interest and a precisely tuned isochronous ring. The mass is determined from the TOF of a particle in the ring and its velocity measured before injected into the ring. In the design study, the expected accuracy of the mass measurement is 10^{-6} when the velocity is measured within the 10^{-4} uncertainty. The momentum acceptance of the Rare RI ring is designed 10^{-2}. Since the kicker magnet can be excited as short as 100 ns, rare RI particles of interest can be selectively injected into the ring if the feedback time 450 ns is achieved, in which the response time of trigger counters for the kicker magnet and the cable-line length are taken into account. The development of the fast kicker magnet is in progress now.

4.5. *Upgrade program of RIPS*

In a RI Spin Lab. project, a beam-transport line from IRC to RIPS is installed.[7,12] Since primary beams are accelerated up to $E \sim 115A$ MeV with IRC[2] in the RIBF configuration, intense beams can be also transported to the former fragment separator RIPS at *intermediate* energies. Beams at $E \sim 115A$ MeV allow for a scheme to produce spin-oriented RIBs[13] and to implant them into sample materials with limited thickness. Based on the technique of fragment-induced spin polarization, many ground-state nuclear moments have been determined so far. In addition to the studies of nuclear structure through electromagnetic moments, β decay, β-γ spectroscopy, and material science studies would be conducted based on spin-related research techniques such as β-NMR and γ-PAD/PAC.

5. Summary

The RIBF project is in progress in RIKEN, in which the cascade acceleration with cyclotrons including three new ones, fRC, IRC, and SRC, boosts beam energies up to $E = 400A$ and $350A$ MeV for lighter- ($A \leq 40$) and heavier-mass ions, respectively, over the whole range of atomic number. The project proceeds through two phases. In the phase-I program, BigRIPS and a ZeroDegree spectrometer are constructed downstream of SRC as well as the three new cyclotrons. The construction and installation of the new cyclotrons and BigRIPS has been finished. The ZeroDegree spectrometer will be soon installed in 2007. In the commissioning, after the first beam of aluminum from the entire cyclotron system was successfully extracted, a ^{238}U beam was also extracted recently. Also, RIBs are successfully produced from the ^{86}Kr and ^{238}U projectiles and isotope-separated with BigRIPS. Besides the new-isotope search, the other approved experiments in the first PAC meeting held in February 2007 will be started soon. The second PAC meeting is scheduled on September 2007. The RIBF project will be fully capitalized in the phase-II program, in which the construction of several experimental key devices, SHARAQ, SLOWRI, SCRIT, Rare RI ring, SAMURAI7, RI Spin Lab., and a new RIBF-LINAC, have been proposed. Although the budget has not yet fully allocated, the installation of SHARAQ will be finished in 2008 and the R&D for the other plans are in progress now.

References

1. Y. Yano, *Proc. 17th Int. Conf. on Cyclotrons and their Applications*, Tokyo Japan, October 18-22, 2004, eds. A. Goto and Y. Yano, Particle Accelerator Society of Japan, 169 (2005).
2. J. Ohnishi *et al.*, *ibid*, p. 197.
3. H. Okuno *et al.*, *ibid*, p. 373.
4. T. Kubo, Nucl. Instr. Meth. B **204**, 97 (2003).
5. Y. Yano, to appear in *Proc. Particle Accelerator Conference 07*, New Mexico U. S. A., June 25-29, 2007.
6. T. Ohnishi and T. Kubo, for the BigRIPS commissioning team, to appear in *Proc. XVth International Conference on Electromagnetic Isotope Separators and Techniques Related to their Applications*, Deauville, France, June 24-29, 2007.
7. Technical information on the phase-II devices are found at `http://ribf.riken.jp/RIBF-TAC05`.
8. T. Kubo *et al.*, Nucl. Instr. Meth. B **70**, 309 (1992).
9. T. Nakamura *et al.*, Phys. Rev. A **74**, 052503 (2006).
10. T.Suda and M.Wakasugi, Prog. Part. Nucl. Phys. **55**, 417 (2005).

11. K. Morita *et al.*, J. Phys. Soc. Jpn. **73**, 2593 (2004).
12. N. Fukunishi *et al.*, RIKEN Accel. Prog. Rep. **35**, 283 (2002).
13. K. Asahi *et al.*, Phys. Rev. C **43**, 456 (1991); K. Asahi *et al.*, Phys. Lett. B **251**, 488 (1990).

POPULATION OF NEUTRON UNBOUND STATES VIA TWO-PROTON KNOCKOUT REACTIONS

N. FRANK,* T. BAUMANN, D. BAZIN, A. GADE, J.-L. LECOUEY,† W.A. PETERS, H. SCHEIT‡, A. SCHILLER, and M. THOENNESSEN

National Superconducting Cyclotron Laboratory, Department of Physics & Astronomy, Michigan State University, East Lansing, MI 48824

J. BROWN

Department of Physics, Wabash College, Crawfordsville, IN 47933

P.A. DEYOUNG

Department of Physics, Hope College, Holland, MI 49423

J.E. FINCK

Department of Physics, Central Michigan University, Mt. Pleasant, MI 48859

J. HINNEFELD

Department of Physics & Astronomy, IUSB, South Bend, IN 46634

R. HOWES

Department of Physics, Marquette University, Milwaukee, WI 53201

B. LUTHER

Department of Physics, Concordia College, Moorhead, MN 56562

The two-proton knockout reaction ^{9}Be(^{26}Ne,^{x}O2p) was used to explore excited unbound states of ^{23}O and ^{24}O. In ^{23}O a state at an excitation energy of 2.79(13) MeV was observed. There was no conclusive evidence for the population of excited states in ^{24}O.

Keywords: two-proton knockout reactions; neutron-unbound states; ^{23,24}O.

*Present address: Department of Physics, Concordia College, Moorhead, MN 56562
†Present address: LPC, IN2P3, 14050 Caen, France
‡Present address: Nishina Center, RIKEN, Wako, Saitama 351-0198, Japan

1. Introduction

Two-proton knockout reactions using intermediate-energy heavy-ion beams have been successfully used to explore the structure of neutron-rich nuclei [1,2]. It has been shown that these reactions can be considered direct reactions. Spectroscopic factors can then be extracted from the longitudinal momentum distribution of the residual fragment. Spectroscopic factors of excited states can be extracted by γ-ray coincidence measurements.

Nuclei close to the dripline have very few if any bound excited states. The structure of these nuclei can be explored with neutron-decay spectroscopy [3]. The evolution of the shell structure along the oxygen isotopes is of particular interest [4]. The two heaviest particle-bound oxygen isotopes ^{23}O and ^{24}O do not have any bound excited states [5].

The MoNA collaboration has recently used the two-proton knockout reaction from ^{26}Ne to populate excited states in ^{24}O and ^{23}O [3,6].

2. Experiment

The experiment was performed at the National Superconducting Cyclotron Laboratory at Michigan State University. The secondary beam of 86 MeV/u was produced from a primary ^{40}Ar beam. The two-proton knockout reaction occurred in a 721 mg/cm^2 Be target located in front of a large-gap sweeper magnet [7]. The oxygen isotopes were detected and identified behind the magnet. Neutrons around zero degrees were measured in coincidence with the Modular Neutron Array (MoNA) [8,9]. The details of the fragment production and separation as well as the experimental setup can be found in references [3,6,10,11].

The decay energies of resonances were reconstructed by the invariant mass method from the measured oxygen and neutron energies and the opening angle between the oxygen and the neutron. The angle and energy of the fragments were ion-optically reconstructed using a novel method which takes into account the position at the target in the dispersive direction [12].

3. Results

Figure 1 shows the reconstructed decay energy spectrum of excited ^{24}O. No obvious resonance structure is apparent. The solid line corresponds to a background contribution simulated by a thermal distribution with a temperature of 1.7 MeV. In contrast, the decay energy spectrum of ^{23}O shown in Figure 2 shows a sharp resonance structure very close to threshold [3,6,10]. The solid line corresponds to a fit with a resonance at 45(2) keV and a

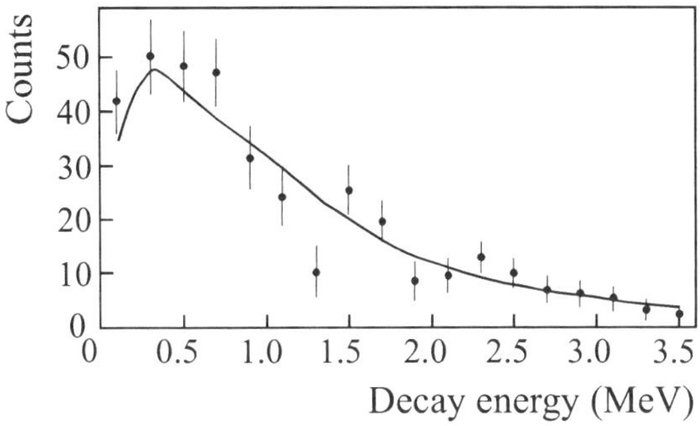

Fig. 1. Decay energy spectrum of ^{24}O (data points). The solid line was calculated from a simulated thermal model.

background contribution with a temperature of 0.7 MeV. With a neutron separation energy of 2.74(13) MeV [13] this corresponds to a state at an excitation energy of 2.79(13) MeV. The state is interpreted as the $5/2^+$ state due to the single particle hole in the $d_{5/2}$ configuration. The $3/2^+$ state recently observed for the first time in the reaction ^{22}O(d,p)^{23}O [14] was not observed.

The absence of excited states in ^{24}O and the selective population of the

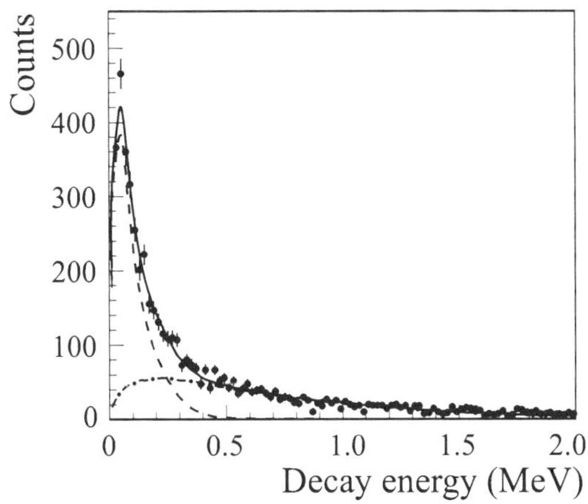

Fig. 2. Decay-energy spectra of ^{23}O* (data points). The solid line corresponds to the sum of the dashed (simulated resonant contribution) and dash-dotted (simulated thermal model) lines [3].

$5/2^+$ state in ^{23}O can be explained with the assumption of a direct knockout reaction and the structure of ^{26}Ne. The knockout of the two valence protons $(\pi(0d_{5/2})^2)$ from ^{26}Ne will populate the particle-bound ground-state of ^{24}O. The spectroscopic factor for the population of the unbound 2^+ state from this reaction is predicted to be quite small [15]. It is not surprising that this state was not observed in the present reaction.

The population of the $5/2^+$ state in ^{23}O is the result of a different two-proton knockout reaction. Instead of the knockout of the two $\pi(0d_{5/2})^2$ protons, the knockout involves one of the core $\pi(0p)^6$ protons in addition to one of the valence $\pi(0d_{5/2})^2$ protons leaving the system in a neutron-unbound excited state. The corresponding one proton (inner shell) knockout from various fluorine isotopes was used to explain the cross section of the population of neutron-rich oxygen isotopes [16].

Figure 3 shows the excitation energy spectrum of this knockout scenario and the subsequent decay to the excited state of ^{23}O, whose neutron decay to ^{22}O is then detected in the present experiment.

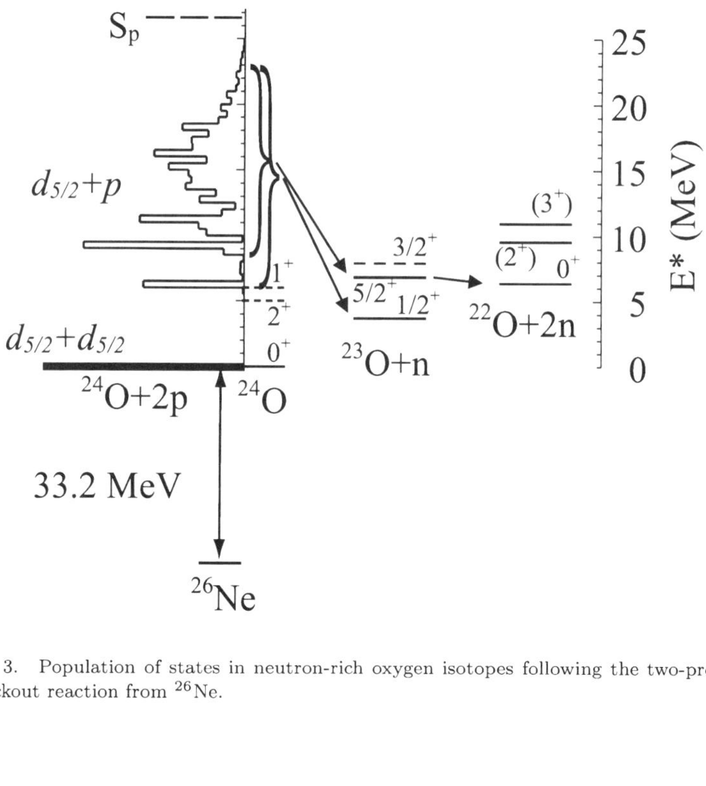

Fig. 3. Population of states in neutron-rich oxygen isotopes following the two-proton knockout reaction from ^{26}Ne.

The selective population of the $5/2^+$ is due to the mixing of the excited proton configuration with neutron excitations. The proton configuration $(\pi(0p) \times \pi(0d_{5/2})^{-1})$ can mix with the neutron excitations of the form $\nu(0p)^{-1} \times \nu(0d_{3/2})^1$, $\nu(1s_{1/2})^{-1} \times \nu(0f1p)^1$, and $\nu(0d_{5/2})^{-1} \times \nu(0f1p)^1$. The first of these has large spectroscopic overlap with high-lying negative-parity excitations in ^{23}O; the other two will excite one neutron (either a $1s_{1/2}$ or a $0d_{5/2}$) to the continuum of the *fp*-shell. The current setup is not geometrically efficient for the detection of these high-energy neutrons which probably contribute to the non-resonant background. The resulting ^{23}O is then in the $1/2^+$ ground state or the $5/2^+$ excited state. The subsequent neutron decay of the $5/2^+$ state is the resonance observed in the present experiment. These different scenarios and the possible direct three-particle $2p1n$ knockout is further described in References [3,6].

In this reaction mechanism the $0d_{3/2}$ configuration is not populated and thus the recently observed $3/2^+$ (particle) state [14] is not populated in our experiment. The experiments are complementary because the present knockout reactions are sensitive to hole states while the single-neutron transfer reaction ^{22}O(d,p)^{23}O* of Reference [14] is predominantly sensitive to particle states.

4. Summary

A two-proton knockout reaction was successfully used to populate neutron-unbound states in ^{23}O. In contrast to the population of bound excited states that are predominantly populated from the knock-out of valence protons, the population of neutron-unbound states involves the knock-out of inner shell protons. The mixing of these deep-hole proton states with neutron states leads to the emission of a continuum neutron. The non-observation of any excited states in ^{24}O and the $3/2^+$ particle state in ^{23}O and the observation of the $5/2^+$ hole state in ^{23}O are consistent with this reaction mechanism.

This new reaction mechanism offers the opportunity to explore excited states in neutron-rich nuclei that might not be accessible with any other method.

Acknowledgments

We would like to thank the members of the MoNA collaboration G. Christian, C. Hoffman, K.L. Jones, K.W. Kemper, P. Pancella, G. Peaslee, W. Rogers, S. Tabor, and about 50 undergraduate students for their contributions to this work. We would like to thank R.A. Kryger, C. Simenel, J.R. Terry, and K. Yoneda for their valuable help during the experiment. Financial support from the National Science Foundation under grant numbers PHY-01-102533, PHY-03-54920, PHY-05-55366 PHY-05-55445, and PHY-06-06007 is gratefully acknowledged. J.E.F. acknowledges support from the Research Excellence Fund of Michigan.

References

1. D. Bazin *et al.*, Phys. Rev. Lett. **91**, 012501 (2003).
2. D. Warner, Nature **425**, 570 (2003).
3. N. Frank *et al.*, Acta. Phys. Hung., in press (2007).
4. T. Otsuka *et al.*, Phys. Rev. Lett. **87**, 082502 (2001).
5. M. Stanoiu *et al.*, Phys. Rev. C **69**, 034312 (2004).
6. A. Schiller *et al.*, subm. for publ. (2007), arXiv:nucl-ex/0612024.
7. M.D. Bird *et al.*, IEEE Trans. Appl. Supercond. **15**, 1252 (2005).
8. B. Luther *et al.*, Nucl. Instrum. Methods Phys. Res. **A505**, 33 (2003).
9. T. Baumann *et al.*, Nucl. Instrum. Methods Phys. Res. **A543**, 517 (2005).
10. N. Frank, Ph.D. thesis, Michigan State University, 2006.
11. W.A. Peters, Ph.D. thesis, Michigan State University, 2007.
12. N. Frank *et al.*, Nucl. Instrum. Methods Phys. Res. **A**, (in press).
13. G. Audi, A.H. Wapstra, and C. Thibault, Nucl. Phys. **A729**, 337 (2003).
14. Z. Elekes *et al.*, Phys. Rev. Lett. **98**, 102502 (2007).
15. B.A. Brown, private communication.
16. M. Thoennessen *et al.*, Phys. Rev. C **68**, 044318 (2003).

STUDIES OF NEUTRON-RICH NUCLEI USING ISOL FACILITIES AT CERN AND JYVÄSKYLÄ*

JUHA ÄYSTÖ
ISOLDE AND IGISOL COLLABORATIONS†
Department of Physics, FI-40014 University of Jyväskylä, Finland
Helsinki Institute of Physics, FI-00014 University of Helsinki, Finland

Neutron-rich nuclei far from stability and near the magic neutron numbers have attained much interest in recent years due to predictions for shell structure modifications when approaching the drip-lines. Some recent experiments at ISOL facilities on binding energies and excited structures studied by high-precision mass spectrometry and post-accelerated beams are discussed in this paper.

1. Introduction

ISOL facilities have played important role in exploring nuclear properties far from the valley of stability. New facilities in Europe with post-acceleration capabilities and modern techniques for production of pure and high-quality beams from lowest to medium-high energies are opening exciting period in the radioactive ion beam science. This paper focuses on two particularly successful areas of research. One deals with the studies carried out at ISOLDE, particularly employing the REX-ISOLDE post accelerator where several succesful experiments employing Coulomb excitation of radioactive neutron-rich beams have been recently performed [1,2]. Another area of research with major successes has been the precision mass measurements of exotic nuclei at ISOL facilities [3,4]. These studies are increasingly depending on novel ion and atom trapping techniques for stopped or low-energy ions as well as production methods and new target techniques. Both low-energy as well as post accelerated radioactive ion beams are beginning to produce new and exciting results on the evolution of shell and collective structures of exotic nuclei, in particular near the "classical" magic neutron numbers at N=20, 50 and 82. In addition to nuclear

* This work has been supported by the EU 6th Framework programme "Integrating Infrastructure Initiative - Transnational Access", Contract Number: 506065 (EURONS) and by the Academy of Finland under the Finnish Centre of Excellence Programme 2006-2011 (Nuclear and Accelerator Based Physics Programme at JYFL).

structure physics the ISOL-facilities are also contributing data to nucleosynthesis calculations. In general, the ISOL method has continued to remain as an important tool providing high-intensity and quality beams which makes it complementary to high-energy in-flight method.

2. Precision mass measurements

2.1. *JYFLTRAP facility*

The JYFLTRAP setup [5] consists of the on-line isotope separator IGISOL [6] coupled to a double Penning trap system via a radiofrequency quadrupole (RFQ) cooler/buncher device. At this facility neutron-rich nuclei of interest are produced in asymmetric and symmetric fission reactions by bombarding a 10 mg/cm^2 thick target of natural uranium typically with 25 MeV protons of 10 μA intensity. Fission fragments recoiling out of the target are thermalized in helium gas at a pressure of about 250 mbar. Because of charge exchange processes with the gas the ions will end up being mainly singly charged. They are then transported with the helium flow in a few ms out of the chamber and are accelerated to 30 keV energy. After acceleration, a 55° bending magnet provided a mass resolving power of 500, sufficient to select ions of only one mass number. Yields of n-rich nuclei in mass measurements range from 10 to several 1000 ions/s depending on the mass number of the studied isotope.

Radioactive ion beam of a selected mass number is injected into a Radio Frequency Quadrupole (RFQ) cooler and buncher placed on a high voltage platform at 30 kV. There the ions are decelerated and cooled by collisions with helium buffer gas [7]. The ions are then released as bunches and transferred into the Penning trap setup on the same high voltage platform [8]. This setup consists of two cylindrical Penning traps housed in the warm bore of a superconducting magnet of 7 T. The magnet has two homogeneous field regions of 1 cm^3 located 20 cm apart from each other. The first trap contains helium buffer gas at a pressure of a few times 10^{-5} mbar and is used for cooling and isobaric mass separation. Dipole excitation is used to increase the radii of all ions. Afterwards, quadrupole excitation is employed to mass-selectively center the ions of the wanted isotopes. Typically, a cycle time of about 400 ms is used in the purification trap and a mass resolving power of about 10^5 was reached with this configuration as demonstrated by the isobar scan shown in Fig. 1 for the A=101 fission products.

After purification, the ion ensemble is injected into the second Penning trap operating in vacuum ($< 10^{-8}$ mbar) for the actual mass measurement. In this measurement ion cyclotron motion is resonantly excited by external quadrupolar RF field. Following this, the radially excited ions are ejected from the trap. They drift towards the microchannel plate (MCP) detector through a set of drift tubes where a strong magnetic field gradient exists. Here the ions experience a force which converts the radial kinetic energy into longitudinal energy additional to the 30 keV acceleration from the high-voltage platform to the ground potential where the MCP is situated. As a result an ion which was in resonance with the applied excitation quadrupolar field moves faster towards the detector than an ion which was on resonance. By measuring the time of flight as

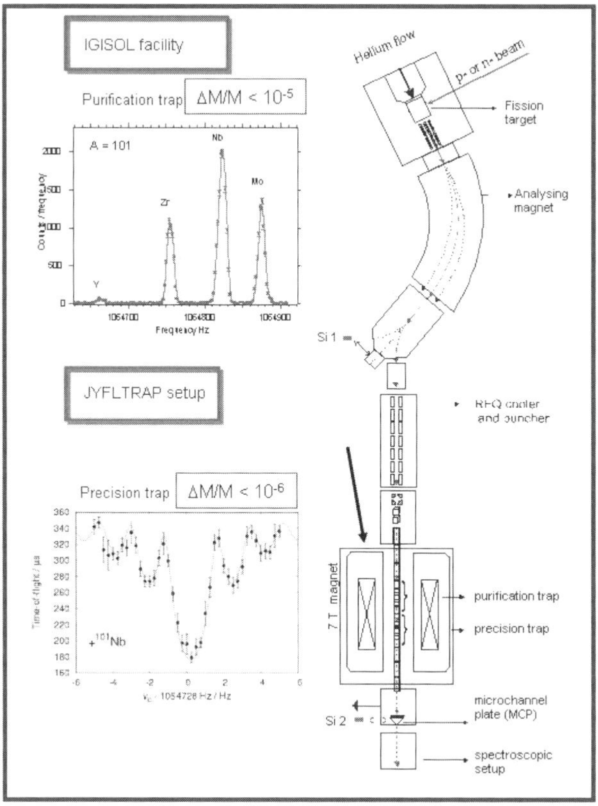

Figure 1. JYFLTRAP setup coupled to the IGISOL facility.

a function of the applied excitation frequency, the cyclotron frequency ν_c was determined. A typical time-of-flight resonance spectrum is shown in Fig.1. This spectrum yields the ion cyclotron frequency

$$\nu_c = \frac{1}{2\pi}\frac{q}{m_{ion}} \cdot B$$

For each isotope measured data are collected in several sets. Each set comprised several pairs of unknown-ion and reference-ion frequency scans taken under the same conditions. Each of the resonance curves was fitted with a realistic function, described in Ref. [9], which yielded values for the resonant frequency and its statistical uncertainty. Fig. 1 shows a typical resonance curve, in this case the one for ^{101}Nb ions that were injected from the purification trap.

2.2. *Mass measurements across the N=50 shell gap*

Precision mass measurements at JYFLTRAP have covered until now a broad range of isotopes produced in fission, as shown in Fig. 2. Complemented by the data from the ISOLTRAP n-rih isotopes of all elements from Ni to Pd and some isotopes of Ag, Cd and Sn have now been measured to a high accuracy with the Penning trap. The data set includes now over 200 n-rich isotopes. These measurements have shown large discrepancies with the previous data [10], in particular at and above the A=100 fission products where deviations as large as 1 MeV have been observed for the most n-rich nuclei studied. The older data is mainly derived from the beta end-point measurements where large systematic errors are easily incorporated due to inadequate knowledge of the decay schemes and unfolding procedures of the continues beta spectra.

An example of the application of the new mass data is shown in Fig. 4 which displays the two-neutron separation energies across the N=50 shell gap from Cu (Z=29) up to Pd (Z=46). The values for n-rich nuclei from Cu to Kr are from Penning trap measurements at ISOLTRAP [4] and JYFLTRAP and p-rich data of Ru, Rh and Pd from experiments at SHIPTRAP and JYFLTRAP [13] Fig. 2 indicates that the shell gap starts to increase towards ^{78}Ni. This finding is somewhat contradictory several other experiments performed on excited states of Z>31, N=50 nuclei. However, as shown later, new Coulomb excitation measurements on the N=50 isotopes of ^{82}Ge and ^{80}Zn are rather suggesting the strengthening of the N=50 shell closure towards ^{78}Ni. However, more experimental data are needed and of particular immediate interest is related to a still unknown mass and two-neutron separation energy of ^{82}Zn. The increase in the shell gap energy is also predicted by a theory calculation employing the HFB method in the frame of the density functional theory in ref. [14].

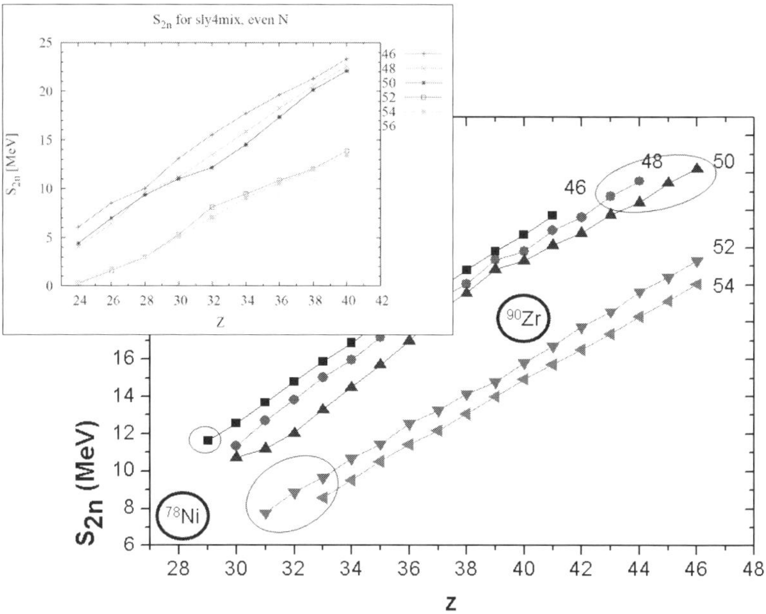

Figure 2. Two-neutron separation energies across the N=50 shell. Data is partly preliminary and is taken from refs. []. Circled values represent the first experimental measurements of the S_{2n} values for these isotopes. Insert shows theoretical S_{2n} values for the same region from ref. [14].

In addition to providing information on nuclear structure also the rapid neutron-capture process is an important topic which can be related to the mass measurements of exotic n-rich nuclei. The r-process has a fundamental importance as it explains the origin of approximately half of the stable nuclei heavier than iron in nature. It occurs at the center of type II supernova at a very high temperature and neutron density. Nuclear masses have the most decisive influence on the reaction flow of the r-process paths, and therefore masses are among the most critical nuclear parameters in nucleosynthesis calculations. The new accurate mass values from penning trap experiments close to the r-process path reported will improve the r-process network calculations to minimize the ambiguity.

2.3. *REX-ISOLDE facility*

A novel post-accelerator complex REX-ISOLDE is opening up an entire new field of research with radioactive ion beams at Coulomb barrier energies. This approach is particularly powerful in structure studies of exotic nuclei using transfer reactions and Coulomb excitation. REX-ISOLDE is now fully operational with energies up to 3.1 MeV/u and has already accelerated more than 40 species of radioactive ions with intensities up to 10^5 ions/s. It has the capability to accelerate heavy-mass ions up to Hg with efficiency of a few percent of the ion source yield. The main work at REX-ISOLDE is performed with the highly efficient and highly segmented gamma-ray MINIBALL array at the secondary target position[15]. In 2003 REX-ISOLDE was integrated into the standard operation of CERN facilities. The upgrade of REX-ISOLDE is planned up to over 5 MeV/u energies within the so called HIE ISOLDE concept.

The first step in REX-ISOLDE shown in Fig. 3 is to inject the singly charged ions from the ISOLDE separator continuosly into a Penning trap where they are accumulated and cooled. Therefore, all beams that are produced at ISOLDE can be fed into the REX facility. After about 20 ms manipulation and purification in the REXTRAP ion bunches are transferred to an electron beam ion source (EBIS). After charge breeding to a charge-to-mass ratio of 1/4.5 the ions are injected into a radio-frequency quadrupole (RFQ) accelerator via a mass separator which is similar to the well known Nier-spectrometer. The accelerator consists of an RFQ, an interdigital H-type (IHS) structure, three seven-gap (7-gap) resonators followed by the 9-gap resonator. The IH-Structure and the 7-gap and 9-gap resonators will allow an energy variation between 0.8 and 3.1 MeV/u to meet the experimental requirements.

The MINIBALL array [18] employed in typical experiments consists of of eight clusters each combining three six-fold segmented HPGe crystals. While the gamma-ray energy iss extracted from the core signal of the individual crystals, the segment with the highest energy deposition determined the emission angle of the gamma-ray. Doppler correction was applied by combining this information with the direction and velocity of the coincident scattered particle detected in the DSSSD detector.

Originally REX-ISOLDE aimed at two main goals: the demonstration of a new concept to bunch, charge-breed and post-accelerate singly charged, low energetic ions in an efficient way and to apply the method in the structure studies of radioactive nuclei in the lower part of the Nuclear Chart due to limitation of kinetic energy. Special focus in the first studies has been on neutron-rich Mg, Cu and Zn isotopes in the vicinity of the closed neutron shells at $N = 20$ and $N = 50$.

Neutron-rich Mg isotopes are particularly interesting for their structure due to rapid shape changes as well as coexistence of vastly different structures in the same nucleus. The first experiment on ^{30}Mg isotope has measured by employing Coulomb excitation the B(E2) value to its first excited 2^+ state [2]. These experiments will soon be extended to heavier Mg isotopes providing important complementary measurements to those made at intermediate energies (> 30 MeV/u), well above the Coulomb barrier where Coulomb-nuclear interference effects and feedings from higher lying states complicate the analysis of the spectra.

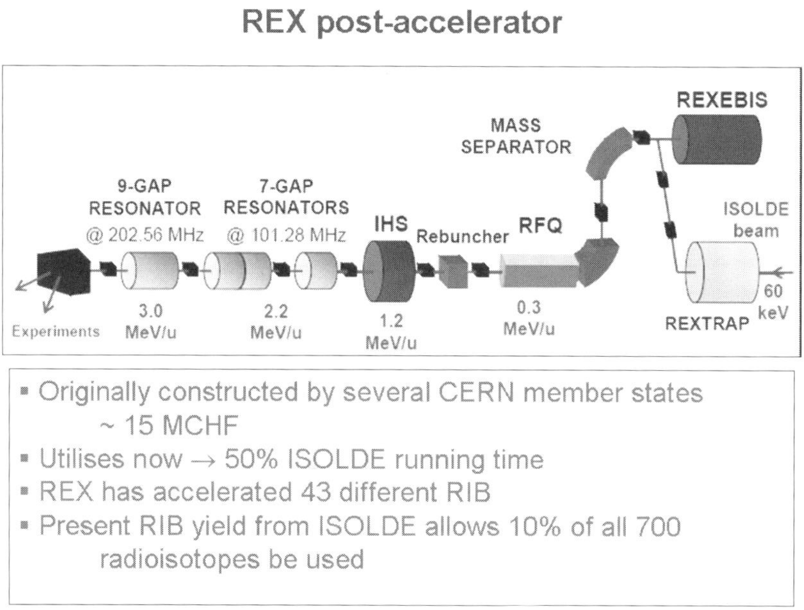

Figure 3. The concept of the REX-ISOLDE facility at CERN.

2.4. *Recent results of experiments on neutron-rich Cu and Zn isotopes*

Currently REX-ISOLDE experiments are being extended to heavier n-deficient as well as n-rich nuclei including ^{70}Se [16], ^{78,80}Zn [17] and ^{108,110}Sn [18]. It is of particular interest to mention the Coulex experiments approaching N=50 and N=82 n-rich nuclei. Some of the key experiments on N=82 nuclei are carried out also at the HRIBF facility of Oak Ridge. Recent experiments measuring the B(E2) values by Coulomb excitation of n-rich Ge [19] at HRIBF and Zn isotopes at REX-ISOLDE could provide the first direct spectroscopic information on the shell closure at N=50. These results together with the analysis based on the standard nuclear shell model suggests the strengthening of the

N=50 shell gap towards ^{78}Ni. These results seem to agree with the newest binding energy data discussed earlier in connection with the mass measurements of neutron-rich nuclei near N=50.

Another major breakthrough in accelerated radioactive beam science, also obtained at REX-ISOLDE, concerns the Coulomb excitation of isomeric beams of ^{68,70}Cu [20]. This has been made possible with isomer selective resonant laser ionization.. In these experiments use was made of Coulomb excitation of post-accelerated 6^- beams of ^{68m}Cu and ^{70}Cu to characterize the states of the $\pi 2p_{3/2}$ $\nu 1g_{9/2}$ multiplet. The beams of ^{68m,70}Cu, post-accelerated up to 2.83 MeV/u, were used to bombard a ^{120}Sn target. Typical beam intensities at the detection setup were 3 10^5 pps [^{68m}Cu 6^-] and 5 10^4 pps [^{70}Cu 6^-]. Scattered projectile and recoiling target nuclei were detected by a DSSSD detector. The detection of gamma-rays was performed with the MINIBALL array, see Fig. 4. The obtained results for the B(E2) values for the 6^- to 4^- transitions indicates weak polarization effects induced by the extra-proton and neutron coupled to the semi-doubly magic ^{68}Ni.

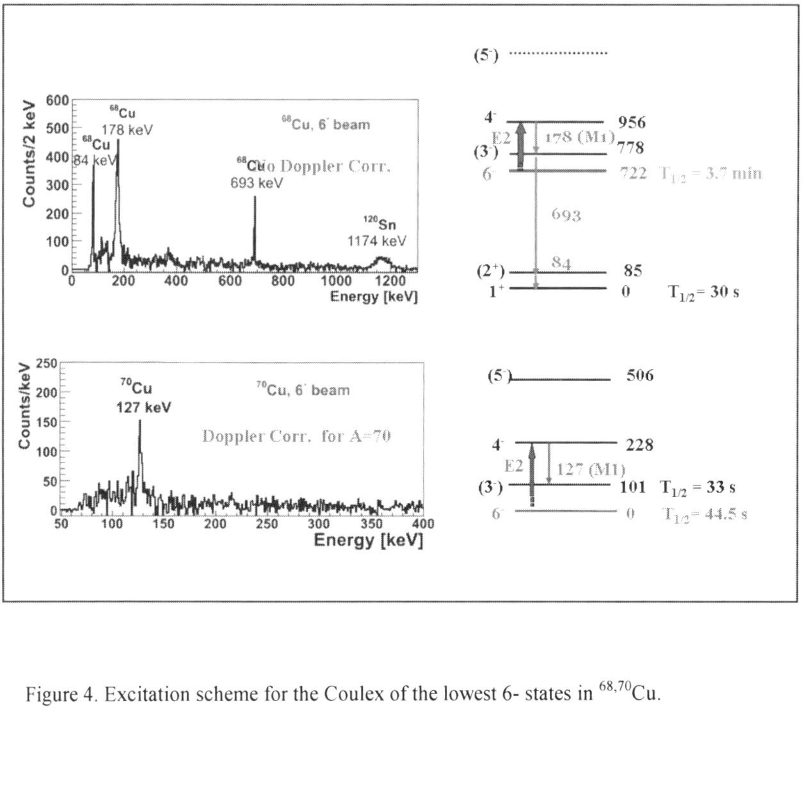

Figure 4. Excitation scheme for the Coulex of the lowest 6- states in ^{68,70}Cu.

3. Future directions

The next major developements in the field are the FAIR and SPIRAL2 construction projects as well as the upgrade of the REX-ISOLDE facility HIE-ISOLDE. The HIE-ISOLDE is planned to consist of both the energy and intensity upgrades. The energy will be boosted by additional superconducting linac sections first to 5.5 MeV/u and later to 10 MeV/u. The intensity upgrade will consist of the PS Booster upgrade with the new linear accelerator injector LINAC4 as well as the target and front-end upgrades. Several other improvements are in preparation as well. They include , for example, the implementation of the RFQ cooler/buncher and the upgrades of REX-TRAP, REX-EBIS and REX-ECR. With these improvements as well as the availability of more than 700 radioisotope beams at ISOLDE, REX-ISOLDE would become the world-leading facility in the field for long time to come.

Acknowledgments

The author wishes to thank several individuilas for fruitful discussions and material for this presentation. In particular, I am indebted to Peter Butler, Ari Jokinen and Piet Van Duppen.

References

1.O. Kester et al., Nucl. Instr. Meth. Phys. Res., B 204, 20 (2003).
2. O. Niedermaier et al., Phys. Rev. Lett. 94, 172501 (2005).
3.U. Hager et al., Phys. Rev. Lett. 96, 042504 (2006)
4.C. Guenaut, Phys. Rev. C 75, 044303 (2007).
5. A. Jokinen et al., Int. J. Mass Spectrometry 251, 204 (2006).
6. J. Äystö, Nucl. Phys. A 693, 477 (2001).
7. A. Nieminen et al., Phys. Rev. Lett. 88 (2002) 094801
8. V. Kolhinen et al., Nucl. Instr. Meth. Phys. Res., A528, 776 (2004).
9. M. König et al., Int. J. Mass Spectrom. Ion processes 142, 95 (1995).
10. G. Audi, A.H. Wapstra and C. Thibault, Nucl. Phys. A729, 337 (2003).
11. S. Rahaman et al., submitted to Eur. Phys. J. A.
12. J. Hakala et al., in preparation.
13. V. Elomaa et al., in preparation.
14. M. Stoitsov et al., Phys. Rev. Lett. 98, 132502 (2007).

15. J. Eberth et al., Prog. Part. Nucl. Phys. **46**, 389 (2001).
16. A . Hurst et al., Phys. Rev. Lett. 98, 072501 (2007).
17. J. Van de Walle et al. Phys. Rev. Lett. to be published
18. J. Cederkäll et al., Phys. Rev. Lett. 98 (2007) 172501
19. E. Padilla-Rodal et al., Phys. Rev. Lett. 94 (2005) 122501
20. I. Stefanescu et al., Phys. Rev. Lett. 98 (2007) 122701

SHELL STRUCTURE EVOLUTION FAR FROM STABILITY: RECENT RESULTS FROM GANIL

F. AZAIEZ

Institut de Physique Nucléaire,
91406, Orsay Cedex
E-mail: azaiez@ipno.in2p3.fr

Shell structure evolution in nuclei situated at the extremes of neutron and proton excess are investigated using in-beam gamma spectroscopy techniques with radioactive beams at GANIL. A selection of results obtained very recently is presented: i) The reduced transition probabilities $B(E2;0_1^+ \rightarrow 2^+)$ of the neutron-rich ^{74}Zn and ^{70}Ni nuclei have been measured using Coulomb excitation at intermediate energy. An unexpected large proton core polarization has been found in ^{70}Ni and interpreted as being due to the monopole interaction between the neutron $g_{9/2}$ and protons $f_{7/2}$ and $f_{5/2}$ spin-orbit partner orbitals. ii) Two proton knock-out reactions has been performed in order to study the most neutron-rich nuclei at the N=28 shell closure. Gamma rays spectra and momentum distribution have been obtained for ^{42}Si and neighboring nuclei. Evidences has been found for a deformed structure for ^{4}2Si and for the disappearance of the spherical N=28 shell effect. iii) The in-beam gamma spectroscopy of ^{36}Ca performed using neutron knock-out reactions revealed that N=16 is as large sub-shell closure as large as Z=16 in ^{36}S. The uniquely large excitation energy difference of the first 2^+ state in these mirror nuclei turns out to be a consequence of the relatively pure neutron (in ^{36}Ca) or proton (in ^{36}S) $1p(d_{3/2})$-$1h(s_{1/2})$ nature.

Keywords: in-beam gamma-spectroscopy, fragmentation reactions, proton and neutron knock-out reactions,Coulomb excitation at intermediate energies, neutron-rich nuclei

1. Introduction

Shell structure predictions towards the neutron and proton drip-lines provide indispensable input to model the rapid neutron and proton capture for nucleosynthesis processes .[1] Furthermore, the evolution of shell structure far off the stability line has become a key topic of nuclear structure studies as it is intimately related to the monopole part of the residual nucleon-nucleon interaction and its origin in one-boson exchange potentials.[2] For instance, shell gap evolutions have been recently ascribed to the strongly attractive (repulsive) tensor force acting between protons and neutrons with opposite (similar) orientation of their intrinsic spin with respect to their angular momentum .[3]

In the following, three recent examples of studies of shell structure evolution away from the valley of stability will be presented. These studies have been performed at GANIL and they are all based on in-beam gamma spectroscopy techniques used with reactions induced by secondary beams at intermediate energies. The three studies concern:

(1) The evolution of the Z=28 shell effect in nuclei with neutron numbers beyond N=40.

(2) The effectiveness of the N=28 shell effect in silicon nuclei.

(3) The difference between the A=36 Ca isotope and its S mirror isotope

2. The evolution of the Z=28 shell effect in nuclei with neutron numbers beyond N=40

Recently the Ni isotopes between the $N = 28$ and 50 shell closures have been the subject of extensive experimental and theoretical studies .[6–12] For $28 \leq N \leq 40$ a parabola-like trend was found[10] in the B(E2;$0^+ \rightarrow 2^+$) values which seemed to indicate a sub-shell closure at $N = 40$. On the other hand, the two-neutron separation energies exhibit a smooth decrease at $N = 40$[8,13] rather than a sharp drop. Beyond $N = 40$ and up to $N = 48$ a striking reduction in the 2^+ excitation energies is observed for the Ni isotopes[14,16–18] which could be due to increasing collectivity due to the neutron-proton tensor force. For that purpose, a direct measurement of the B(E2;$0^+ \rightarrow 2^+$) value in the Ni and Zn isotopes have been carried out at GANIL, using Coulomb excitation at intermediate energies.

${}^{74}_{30}Zn_{44}$ and ${}^{70}_{28}Ni_{42}$ were produced via reactions of a 60·A MeV ${}^{76}Ge^{30+}$ beam with an average intensity of 1.2 eμA in a 500 μm-thick Be target. Two settings of the LISE3 spectrometer[20] were used to select these isotopes. Mean rates of 2800 and 800 per second were obtained for the ^{74}Zn and ^{70}Ni isotopes, respectively. Another spectrometer setting was set to transmit the $Q = 28^+$ charge state of the primary beam at a rate of $10^4 s^{-1}$ in the same optical conditions. The nuclei produced were identified by means of their energy loss and time of flight measured in a "removable" Si detector placed downstream from the spectrometer at the entrance of the target chamber.

Two annular Si detectors were used in order to identify the deflected nuclei. The angular coverage of the silicon detector (from 1.5 to 6.1 degrees) in order to reduce as much as possible nuclear interaction with the target.

A ^{208}Pb target with 120 mg/cm^2 thickness was located at the focal plane of LISE3 and was surrounded by four segmented EXOGAM clover Ge detectors with a total photo-peak efficiency of ϵ_γ=5.0 % at 1.3 MeV. The segmentation of the clover detectors allowed to reduce the Doppler broadening by 40%, leading to an energy resolution (FWHM) of 75 keV at 1 MeV. The Doppler-corrected spectra for the 3 studied nuclei ^{76}Ge, ^{74}Zn and ^{70}Ni are shown in Fig. 1. They exhibit photo-peaks associated with the Coulomb excitation of the 2^+ energy level.

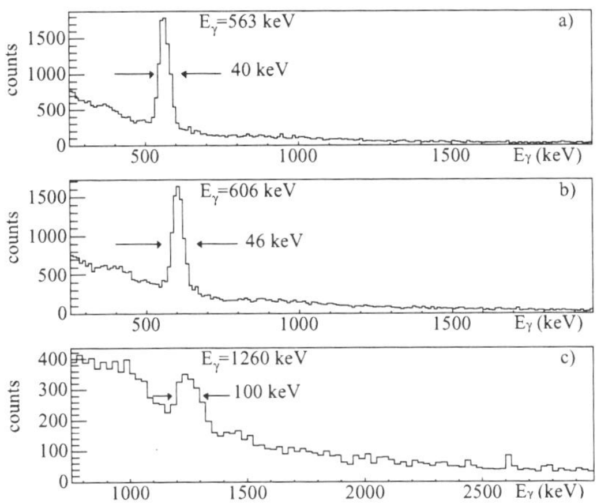

Fig. 1. Doppler-corrected γ-energy spectra obtained in the Ge clover detectors from the Coulomb excitation of the ^{76}Ge, ^{74}Zn and ^{70}Ni isotopes.

The measured total Coulomb excitation cross-section for ^{76}Ge, was found to be 1.09(0.1) b. By comparing the calculated cross-section to the experimental value, we find a B(E2) of 2720(250) e^2fm^4 for ^{76}Ge, which is in close agreement with the value of 2680(80) e^2fm^4[19] obtained using a low-energy Coulomb excitation. This gives more confidence in the values measured for new isotopes such as ^{70}Ni and ^{74}Zn which were found to be 860(170) e^2fm^4 and 1960(140) e^2fm^4, respectively.

The behavior of the B(E2;$0^+ \rightarrow 2^+$) values ,as shown in Fig. 2, provides nuclear structure information on the existence of the $N = 40$ sub-shell closure and the evolution of collectivity in the Ni and Zn isotopic chains. In the Ni isotopic chain the $\pi f_{7/2}$ orbital is separated from the remaining proton orbitals of the fp shell by the $Z = 28$ gap. In the Zn isotopic chain, valence protons in the $\pi p_{3/2}$ and $\pi f_{5/2}$ orbitals (above the $Z = 28$ gap) also add to polarization by proton-neutron interaction besides their direct contribution to the E2 strength. The large scale shell model (LSSM) reproduce the Ni and Zn B(E2) curves below mid-shell $N = 34$ within a fp model space [24,25] while for N≥34 the $g_{9/2}$ neutron orbit is essential for a correct description.[10,25]

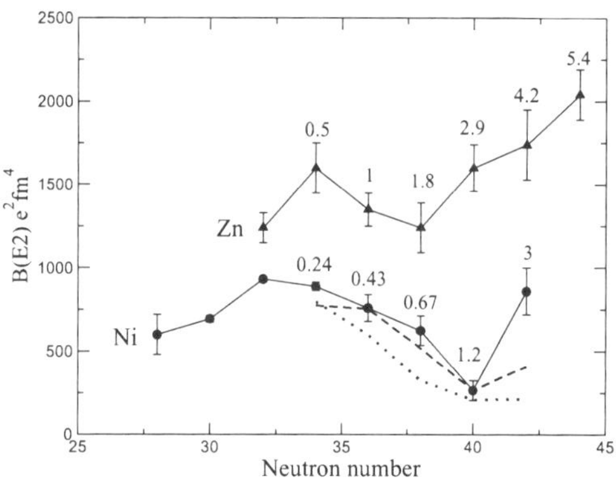

Fig. 2. Experimental B(E2;$0^+ \rightarrow 2^+$) values in units of e^2fm^4 in the Ni and Zn isotopic chains. Results on ^{70}Ni and ^{74}Zn nuclei are from the present study. Other values are taken from .[19,26] The number of neutrons in the $g_{9/2}$ orbital (written on top of each B(E2) curve) and the B(E2) values in Ni are calculated with the shell model of Ref.[10] (dashed line) or with the QRPA model of Ref.[12] (dotted line).

Beyond $N = 40$, the B(E2) values of the Zn isotopes continue to increase at least towards the $g_{9/2}$ mid-shell ($N \simeq 44$). the results of the present work indicate indicate also an increase of the B(E2) from ^{68}Ni to ^{70}Ni which is the result of the Z=28 core polarization induced by the extra two neutron out the N=40 shell closure. The amount of the proton core polarization of the 2^+ state in ^{70}Ni was inferred through the evolution of the B(E2;$J \rightarrow J-2$) values along the 8^+, 6^+, 4^+, 2^+ components of the $(\nu g_{9/2})^2$ multiplet. In this approach the experimental B(E2;$8^+ \rightarrow 6^+$) = 19(4),[28] B(E2;$6^+ \rightarrow 4^+$) = 43(1)[29,30] and the present B(E2;$2^+ \rightarrow 0^+$)= 172(34)e^2fm^4 are calculated as 17.3, 44.6 and 92.2 e^2fm^4 using an effective neutron charge e_ν= 1.2 e. The good agreement for the high-spin states breaks down for the $2^+ \to 0^+$ transition which is a clear signature for an enhanced proton core polarization at low excitation energy.

The strong polarization in the Ni and Zn isotopes beyond N=40 could be due to the attractive $\pi f_{5/2}$ - $\nu g_{9/2}$ monopole interaction,[34,35] ascribed to the tensor force of the in-medium nucleon-nucleon interaction.[3] This force also acts through the repulsive $\pi f_{7/2}$ - $\nu g_{9/2}$ interaction to reduce the apparent $\pi f_{7/2}$ - $\pi f_{5/2}$ spin-orbit splitting, provoking the crossing of the $\pi f_{5/2}$ and the $\pi p_{3/2}$ levels.[34,36] Eventually the effective $Z = 28$ shell gap is decreased and enhanced *ph* excitations across the $Z = 28$ shell closure are generated. It is foreseen that this interaction will continue to weaken the Z=28 gap with the complete filling of the $\nu g_{9/2}$ at the doubly magic $^{78}_{28}Ni_{50}$.

3. The effectiveness of the N=28 shell effect in silicon nuclei

The aim of this studies is to measure the energy of the first 2^+ level in ^{42}Si and to perform spectroscopy in the neighbors nuclei through in beam γ-spectroscopy using double fragmentation. This method has been already successfully applied to the spectroscopy of lighter nuclei around N=16[38] whereas ^{44}S was also studied but using a single fragmentation.[39] A first fragmentation in the SISSI device of a 4μA ^{48}Ca beam at 60MeV/u produced a cocktail beam optimized for the production of nuclei around ^{44}S. An average ^{44}S production of 100 pps was obtained. These radioactive ions, selected by the alpha spectrometer were transported at the target point of the SPEG and identified event by event by their energy loss (ΔE) in a Si detector located before the secondary target and their time-of-flight from the exit of the alpha spectrometer to the secondary target. The cocktail beam impinged the secondary Be target (185 mg/cm^2) and the nuclei produced in the second fragmentation were transmitted through the SPEG spectrometer with a Bρ value adjusted to maximize the collection of the ^{42}Si. They were identified through their energy loss in an Ionization Chamber (IC) and their masse over charge (A/Q) ratio deduced from time-of-flight and Bρ measurement.

The secondary target was surrounded by the 4π gamma array 'Chateau de Crystal' consisting in 74 BaF_2 scintillators. The time resolution, around 1 nsec, allows to remove the neutrons and light charged particles emitted in the fragmentation process. The gamma efficiency and energy resolution have been checked with sources and were close to 50% and 10% at 1 MeV, respectively. The Doppler correction has been performed using the proper velocity of final fragments detected at the focal plan of SPEG. The spectrum extracted in the $^9Be(^{44}S,^{42}Si)$ reaction is displayed in figure 3.

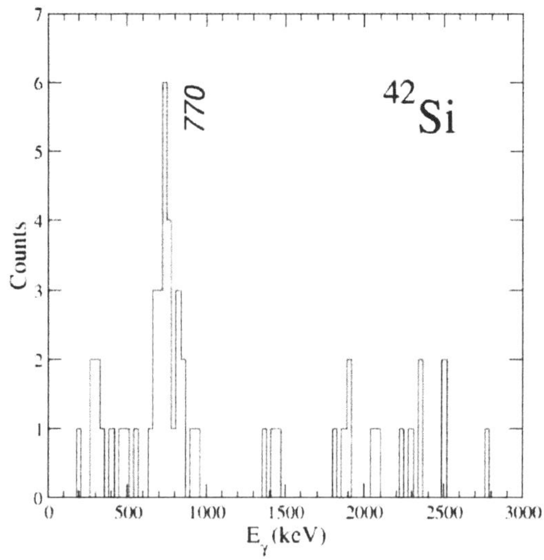

Fig. 3. γ-ray spectrum obtained for the ^{42}Si in the (2p) removal reactions.

The ^{42}Si spectrum exhibits a clear peak 765±20 keV with 23 counts which is attributed to the $2^+ \rightarrow 0^+$ transition. The ^{40}Si spectrum shows two peaks located at 624±10 MeV and 991±10 MeV. The ^{38}Si spectrum exhibits a single peak located at 1081±8 MeV which agree with the known $2^+ \rightarrow 0^+$ transition. We have display in figure 4 the energies of the 2^+ state in the Si isotopes, from N=18 to N=28. The case of the Ca isotopes is also displayed for comparison. For Si and Ca isotopes, the 2^+ excitation energy reaches a maximum at N=20 illustrating the N=20 spherical shell effect and the 'magic' character of the N=20. For the isotone ^{32}Mg located in the island of inversion where intruder configurations dominate the ground state structure leading to a deformed nucleus, the 2^+ energy is known to drop at 885 keV. A similar drop is clearly observed for the ^{42}Si at N=28. From this observation, we do conclude that ^{42}Si is a collective nucleus which should be very deformed.[41] From this similarity we are tempted to call ^{42}Si a first nucleus of the isle of inversion around N=28. Shell model calculations and mean field calculations so far also support a similar description of ^{42}Si with an oblate deformation resulting from a reduction of the N=28 shell closure.[42–45]

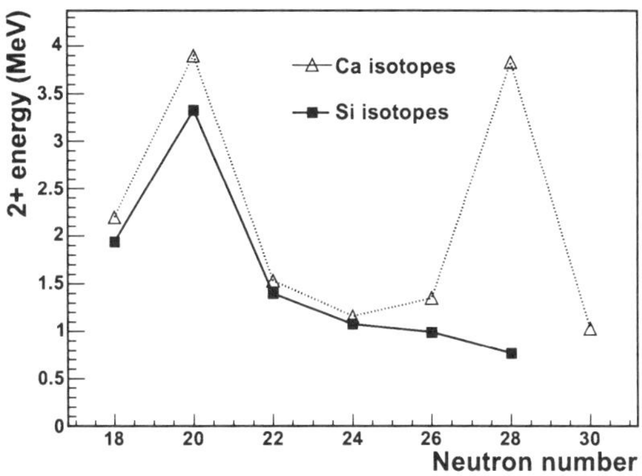

Fig. 4. Energies of the 2^+ states measured in the Ca and Si isotopes from N=18 to N=28. The drop of the energy for the ^{42}Si is a clear evidence of the absence of magicity in that nucleus.

4. The difference between the A=36 Ca and S mirror isotopes

The so-called isle of inversion is currently understood [3] as due to the neutron-proton interaction between the ν $d_{3/2}$ and the π $d_{5/2}$. This n-p interaction between the l=2 spin-orbit partners is also responsible of the large excitation energy of the 2^+ states in ^{34}Si and ^{36}S. The aim of the study of ^{36}Ca is primary to check if this interaction acts similarly in the ^{36}S mirror nucleus. For that purpose an experiment was performed at GANIL in order to measure the excitation energy of the first 2^+ state in ^{36}Ca and to make a comparison with the mirror nucleus, ^{36}S.

For this experiment the same technique of double step fragmentation (see previous section)[38] was used. From a ^{40}Ca beam at around 95 $A\times$MeV, the secondary ^{37}Ca beam was produced. The ^{37}Ca beam was guided to the SPEG experimental area. There, it first passed through a detector for time-of-flight measurements. Then it impinged on the secondary target, a ^{9}Be foil of 1072 μm thickness at an energy of around 60 $A\times$MeV. The identification of produced fragments after the target was done using the SPEG spectrometer as it was explained in the previous section. In order to measure the γ-ray energies, the same array of 74 BaF_2 scintillators was used. The doppler correction for the spectra was done assuming that the nucleon removal takes place in the middle of the secondary target. The spectra obtained for the two most proton rich nuclei ^{36}Ca and ^{28}S are shown in fig. 5. Using gaussian fits for the peaks, the energy of the 2^+ state in ^{36}Ca has been determined to be $E(2^+) = 3025(30)$ keV. The energy of the first 2^+ state in ^{28}S has been also measured for the first time to be $E(2^+) = 1525(30)$ keV. The result obtained for ^{36}Ca indicates that the N=16 neutron gap in Ca is as large as the Z=16 proton gap in ^{36}S and that the tensor monopole interaction between the neutron and proton d orbits acts similarly when a proton and neutrons are exchanged.

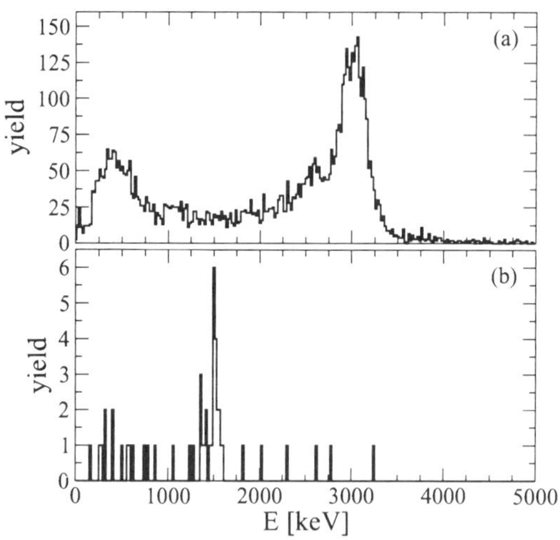

Fig. 5. γ-ray spectra obtained for the proton-rich (a)^{36}Ca and (b)^{28}S isotopes.

When compared to the mirror nucleus ^{36}S, one can see in figure 6, a relatively large excitation energy difference for the first 2^+ states of ^{36}Ca. This implies an MED (Mirror Energy Difference) of 266(30) keV which is one of the largest value observed so far in $T = 2$ mirror nuclei.

This is understood in a simple way as due to the combination of different effects. Because of parity non-conservation, 1p-1h proton excitations across the large Z=20 and 1p-1h neutron excitations across the large N=20 do not contribute to the first 2^+ state in respectively ^{36}Ca and ^{36}S. In addition, because of the N=16 and Z=16 sub-shell closures in respectively ^{36}Ca and ^{36}S, the 2^+ state in these two nuclei is relatively pure 1p($s_{1/2}$)-1h($d_{3/2}$)

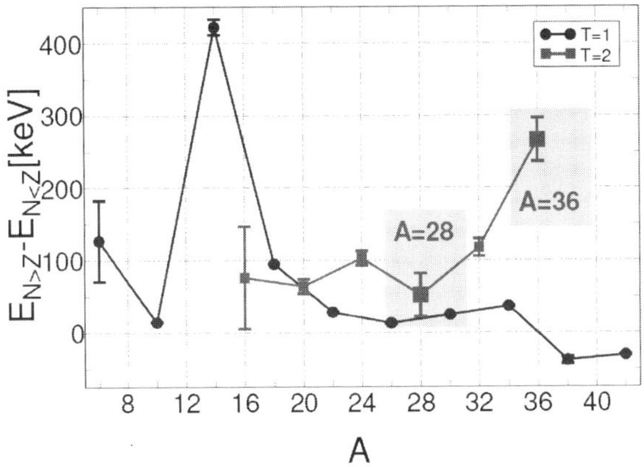

Fig. 6. Excitation energy differences of the first 2^+ states in the chain of T=2 mirror nuclei, completed with the present measurements for ^{36}Ca and ^{28}S.

neutron and proton excitations. Consequently the difference in the Coulomb energy between the valence s and d orbits, due to the large difference in the distributions of their radial wave functions, generates a large MED between these mirror nuclei.

5. Acknowledgments

I am thankful to all my collaborators who helped preparing and performing the three experiments discussed in this paper. Special thanks to my close colleagues A. Bürger, Zs. Dombradi, S. Franchoo, S. Grevy, O. Sorlin and M. Stanoiu with whom I had many illuminating discussions about the analysis and interpretation of the data.

6. References

References

1. B. Pfeiffer et al., Nucl. Phys. A 693, 282 (2001)
2. T. Otsuka et al., Phys. Rev. Lett. 87, 082502 (2001)
3. T. Otsuka et al., Proc. XXXIX Zakopane School of Physics, Acta Physica Polonica B36, 1213 (2005)
4. L.S. Kisslinger and R.A. Sorensen, Mat. Fys. Medd. Dan. Vid. Selsk. 32, No. 9 (1960)
5. M.Hjorth-Jensen et al., Phys. Rep. 261, 125 (1995)
6. O. Kenn et al., Phys. Rev. C63, 064306 (2001)

7. T. Ishii et al., Phys. Rev. Lett 84, 39 (2000)
8. H. Grawe et al., Tours Symposium on Nuclear Physics IV, Tours 2000, AIP Conf. Proc. 561, 287 (2001)
9. A.M. Oros-Peusquens and P.F. Mantica, Nucl. Phys. A 669, 81 (2001)
10. O. Sorlin et al., Phys. Rev. Lett. 88, 092501 (2002)
11. P.G. Reinhard et al., RIKEN Review 26, 23 (2000)
12. K.H. Langanke et al., Phys. Rev. C67, 044314 (2003)
13. C. Guénaut et al., ENAM04 proceedings, Eur. Phys. J A Direct in print.
14. R. Grzywacz et al., Phys. Rev. Lett. 81, 766 (1998)
15. M. Hannawald et al., Phys. Rev. Lett. 82 (1999) 1391
16. M. Sawicka et al., Phys. Rev. C68, 044304 (2003)
17. R. Grzywacz et al., ENAM04 proceedings, Eur. Phys. J A Direct in print.
18. M. Sawicka et al., Eur. Phys. J. A20, 109 (2004)
19. S. Raman et al., At. Data Nucl. Data Tab. 36, 1 (1987)
20. R. Anne and A.C. Mueller, Nucl. Inst. and Meth. B70 (1992) 276
21. A. Winther and K. Alder, Nucl. Phys. A 319, 518 (1979)
22. R. Ardill, K. Moriarty, and P. Koehler, Comput. Phys. Commun. 22 (1981) 419
23. G. Kraus et al., Phys. Rev. Lett. 73, 1773 (1994)
24. O. Kenn et al., Phys. Rev. C63, 021302(R) (2001)
25. O. Kenn et al., Phys. Rev. C65, 034308 (2002)
26. S. Leenhardt et al., Eur. Phys. J A14, 1 (2002)
27. I. Deloncle and B. Roussière, arXiv:nucl-th/0309050
28. M. Lewitowicz et al., Nucl. Phys. 654, 687c (1999)
29. H. Mach et al., Nucl. Phys. A 719, 213c (2003)
30. M. Stanoiu PhD Thesis, Caen (2003) GANIL T03-01
31. A. Lisetskiy et al., Phys. Rev. C70, 044314 (2004)
32. C.M. Baglin, Nucl. Data Sheets 91, 423 (2000)
33. D. Rudolph et al., Nucl. Phys. A597, 298 (1996)
34. S. Franchoo et al., Phys. Rev. Lett. 81, 3100 (1998)
35. H. Grawe, Lecture Notes in Physics 651, Springer Berlin - Heidelberg 2004, p. 33
36. N. Smirnova et al., Phys. Rev. C69, 044306 (2004)
37. R. W. Ibbotson et al., Phys. Rev. Lett. 80, 2081 (1998)
38. M. Stanoiu et al., Phys. Rev. C69, 034312 (2004).
39. D. Sohler et al., Phys. Rev. C66, 054302 (2002).
40. J. Fridmann et al., Nature (London) C435, 932 (2005).
41. B. Bastin et al., Phys.Rev.Lett. 99, 022503 (2007).
42. T. R. werner et al., Nucl. Phys. A 597, 327 (1996).
43. G. A. Lalazissis, Phys. Rev. C60, 014310 (1999).
44. R. Rodriguez-Guzman, Phys. Rev. C65, 024304 (2002).
45. E. Caurier, Nucl. Phys. A 742, 14 (2004).

MAGNETIC MOMENT MEASUREMENTS: PUSHING THE LIMITS

N. BENCZER-KOLLER

Department of Physics, Rutgers University,
New Brunswick, NJ 08903, USA
**E-mail: nkoller@physics.rutgers.edu*

The last three years have seen developments in both experimental technology and theoretical capabilities. Thus it has become possible to measure the magnetic properties of weakly excited nuclear states with shorter half-lives and in nuclei further away from stability. In particular, two examples will be given concerning the structure of 4^+ states in the mass 70 nuclei and the nature of the excitation of mixed-symmetry states in Zr nuclei. In addition, progress has been made in experiments involving radioactive beams. While these experiments are very difficult, successful experiments have been carried out in light Ar and S nuclei as well as in heavier isotopes in the A=132 region. The experiments will be described and the latest theoretical approaches will be discussed.

Keywords: 21.10.Ky,25.70.De

1. Introduction

Over the last decade experiments have progressed hand in hand with theoretical work with the common task of describing the wave functions of excited states of low-lying states of nuclei. In this paper, certain advances in experimental techniques designed to measure magnetic moments, or g factors, of states in nuclei that are difficult to excite are presented. In particular, the magnetic moments of 4_1^+ states in some even-even Ge and Zn isotopes were measured and compared with new shell model calculations. In the ^{92,94}Zr isotopes the second 2_1^+ states, which had been confirmed as being mixed-symmetry states, were studied. Finally, new results of magnetic moment measurements in radioactive S and Ar, on the one hand and ^{132}Te on the other, were completed.

2. Experimental techniques

The main technique used for these experiments involves the application of the transient field hyperfine interaction between a nuclear moment and the effective field produced by polarized electrons captured while the nucleus traverses a ferromagnetic target. The details have been amply described in two review papers Refs.[1,2] It turns out that, contrary to the experiments carried out in this field over the last decade in which the experimental setup are very similar for all species studied, in the realm of radioactive beams, each isotope poses very special challenges and experiments have to be tailored to these particular needs. Only details relevant to the current experiments will be highlighted.

3. The $N = 38$ isotones: ^{70}Ge and ^{68}Zn

The issue in this region was that the g factor of the 4_1^+ state in ^{68}Zn had been found to be negative,[3] contrary to theoretical expectations by the authors themselves and other shell model calculations. In order to clarify this issue, a measurement of the g factor of the 4_1^+state in the isotone ^{70}Ge was carried out. The experiment was performed at the Wright Nuclear Structure Laboratory at Yale. Beams of ^{70}Ge at 225 MeV were focused on multilayered targets faced by a C layer and backed by gadolinium/copper layers. Forward scattered C ions were detected in coincidence with the γ rays recorded in four Clover detectors. The g factor was found to be indeed positive, $g(^{70}\text{Ge};4_1^+) = +\ 0.5\ (3)$.[4]

Consequently, this result stimulated a remeasurement of the g factor of the 4_1^+ state in ^{68}Zn done under similar conditions. Again, the result yielded a positive g factor, $g(^{68}\text{Zn};4_1^+0 = +0.6(3)$.[5]

These results are consistent with the systematics of the g factors of the 4_1^+ states in the region, as shown in Fig. 1. Extensive shell model calculations were carried out in these two papers to understand the composition of the wave function of the low excited states in these isotopes and, in particular, to examine the possible contribution of the $g_{9/2}$ orbital. It appears that the measured values are in good agreement with full fp shell model calculations carried out with several commonly used effective interactions. The $g_{9/2}$ does not play a significant role in the wave function.[4]

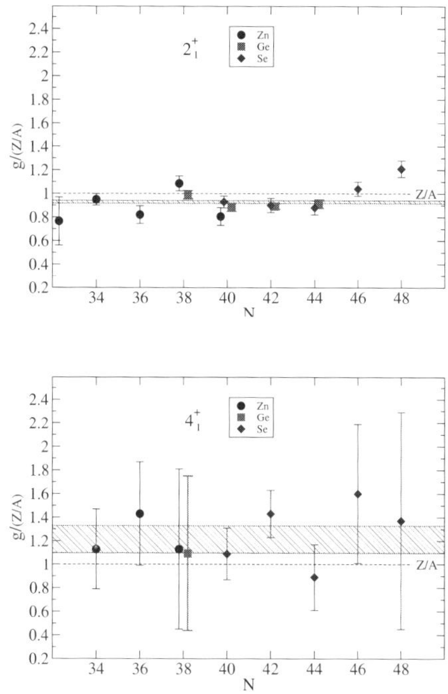

Fig. 1. Systematics of $g(2_1^+)$ and $g(4_1^+)$ factors in Zn, Ge and Se isotopes.

4. Magnetic moments and the wave function of the mixed-symmetry states in ^{92,94}Zr.

The Zr isotopes span a region of nuclei where the protons fill the $Z = 40$ shell and, as N increases, the neutrons gradually fill the $d_{5/2}$ orbit. The magnetic moments of the the first 2^+ and 4^+ states are negative. These states were reasonably described by a wave function that contains a strong component of $d_{5/2}$ neutrons. The second 2^+ states are different. They have been postulated as not being fully symmetric with respect to an exchange of protons and neutrons. So, while the first 2^+ states are the fully symmetric states, the second 2^+ states represent the mixed-symmetry states. The magnetic moments of the second 2^+ states in ^{92,94}Zr isotopes were measured with the aim of determining whether this characterization is indeed correct.[6]

Experiments were also carried out at Yale, in inverse kinematics, with beams of ^{92,94}Zr at 275 and 290 MeV impinging on C target deposited on gadolinium backed by copper. The transient field technique was used. The g factors of the 2^+ and 4^+ states in the even Zr isotopes are displayed on Fig. 2.

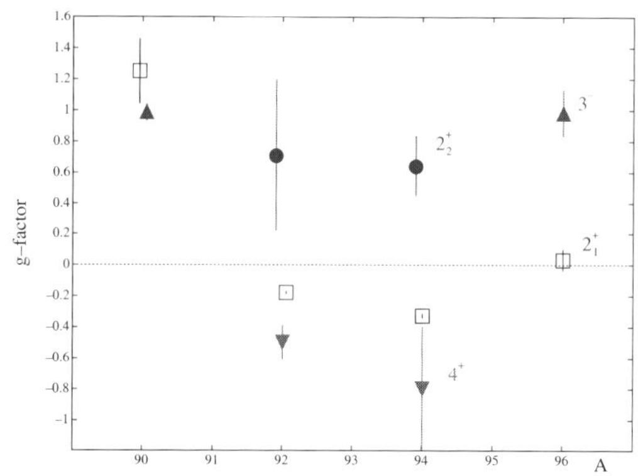

Fig. 2. Systematics of $g(2^+)$ and $g(4^+)$ factors in the even Zr isotopes. The g factors of the 2_2^+ states are preliminary.

The preliminary results yield g factors that are large and positive in agreement with shell model calculations with surface delta (SDI) and the V_{lowK} interactions as well as with calculations based on the quasiparticle phonon model (QPM) .[7]

5. Magnetic moment measurements in a radioactive beam environment: ^{132}Te.

The B(E2) values of several neutron rich Te and Sn isotopes around $N = 82$ have been measured recently at ORNL.[8,9] The interest in finding deviations from the structures predicted by standard shell model calculations has stimulated experiments to determine other properties. In particular magnetic moments usually offer a sharp picture of the nature of the wave functions that describe the low lying states.

Two approaches have been used to measure the magnetic moment of the first excited of ^{132}Te. The first results were obtained by the recoil-in-vacuum technique (RIV). This technique was pioneered by G. Goldring at the Weizmann Institute in the 1970's .[10] Ions recoiling from a thin target into vacuum experience hyperfine interactions with an ensemble of atomic electrons in a variety of charge and angular momentum states. These interactions can be quantified and calibrated via series of experiments on stable nuclei whose excited state lifetimes and g factors are known. The measurements were carried out at the Australian National University on stable Te isotopes as well as on ^{132}Te[11] at ORNL using the CLARION array for γ-ray detection operated in coincidence with particle detectors from the Hyball array.

The resulting g factor, $|g| = 0.35(5)$ [12] agrees well with three shell model calculations that yield $g = +0.488$,[13] $+0.491$ [14] and $+0.35$.[15]

The second approach involves measuring the magnet moment by the transient field technique. This experiment was also carried out at ORNL. Because of the problems that arise from working in a radioactive environment, the standard configuration of the experimental set-up had to be modified. The use of thick targets commonly prepared for such measurements had to be curtailed because of scattering of the radioactive beam into the chamber surroundings.

First, the target had to be thinner than those used for the conventional experiments in inverse kinematics. A 1.4 mg/cm^2 of C was deposited on a gadolinium foil of moderate thickness, 4 mg/cm^2. Second the copper backing, generally required for cooling the ferromagnetic target, needs to be drastically reduced. In this particular case, since the beam is contaminated by 10% of ^{132}Sb which decays to the same 2_1^+ state in ^{132}Te that is under study via Coulomb excitation of the ^{132}Te beam, the recoiling excited Te ions had to be allowed to recoil in vacuum. By using this particular target, the γ rays of interest are Doppler shifted and easily distinguishable from the γ rays that are emitted from the ^{132}Sb background decaying at rest, as shown in Fig. 3. The target did not heat up because the radioactive beam intensity is a factor of a 100 lower than that usually employed with stable beams.

The particle detector consisted of two solar cells 15 x 15 mm located 25 mm away from the target, 10 mm above and below the beam axis. Thus, the unscattered beam was allowed to traverse the chamber and stop downstream, far from target.

These modifications made the experiment feasible. While the beam during the experiment was rather weak, a result could nevertheless be obtained. The final analysis has not yet been completed, nevertheless, the method proved to be applicable to such an environment. In general, each isotope has its own peculiarities and targets and detector systems need to be very carefully tailored to match the requirements of each experiment.

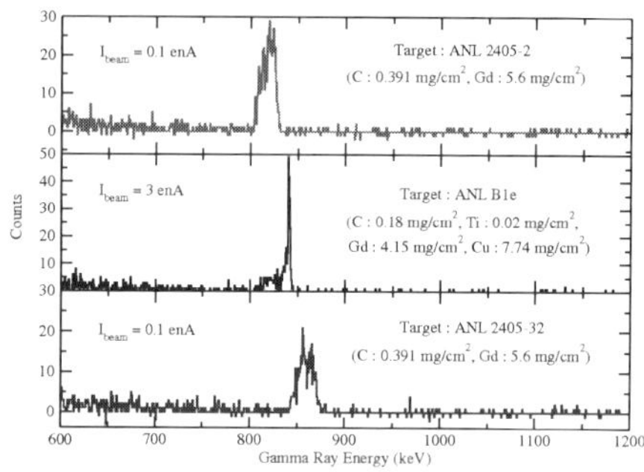

Fig. 3. Doppler-shifted γ spectra obtained with ^{130}Te with beam intensities comparable to those available for radioactive ^{132}Te. The central panel show a γ spectrum obtained with a thick target.

6. From ISOL to fragmentation beams: Ar and S isotopes and the *sd* shell.

The production of radioactive beams in fragmentation reactions has stimulated new developments in the TF technique to accommodate the intermediate and high ion velocities that pertain to these reactions. The Ar experiments described in this section have been carried out at low energies by the conventional TF method ,[16–18] but the S-beams experiments represent new directions.

High velocity ^{38,40}S, 40 Mev/u, beams were produced at the National Superconducting Cyclotron Laboratory at Michigan State University. The work was carried jointly by Michigan State University and a group from the Australian National University .[19,20] These experiments necessitate thick, high Z, scatterers such as Au to increase the Coulomb excitation probability and, at the same time, reduce the excited nuclei velocity in order to be in the range of effectiveness of the transient field, and thick ferromagnetic foils to increase the magnitude of the spin precession. The scattered S recoils were detected in a plastic scintillator phoswich detector while the γ rays were detected in the Segmented Germanium Array (SEGA).

The g factors of the 2_1^+ in these Ar and S isotopes are shown in Fig. 4 where they are compared with shell model calculations.

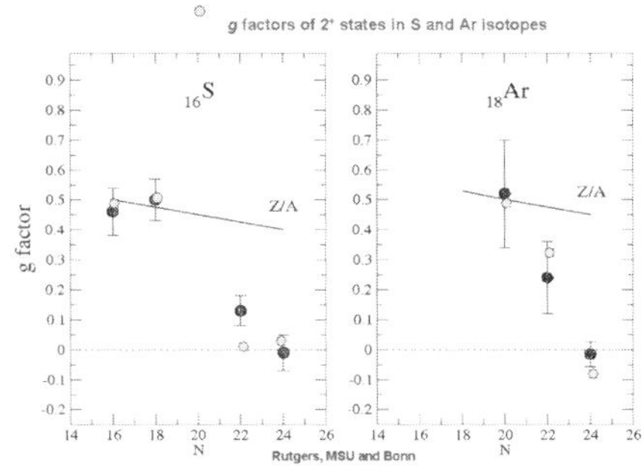

Fig. 4. g factors of the 2_1^+ states in S and Ar isotopes. The stars are the results of shell model calculations (Refs.17 and 18).

Shell model calculations within the (full $sd)_\pi$(full $fp)_\nu$ space without including core excitations have been performed for $^{36,38,40}S_{20,22,24}$ and $^{48,40,42}Ar_{20,22,24}$ nuclei for which valence neutrons are restricted to the fp shell and the valence protons to the sd shells .[17,18] Both WBT and FSZM interactions were used and results were obtained with the free nucleon g factors as well as with effective g factors. Both calculations discuss possible effects of collectivity and admixtures of deformed components in the wave functions. The results of these calculations are in good agreement with energy levels, quadrupole properties and magnetic moments.

7. Future measurements

In this presentation it was shown that magnetic moment of weakly excited states in stable nuclei or strongly excited states in nuclei prepared as radioactive beams can be measured with the existing techniques with sufficient accuracy for reasonable comparison with theoretical calculations. In the future, both the experimental techniques and the availability, quality and intensity of the beams will have to be developed further.

References

1. K.-H. Speidel, O. Kenn and F. Nowacki, *Prog. Part. Nucl. Phys.* **49**, p. 91 (2003).
2. N. Benczer-Koller and G. Kumbartzki, *J. Phys. G* (2007).
3. J. Leske, K.-H. Speidel, S. Schielke, J. Gerber, P. Maier-Komor, T. Engeland and M. Hjorth-Jensen, *Phys. Rev. C* **72**, p. 044301 (2005).
4. P. Boutachkov, S. J. Q. Robinson, A. Escuderos, G. Kumbartzki, N. Benczer-Koller, E. Stefanova, Y. Y. Sharon, L. Zamick, E. A. McCutchan, V. Werner,

H. Ai, A. B. Garnsworthy, G. Gürdal, A. Heinz, J. Qian, N. J. Thompson, E. Williams and R. Winkler, *Phys. Rev.* **C** (2007), to be published.

5. P. Boutachkov, N. Benczer-Koller, G. J. Kumbartzki, A. Escuderos, L. Zamick, S. J. Q. Robinson, H. Ai, M. Chamberlain, G. Gürdal, A. Heinz, E. A. McCutcheon, J. Qian, V. Werner, E. Williams, K. Aleksandrova, C. A. Copos, D. A. Kovacheva and P. Manchev, *Phys. Rev. C* **75**, p. 021302 (2007).
6. N. L. Iudice and C. Stoyanov, *Phys. Rev. C* **73**, p. 037305 (2006).
7. V. Werner, N. Benczer-Koller, P. Boutachkov, G. Kumbartzki, J. Holt, M. Perry, N. Pietralla, E. Stefanova, H. Ai, K. Aleksandrova, G. Anderson, R. B. Cakirli, R. F. Casten, M. Chamberlain, C. Copos, B. Darakchieva, S. Eckel, M. Evtimova, C. R. Fitzpatrick, A. B. Garnsworthy, G. Gürdal, A. Heinz, D. Kovacheva, C. Lambie-Hanson, X. Liang, P. Manchev, E. A. McCutchan, D. A. Meyer, J. Qian, A. Schmidt, N. J. Thomson, E. Williams and R. Winkler, *Phys. Rev.* **C** (2007), to be published.
8. D. Radford, J. Beene, R. Varner and collaborators, in *Proceedings of the XLI Zakopane Conference in Nuclear Physics*, 2006.
9. C. Baktash, (2007), private communication and contribution to this Seminar.
10. G. Goldring, *Heavy Ion Collisions* 1982. edited by R. Bock (North Holland, Amsterdam.
11. V. Zamfir, (2002), private communication.
12. N. J. Stone, A. E. Stuchbery, M. Danchev, J. Pavan, C. Timlin, C. Baktash, C. Barton, J. Beene, N. Benczer-Koller, C. R. Bingham, J. Dupak, A. Galindo-Uribarri, C. J. Gross, G. Kumbartzki, D. C. Radford, J. R. Stone and N. V. Zamfir, *Phys. Rev. Lett.* **94**, p. 192501 (2005).
13. B. A. Brown, N. J. Stone, J. R. Stone, I. S. Towner and M. Hjorth-Jensen, *Phys. Rev. C* **71**, p. 044317 (2005).
14. J. Terasaki, J. Engel, W. Nazarewicz and M. Stoitsov, *Phys. Rev. C* **66**, p. 054313 (2002).
15. A. Covello, (2004), private communication.
16. J. Cub, U. Knopp, K.-H. Speidel, H. Bush, S. Kremeyer, H.-J. Wollersheim, N. Martin, X. Hong, K. Vetter, N. Gollwitzer, A. I. Levon and A. Booten, *Nucl. Phys.* **A 549**, p. 304 (1992).
17. E. A. Stefanova, N. Benczer-Koller, G. J. Kumbartzki, Y. Y. Sharon, L. Zamick, S. J. Q. Robinson, L. Bernstein, J. R. Cooper, D. Judson, M. J. Taylor, M. A. McMahan and L. Phair, *Phys. Rev. C* **72**, p. 014309 (2005).
18. K.-H. Speidel, S. Schielke, J. Leske, J. Gerber, P. Maier-Komor, S. J. Q. Robinson, Y. Y. Sharon and L. Zamick, *Phys. Lett. B* **632**, p. 207 (2006).
19. A. D. Davies, A. Stuchbery, P. F. Mantica, P. M. Davidson, A. N. Wilson, A. Becerril, B. A. Brown, C. M. Campbell, J. M.Cook, D. C. Dinca, A. Gade, S. N. Liddick, T. J. Mertzimekis, W. F. Mueller, J. R. Jerry, B. E. Tomlin, K. Yoneda and H. Zwahlen, *Phys. Rev. Lett.* **96**, p. 112503 (2006).
20. A. Stuchbery, A. D. Davies, P. F. Mantica, P. M. Davidson, A. N. Wilson, A. Becerril, B. A. Brown, C. M. Campbell, J. M.Cook, D. C. Dinca, A. Gade, S. N. Liddick, T. J. Mertzimekis, W. F. Mueller, J. R. Terry, B. E. Tomlin, K. Yoneda and H. Zwahlen, *Phys. Rev. C* **74**, p. 054307 (2006).

TECHNIQUE FOR MEASURING ANGULAR CORRELATIONS AND g-FACTORS OF EXCITED STATES WITH LARGE MULTI-DETECTOR ARRAYS: AN APPLICATION TO NEUTRON RICH NUCLEI PRODUCED IN SPONTANEOUS FISSION

A.V. RAMAYYA*, C. GOODIN, K. LI, J.K. HWANG, J.H. HAMILTON, Y.X. LUO

Department of Physics, Vanderbilt University, Nashville, TN 37235, USA
**E-mail: a.v.ramayya@vanderbilt.edu*

A.V. DANIEL, G.M. TER-AKOPIAN

Flerov Laboratory for Nuclear Reactions, JINR, Dubna, Russia

N.J. STONE

Department of Physics, Oxford University, Oxford OX1 3PU, United Kingdom
Department of Physics, University of Tennessee, Knoxville, Tennessee 37996, USA

J.R. STONE

Department of Physics, University of Tennessee, Knoxville, Tennessee 37996, USA
Department of Chemistry and Biochemistry, University of Maryland, College Park, Maryland 20742, USA

J.O. RASMUSSEN, I.Y. LEE

Lawrence Berkeley National Laboratory, Berkeley, CA 24720, USA

S.J. ZHU

Department of Physics, Tsinghua University, Beijing 100084, Peoples Republic of China

M. STOYER

Lawrence Livermore National Laboratory, Livermore, CA 24550, USA

Triple coincidences between prompt γ-rays emitted in the spontaneous fission of ^{252}Cf were measured with Gammasphere. These data are used to measure the angular correlation of cascades of γ-rays from excited states of neutron rich fission fragments stopped in an unmagnetized iron foil. The hyperfine fields in the iron lattice cause attenuations of the angular correlations between γ rays emitted from the excited states which have sufficiently long lifetimes. This attenuation is measured and used to calculate the g-factors of excited states in many neutron rich nuclei.

Keywords: g-factor, ^{252}Cf, Angular correlation, Gammasphere

1. Introduction

Experiments measuring the angular correlations between successive nuclear de-excitations can be used to determine certain properties of the nuclear structure. From unperturbed angular correlations one can extract information on the spins of the energy levels and the multipole mixing ratios of the transitions. Another important property that can be measured with perturbed angular correlations is the g-factor of an excited state. The g-factors of excited states in even-even nuclei give some sense of their degree of collectivity; for odd A nuclei, the g-factor is related to the purity of the single particle state that characterizes the level. Therefore, we have developed a method for measuring the angular correlations and g-factors of the excited states in nuclei with large γ-ray detector arrays. The principle of the method is fully described in Ref. 1. We use this technique to measure the g-factors of excited states in neutron rich nuclei. There are many cases in our previous nuclear structure studies where the spins were assigned based only on the systematics. Although the g-factors of 2^+ states for many of the isotopes produced in the fission of ^{252}Cf have been previously measured,[2] our high statistics data set may allow us to measure the g-factors of many new states.

The advantage of doing angular correlation experiments with large detector arrays such as Gammasphere as opposed to traditional setups with 2-4 detectors is a major increase in angular resolution and detection efficiency. However, there are many difficulties for measuring angular correlations with the large detector arrays. Specifically, the number of pairs increases as the square of the number of detectors. Since the detector efficiency and the solid angle correction factors are different for each pair, handling a problem of this magnitude requires careful analysis techniques, and we describe these below. Following the description of our method, we show results for unattenuated angular correlations, and then for some attenuated angular correlations and g-factors.

2. Theory

For a cascade of two γ-rays, the angular correlation function is given by

$$W(\theta) = 1 + A_2(\delta)P_2(cos\theta) + A_4(\delta)(P_4(cos\theta) \quad (1)$$

where δ is the mixing ratio for γ_1 or γ_2.

For a state with a known lifetime, τ, the average precession angle ϕ is given by

$$\phi = -\frac{\mu_N g B \tau}{\hbar} \quad (2)$$

If the excited nucleus is somehow implanted into the lattice of a ferromagnetic material, as in the case of fission fragments stopping in an iron foil, then the nucleus will be subject to a hyperfine field (HFF), and will precess about this field. For our analysis, we have assumed that the HFFs are those of an ion implanted in a substitutional site in the lattice.[3] We have also assumed that the domains in the foil are randomly oriented and small enough that there is no preferential direction in the foil.

One further point is that the following discussion assumes that the lifetime, τ, is much longer than the stopping time, which is a few picoseconds. If this is not the case, then the observed angular correlation will be affected by transient field effects. If the HFFs in iron are randomly oriented, then for an ensemble of excited nuclei implanted throughout the lattice, the net effect on the experimentally measured angular correlation is an attenuation ($A_k \rightarrow 0$) of the correlation function $W(\theta)$. Therefore $W(\theta)$ must be modified to include the effect of the attenuation

$$W(\theta) = 1 + A_2^{theory} G_2 P_2(cos\theta) + A_4^{theory} G_4 P_4(cos\theta) \quad (3)$$

where G_2 and G_4 are the attenuation factors, defined as

$$G_k = \frac{A_k^{exp}}{A_k^{theory}} \quad (4)$$

In practice, the experimental angular distribution function is given by

$$N(\theta_n) = A_0^{exp}(1 + A_2^{exp} P_2(cos\theta_n) + A_4^{exp} P_4(cos\theta_n)) \quad (5)$$

for some set of discrete angles, θ_n.

The attenuation factor G_k is found by measuring A_k^{exp}. The average precession angle ϕ can then be found by solving[4]

$$G_k = \frac{1}{2k+1}(1 + 2\sum_{q>0}^{k} \frac{1}{1+q^2\phi^2}) \quad (6)$$

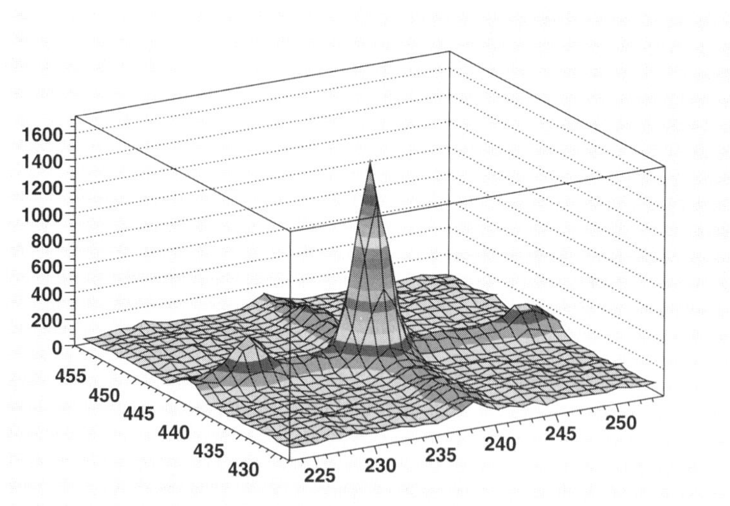

Fig. 1. Two-dimensional histograms showing the number of coincidences on the z-axis between E_1 and E_2 on the x and y axes. The histograms are constructed for the $4^+ \rightarrow 2^+ \rightarrow 0^+$cascade in ^{148}Ce in coincidence with at least one of 9 γ-rays corresponding to coincident transitions in ^{148}Ce or in ^{100,102}Zr.

3. Experiment and Analysis

The detailed description of the experiment can be found in Ref. 5. It should be noted that in this arrangement, the fission fragments were fully stopped in the iron foils. Therefore, the implanted fission fragments were subject to the hyperfine fields (HFFs) caused by their implantation in substitutional sites in the iron lattice. The strength of these fields can be quite high ($\sim$30 T), causing a significant attenuation in many cases. One difficulty in this procedure is to determine the HFF in iron for each ion. These difficulties are discussed in Ref. 2. For example, we have found evidence that the HFF field for barium implanted in iron might be quite different than that given in Ref. 6 or 7.

For a typical angular correlation measurement, it is necessary to calculate a solid angle correction factor Q_k for each parameter A_k. However, for very low intensity transitions, the sensitivity of the angular correlation measurement can be improved by determining the detector response function $R_n(\theta, E_1, E_2)$ for each pair of detectors. For a given detector pair, the response function describes the distribution of possible angles about the central angle of the pair as a function of energy. The response functions for each pair can then be summed to find the response function of each angle bin.

We calculated the response function using a simple Monte Carlo simulation, with the γ ray transport simulated up to the first collision. This is equivalent to the traditional calculation of Q_k.[1] The mean free path, $\lambda(E)$, of γ-rays was calculated using the known Gammasphere detector properties. The energy dependence of $R_n(\theta,E_1,E_2)$ is negligible, and so only $R_n(\theta)$ was calculated.

To determine the angular correlation, each of the 64 histograms was fit to find the intensity of the peak of interest, N_n. Our fitting method is based on the analysis of two dimensional $\gamma-\gamma$ coincidence spectra given in.[8] To fit a given histogram, a window within the histogram is selected. Within this window, the positions of the peaks are defined by using projections of the histogram on the two axes. The surface is then approximated by three types of background and the sum of the two dimensional γ peaks found in the projections. The three types of background consist of a smooth background and two series of ridges parallel to the axes corresponding to γ lines in the x and y direction. The last part of the fitting procedure is to solve a well known NNLS (Non-Negative Least Square) problem.[9]

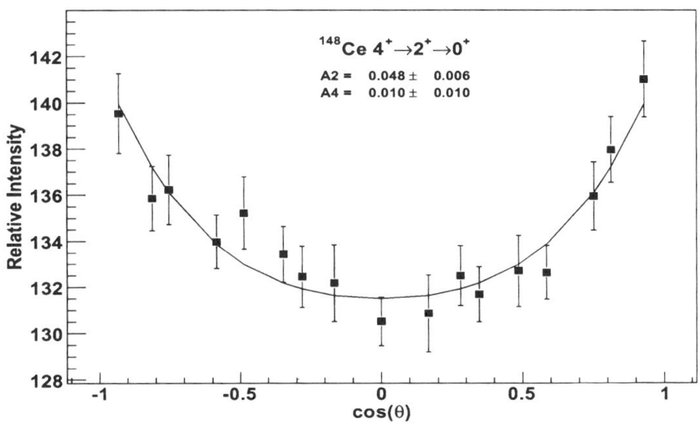

Fig. 2. Angular correlation of the $4^+ \to 2^+ \to 0^+$ cascade in ^{148}Ce. The solid black line is the result of the fitting procedure.

As an example of an attenuated correlation, Figs. 1 and 2 show the

Table 1. Calculated g-factors of 2^+ states in some even-even nuclei compared with previously published values.

Fragment	τ (ns)	B (Tesla)	g	g(refs.)	Reference
^{100}Zr	0.78(3)	27.4(4)	0.32(5)	0.30(3)	2
^{102}Zr	2.76(36)	27.4(4)	0.25(4)	0.22(5)	2
^{104}Mo	1.040(59)	25.6(1)	0.28(6)	0.27(2)	2
^{106}Mo	1.803(43)	25.6(1)	0.24(3)	0.21(2)	2
^{108}Mo	0.72(43)	25.6(1)	0.3(3)	0.5(3)	2
^{146}Ce	0.36(4)	41(2)	0.30(6)	0.24(5),0.46(34)	7,10
^{148}Ce	1.46(9)	41(2)	0.37(5)	0.37(6)	7

result of the angular correlation measurement for the $4^+ \rightarrow 2^+ \rightarrow 0^+$ cascade in ^{148}Ce. The 2^+ state has a relatively long lifetime of 1.46(9) ns, resulting in a reasonably large attenuation; G_2=0.47(6). This example is typical of the cases listed in Table 1, which gives a comparison of our results to known g-factors and shows the good agreement with previously measured values. The lifetimes listed in the table come from the Brookhaven website.[11] This agreement demonstrates the validity of our method. Note that for ^{146}Ce, we are in agreement with,[7] but not with,[10] which we believe is likely due to some discrepancy in the value taken for the hyperfine field. In fact, the HFF was not measured in,[10] but was calibrated based on the value for the g-factor of the 2^+ state in ^{148}Ce given in Ref. 7.

Another possible use of this method is the determination of the multipole mixing of transitions. Fig. 3 shows how we can use correlations with states of known spin to determine δ, with the $3^+ \rightarrow 2^+$, 732.5 keV transition in ^{108}Ru as an example. First, we use the $4^+ \rightarrow 2^+ \rightarrow 0^+$ cascade to determine the attenuation factor G_2 for the 2^+ level. We then use this value of G_2 to correct A_2 in the $3^+ \rightarrow 2^+ \rightarrow 0^+$ cascade and determine the mixing ratio with our software. From this, we find two possible values for the mixing ratio, δ=-0.15$^{+0.02}_{-0.07}$ and δ=80(65). We adopt the second value, which corresponds to an almost pure quadrupole transition. This is in rough agreement with the result found by[13] and is the expected result for the transition from a 3^+ state in the γ-band to a state in the ground state band.[12] The angular correlation coefficients A_2^{exp} and A_4^{exp} for the 174.5 and 1156.7 keV cascade in ^{138}Cs were measured to be -0.07(1) and -0.02(2).[14] They are consistent with the theoretical coefficients of A_2^{exp} = -0.071 and A_4^{exp} = 0.0[14] for pure quadrupole and pure dipole transitions and for the proposed spin assignments of (9^-) and (7^-) to the 1411.3 and 254.4 keV levels. These examples demonstrate the validity of our method in determining mixing ratios.

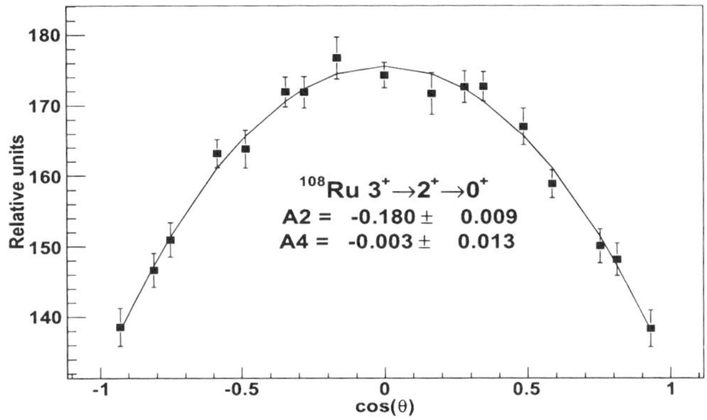

Fig. 3. Angular correlation of the $3^+ \to 2^+ \to 0^+$ cascade in ^{108}Ru.

As mentioned above, our method requires that the spins and parities of the three levels be known, as well as the multipolarities of the transitions and the lifetime of the state of interest. Fortunately, the isotopes produced in the fission of ^{252}Cf have been studied extensively,[15] and there are many cases satisfying these requirements. While we have shown results only for 2^+ states in even-even nuclei, our method is also valid for odd-A nuclei, as well as for high spin excited states. Our high statistics data should allow us to determine the g-factors for many previously unmeasured states. The more details are published in Ref. 16.

4. Summary

We have developed a novel method to measure angular correlations and in some cases determine the magnitudes of the g-factors of excited states in the neutron-rich nuclei produced in the fission of ^{252}Cf. The large angular resolution of Gammasphere allows us to measure the attenuation in the angular correlation of a cascade of γ rays caused by the implantation of highly excited fission fragments into an unmagnetized iron lattice, and the subsequent interaction of those excited fragments with the randomly oriented hyperfine fields of the lattice. This method has been validated by comparison to previously measured g-factors and will be used to measure the g-factors of other excited states.

5. Acknowledgments

The work at Vanderbilt University, Lawrence Berkeley National Laboratory, Lawrence Livermore National Laboratory, and Idaho National Engineering and Environmental Laboratory are supported by U.S. Department of Energy under Grant No. DE-FG05-88ER40407 and Contract Nos. W-7405-ENG48, DE-AC03-76SF00098, and DE-AC07-76ID01570.

References

1. K. Siegbahn, *Alpha-,Beta-, and Gamma-Ray Spectroscopy (Part2)*, North-Holland, Amsterdam, 1968, p. 1192.
2. A.G. Smith et al., Phys. Lett. **B591**, 55 (2004).
3. G.N. Rao,G, Hyperfine Interactions **7**, 141 (1979).
4. E. Matthias, S.S. Rosenblum and D.A. Shirley, Phys. Rev. Lett. **14**, 46 (1965).
5. Y.X. Luo et al., Phys. Rev. **C64**, 054306 (2001).
6. A.G. Smith et al., AIP Conf. Proc., Nuclear Structure 98. **481**, 283 (1999).
7. R.L. Gill et al., Phys. Rev. **C33**, 1030 (1986).
8. G.M. Ter-Akopian et al., Phys. Rev. **C55**, 1146 (1997).
9. C.L. Lawson and R.J. Hanson, *Solving Least Square Problems*, Prentice-Hall, Englewood Cliffs, NJ, 1974.
10. A.G. Smith et al., Phys. Lett. **B453**, 206 (1999).
11. http://www.nndc.bnl.gov/.
12. L.C. Whitlock et al., Phys. Rev. **C3**, 313 (1971).
13. J. Stachel et al., Z. fur Phys. **A316**, 105 (1984).
14. K. Li et al., Phys. Rev. **C75**, 044314 (2007).
15. J.H. Hamilton, Prog. in Part. and Nucl. Phys. **35**, 635 (1995).
16. A.V. Daniel et al., Nucl. Inst. and Meth. **A** (2007) in press.

ISOSPIN SYMMETRY AND PROTON DECAY: IDENTIFICATION OF THE 10^+ ISOMER IN ^{54}Ni

C. FAHLANDER*[1], R. HOISCHEN[1,2], D. RUDOLPH[1], M. HELLSTRÖM[1], S. PIETRI[3], ZS. PODOLYÁK[3], P.H. REGAN[3], A.B. GARNSWORTHY[3,4], S.J. STEER[3], F. BECKER[2], P. BEDNARCZYK[2,5], L. CACERES[2,6], P. DOORNENBAL[2], J. GERL[2], M. GÓRSKA[2], J. GRĘBOSZ[5,2], I. KOJOUHAROV[2], N. KURZ[2], W. PROKOPOWICZ[2,5], H. SCHAFFNER[2], H.J. WOLLERSHEIM[2], L.-L. ANDERSSON[1], L. ATANASOVA[7], D.L. BALABANSKI[7,8], M.A. BENTLEY[9], A. BLAZHEV[10], C. BRANDAU[2,3], J. BROWN[9], E.K. JOHANSSON[1], A. JUNGCLAUS[6],

[1]*Department of Physics, Lund University, SE-22100 Lund, Sweden*
**E-mail: claes.fahlander@nuclear.lu.se*
wwwnsg.nuclear.lu.se
[2]*Gesellschaft für Schwerionenforschung mbH, D-64291 Darmstadt, Germany*
[3]*Department of Physics, University of Surrey, Guildford, GU2 7XH, UK*
[4]*WNSL, Yale University, New Haven, CT 06520-8124, USA*
[5]*The Henryk Niewodniczański Institute of Nuclear Physics (IFJ PAN), PL-31-342 Kraków, Poland*
[6]*Departamento de Fisica Teórica, Universidad Autonoma de Madrid, E-28049 Madrid, Spain*
[7]*Faculty of Physics, University of Sofia, BG-1164 Sofia, Bulgaria*
[8]*Institute for Nuclear Research and Nuclear Energy, Bulgarian Academy of Sciences BG-1784 Sofia, Bulgaria*
[9]*Department of Physics, University of York, York, Y01 5DD, UK*
[10]*Institut für Kernphysik, Universität zu Köln, D-50937 Köln, Germany*

Time-correlated γ decays from individually separated ^{54}Ni nuclei have been measured in an experiment within the stopped-beam campaign of the RISING project at GSI. An isomeric 10^+ state was identified and its lifetime measured to 218(4) ns. The isomer decays by $E2$ and $E4$ γ-ray transitions to the 8^+ and 6^+ states, respectively, and by proton emission to the first excited state of ^{53}Co. The low-lying level structure of ^{54}Ni is discussed within the shell model in comparison to its mirror nucleus ^{54}Fe.

1. Introduction

Isomer spectroscopy is a powerful tool for studies of nuclear structure at extreme values of isospin. It requires an efficient identification of the radioactive nuclear beam species produced in the reaction in combination with high efficiency for γ-ray detection. In the RISING project (*Rare ISotope INvestigations at GSI*) this is acheived by the *FRagment Separator* (FRS)[1] and fifteen CLUSTER germanium detectors.[2] RISING has been used in several experiments to study a wide range of exotic nuclei both at the proton drip-line and in heavy, neutron-rich systems. Information on the results of the first stopped-beam RISING campaign is provided in Refs.[3–5] The present contribution focuses on ^{54}Ni. The aim of the experiment is to reveal information on isospin symmetry aspects of the effective nuclear force by comparing ^{54}Ni with its mirror partner ^{54}Fe.

2. Experiments

The ^{54}Ni nuclei were produced in projectile fragmentation reactions using a primary beam of ^{58}Ni ions provided by the SIS accelerator at GSI, and a beryllium target with a thickness of 1 g/cm^2. The reaction products were selected by means of a $B\rho - \Delta E - B\rho$ technique in the FRS.[1] The identification in terms of mass A and proton number Z of each transmitted ion was performed with a set of detector elements placed at the intermediate S2 and final S4 focus of the FRS. The ions were then slowed down in an aluminum degrader to achieve proper implantion into a stopper foil surrounded by the RISING Ge-array. It comprises fifteen CLUSTER germanium detectors,[2] i.e., a total of 105 Ge crystals arranged in three rings of five detectors at central angles of 51°, 90°, and 129° with respect to the secondary beam direction. The distance of the crystals from the center of the S4 focal plane was $\sim$ 21 cm. The array combines high efficiency with high granularity. The digital processing of the energy and timing signals was done using thirty 4-channel XIA Digital Gamma Finder (DGF) modules. The γ-ray events were time stamped in steps of 25 ns, and a conventional timing branch is installed in parallel and digitized with a short-range ($t \leq 1$ μs) and a long-range ($t \leq 0.8$ ms) VME TDC. If the isomer survives the $\sim$ 200 ns flight time through the FRS, the γ-ray set-up is sensitive to isomeric decays in the range of about 10 ns to 1 ms. Further details of the experiment is found in Refs.[3,4]

3. Results

About 5 million ^{54}Ni isotopes were rigorously identified by the ionization chambers in the FRS, which determines Z, and by time-of-flight between the S2 and S4 areas, which determines A/Z. This is shown in a preliminary scatter plot in Fig. 1(a). Figure 1(b) shows the correlation matrix constructed between γ-ray energy and time of the observation of the γ ray relative to the implantation of the ions in the stopper. The vertical line marks the "prompt flash", i.e., implantation time $t = 0$, while fading, horizontal lines indicate decays from isomeric states.

From the energy-time correlation matrix time spectra can be produced for selected γ rays, which are then used to derive the half-life of the isomer. In turn, clean γ-ray energy spectra can be produced by choosing proper timing windows. This is illustrated in Fig. 2(a), which reveals six delayed γ-ray transitions at 146, 451, 1227, 1327, 1392, and 3241 keV, all having the same lifetime of $\tau \sim 220$ ns. Note that the 451, 1227, and 1392 keV lines are known to belong to ^{54}Ni, representing the $6^+ \to 4^+ \to 2^+ \to 0^+$ cascade in that nucleus.[6–8] The 146 and 3241 keV lines are suggested to be the $10^+ \to 8^+$ and $8^+ \to 6^+$ transitions, at first based on mirror symmetry to ^{54}Fe. More reliably they are found in coincidence with the known cascade, which is proven by the γ-γ coincidence spectra of Fig. 2(b). Here, another weak but distinct line can be discriminated at 3386 keV, which marks the

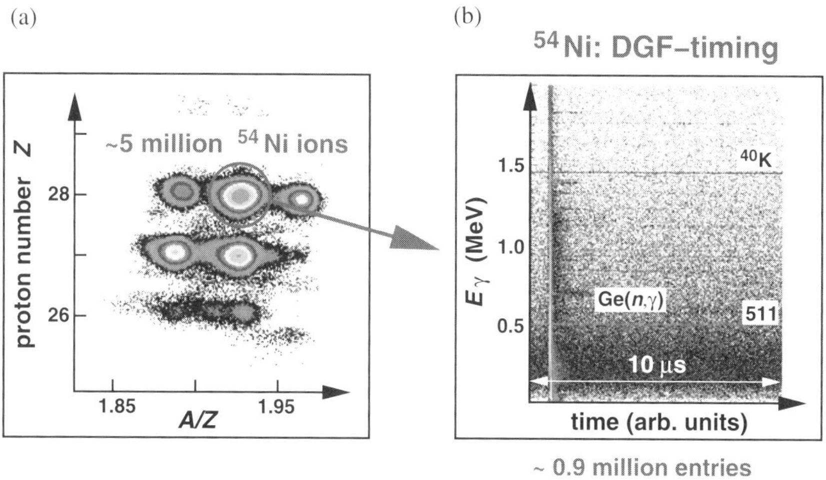

Fig. 1. Identification of the ^{54}Ni ions in the S4 focal-plane of the FRS, and the energy-time correlated γ-rays associated with ^{54}Ni.

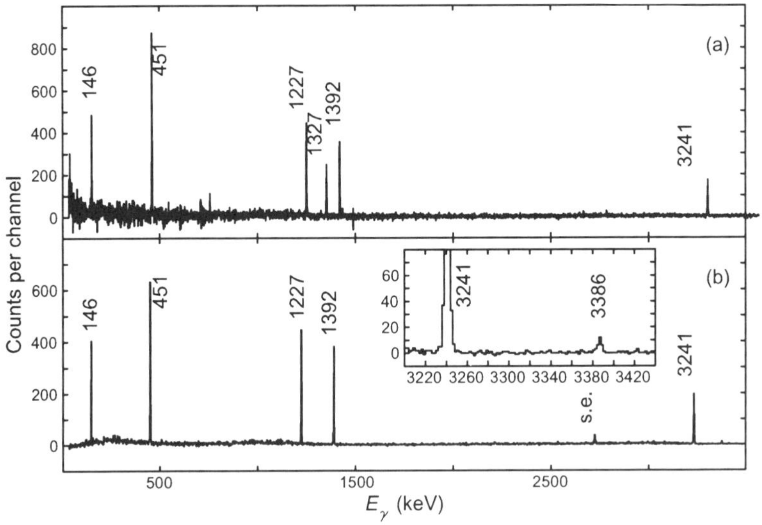

Fig. 2. Panel (a) provides a γ-ray singles spectrum in the time range $0.05\ \mu s \leq t \leq 1.0\ \mu s$ following the implantation of ^{54}Ni ions in the stopper. Panel (b) is a $\gamma\gamma$ correlation spectrum in coincidence with one of the 451, 1227, or 1392 keV transitions in ^{54}Ni. The same timing conditions as for panel (a) were applied, but the FRS gates were less restrictive.

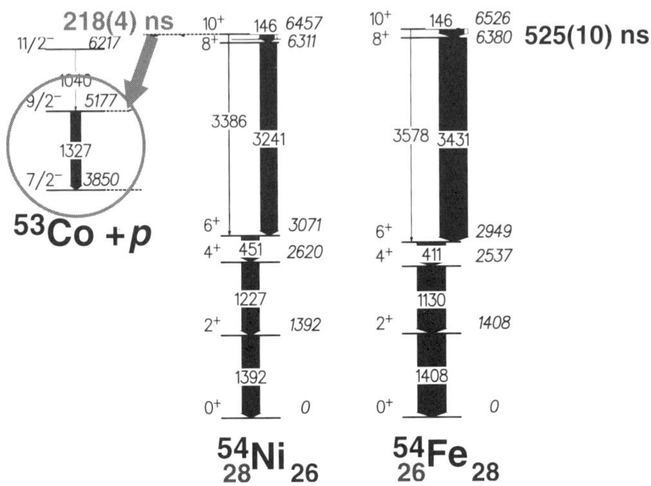

Fig. 3. Preliminary decay scheme of ^{54}Ni from the present experiment.

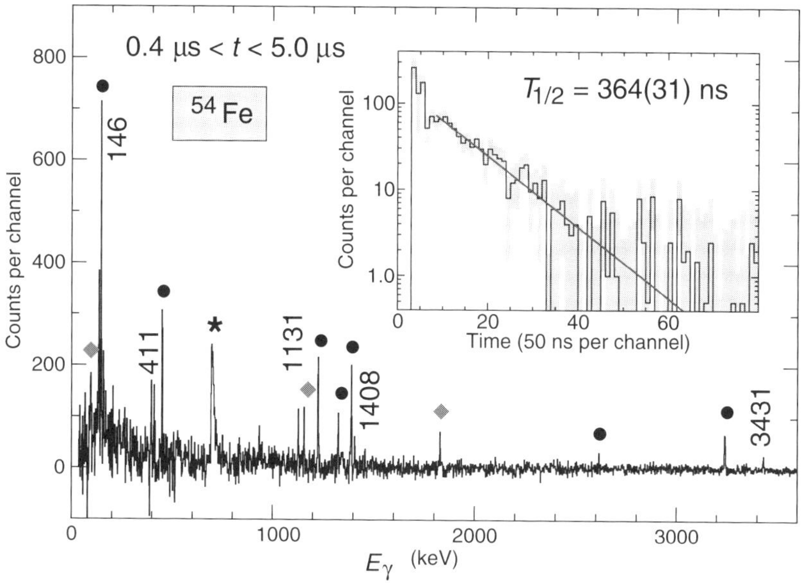

Fig. 4. Gamma-ray transitions in ^{54}Fe (energy labels in keV) and ^{43}Sc (diamonds) produced in secondary reactions in the stopper. Filled circles are ^{54}Ni and * is Ge(n,n'γ). See text for explanation.

$10^+ \rightarrow 6^+$ $E4$ branch of the isomeric decay in ^{54}Ni. These γ rays define the level structure shown in Fig. 3.

^{54}Fe has a well established isomeric 10^+ state with $\tau = 525(10)$ ns.[9] The lifetime of the isomeric state in ^{54}Ni was measured to 218(4) ns, which at first is somewhat puzzeling since the lifetimes of the two mirror states are expected to be similar. Interestingly, however, the 1327 keV line seen clearly in the singles spectrum [Fig. 2(a)] is absent in the γ-γ coincidence spectrum [Fig. 2(b)]. Since it exhibits (within uncertainties) the same half-life as the other transitions and because it fits in energy, we associate it with the $9/2^- \rightarrow 7/2^-$ ground-state transition in ^{53}Co. The $9/2^-$ state in ^{53}Co can be populated with a direct proton decay of 1.28 MeV from the 10^+ isomer in ^{54}Ni. This would be the first (indirect) evidence for proton emission following a fragmentation reaction and brings the associated research field back to its roots, since direct proton decay was first observed in ^{53m}Co in 1970.[10]

The primary goal of finding the predicted isomeric 10^+ state and its corresponding decay branches in ^{54}Ni has been achieved. An interesting secondary outcome of the present experiment is visualized in the γ-ray spectrum of Fig. 4. It is obtained by gating on the ^{54}Ni ions (see Fig. 1(a)), but with less restrictive conditions of the FRS settings such that $\sim$ 14 million events were allowed to pass through. It is further gated on delayed time between 0.4 and 5.0 μs. In addition to the γ rays of ^{54}Ni (lines marked with filled circles) the spectrum also includes γ rays from ^{43}Sc (lines marked with diamonds) and from ^{54}Fe (the 146, 1131, 1408, and 3431 keV lines). The only possibility to account for their presence is that they are produced by secondary nuclear reactions of the ^{54}Ni ions while they are implanted in the passive stopper. Both ^{54}Fe and ^{43}Sc have known isomers in the relevant time range, a 10^+ state and a $19/2^-$ state, respectively. The inserted time spectrum of Fig. 4 was obtained by gating on the ^{54}Fe transitions. The half-life of its 10^+ state is re-measured with this new type of method and is in nice agreement with the previously measured half-life.[9]

4. Discussion

The level structures of ^{54}Ni and ^{54}Fe (see Fig. 3) are similar to each other as expected from the underlying charge symmetry of the nuclear force. The energy difference between their states, as plotted in Fig. 5 as a function of angular momentum in a so called mirror energy difference (MED) diagram, is a measure of the extent to which this symmetry is broken. The asymmetry is mainly due to the Coulomb interaction, which is different in the two nuclei.

Both ^{54}Ni and ^{54}Fe have very simple structures with only two neutron and two proton holes, respectively, in the $1f_{7/2}$ shell. When the proton-hole pair in ^{54}Fe is aligned first to angular momentum 2, then to 4, and finally to 6, the Coulomb energy is expected to increase, the binding energy to decrease, and the states in ^{54}Fe are expected to move up in energy with respect to ^{54}Ni causing a gradual increase of the MED. This behaviour is reproduced by the shell model shown in Fig. 5 (the dashed curve). The calculation is based on the KB3G interaction using the code ANTOINE.[11] However, the experimental MED diagram (the solid curve) does not reproduce this behaviour for the 2^+ states as their energy difference decreases when the particles align.

The same effect has been observed in the mirror nuclei ^{42}Ti and ^{42}Ca.[12] They are cross-conjugate to ^{54}Ni and ^{54}Fe and have the same simple structure, but instead of two holes in the $1f_{7/2}$ shell they have two particles. The obeserved effect is the so-called 2^+ anomaly, which Zuker et al.[12] suggest is due to isospin breaking of the nucleon-nucleon interaction.

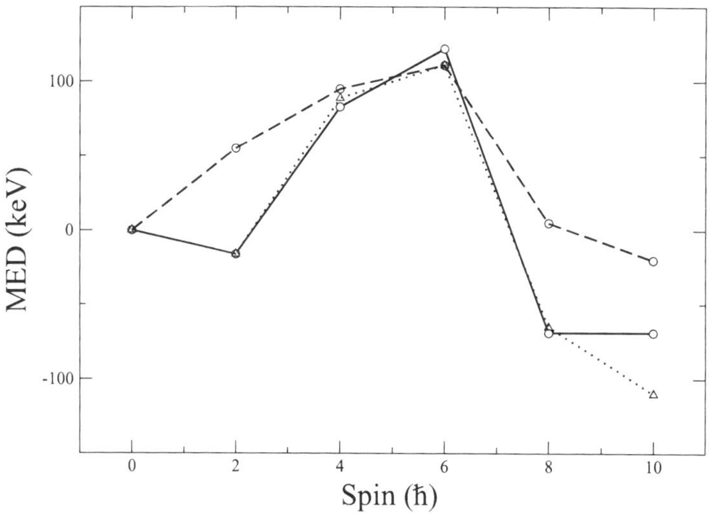

Fig. 5. Mirror Energy Difference diagram of the mass $A = 54$ mirror pair.

To account for the effect they proposed an ad hoc adjustment of the two-body matrix element of the 2^+ coupling. If we adjust this matrix element by about 100 keV in the KB3G interaction above the result of the shell model calculation is the dotted curve in Fig. 5, which now nicely reproduce the 2^+ state.

The original calculation failed to reproduce also the 8^+ and 10^+ MED values, but interestingly with the modified interaction of the 2^+ coupling they are much better reproduced (see Fig. 5). To make the 8^+ and 10^+ states it is necessary to break two pairs in the $1f_{7/2}$ shell, and one of those pairs couple to 2^+. In fact both the 8^+ and 10^+ states should have large 2^+ partitions in their wave functions, so it appears as if isospin breaking of the nucleon-nucleon interaction indeed mainly affects the 2^+ coupling.

5. Summary

^{54}Ni was produced in an experiment within the RISING stopped beam campaign at GSI in Darmstadt. The aim was to study isospin symmetry aspects of the effective nuclear interaction. The expected 10^+ isomeric state of ^{54}Ni was successfully observed. Unexpectedly though the 10^+ level was found to decay not only via $E2$ and $E4$ γ-ray transitions, but also via direct proton emission. A new facet of the prompt proton decay has thus been discovered following a fragmentation reaction, where the initial state in ^{54}Ni is isomeric with $\tau = 218(4)$ ns. The experiment reveals new and interesting information on isospin symmetry of the effective nuclear force.

Acknowledgements

The authors gratefully acknowledge the outstanding work of the GSI accelerator and ion-source crews in providing the experiments with the envisaged high beam intensities. This work is supported by the European Commission contract no. 506065 (EURONS), the Swedish Research Council, EPSRC (United Kingdom), the German BMBF, the Polish Ministry of Science and Higher Education, the Bulgarian Science Fund, the Spanish, Italian, and French science councils, and the U.S. Department of Energy.

References

1. H. Geissel *et al.*, Nucl. Instr. Meth. **B70**, 286 (1992).
2. J. Eberth *et al.*, Nucl. Instr. Meth. **A369**, 135 (1996).
3. D. Rudolph *et al.*, Proc. Seventh International Conference on Radioactive Nuclear Beams, RNB7, Cortina d'Ampezzo, Italy, 2006; Ed. C. Signorini.
4. S. Pietri *et al.*, Proc. Seventh International Conference on Radioactive Nuclear Beams, RNB7, Cortina d'Ampezzo, Italy, 2006; Ed. C. Signorini.
5. Zs. Podolyák *et al.*, Proc. Seventh International Conference on Radioactive Nuclear Beams, RNB7, Cortina d'Ampezzo, Italy, 2006; Ed. C. Signorini.
6. K.L. Yurkewicz *et al.*, Phys. Rev. C **70**, 054319 (2004).
7. K. Yamada *et al.*, Eur. Phys. J. **A25** S1, 409 (2005).
8. A. Gadea *et al.*, Phys. Rev. Lett. **97**, 152501 (2006)
9. J. Huo *et al.*, Nucl. Data Sheets **68**, 887 (1993).
10. K.P. Jackson *et al.*, Phys. Lett. **33B**, 281 (1970); J. Cerny *et al.*, Phys. Lett. **33B**, 284 (1970).
11. E. Caurier, shell model code ANTOINE, IReS Strasbourg 1989, 2002.
12. A. Zuker *et al.*, Phys. Rev. Lett. **89**, 142502 (2002)

EXPLORING THE EVOLUTION OF THE SHELL STRUCTURE BY MEANS OF DEEP INELASTIC REACTIONS

G. DE ANGELIS

INFN Laboratori Nazionali di Legnaro,
Legnaro (Pd) I-35020, Italy
**E-mail: ab_giacomo.deangelis@lnl.infn.it*

The study of nuclear structure far from stability, which mainly rely on the availability of radioactive nuclear beams, can complementary be addressed by means of high intensity beams of stable ions. Deep-inelastic and multi-nucleon transfer reactions are a powerful tool to populate yrast and non yrast states in neutron-rich nuclei. Particularly successful is here the combination of large acceptance spectrometers with highly segmented γ-detector arrays. Such devices, eventually complemented by large coverage particle detectors, can provide the necessary channel selectivity to identify very rare signals. An example is the CLARA γ-ray detector array coupled with the PRISMA spectrometer at the Legnaro National Laboratories (LNL). Large data sets have been recently collected for nuclei close to the N=28, 40 and 50 shell closures. The obtained results complement studies performed with current radioactive beam (RIB) facilities. The data clearly show the evolution of the effective single particle energies in very good agreement with the predictions of the mean field model with tensor interaction. New experimental information has been obtained on a wide range of nuclei close to the N=28, 40 and 50 shell closures allowing the population of medium and high-spin yrast states. The excited states of the N = 50 isotones, extended down to Z=31, and of N=51 isotones, extended down to Z=34, have been used to test the predictions of the shell evolution based on the effects of the tensor interaction as well as of the different effective interactions. As future perspective the development of a γ-ray detection system capable of tracking the location of the energy deposited at every γ-ray interaction point will also provide an unparallel level of detection sensitivity, and will open new revenues for nuclear structure studies.

Keywords: Gamma Spectroscopy, Deep Inelastic Reactions, Neutron Rich Nuclei.

1. Introduction

Magic numbers are a key feature in finite Fermion systems since they are strongly related to the underlying mean field. The study of the evolution of the shells far from stability is therefore of high interest since such information can be linked to the shape and symmetry of the nuclear mean field. The study of nuclei with large neutron/proton ratio allow to probe the density dependence of the effective interaction. Changes of the nuclear density and size in nuclei with increasing N/Z ratios are expected to lead to different nuclear symmetries and excitations. Recently it has also been shown that the tensor force play an important role in breaking and creating magic numbers being a key element of the shell evolution along the nuclear chart. A tremendous effort is presently going on through a systematic exploration of new regions of the chart of nuclei as well as through the developments of high intensity stable and radioactive ion beam facilities. In this contribution I will discuss some selected examples which show the big potential of stable beams using "unconventional" reactions for the study of the properties of the nuclear many body system at large neutron excess.

2. Shell structure evolution far from stability

One of the most critical ingredient in determining the properties of a nucleus from a given effective interaction, is the overall number of nucleons and the ratio N/Z of neutrons to protons. One aspect which is presently strongly discussed concerns the modification of the average field experienced by a single nucleon due to the changes in size and diffusivity for nuclei with strong neutron excess.[1,3] For large neutron excess the softening of the Woods-Saxon shape of the neutron potential is expected to cause a reduction of the spin-orbit interaction and therefore a migration of the high-l orbitals with a large impact on the shell structure of nuclei far from stability.[3–6]

A different scenario has been recently suggested where the evolution of the shell structure, in going from stable to exotic nuclei, can be related to the effect of the tensor part of the nucleon-nucleon interaction.[7,8] The tensor-force, one of the most direct manifestation of the meson exchange origin of the nucleon-nucleon interaction, is responsible of the strong attraction between a proton and a neutron in the spin-flip partner orbits. A recent generalization of such mechanism foresees a similar behaviour also for orbitals with non identical orbital angular momenta. An attraction is expected for orbitals with antiparallel spin configuration whereas orbitals

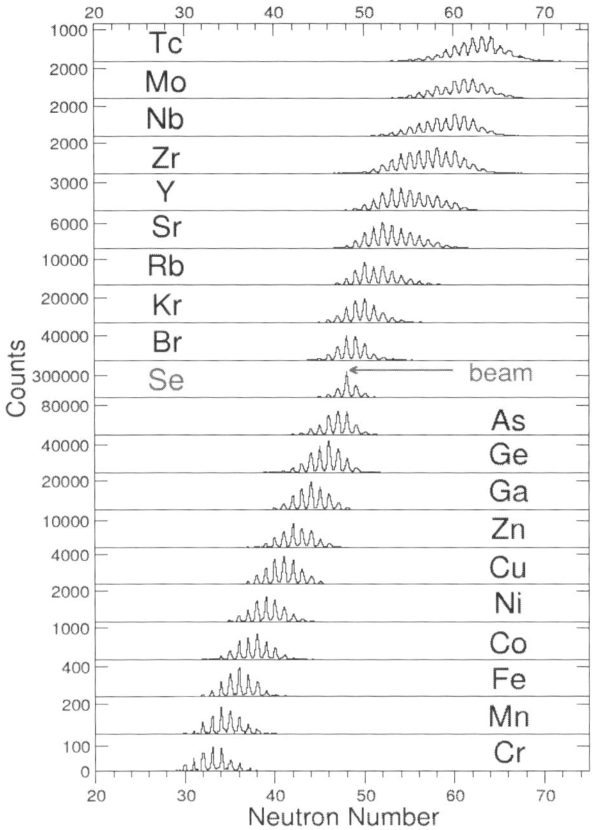

Fig. 1. Mass distributions for the different elements populated in the ^{82}Se + ^{238}U reaction at 505 MeV of beam energy.

with parallel spin configuration should repell each other.[8] In most of the cases one is dealing with a combined effect of the attraction among orbitals with antiparallel spins and repulsion between orbitals with parallel spins. Those effects become particularly visible when moving away from the line of stability. In such cases the proton-neutron interaction is changed by empting of the partner orbit causing a modification into the effective single particle energies (evolution of the shell structure). The change of the shell structure based on such mechanism has been recently discussed in different

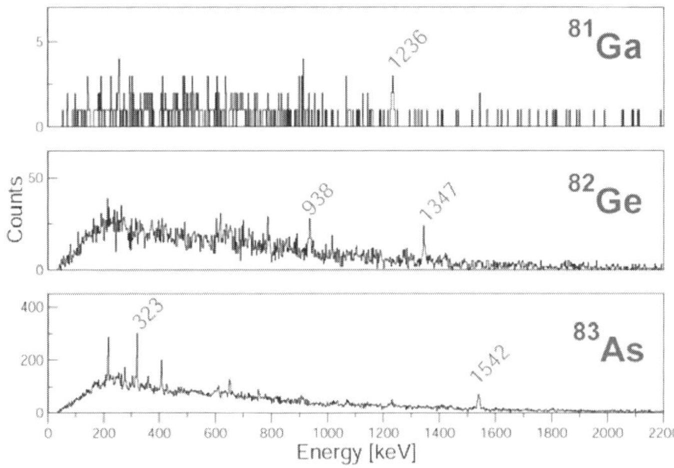

Fig. 2. Doppler corrected γ-ray spectra measured with the CLARA Ge-detector array for Z=33, 32, 31 and N=50 selected with the PRISMA Specrometer.

mass regions of the nuclear chart[8,10] . In such contest neutron-rich nuclei close to shell gaps are particularly interesting since, when compared with the shell model prediction, they allow to search for anomalies into the shell structure.[3,10] It is predicted, for example, that the Z=28 gap for protons in the pf-shell becomes smaller moving from ^{68}Ni to ^{78}Ni as consequence of the attraction between particles in the proton $f_{5/2}$ and neutron $g_{9/2}$ orbits and the repulsion between particles in the proton $f_{7/2}$ and the neutron $g_{9/2}$ configurations. The same argument also predicts a weakening of the N=50 shell gap when approaching the ^{78}Ni nucleus due to the attraction between particles in the neutron $g_{9/2}$ and $d_{5/2}$ configurations with the proton $f_{5/2}$ state and the repulsion between particles in the neutron $g_{7/2}$ with the proton $f_{5/2}$ states.

3. Experiment

Experimental information has been obtained on a wide range of nuclei close to the N=50 shell closure by means of multi-nucleon transfer and deep-inelastic collisions.[11,12] The reaction mechanism allows the population of medium and high spin yrast states. In most of the cases previous information on the excited states comes mainly from β-decay studies

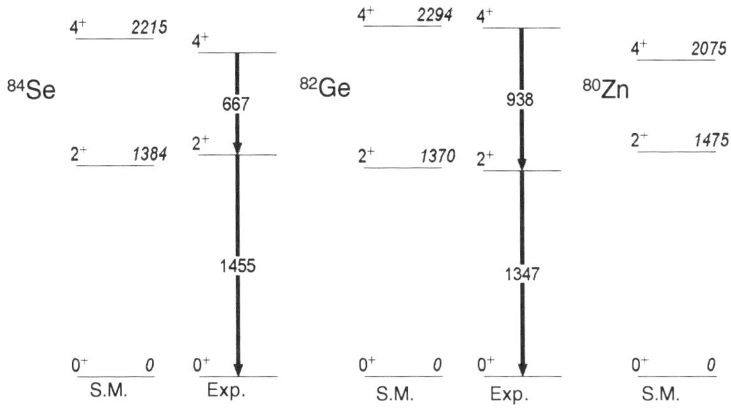

Fig. 3. Level scheme of the N=50 ^{84}Se, ^{82}Ge nuclei compared to Shell model calculations.

and direct reactions.[13] Excited states of previously unknown isotopes have been identified using the CLARA γ-detector array[14] in coincidence with the PRISMA spectrometer.[15] The N≈50 nuclei have been populated using the reaction ^{82}Se + ^{238}U. The combination of the Tandem-XTU and the superconductive LINAC ALPI accelerators at the Laboratori Nazionali di Legnaro, Italy, was used to accelerate a beam of ^{82}Se ions at an energy of 505 MeV. The target, isotopically enriched, was of a thickness of 400 μg/cm2. Projectile-like nuclei, produced following multinucleon transfer, were detected by the PRISMA spectrometer,[15–17] placed at an angle of 64 degrees, covering an angular region around the grazing angle of the reaction. PRISMA is a magnetic spectrometer of large angular (≈80 msr) and energy (20%) acceptance, consisting of a quadrupole singlet followed by a dipole magnet. The position of the incoming ions entering the spectrometer were measured using a position sensitive MCP detector placed at 25 cm from the target. After the magnetic elements, the position of the ions were determined again at the focal plane of the spectrometer using a 10 elements, 100 cm long, multi-wire PPAC detector. Finally, the ions were stopped in a 10×4 elements ionization chamber measuring the time of flight (TOF). Such information allow a reconstruction of the trajectories, the identification of the atomic number Z, of the mass A and of the absolute value of the velocity. The mass resolution obtained in the present experiment,

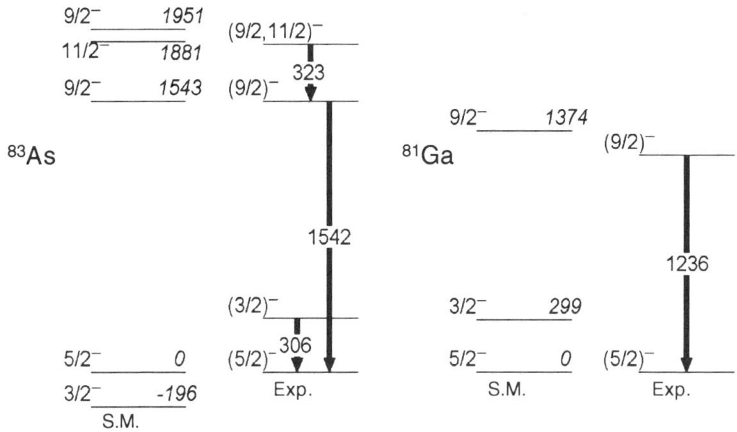

Fig. 4. Level scheme of the N=50 ^{83}As and ^{81}Ga compared to Shell model calculations.

shown in fig.1, is of 1/180, allowing a good identification of all isotopes. Gamma ray spectra for selected N=50 isotones are shown in fig.2. For the construction of the level schemes as well as for the spin and parity assignment we have assumed that the population is mainly yrast and used, where possible, γ-coincidence data obtained in a second experiment at the GASP detector array.[18] The GASP apparatus consists of 40 Compton-suppressed large-volume germanium detectors and of an inner BGO ball acting as a total energy and multiplicity filter. Neutron-rich nuclei where populated using the reaction ^{82}Se + ^{192}Os at 460 MeV of beam energy. The Se beam was provided by the LNL accelerator complex. Triple and higher-fold γ-γ coincidences have been acquired in a "thick target" measurement. Here the target, isotopically enriched, was of a thickness of 60 mg/cm2, sufficient to stop all reaction fragments. Since all recoiling fragments were stopped in the target, Doppler broadening prevented the observation of transitions deexciting short lived states and only γ-decay with lifetime longer than the slowing-down time of the recoiling nuclei ($\approx$1 ps) could be resolved. The spin and parities of the levels were deduced, where possible, from angular distribution ratios from oriented states (ADO) as well as from the decay branchings. Details of the measurements as well as of the data reduction are reported in Ref.[19]

4. Results and Discussion

Results obtained for the N=50 isotones are shown in fugures 3 and 4 together with the predictions of shell model calculations. Within the shell-model, the description of the nuclear mean field is conventionally obtained by considering the so-called monopole Hamiltonian constructed from the centroids of the two-body interaction. The eigenvalues of this Hamiltonian, usually referred as effective single particle energies, provide the average energies of the specific spherical configurations. They should reproduce the energies of single-proton or single neutron states in odd-A nuclei with Z or N equal to a magic number plus or minus one proton or one neutron. The nucleon-nucleon residual interaction, and in particular the tensor force part, changes the spherical single particle energies through the nuclear chart. As reported in Ref.[7–9] the monopole effect of the tensor force shifts systematically the single-particle levels as protons and neutrons fill certain orbits, the proton-neutron part being the dominant. This works attractively for anti-parallel spin configurations and repulsively for parallel spin configurations. The radial wave functions of the two orbits must also be similar in order to have a large overlap in the radial direction. For the same radial condition, larger orbital angular momenta enhance the tensor monopole effect.[9] For the N=50 isotones, due to the strong attraction between particles filling the $\pi f_{5/2}$ and the $\nu g_{9/2}$ orbits and the repulsion between the particles filling the $\pi d_{3/2}$ and the $\nu g_{9/2}$ orbits one expects an inversion of the relative position of the $\pi f_{5/2}$ - $\pi d_{3/2}$ effective single particle states respect to the usual order where the $\pi d_{3/2}$ state is the lowest in excitation enery. Such inversion is confirmed by the experimental data (figure 4). It is interesting to mention here the good agreement obtained between the excitation energies of the newly identified states and the predicted values of the shell model calculation using the effective interaction reported in ref.[20]

5. Conclusion and Perspectives

The study of neutron rich nuclei at medium angular momenta can provide basic information on the stability and evolution of the nuclear many body system. High-intensity beams of heavy stable ions and binary reactions are powerful tools to access the neutron rich side of the chart of the nuclei. Within this perspective the accelerator complex at LNL is going to be upgraded with a new ECR ion source and with additional accelerator modules, the main aim being a sensible increase of the beam intensities for the very heavy stable ions. Future perspectives on a loger time scale are also based on the realization of a new radioactive ion beam facility based on fission fragments[21,22] . Among the future perspectives concerning detector technology, the development of a γ-ray detection system (AGATA) capable of tracking the location of the energy depositet at every interaction point will costitute a major advance. This will provide an unparalleled level of detection sensitivity and will open new avenues for nuclear structure studies.

6. Acknowledgments

I would like to thank E. Sahin, T. Faul, G. Duchene, A. Gadea, for providing most of the material reported here and for the useful discussions on the results. Many thanks are due to all the colleagues of the PRISMA, CLARA and GASP groups for running the apparatus. This research has been supported by the European Commission through the contract number HPRI-CT-1999-00078.

References

1. G.A. Lalazissis et al., Phys. Lett. B 418, 7 (1998)
2. B. Blank, P.H. Regan, Nucl. Phys. News Internat. 11 (1) (2001) 15.
3. J. Dobaczewski et al., Phys. Scr. T56, 15 (1995).
4. R.C. Nayak. Phys. Rev. C 60 (1999) 064305.
5. X. Campi, et al. Nuclear Phys. A 251 (1975) 193.
6. N. Fukunishi et al., Phys. Lett. B 296 (1992) 279.
7. T. Otsuka et al., Phys. Rev. Lett. 11 (1964) 145.
8. T. Otsuka, Progr. Theoret. Phys. Suppl. 146 (2002) 6.
9. T. Otsuka et al., Phys. Rev. Lett. 97 (2006) 162501.
10. H. Grave, Springer Lect. Notes in Phys. 651 (2004) 33.
11. R. Broda et al., Phys. Rev. Lett. 74, 868 (1995).
12. I. Ishii et al., Phys. Rev. Lett. 81, 4100 (1998) and ref. therein.
13. R.B. Firestone, Table of Isotopes, 8th. ed., Wiley, 1996.
14. A. Gadea et al., Eur. Phys. J. A 20, 193 (2004)
15. A. M. Stefanini et al., Nucl. Phys. A 701, 217c (2002).
16. G. Montagnoli et al., Nucl. Instrum. Methods Phys. Res. A 547, (2005) 455.
17. S. Begini et al., Nucl. Instrum. Methods Phys. Res. A 551 (2005) 364.
18. D. Bazzacco, Proceedings of the Intern. Conference on Nuclear Structure at high Angular Momentum, vol II, Ottawa, 1992, p.376. Rep. N. AECL10613.
19. Y.H. Zhang et. al. Phys. Rev. C. 70 (2004) 024301.
20. Lisetskiy et al., Phys. Rev. C 70, 044314 (2004).
21. G. Prete, see contribution to these preceedings.
22. SPES Thechnical Design for an Advanced Exotic Ion Beam Facility at LNL, LNL-INFN (REP) 181/02, June 2002

STUDIES ON THE EXOTIC STRUCTURE OF ^{23}Al BY MEASUREMENTS OF σ_R AND $P_{//}$

D. Q. FANG[1], Y. G. MA[1], W. GUO[1], C. W. MA[1], K. WANG[1], T. Z. YAN[1], X. Z. CAI[1], Z. Z. REN[2], Z. Y. SUN[3], J. G. CHEN[1], W. D. TIAN[1], C. ZHONG[1], W. Q. SHEN[1], M. HOSOI[4], T. IZUMIKAWA[5], R. KANUNGO[6], S. NAKAJIMA[4], T. OHNISHI[7], T. OHTSUBO[5], A. OZAWA[8], T. SUDA[7], K. SUGAWARA[4], T. SUZUKI[4], A. TAKISAWA[5], K. TANAKA[7], T. YAMAGUCHI[4], I. TANIHATA[6]

[1]*Shanghai Institute of Applied Physics, Chinese Academy of Sciences, Shanghai 201800, China*

[2]*Department of Physics, Nanjing University, Nanjing 210008, China*

[3]*Institute of Modern Physics, Chinese Academy of Sciences, Lanzhou 730000, China*

[4]*Department of Physics, Saitama University, Saitama 338-8570, Japan*

[5]*Department of Physics, Niigata University, Niigata 950-2181, Japan*

[6]*TRIUMF, 4004 Wesbrook Mal, Vancouver, British Columbia, V6T 2A3, Canada*

[7]*Institute of Physical and Chemical Research (RIKEN), Wako, Saitama 351-0198, Japan*

[8]*Institute of Physics, University of Tsukuba, Ibaraki 305-8571, Japan*

The longitudinal momentum distribution ($P_{//}$) of fragments after one-proton removal from ^{23}Al and reaction cross sections (σ_R) for ^{23,24}Al on carbon target at 74A MeV have been measured using 135A MeV ^{28}Si primary beam on RIPS in RIKEN. $P_{//}$ is measured by a direct time-of-flight (TOF) technique, while σ_R is determined using a transmission method. An enhancement in σ_R is observed for ^{23}Al compared with ^{24}Al. The $P_{//}$ for ^{22}Mg fragments from ^{23}Al breakup has been obtained for the first time. FWHM of the distributions has been determined to be 232±28 MeV/c. The experimental data are discussed by use of the Few-Body Glauber Model (FBGM). Analysis of $P_{//}$ indicates a dominant d-wave configuration for the valence proton in the ground state of ^{23}Al. The exotic structure in ^{23}Al is discussed.

Keywords: proton-rich nuclei; reaction cross section; longitudinal momentum distribution

1. Introduction

Since the pioneering measurements of the interaction cross sections (σ_I) and observation of an remarkably large σ_I for ^{11}Li,[1] it has been shown that there is exotic structure like neutron halo or skin in light neutron-rich nuclei. Measurements of σ_R, $P_{//}$ of one or two nucleons removal reaction, quadrupole moment and Coulomb dissociation have been demonstrated to be very effective methods to identify and investigate the structure of halo nuclei. The neutron skin or halo nuclei ^{6,8}He, ^{11}Li, ^{11}Be, ^{19}C etc.,[1–3] have been identified by these experimental methods. Due to the centrifugal and Coulomb barriers, the identification of a proton halo is more difficult compared to a neutron halo. The quadrupole moment, $P_{//}$ and σ_R measurements indicate a proton halo in ^{8}B,[4,5] whereas no enhancement is observed in the measured σ_I at relativistic energies.[1,6] The proton halo in ^{26,27}P and ^{27}S has been predicted theoretically.[7] And the measurements of $P_{//}$ have shown a proton halo character in ^{26,27,28}P.[8]

Proton-rich nucleus ^{23}Al has a very small separation energy (S_p = 0.125 MeV)[9] and is a candidate of proton halo. An enhanced σ_R for ^{23}Al has been observed previously.[10] To reproduce the σ_R for ^{23}Al using the Glauber model, the assumption of a considerable $2s_{1/2}$ component for the valence proton is necessary.[10] Thus a long tail in the proton density distribution has been extracted for ^{23}Al which indicates an exotic structure. While the spin and parity (J^π) for ^{23}Al has been deduced to be $5/2^+$ recently.[11] This result favors the d-wave configuration for the valence proton in ^{23}Al. But it does not eliminate the possibility of a s-wave valence proton if the ^{22}Mg core is in the excited state ($J^\pi = 2^+$). Therefore it will be very important to determine the configuration of the valence proton for ^{23}Al. In this paper we will report the simultaneously measurement of σ_R and $P_{//}$ for further investigation on the exotic structure of ^{23}Al.

2. Experiment

The experiment was performed at RIPS in RIKEN. Secondary beams were generated by fragmentation reaction of 135A MeV ^{28}Si beam on a ^{9}Be target in F0 chamber. At the first dispersive focus F1, an Al wedge-shape degrader was installed. And a Parallel Plate Avalanche Counter (PPAC) was placed to measure momentum broadening of the beam. Then the secondary beam was focused onto the achromatic focus F2. Two PPACs were installed to determine the beam position and angle. An ion chamber was used to measure the energy loss (ΔE).[12] And an ultra-fast plastic scintillator was placed before a carbon reaction target to measure the time-of-flight (TOF) from the PPAC at F1. The particle identification before the reaction target was done by means of $B\rho - \Delta E -$TOF method. After the reaction target, a quadrupole triplet was used to transport and focus the beam onto F3 ($\sim$ 6 m from F2). Two PPACs were used to monitor the beam size and emittance angle. Another plastic scintillator gave a stop signal of the TOF from F2 to F3. Another ion chamber was used to measure the energy loss (ΔE). The total energy (E) was measured by a NaI(Tl) detector. The particles were identified by TOF$-\Delta E-E$ method.

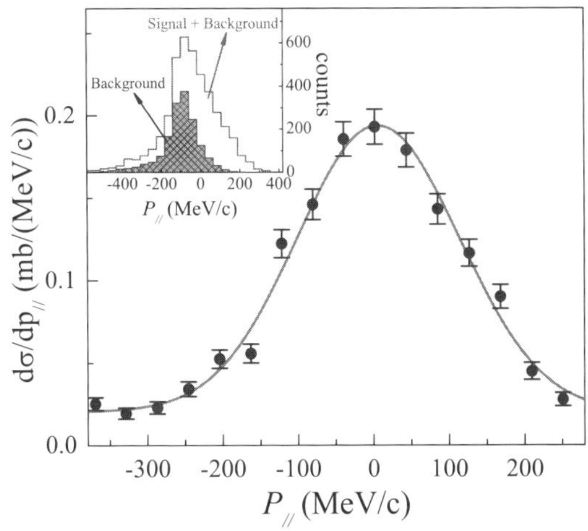

Fig. 1. $P_{//}$ distribution of fragment ^{22}Mg after one-proton removal from ^{23}Al. The closed circles with error bars are the present experimental data, the solid curve is a Gaussian fit to the data. Spectra in the inset are the original $P_{//}$ of ^{22}Mg from target-in (solid line) and target-out (shadowed area) measurements, respectively.

For one-proton removal reactions of ^{23}Al, the signal (reactions in the carbon target) is mixed up with the background (reactions in the detectors and material other than the reaction target) in the $P_{//}$ spectrum of the fragment from target-in measurement. This background was carefully reconstructed based on the ratio of fragments to unreacted projectile identified in the target-out measurement and the broadening effect of the carbon target on $P_{//}$. Then the estimated background was subtracted from the mixed spectrum. The obtained momentum distribution of the ^{22}Mg fragments from ^{23}Al breakup in the carbon target at 74A MeV is shown in Fig. 1. We normalized the experimental counts to the measured one-proton removal cross section (σ_{-1p}) so that $\sum N(p_{\mathrm{i}})\Delta p_{||}$ equals σ_{-1p}. A Gaussian function was used to fit the distributions. The full width at half maximum (FWHM) was determined to be 232±28 MeV/c after unfolding the Gaussian-shaped system resolution. The FWHM is consistent with the Goldhaber model's prediction (FWHM=2[illegible]2 MeV/c with $\sigma_0 = 90$ MeV/c) within the error bar.[13] Since the magnetic fields of the quadruples between F2 and F3 were optimized for the projectile in the measurement, the momentum dependence of the transmission from F2 to F3 for fragments was simulated by the code MOCADI.[14] The effect of transmission on the width of $P_{//}$ distribution was found to be negligible which is similar with the conclusion for neutron-rich nuclei.[15] Using the estimated transmission value, the one-proton removal cross sections for ^{23}Al was obtained to be 63±9 mb.

Reaction cross section is determined using the transmission method, by events of projectile before (incident) and after (unreacted) the reaction target for target-in and target-out measurements:

$$\sigma_R = \frac{1}{t} \ln \left(\frac{\gamma_0}{\gamma} \right) \tag{1}$$

where γ and γ_0 denote ratio of the unreacted outgoing and incident projectiles for target-in and target-out cases, respectively; t the thickness of the reaction target, i.e., number of particle per unit area.

The σ_R of ^{23,24}Al at 74A MeV were obtained to be 1609±79 mb and 1527±112 mb, respectively. The errors include the statistical and systematic uncertainties. The probability of inelastic scattering reaction was estimated to be very small ($< 1\%$), e.g., the inelastic cross section is only around 11 mb for ^{23}Al which is much smaller than the error of σ_R. And we observed an enhanced σ_R for ^{23}Al in our data again as in the previous experiment.[10]

3. Discussion

To interpret the measured reaction cross section and momentum distribution data, we performed a Few-Body Glauber Model (FBGM) analysis for $P_{//}$ of $^{23}\text{Al} \rightarrow {}^{22}\text{Mg}$ processes and σ_R of ^{23,24}Al.[17] In this model, a core plus proton structure is assumed for the projectile. For the core, HO-type functions were used for the density distributions. The wavefunction of the valence neutron was calculated by solving the eigenvalue problem in a Woods-Saxon potential. The separation energy of the last proton is reproduced by adjusting the potential depth.

In the recent g factor measurement using a β-NMR method, the spin and parity for the ground state of ^{23}Al is shown to be $5/2^+$. This gives a strong restriction on the possible nuclear structure of nucleus. Assuming the ^{22}Mg+p structure, three different configurations are possible for $J^\pi = 5/2^+$ of ^{23}Al: $0^+ \otimes 1d_{5/2}$, $2^+ \otimes 1d_{5/2}$ and $2^+ \otimes 2s_{1/2}$.[11] The s-wave configuration is also possible for the core in the excited state.

The momentum distributions for the valence proton in the s or d-wave

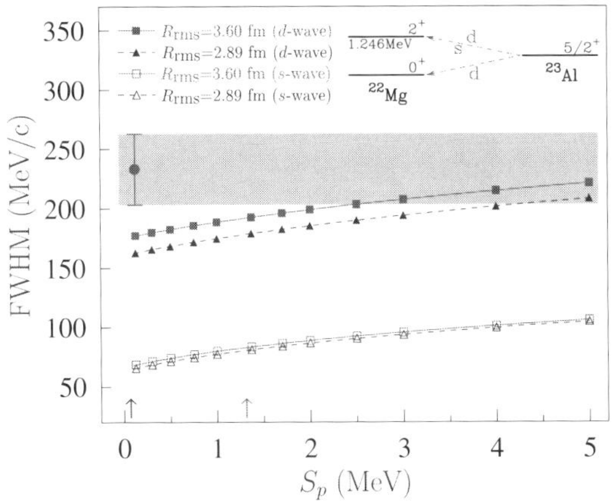

Fig. 2. The dependence of FWHM for the $P_{//}$ distribution after one-proton removal of ^{23}Al on the separation energy of the valence proton. The solid circles with error bars is result of the present experiment, the shaded area refers to error of the data. The solid and open squares are the FBGM calculations for the d and s-wave configuration of the valence proton with the core $R_{\rm rms}$ = 3.6 fm. The solid and open triangles are for the core $R_{\rm rms}$ = 2.89 fm. The lines are just for guiding the eyes. The two arrows refer to the separation energy of 0.125 MeV and 1.37 MeV (the excitation energy for the first excited state of ^{22}Mg plus the experimental one proton separation energy of ^{23}Al).

configuration are calculated by use of the FBGM. To fix the core size, the width parameters in the HO density distribution of ^{22}Mg were adjusted to reproduce the experimental σ_I data at around 1A GeV.[18] The extracted effective root-mean-square matter radii ($R_{\rm rms} \equiv< r^2 >^{1/2}$) for ^{22}Mg is 2.89 ± 0.09 fm. To see the separation energy dependence, the $P_{//}$ is calculated assuming an arbitrary separation energy in calculation of the wavefunction for the valence proton in ^{23}Al and shown in Fig. 2. And, if we adopt a larger radii of $R_{\rm rms}$ = 3.6 fm for ^{22}Mg to see the core size effect on $P_{//}$, we obtained solid and open squares of FWHM in Fig. 2. The one proton separation energies for ^{22}Mg in the ground and excited ($J^\pi = 2^+$, $E_x = 1.25$ MeV) states are taken as 0.125 MeV and 1.375 MeV ($E_x + 0.125$ MeV). Those two values are marked by two arrows in Fig. 2. In this figure, we can see that the width for the s and d-wave are obviously separated. The width for the s-wave is much lower than the experimental data, while that of the d-wave is close to the experimental FWHM. With the increase of S_p, the width of $P_{//}$ increases slowly. That means $P_{//}$ will become wider for ^{22}Mg in the excited state. The effect of the core size on $P_{//}$ is negligible for the s-wave but not for the d-wave configuration. The larger sized core will give a wider $P_{//}$ distribution. From comparison of the

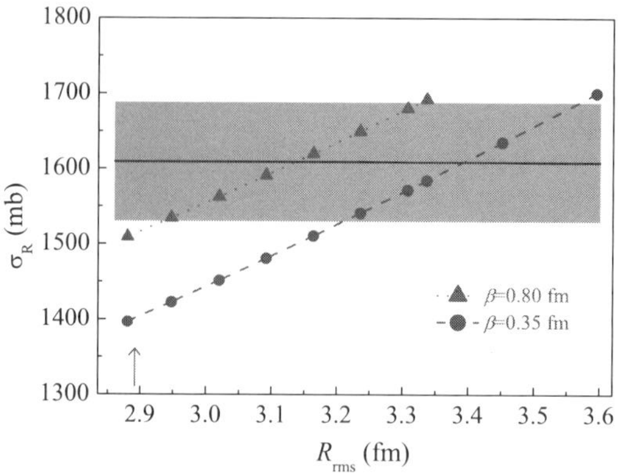

Fig. 3. The dependence of σ_R at 74A Mev on the core size ($R_{\rm rms}$). The horizontal line is the experimental σ_R value, the shadowed area is the error of σ_R. The solid circles and triangles denote the FBGM calculations with the range parameter $\beta = 0.35$ fm and $\beta = 0.80$ fm, respectively. The size of ^{22}Mg obtained by fitting the σ_I data at around 1A GeV is marked by an arrow.

FBGM calculation with the experimental data in Fig. 2, it clearly indicates that the valence proton in ^{23}Al is dominantly in the d-wave configuration. And the possibility for the s-wave should be very small. This is consistent with the shell model calculations and also the Coulomb dissociation measurement.[11,19]

From above discussions of $P_{//}$, the valence proton in ^{23}Al is determined to be in the d-wave configuration, which is used in the following calculations. For the calculation of reaction cross section using the FBGM, $R_{\rm rms} = 2.89 \pm 0.09$ fm is used for ^{22}Mg by reproducing the σ_I data as described above. But the calculated σ_R for ^{23}Al is much lower than the obtained σ_R data. Similar puzzle is also encountered for some neutron-rich nuclei. The large σ_I cannot be reproduced by the FBGM even for the valence neutron in the s-wave for ^{19}C and ^{23}O. One way is to enlarge the core size to reproduce the experimental reaction cross section.[20] Here we changed the core size by adjusting the width parameters in the HO density distribution of ^{22}Mg. The dependence of σ_R on $R_{\rm rms}$ of the core is shown in Fig. 3. In the FBGM calculations, the parameters α and σ_{NN} in the profile function $\Gamma(b) = \frac{1-i\alpha}{4\pi\beta^2}\sigma_{NN}\exp(-\frac{b}{2\beta^2})$ (b is the impact parameter) are taken from Ref.[17] The range parameter (β) is calculated by the formula which is determined by fitting the σ_R of ^{12}C + ^{12}C from low to relativistic energies.[21] β is 0 and 0.35 fm for 1A GeV and 74A MeV, respectively. The σ_R of ^{23}Al is very sensitive to the size of ^{22}Mg core. To reproduce the measured σ_R of ^{23}Al, the calculated results indicates an enlarged ^{22}Mg core with $R_{\rm rms} = 3.37 \pm 0.18$ fm. It is 17±7% larger than the size of ^{22}Mg extracted using the σ_I at around 1A GeV.[18]

But there may be another possibility for the Glauber model. Global underestimation of the reaction cross section (10-20% for stable nuclei and 30-50% for exotic nuclei) was found at intermediate energies in the Glauber model when the densities of the projectile are determined by fitting the σ_I at relativistic energies.[22] To correct the underestimation in the Glauber model, different method has been tried. The energy dependent phenomenological correction factor and finite-range effect were introduced into the Glauber model.[2,21,23] But these corrections are performed for almost light stable nuclei and no systematic study has been done for proton-rich nuclei with $A > 20$. The reaction cross section of ^{24}Al is calculated with the size of ^{23}Mg core determined by fitting σ_I at around 1A GeV.[18] But the calculated σ_R for ^{24}Al is only 1430 mb and underestimation still exists ($\sim$ 10%). Since scope of the underestimation in the Glauber model is large even for stable nuclei, underestimation may still exist in the finite range Glauber calculations with $\beta = 0.35$ fm at 74A MeV. To correct the possible underestimation, we adjusted the range parameter to fit the σ_R of ^{24}Al from the present measurement. And $\beta = 0.8$ fm is obtained when the σ_R of ^{24}Al is reproduced. Using this range parameter, the σ_R of ^{23}Al is calculated and shown in Fig. 3. The calculated results indicates the core size of $R_{\rm rms} = 3.13 \pm 0.18$ fm (8±7% larger than the size of ^{22}Mg deduced by the σ_I data). Even if a larger range parameter is used in this calculation, a relatively larger core is also obtained for ^{23}Al compared to ^{24}Al. The obtained size of ^{22}Mg is different for the two range parameters, but both calculations suggest an enlarged core inside ^{23}Al.

4. Conclusion

In summary, the $P_{//}$ after one-proton removal for ^{23}Al and σ_R for ^{23,24}Al were measured. An enhancement was observed for the σ_R of ^{23}Al. The $P_{//}$ distributions were found to be wide. The experimental $P_{//}$ and σ_R results were discussed within a Few-Body Glauber Model. We determined the valence proton to be a dominant d-wave configuration in the ground state of ^{23}Al. An enlarged ^{22}Mg core was revealed in order to explain both the σ_R and $P_{//}$ distributions.

Acknowledgements

This work was partially supported by the National Natural Science Foundation of China (NNSFC) under Grant No. 10405032, 10535010, 10405033 and 10475108, Shanghai Development Foundation for Science and Technology under contract No. 06QA14062, 06JC14082 and 05XD14021, the Major State Basic Research Development Program in China under Contract No. G200077404 and the Knowledge Innovation Project of Chinese Academy of Sciences under Grant No. KJCX-SYW-NZ.

References

1. I. Tanihata *et al.*, *Phys. Rev. Lett.* **55**, 2676 (1985);
   I. Tanihata *et al.*, *Phys. Lett. B* **287**, 307 (1992).
2. M. Fukuda *et al.*, *Phys. Lett. B* **268**, 339 (1991).
3. D. Bazin *et al.*, *Phys. Rev. Lett.* **74**, 3569 (1995);
   T. Nakamura *et al.*, *Phys. Rev. Lett.* **83**, 1112 (1999);
   A. Ozawa *et al.*, *Nucl. Phys. A* **691**, 599 (2001).
4. T. Minamisono *et al.*, *Phys. Rev. Lett.* **69**, 2058 (1992);
   W. Schwab *et al.*, *Z. Phys. A* **350**, 283 (1995).
5. R.E. Warner *et al.*, *Phys. Rev. C* **52**, R1166 (1995);
   F. Negoita, C. Borcea, F. Carstoiu, *Phys. Rev. C* **54**, 1787 (1996);
   M. Fukuda *et al.*, *Nucl. Phys. A* **656**, 209 (1999).
6. M.M. Obuti *et al.*, *Nucl. Phys. A* **609**, 74 (1996).
7. B.A. Brown, P.G. Hansen, *Phys. Lett. B* **381**, 391 (1996);
   Z.Z. Ren *et al.*, *Phys. Rev. C* **53**, R572 (1996).
8. A. Navin *et al.*, *Phys. Rev. Lett.* **81**, 5089 (1998).
9. G. Audi, A.H. Wapstra, *Nucl. Phys. A* **565**, 66 (1993).
10. X.Z. Cai *et al.*, *Phys. Rev. C* **65**, 024610 (2002);
    H.Y. Zhang *et al.*, *Nucl. Phys. A* **707**, 303 (2002).
11. A. Ozawa *et al.*, *Phys. Rev. C* **74**, 021301R (2006).
12. K. Kimura *et al.*, *Nucl. Inst. Meth. A* **538**, 608 (2005).
13. A.S. Goldhaber, *Phys. Lett. B* **53**, 306 (1974).
14. N. Iwasa *et al.*, *Nucl. Instrum. Methods B* **126**, 284 (1997).
15. D.Q. Fang *et al.*, *Phys. Rev. C* **69**, 034613 (2004);
    T. Yamaguchi *et al.*, *Nucl. Phys. A* **724**, 3 (2003).
16. W.Q. Shen *et al.*, *Nucl. Phys. A* **491**, 130 (1989).
17. Y. Ogawa *et al.*, *Nucl. Phys. A* **543**, 722 (1992);
    B. Abu-Ibrahim *et al.*, *Comput. Phys. Comm.* **151**, 369 (2003).
18. T. Suzuki *et al.*, *Nucl. Phys. A* **630**, 661 (1998).
19. T. Gomi *et al.*, *Nucl. Phys. A* **758**, 761c (2005).
20. R. Kanungo *et al.*, *Nucl. Phys. A* **677**, 171 (2000);
    R. Kanungo *et al.*, *Phys. Rev. Lett.* **88**, 142502 (2002).
21. T. Zheng *et al.*, *Nucl. Phys. A* **709**, 103 (2002).
22. A. Ozawa *et al.*, *Nucl. Phys. A* **608**, 63 (1996).
23. M. Takechi *et al.*, *Eur. Phys. J. A* **25**, s01, 217 (2005).

EXTENDED CLUSTER MODEL FOR LIGHT AND MEDIUM NUCLEI

M. TOMASELLI*, T. KÜHL, D. URSESCU, AND S. FRITZSCHE

GSI, Gesellschaft für Schwerionenforschung, Darmstadt
D24691 Germany
**E-mail: m.tomaselli@gsi.de*

The structures, the electromagnetic transitions, and the beta decay strengths of exotic nuclei are investigated within an extended cluster model. We start by deriving an effective nuclear Hamiltonian within the S_2 correlation operator. Tensor forces are introduced in a perturbative expansion which includes up to the second order terms. Within this Hamiltonian we calculate the distributions and the radii of A=3, 4 nuclei. For exotic nuclei characterized by n valence protons/neutrons we excite the structure of the closed shell nuclei via mixed modes formed by considering correlations operators of higher order. Good results have been obtained for the calculated transitions and for the beta decay transition probabilities.

Keywords: Cluster model, exotic nuclei, electromagnetic transitions, beta decay.

1. Introduction

The data obtained at RIA, Riken and GSI have given new life to the old Nuclear Physics. Nuclear structures of proton and neutron rich nuclei are and will be investigated giving a new insight to the fundamental observable such as the nuclear forces in the proton-proton and neutron-neutron components, shell closure far from stability, magnetic properties of weakly excited nuclear states, and many others. The theoretical analysis of these data requires a reliable nuclear model which can reproduce the data of stable nuclei and be extrapolated to predict or at least reproduce the experimental results. The Cluster Correlation Model offers effective tools in this direction. We have to depart, however, from a perturbative scheme which is generally used to treat the two body correlation. Generalization to the perturbative method to include the many body correlation is in this paper realized within an extended, non-perturbative Cluster Correlation Model. One of the central challenges of theoretical nuclear physics is the attempt to describe unknown properties of the exotic systems in terms of a realistic nucleon-nucleon (NN) interaction. In order to calculate matrix elements with the singular interaction (hard core) we have to define effective correlated Hamiltonians.

Correlation effects in nuclei have been first introduced in nuclei by Villars,[1] who proposed the unitary-model operator (UMO) to construct effective operators. The method was implemented by Shakin[2] for the calculation of the G-matrix from hard-core interactions. Non perturbative approximations of the UMO have been recently applied to odd nuclei in Ref. [3] and to even nuclei in Ref. [4]. The basic formulas of the Dynamic Correlation Model and of the Boson Dynamic Correlation Model (BDCM) presented in the above quoted papers have been obtained by separating the n-body correlation operator in short- and long-range components. The short-range component is considered up to the two body correlation while for the long range component the three and four body correlation operators have been studied. The extension of the correlation operator to high order diagrams is especially important in the description of exotic nuclei (open shell). In the short range approximation the model space of two interacting particles is separated in two subspaces: one which includes the shell model states and the other (high momentum) which is used to compute the G-matrix of the model. The long range component of the correlation operator has the effect of generating a new correlated model space (effective space) which departs from the originally adopted one (shell model). The amplitudes of the model wave functions are calculated in terms of non linear equation of motions (EoM). The derived systems of commutator equations, which characterize the EoM, are finally linearized. Within these generalized linearization approximations (GLA) we include in the calculation presented in the paper up to the ((n+1)p1h) effective diagrams. The linearized terms provide, as explained later in the text, the additional matrix elements that convert the perturbative UMO expansion in an eigenvalue equation. The n-body matrix elements needed to diagonalize the resulting eigenvalue equations are calculated exactly via the Cluster Factorization Theory (CFT).[5]

2. The S_2 correlated Hamiltonian

In order to describe the structures and the distributions of nuclei we start from the following Hamiltonian:

$$H = \sum_{\alpha\beta} \langle \alpha | t | \beta \rangle \, a^\dagger_\alpha a_\beta + \sum_{\alpha\beta\gamma\delta} \langle \Phi_{\alpha\beta} | v_{12} | \Phi_{\gamma\delta} \rangle \, a^\dagger_\alpha a^\dagger_\beta a_\delta a_\gamma \tag{1}$$

where v_{12} is the singular nucleon-nucleon two body potential. Since the two body states $|\alpha\beta\rangle$ are uncorrelated the matrix elements of v_{12} are infinite. This problem can be avoided by taking matrix elements of the Hamiltonian between correlated states. In this paper the effect of correlation is introduced via the e^{iS} method. In dealing with very short range correlations only the S_2 part of the correlation operator needs to be considered.

Following Ref. [2] we therefore calculate an "effective Hamiltonian" by using only the S_2 correlation operator obtaining:

$$\begin{aligned} H_{eff} &= e^{-iS_2} H e^{iS_2} = \sum_{\alpha\beta} \langle\alpha|t|\beta\rangle a^\dagger_\alpha a_\beta + \sum_{\alpha\beta\gamma\delta} \langle\Psi_{\alpha\beta}|v^l_{12}|\Psi_{\gamma\delta}\rangle a^\dagger_\alpha a^\dagger_\beta a_\delta a_\gamma \\ &= \sum_{\alpha\beta} \langle\alpha|t|\beta\rangle a^\dagger_\alpha a_\beta + \sum_{\alpha\beta\gamma\delta} \langle\Psi_{\alpha\beta}|v|\Psi_{\gamma\delta}\rangle a^\dagger_\alpha a^\dagger_\beta a_\delta a_\gamma \end{aligned} \tag{2}$$

where v^l_{12} refers to the long-range part of the nucleon-nucleon force diagonal in the relative orbital angular momentum, after the separation:[6]

$$v_{12} = v^s_{12} + v^l_{12} \tag{3}$$

The separation is made in such a way that the short range part produces no energy shift in the pair state.[6] In doing shell model calculation with the Hamiltonian Eq. (2), we remark: a) only the long tail potential plays an essential role in the calculations of the nuclear structure i.e.: the separation method and the new proposed v_{low-k}[7] method show a strong analogy and b) the v^{od}_T must be included as an additional re-normalization of the effective Hamiltonian Eq. (2).

In Eq. (2) the $\Psi_{\alpha\beta}$ is the two particle correlated wave function:

$$\Psi_{\alpha\beta} = e^{iS_2}\Phi_{\alpha\beta} \tag{4}$$

In order to evaluate the effect of the tensor force on the $\Psi_{\alpha\beta}$ we calculate:

$$w(r) = V^{od}_T \frac{Q}{\Delta E} u(r) = V^{od}_T \frac{Q}{\Delta E} |(\tilde{nl}S), J' : NL, J\rangle \tag{5}$$

where Q is a momentum dependent projection operator, $\Delta E(k_1, k_2)$ the energy denominator and $\tilde{nl}$ the correlated two particle state in the relative coordinates. In Eq. (5) u(r) is generated as in Ref.[2] by a separation distance calculation for the central part of the force in the $^3S_1$ state. The wave function obtained in this way (full line) heals to the harmonic-oscillator wave function (dashed line) as shown in Fig. 1. The result obtained for Eq. (5) calculated with the tensor force of the Yale potential[8] is given also in Fig. 1 left where we plot for the harmonic oscillator size parameter b=1.41 fm:

$$\Psi(\vec{r}) = [u(r)Y^1_0(\Omega_{\vec{r}}) + w(r)Y^1_2(\Omega_{\vec{r}})] \tag{6}$$

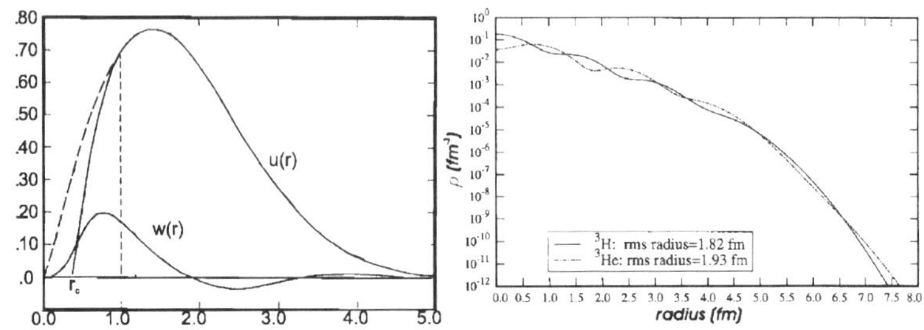

Fig. 1. Left: The u(r) and w(r) wave functions of the deuteron, with quantum numbers $^3S_1$ and $^3D_1$, plotted as function of r; Right: Distributions of ^{3}H and ^{3}He.

Being the admixture of the two components, circa 4%, the wave function Eq. (6) can be associated to the deuteron wave function. Let us use then the Hamiltonian Eq. (2) to calculate the structure of the A=3 nuclei. Here we propose to calculate the ground state of ^{3}H, ^{3}He, and ^{4}He within the EoM method which derive the eigenvalue equations by working with the e^{iS_2} operator on the wave functions of the A=3, 4 nuclei Ref. From the diagonalization of the eigenvalue equation of the three particles, we obtain an energy difference $\Delta E(^3H -^3 He)$=0.78 MeV and the distributions and radii given in Fig. 1 Right. By extending the commutator to a four particle state we obtain for the ground state of ^{4}He the binding energy of E=28.39 Mev and the rms radius of 1.709 fm. In dealing with complex nuclei however the $(S_i,\ i = 3 \cdots n)$ correlations should also be considered. The evaluation of these diagrams is, due to the exponentially increasing number of terms, difficult in a perturbation theory. We note however that one way to overcome this problem is to work with $e^{i(S_1+S_2+S_3+\cdots+S_i)}$ operator on the Slater's determinant by keeping the n-body Hamiltonian uncorrelated. Via the long tail of the nuclear potential the Slater determinant of the "n" particle systems are interacting with the excited Slater's determinants formed by the ("n" particles+(mp-mh) mixed-mode excitations). The amplitudes of the different determinants are calculated via the EoM method. After having performed the diagonalization of the n-body Hamilton's operator we can calculate the form of the effective Hamiltonian which, by now, includes the complete set of the commutator equations. The method is here applied to ^{6}He, ^{11}Be, ^{14}C, ^{15}O, and ^{17}O. A detailed formulation of the model my be found in Ref.[9]

3. Results

In order to perform structure calculations for complex nuclei, we have to define the CMWFs base, the "single-particle energies" and to choose the nuclear two-body interactions. The CMWFs are defined as in Ref.[9] by including mixed valence modes and core-excited states. The base is then orthonormalized and, since the single particle wave functions are harmonic oscillators, the center-of-mass (CM) is removed. The single-particle energies of these levels are taken from the known experimental level spectra of the neighboring nuclei. For the particle-particle interaction, we use the G-matrix obtained from Yale potential.[10] These matrix elements are evaluated by applying the e^S correlation operator, truncated at the second order term of the expansion, to the harmonic oscillator base with size parameter b=1.76 fm. As elucidate in Refs. [3] and [4] the effective two-body potential used by the DCM and the BDCM models is separated in low and high momentum components. Therefore, the effective model matrix elements calculated within the present separation method and those calculated by Kuo[7] in the v_{low-k} approximation are pretty similar. The adopted separation method and the v_{low-k} generate two-body matrix elements which are almost independent from the radial shape of the different potentials generally used in structure calculations.

The particle-hole matrix elements could be calculated from the particle-particle matrix elements via a re-coupling transformation. In this contribution we present application of the S_n correlated model to the charge distributions of ^{6}He, ^{11}Be, and to the electromagnetic transitions of neutron rich Carbon and Oxygen isotopes. The beta decay strengths from the ground state ^{14}N to the excited states of ^{14}C are also calculated. In Fig. 2) Left three distributions are given for ^{6}He: 1) the correlated charge distribution calculated with the full S_3 operator, 2) the correlated charge distribution calculated with the partial S_3 operator obtained by neglecting the folded diagrams, 3) the charge distribution calculated for two correlated protons in the $1s_{\frac{1}{2}}$ shell. The full S_3 correlation operator therefore increases the calculated radii. In Fig. 2) Right the charge distribution for ^{11}Be is given. A charge radius of 3.12 fm has been obtained. Calculations are performed in a mixed S_3 and S_5 system. The results obtained for the Carbon and Oxygen isotopes are in the following presented as function of the increasing valence neutrons. It is worthwhile to remark that the high order correlation operators generate the interaction of the valence particles with the closed shell nucleus. The correlation model treats therefore consistently the "A" particles of the isotopes.

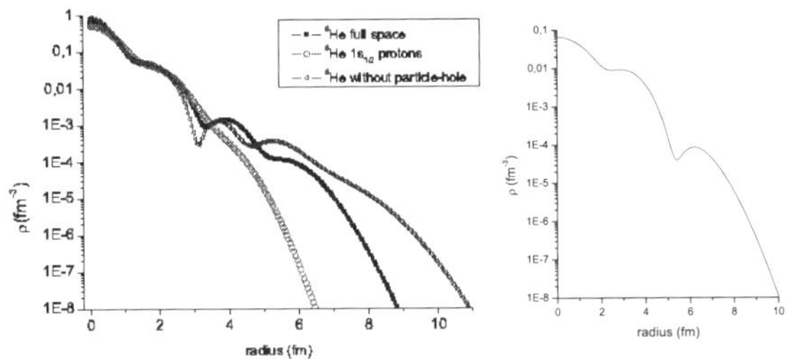

Fig. 2. Left:Charge distributions of ^{6}He calculated in different approximations; Right: Charge distribution of ^{11}Be.

By using generalized linearization approximations and cluster factorization coefficients[5] we can perform exact calculations. In following Tables an over all b=1.76 fm has been used.

In Table 1, 3) we give the calculated magnetic moments and rms radii for one-hole and for one-particle in ^{16}O. The energy splitting between the ground- and the second (first) excited states and the electromagnetic transitions for the two isotopes are given in Tables 2, 4).

Table 1: Magnetic moment (nm) and rms (fm) of the ground state of ^{15}O with $J = \frac{1}{2}^-; T = \frac{1}{2}$

	DCM	Exp.[11]
Magnetic Moment (mm)	.70	.7189
	DCM	Exp.[12]
rms (fm)	2.74	2.73(3)

Table 2: Energy splitting between the ground and the second excited states and the corresponding electromagnetic transitions for ^{15}O.

Energy (MeV)	DCM	Exp.[11]
$\Delta E_{\frac{1}{2}^- \frac{5}{2}^+}$	5.41	5.24
Ratio	DCM	Exp.[11]
$\frac{BE(E3;\frac{5}{2}^+ \to \frac{1}{2}^-)}{BE(M2;\frac{5}{2}^+ \to \frac{1}{2}^-)}$	.15	.10

Table 3: Magnetic moment (nm) and rms (fm) of the ground state of ^{17}O with $J = \frac{5}{2}^+; T = \frac{1}{2}$

	DCM	Exp.[11]
Magnetic Moment (nm)	-1.88	-1.89
	DCM	Exp.[12]
rms (fm)	2.73	2.72(3)

Table 4: Energy splitting between the ground- and the first excited states and the E_2 transition for ^{17}O.

Energy (MeV)	DCM	Exp.[11]
$\Delta E_{\frac{1}{2}^+\frac{5}{2}^+}$	0.87	0.89
Transition(e^2fm^4)	DCM	Exp.[11]
$BE(E2;\frac{1}{2}^+\rightarrow\frac{5}{2}^+)$	2.10	2.18±0.16

In Table 5) we give the calculated results for the energy splitting between the ground- and the 2^+ excited state and the corresponding electromagnetic transition for the ^{14}C. The commutator equations involve S_2 and S_3 diagrams.

Table 5: Calculated energy splitting and $BE(E2;2^+\rightarrow0^+)$ transition for ^{14}C

Energy (MeV)	Ref.[13]	BDCM	Exp.[14]
$\Delta E_{0^+2^+}$		8.38	8.32
Transition(e^2fm^4)	Ref.[13]	BDCM	Exp.[14]
$BE(E2;2^+\rightarrow0^+)$	3.38	3.65	3.74 ± .50

In Table 6) preliminary results for the calculated reduced transition probabilities from the ground state of ^{14}N to the $0^+,1^+,2^+$ excited states of ^{14}C are given. The calculated strengths reproduce reasonably well the experimental values.[15]

Table 6: Calculated energies of the low-lying states (MeV) $0^+,1^+,2^+$ of ^{14}C and the associated reduced transition probabilities $B(GT)$ from the J=1^+ T=0 ground state of ^{14}N.

^{14}N	J_i^+T	^{14}C	J_f^+T	Energy (MeV)	B(GT)
	1^+0		0^+1	0.0	0.06
			0^+1	7.81	0.15
			1^+1	12.17	0.12
			2^+1	7.38	0.42
			2^+1	8.38	0.50
			2^+1	10.91	0.35

Good results have been overall obtained for the transitions with a neutron effective charge varying between 0.1- to 0.12-e_n.

4. Conclusion and Outlook

In this contribution we have investigated the effect of the microscopic correlation operators on the exotic structure of the Carbon and Oxygen isotopes. The microscopic correlation has been separated in short- and long-range correlations according to the definition of Shakin. The short-range correlation has been used to define the effective Hamiltonian of the model while the long-range correlation is used to calculate the structures and the distributions of exotic nuclei. As given in the work of Shakin, only the two-body short-range correlation need to be considered in order to derive the effective Hamiltonian especially if the correlation is of very short range. For the long range correlation operator the three body component is important and should not be neglected. Within the three body correlation operator, one introduces in the theory a three body interaction which compensates for the use of the genuine three body interaction of the no-core shell model. Within the S_2 effective Hamiltonian, good results have been obtained for the ground state energies and the distributions of ^{3}H, ^{3}He, and ^{4}He. The higher order correlation operators $S = 3 \cdots n$ have been used to calculate the structure and the electromagnetic transitions of ground and first excited states for the isotopes of Carbon and Oxygen. By using generalized linearization approximations and cluster factorization coefficients we can perform expedite and exact calculations.

(1) F. Villars, *Proc. Enrico Fermi Int. School of Physics XXII*, Academic Press N.Y. (1961).
(2) C.M. Shakin and Y.R. Waghmare, *Phys. Rev. Lett.* **16** (1966) 403; C.M. Shakin, Y.R. Waghmare, and M.H. Hull, *Phys. Rev.* **161** (1967) 1006.
(3) M. Tomaselli, L.C. Liu, S. Fritzsche, T. Kühl, and D. Ursescu, *Nucl. Phys.* **A738** (2004) 216; M. Tomaselli, Ann. Phys. **205** (1991) 362.
(4) M. Tomaselli, L.C. Liu, S. Fritzsche, T. Kühl, *J. Phys. G: Nucl. Part. Phys.* **30** (2004) 999.
(5) M. Tomaselli, T. Kühl, D. Ursescu, and S. Fritzsche, Prog. Theor. Phys. **116** (2006) 699.
(6) S.A. Moszkowski and B.L. Scott, Ann. of Phys. (N.Y.) **11** (1960) 55.
(7) S. Bogner, T.T.S. Kuo, L. Coraggio, A. Covello, and N. Itaco, Phys. Rev. **C65** (2002) 051301(R).
(8) K.E. Lassila, M.H. Hull, H.M. Ruppel, F.A.McDonald, and G. Breit, Phys. Rev. **126** (1962) 881.
(9) M. Tomaselli, T. Kühl, and D. Ursescu, Prog. Part. Nucl. Phys. **59** (2007) 455; M. Tomaselli, T. Kühl, and D. Ursescu, Nucl. Phys. **A790** (2007) 246.
(10) C.M. Shakin, Y.R. Waghmare, M. Tomaselli, and, M.H. Hull, Phys. Rev. **161** (1967) 1015.
(11) http://www.tunl.duke.edu/nucldata, and references therein quoted.
(12) C.W. de Jager, H. de Vries, and C. de Vries, Atomic Data and Nuclear Data Tables **14** (1974), 479.
(13) S. Fujii, T, Misuzaki, T. Otsuka, T. Sebe, A. Arima, arXiv:nucl-th/0602002, 2006.
(14) S. Raman *et al.* Atomic Data and Nuclear Data Tables **31** (1987) 1.
(15) A. Negret, et al., Phys. Rev. Lett. **97** (2006) 062502.

NUCLEAR STRUCTURE STUDIES ON EXOTIC NUCLEI WITH RADIOACTIVE BEAMS – PRESENT STATUS AND FUTURE PERSPECTIVES AT FAIR

PETER EGELHOF

Gesellschaft für Schwerionenforschung, GSI, Planckstr.1, D-64291 Darmstadt, Germany

The investigation of nuclear reactions using radioactive beams in inverse kinematics gives access to a wide field of nuclear structure studies in the region far off stability. The basic concept and the methods involved are briefly discussed, and an overview including some selected examples of the most recent results obtained with radioactive beams from the present fragment separator at GSI Darmstadt is presented. The experimental conditions expected at the future international facility FAIR will allow for a substantial improvement in intensity and quality of radioactive beams as compared to present facilities. Therefore it is expected that FAIR will provide unique opportunities for nuclear structure studies on nuclei far off stability, and will allow to explore new regions in the chart of nuclides of high interest for nuclear structure and nuclear astrophysics. A brief overview on the new facility, and on the experimental setups planned for nuclear structure research with radioactive beams is given.

1. Introduction

The exploration of nuclear matter under extreme conditions is one of the major goals of modern nuclear physics. The opportunities offered by beams of exotic nuclei for research in the areas of nuclear structure physics and nuclear astrophysics are exciting, and worldwide activity in the construction of different types of radioactive beam facilities reflects the strong scientific interest in the physics that can be probed with such beams.

One of the most powerful classical methods for obtaining spectroscopic information on the structure of nuclei is the investigation of direct reactions. The study of such reactions, as for example elastic and inelastic scattering, one- and few-nucleon transfer reactions, charge exchange, knock-out, or breakup reactions, etc., contributed in the past substantially to our present knowledge about stable nuclei. Of course, before the availability of radioactive ion beams, this method was limited to stable or very long-lived nuclei, which allow to produce targets. In the recent past the use of good-quality secondary radioactive ion beams, now available at several major accelerator facilities world wide, enabled to extend such studies on exotic nuclei by using the method of inverse kinematics. Despite the experimental challenge in performing such experiments with relatively low secondary beam intensities, investigations with radioactive beams have opened new territories of nuclei in the nuclear chart to be explored and lead to the discovery of new exciting phenomena of nuclear structure.

The present article is intended to focus on nuclear structure studies with radioactive beams presently or in the recent past performed at GSI Darmstadt, and on perspectives at the future international facility FAIR (Facility for Antiproton and Ion Research) . An overview on the layout of the present GSI facility and the future FAIR facility is presented in figure 1. At the present GSI facility powerful dedicated experimental setups, as for example LAND (Large Area Neutron Detector), the active target setup IKAR, the RISING (Rare ISotope INvestigations at GSI) setup with a highly efficient large solid angle array of Ge crystals, allowed to investigate new and exciting phenomena, like, for example, skin and halo structures in light neutron-rich dripline nuclei, new collective excitation modes, like the Pygmy resonance, and new shell closures far off stability.

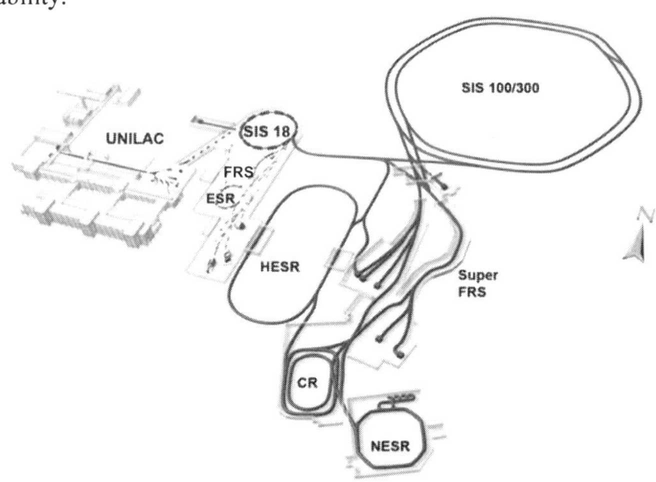

Figure 1. Schematic layout of the present GSI facility (left) and the FAIR facility (right). At FAIR the existing GSI accellerators UNILAC and SIS18 serve as an injector for the two large synchrotron rings SIS-100/300. The beams are distributed to various experimental areas or to the production targets for the anti-proton or radioactive-isotope beam facilities. For beam storage and manipulation several rings are foreseen including the high-energy storage ring HESR for anti-protons, the collector ring CR, and the experimental storage ring NESR for ions. The storage-ring complex provides possibilities of beam cooling as well as target areas for scattering experiments. The present fragment separator FRS and the new superconducting fragment separator Super-FRS for preparation of rare-isotope beams are indicated as well.

The experimental conditions at the future facility FAIR (see [1,2,3,4,5] for an overview) will provide unique opportunities for nuclear structure studies on nuclei far off stability, and will allow for exploring new regions in the chart of nuclides of high interest for nuclear structure and astrophysics. Here, in particular completely new and powerful experimental methods, like, for example, the investigation of direct reactions with stored radioactive beams interacting with internal targets of the storage ring within the EXL project [5], and electron scattering on exotic nuclei at the eA collider within the ELISe project [5] will provide new and challenging perspectives.

2. Nuclear Physics with Radioactive Beams at the Present GSI Facility – Some Selected Examples

The layout of the present GSI facility including its radioactive beam facility is displayed in figure 1 (left side). Heavy ion beams from the UNILAC linear accelerator are injected into the heavy ion synchrotron SIS18, which feeds the fragment separator FRS [6]. Good quality secondary radioactive beams with intermediate energies of about 100 – 1000 MeV/u, produced by projectile fragmentation or fission, the latter process leading to relatively high intensities at the neutron-rich side of heavy isotopes, are extracted and guided to the experimental areas. An overview on various kinds of investigations performed at the GSI radioactive beam facility is displayed in figure 2. In the following a few selected examples of most recent results are presented.

2.1. *The Dipole Response of Neutron-Rich Nuclei – Evidence for the Pygmy Dipole Excitation*

The multipole response of exotic nuclei, and in particular the question how the excitation spectra qualitatively depend on the proton-neutron asymmetry, is of high interest for our understanding of the structure of exotic nuclei. Such investigations are performed by intermediate energy inelastic scattering of radioactive beams at the LAND setup of GSI [7]. Data recently obtained [8] by the LAND collaboration on the dipole response of the neutron-rich ^{130,132}Sn isotopes are displayed in figure 3. Beams of ^{130,132}Sn isotopes with energies of about 500 MeV/u were produced by inflight fission of a primary ^{238}U beam in the fragment separator, and excited by interaction with a ^{208}Pb target. The excitation energy E* is finally determined from a kinematically complete measurement of the four-momenta of all decay products, i.e. heavy fragments, neutron(s) and γ-rays, applying the invariant mass method (for details see [8]). The good quality of the deduced data demonstrates the high sensitivity of the

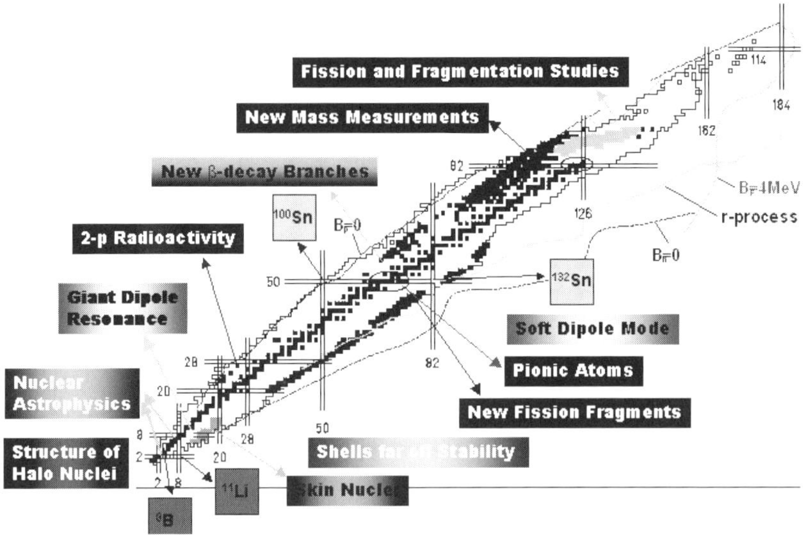

Figure 2. Overview on fields of research at the present radioactive beam facility at GSI.

method when taking into account the relatively low secondary beam intensity of only about 10 sec^{-1} for ^{132} Sn.

The differential cross section for electromagnetic excitation of ^{130,132}Sn on a lead target is displayed in fig. 3 (left frames). The right frames display the corresponding photo-absorption (γ,n) cross sections. The upper right panel shows the result from a real-photon experiment [9] for the stable isotope ^{124}Sn for comparison. The spectrum is dominated by the excitation of the giant dipole resonance (GDR). A fit of a Lorentzian plus a Gaussian parametrization for the photo-absorption cross section (solid curves) yields a position and width of the GDR comparable to those known for stable nuclei in this mass region. The GDR exhausts almost the energy-weighted dipole sum rule. An additional peak structure is clearly visible below the GDR energy region. The position of around 10 MeV is close to the predicted [10] energy of the collective soft mode (Pygmy resonance), which is commonly interpreted as a dipole vibration of skin neutrons versus the isospin-saturated core. The experimentally observed strength in this peak corresponds to about 4% of the TRK sum rule, also in good agreement with the QRPA prediction [10] as well as with non-relativistic QRPA calculations [11].

It is interesting to note that the strength of the low lying Pygmy resonance mode turns out to be strongly related to the neutron-skin thickness of neutron-rich nuclei, and to the parameters of the equation of state of neutron-rich nuclear

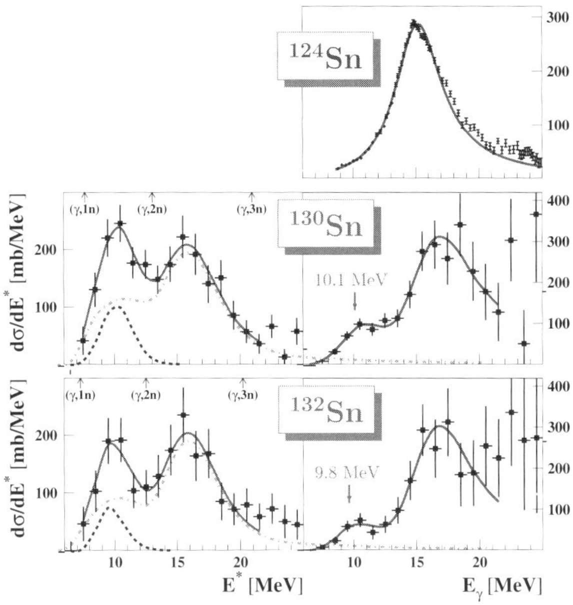

Figure 3. Left frames: Differential cross section for the electromagnetic excitation of ^{130,132}Sn on a lead target at around 500 MeV/u (from ref. [8]). Corresponding photo-neutron cross sections are shown in the right panels. For comparison, photo-neutron cross sections are shown in the upper right frame for the stable isotope ^{124}Sn measured in a real-photon experiment [9].

matter [12]. On theoretical grounds, employing relativistic mean field calculations [10,13], a one-to-one correlation is found between the Pygmy strength and parameters a_4 and p_0 of the nuclear symmetry energy, and in turn with the thickness R_n-R_p of the neutrons skins. On this basis, by using the experimental Pygmy strength obtained for ^{132}Sn, as discussed above, preliminary values for $a_4 = 32.0 \pm 1.8$ MeV, $p_0 = 2.3 \pm 0.8$ MeV/fm^3, and R_n-$R_p = 0.24 \pm 0.04$ fm for ^{132}Sn were deduced [12].

2.2. *Investigation of Nuclear Matter Distributions of Halo Nuclei from Intermediate Energy Elastic Proton Scattering*

The size of nuclei and the radial shape of the distribution of nuclear matter and charge are fundamental properties of nuclei, and therefore of high interest for various fields in nuclear physics. Proton-nucleus elastic scattering at intermediate energies around 700 - 1000 MeV is known to be a method well established for obtaining accurate nuclear matter distributions [14]. This method was applied at GSI Darmstadt for the first time for the investigation of exotic nuclei by using the technique of inverse kinematics. Simulation calculations [15,16] have clearly demonstrated that proton scattering in the region of small momentum transfer is particularly sensitive to the nuclear matter radius, and the halo structure of nuclei.

Differential cross sections for elastic proton scattering at small momentum transfer were measured at GSI Darmstadt by the IKAR collaboration at energies around 700 MeV/u in inverse kinematics for the neutron-rich helium isotopes ^{6}He, ^{8}He [16,17], for the neutron-rich lithium isotopes ^{8}Li, ^{9}Li, and ^{11}Li [18,19], and most recently for ^{12,14}Be and ^{8}B. The experimental method is based on the high-pressure hydrogen-filled ionization chamber IKAR, which serves simultaneously as gas target and recoil proton detector. The use of this dedicated active-target technique enabled to detect the low energy recoil protons with sufficient angular and energy resolution covering the full solid angle of interest (for details see [17]).

The differential cross sections deduced [20] for the example of the most recently investigated isotopes ^{12}Be and ^{14}Be are displayed in figure 4. Note that the data shown are still preliminary. For establishing the nuclear matter density distributions from the measured cross sections, the Glauber multiple scattering theory was applied. For modelling the nuclear matter distributions, parametrical functions were used. The free parameters of these density distribution parametrizations were then determined by a least-square fit of the calculated cross sections to the experimental ones. A reasonably good description of the cross section data is obtained as may be seen in fig. 4 (left part).

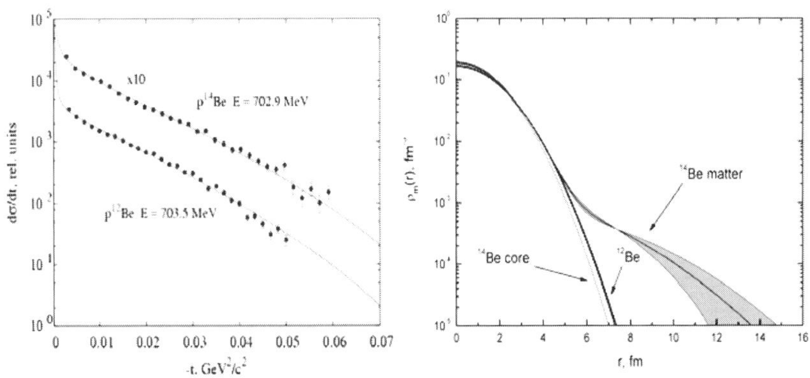

Figure 4. Differential cross sections versus the four-momentum transfer squared -t for p^{12,14}Be elastic scattering (left) and the nuclear matter and nuclear core density distributions deduced from the experimental cross section data. Note that the data are still preliminary.

The present results (still preliminary!) on the nuclear matter densitiy distributions obtained for the ^{12,14}Be isotopes are displayed in figure 4 (right side). As compared to the nuclear matter distribution deduced for ^{12}Be, which exhibits a behaviour of "normal nuclei", a rather extended nuclear matter distribution is obtained for ^{14}Be, the matter density decreasing much slower with the radius parameter than the one for ^{12}Be, and therefore representing a clear signature for a pronounced neutron halo in ^{14}Be. This is also reflected in the deduced nuclear matter radii, where the result for ^{14}Be: R_m = 3.78 (37) fm has to be compared with R_m = 2.88 (7) for ^{12}Be. The close resemblence of the distribution deduced for the core of ^{14}Be (with R_{core} = 2.76 (4)) with the matter distribution of ^{12}Be supports the picture of a two-neutron halo structure in ^{14}Be as predicted from various theoretical calculations.

With ^{8}B for the first time an isotope supposed to be a candidate for a proton halo was investigated by the present method. From the preliminary results it can be concluded that the halo structure of ^{8}B is confirmed. The deduced matter radius of ^{8}B amounts to R_m = 2.88 (13) fm. The extended nuclear matter distribution of ^{8}B, and in particular its long range tail may also have a significant relevance for the astrophysical S (E) factor for E $\Rightarrow$ 0 for the ^{7}Be (p,γ) ^{8}B reaction. This needs to be investigated in more detail in near future when final results from the present experiment are available.

2.3. *Gamma Spectroscopy with Fast and Stopped Radioactive Beams – Recent Results from RISING at GSI*

Besides the investigation of direct reactions, γ-ray spectroscopy is another important tool for nuclear structure studies far off stability. Since 2003 the powerful RISING setup (Rare ISotope INvestigations at GSI) [21] was installed at the focal plane of the FRS, and several campaigns with fast and stopped beams were performed, addressing various topics concerning the shell structure far off stability, nuclear shapes, collective modes, nuclear symmetries, etc. Meanwhile the RISING setup consists of 105 Ge crysals with an energy resolution (FWHM) of 1.24%, and a total efficiency of 2.9% for E_γ =1.3 MeV at 100 MeV/u.

One of the recent results concerns the relativistic Coulomb excitation of neutron-rich ^{54,56,58}Cr isotopes [22]. This experiment was intended to investigate the existence of a new sub-shell closure at N=32 for Z=24 by measuring the B(E2, $2_1^+ \Rightarrow 0^+$) values of the neutron-rich Cr isotopes which are located at a key point on the pathway from the N=40 sub-shell closure via the deformed region to the spherical nuclei at N=28. The first excited 2^+ states in ^{54,56,58}Cr were populated by Coulomb excitation at relativistic energies and γ-rays were measured using the RISING setup. For ^{56}Cr and ^{58}Cr the $B(E2, 2_1^+ \Rightarrow 0^+)$ values relative to the previously known B(E2) value for ^{54}Cr are determined as 8.7(3.0) and 14.8(4.2) W.u., respectively [22]. The results confirm a sub-shell closure at N=32 which was already indicated by the higher energy of the 2_1^+ state in ^{56}Cr. Recent large-scale shell model calculations using effective interactions reproduce the trend in the excitation energies, but fail to account for the minimum in the B(E2) values at N=32 [22].

3. Nuclear Physics with Radioactive Beams at the Future International Facility FAIR

FAIR, the international Facility for Antiproton and Ion Research, will provide the european and international science community with a technically novel and in many respects unique accelerator system for a multidisciplinary scientific program (see [1,2,3,4,5] for an overview). The layout of the FAIR accelerator and storage ring complex is displayed in fig. 1 (right side). It constitutes a substantial extension of the capabilities of the present GSI facility by delivering ion beams of much increased intensity, of higher energy, and of improved beam quality. Ions of all elements from hydrogen to uranium are accelerated and intense antiproton beams will be provided in addition. Two synchrotrons SIS100 and SIS300 are the heart of the FAIR accelerator system. The existing accelerators UNILAC and SIS18 and a proton linac will serve as injectors. With the high bending power of the SIS100/300 super-conducting magnets high beam energies are reached, e.g., 35 GeV/u for U^{92+}. Highest beam intensities around 10^{12} ions per second, with beam energy restricted to about 1.5 GeV/u, are achieved by a high cycle frequency and by accelerating ions of lower atomic charge state. Adjacent to the synchrotrons is a complex system of storage-cooler rings HESR, CR, NESR, and various experimental stations, which, in conjunction with production targets and separators, provide high-quality secondary beams of antiprotons and radioactive nuclei. Radioactive beams are produced and separated in the Super-conducting Fragment Separator (Super-FRS), and are delivered to experimental stations described below.

FAIR will open up unique opportunities for a multifaceted forefront physics program, which can be grouped into the following specific fields of research: nuclear structure physics and nuclear astrophysics with radioactive ion beams; hadron physics with antiproton beams; physics of compressed nuclear matter with relativistic nuclear collisions; plasma physics with highly bunched laser- and ion-beams; atomic physics with highly charged ions and antimatter; and various fields of applied science. A recent collection of articles on the FAIR physics program can be found in [2].

The present article is focused on the part of FAIR which deals with the investigation of properties of exotic nuclei with intense rare-isotope beams, which is an integral part of the FAIR project, and in which advanced experimental concepts promise a new quality in investigations of nuclei far from stability. The experimental conditions at the future facility FAIR will substantially improve qualitatively and quantitatively our research potential on the physics of exotic nuclei, and will allow for exploring new regions in the chart of nuclides, of high interest for nuclear structure and astrophysics. FAIR will therefore provide world-wide unique opportunities for nuclear structure studies on nuclei, mainly due to two major achievements, as compared to present radioactive beam facilities, which are essential for reaction experiments with exotic nuclei. Firstly, secondary beam intensities will be superior by about 4 orders of magnitude compared to those presently available, and secondly, new experimental concepts and highly advanced instrumentation will allow to cope with the low production cross sections for exotic nuclei far off stability.

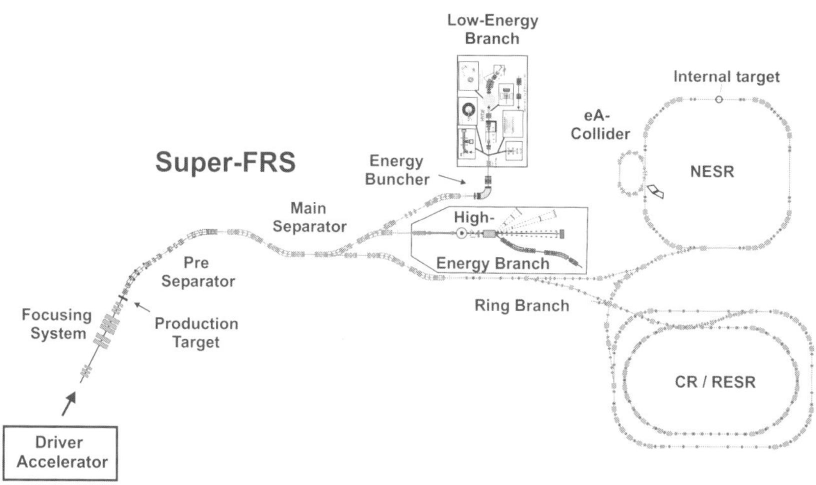

Figure 5. Schematic view of the rare-isotope beam facility at FAIR. The two-stage large-acceptance superconducting fragment separator Super-FRS consists of a pre-separator and a main separator. The super-FRS serves three experimental areas, the low-energy branch, the high-energy branch, and the storage ring complex (ring branch) comprising a large-acceptance collector ring (CR), an accumulator ring (RESR), and the experimental storage ring NESR including an internal target and the electron-ion (eA) collider.

The layout of the rare-isotope beam facility at FAIR is displayed in figure 5 (for an overview see [2,5,23]). It consists of the new two-stage large-acceptance superconducting fragment separator Super-FRS [24], and three experimental areas, the low-energy branch for experiments with radioactive beams slowed down to energies ranging from 100 MeV/u down to a few MeV/u, and with stopped or reaccelerated beams, the high energy branch for experiments with relativistic energy radioactive beams, and the ring branch. The ring branch includes several storage rings, first the Collector Ring CR, which has a large acceptance to allow efficient injection of secondary beams. Before injection into the New Experimental Storage Ring NESR, the beam is stochastically precooled. Further electron cooling is provided in the NESR, which is equipped with an internal target for reaction studies. A smaller intersecting electron storage ring will allow for the first time electron-scattering experiments off radioactive isotopes to be performed in a collider mode (eA-collider).

The novel instruments and experimental opportunities have attracted a large community of nuclear physicists who have, under the umbrella collaboration NUSTAR (NUclear STructure, Astrophysics and Reactions) formed several individual collaborations, and defined a considerable number of projects spanning a huge range of physics interest and experimental methods, including the investigation of direct reactions, γ-spectroscopy, laser spectroscopy, direct mass measurements, and many more (see [2,5,23] for an overview). The present article will focus on the investigation of direct reactions at relativistic energies, which will be performed at the high energy branch and the ring branch.

To address the key physics issues (formulated in detail in [2,5,23]) such as the investigation of: nuclear matter distributions near the drip lines (halo -, skin - structures, etc.), the isospin-dependence and evolution of the single-particle shell structures (new magic numbers, new shell gaps, spectroscopic factors), nucleon-nucleon correlations, and pairing and clusterization phenomena, new collective modes (different deformations for protons and neutrons; giant resonances and soft excitation modes), parameters of the nuclear equation of state, in-medium interactions in asymmetric and low-density nuclear matter, the astrophysical r- and rp-processes, and the understanding of supernova evolution (Gamow-Teller transitions; dipole strength; neutron-capture; weak decay rates), and many more, a large variety of direct reactions need to be investigated.

At the high-energy branch, an experimental area has been foreseen for high-energy reaction studies in inverse kinematics employing an apparatus of highest efficiency and full solid-angle coverage applicable to a wide class of reactions. For this, the R^3B collaboration (Reactions with Relativistic Radioactive Beams) has designed an experimental setup capable of fully benefiting from the high energies and intensities of the beams from the Super-FRS. Located at the focal plane of the high-energy branch of the Super-FRS, R^3B is a versatile fixed-target detector with high efficiency, acceptance, and resolution for kinematically complete measurements of reactions with high-energy radioactive beams, even at very low beam intensities down to one ion per second (for details see [2,23]).

The objective of the EXL-project (EXotic nuclei studied in Light-ion induced reactions at the NESR storage ring), is to capitalize on light-ion induced direct reactions in inverse kinematics by using novel storage ring techniques and a universal detector system (for details see [2,5,25]. Due to their spin-isospin selectivity, light-ion induced direct reactions at intermediate to high energies are an indispensable tool in nuclear structure investigations as evident from investigations of stable nuclei in the past. For many cases of direct reactions the essential nuclear structure information is deduced from high-resolution measurements at low-momentum transfer. This is in particular true for example for the investigation of nuclear matter distributions by elastic proton scattering at low q, for the investigation of giant monopole resonances by inelastic scattering at low q, and for the investigation of Gamow-Teller transitions by charge exchange reactions at low q. Because of the kinematical conditions of inverse kinematics in case of beams of unstable nuclei, low-momentum transfer measurements turn out to be an exclusive domain of storage ring experiments. Here luminosities are superior by orders of magnitude compared to experiments with external targets. The EXL program thus utilizes one of the outstanding

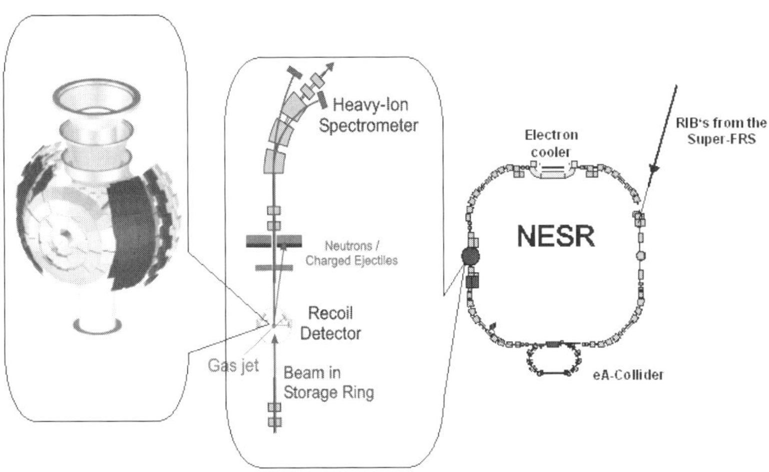

Figure 6. Layout of the EXL setup at the NESR storage ring. The insert on the left displays a view of the target recoil detector surrounding the internal target (for details see text).

features of FAIR, i.e., the availability of a multi-storage-ring complex coupled to the superconducting fragment separator (Super-FRS).

Since the domain of low momentum transfer is of interest, extremely thin targets are requested, resulting in too low luminosities if external targets would be used. Likewise, due to their production mechanism, a large momentum spread and large emittance are inherent to the secondary ion beams which would deteriorate a measurement of the target-recoil momenta and kinetic energies if not counteracted. These problems can be overcome using stored and cooled secondary beams of unstable nuclei interacting with thin internal gas-jet targets. This method provides: high luminosities due to the continuous beam accumulation and beam recirculation; high resolution detection of low-energy recoil particles due to beam cooling, and thin targets; and low-background conditions due to pure, windowless H, He, etc. targets.

Within the EXL project the design of a complex detection set-up was investigated with the aim to provide a highly efficient, high-resolution universal detection system, applicable to a wide class of reactions (see figure 6). The apparatus is foreseen being installed at the internal target of the NESR storage cooler ring. Since a fully exclusive measurement is envisaged, the detection system includes: a recoil detector consisting of a Si-strip and Si(Li) detector array for recoiling target-like reaction products, completed by a scintillator array of high granularity for γ-rays and for the total-energy measurement of more energetic target recoils; a detector in forward direction for fast ejectiles from the excited projectiles, i.e., for neutrons and light charged particles; and heavy-ion detectors for the detection of beam-like reaction products. All detector components will practically cover the full relevant solid angle and have detection efficiency close to unity.

The ELISe project (ELectron Ion Scattering in a Storage ring e-A collider) aims at elastic, inelastic and quasifree electron scattering, which will be possible for the first time for short lived nuclei (for details see [2,5,23]. It uses the intersecting ion and electron storage rings (see figure 6), and an electron spectrometer with high resolution and large solid angle coverage. The investigation of charge densities, transition densities, etc. on nuclei far off stability will complement the investigations of the EXL project, and allow in many cases for a more complete and model independent information of the structure of such nuclei.

References

1. Conceptual Design Report, GSI, 2001, http://www.gsi.de/GSI-Future/cdr/.

2. Nuclear Physics News (ISSN: 1050-6896) 16 No. 1, 5-14 (2006).
3. W.F. Henning, *Nucl. Phys.* **A734**, 654 (2004).
4. H.H. Gutbrod, *Nucl. Phys.* **A752**, 457c (2005).
5. Baseline Technical Report, Vol. 4, July 2006, http://www.gsi.de/fair/reports/btr.html.
6. H. Geissel et al., *Nucl. Instr. Meth.* **B70**, 286 (1992).
7. T. Aumann, *Eur. Phys. J.* **A26**, 441 (2005).
8. P. Adrich et al., *Phys. Rev. Lett.* **95**, 132501 (2005).
9. S.C. Fultz, *Phys. Rev.* **186**, 1255 (1969).
10. N. Paar et al., *Phys. Rev. C* **67**, 034312 (2003).
11. D. Sarchi et al., *Phys. Lett. B* **601**, 27 (2004).
12. A. Klimkiewicz, *Phys. Rev. C* (2007) to be published.
13. D. Vretenar et al., *Phys. Rep.* **409**, 101 (2005).
14. G.D. Alkahzov et al., *Phys. Rep. C* **42**, 89 (1978).
15. P. Egelhof et al., *Prog. Part. Nucl. Phys.* **46**, 307 (2001).
16. G.D. Alkahzov et al., *Nucl. Phys. A* **712**, 269 (2002).
17. S.R. Neumaier et al., *Nucl. Phys. A* **712**, 247 (2002).
18. P. Egelhof et al., *Eur. Phys. J.* **A15**, 27 (2002).
19. A.V. Dobrovolsky et al., *Nucl. Phys. A* **766**, 1 (2006).
20. S. Ilieva, A. Inglessi et al., to be published.
21. H. Wollersheim et al., *Nucl. Instr. Meth..* **A537**, 637 (2004).
22. A. Bürger et al., *Phys. Lett. B* **622**, 29 (2005).
23. T. Aumann, *Progr. Part. Nucl. Phys.* **59**, 3 (2007).
24. H. Geissel et al., *Nucl. Instr. Meth.* **B204**, 71 (2003).
25. P. Egelhof et al., *Phys. Scripta* **T104**, 151 (2003).

THE SPES DIRECT TARGET PROJECT AT THE LABORATORI NAZIONALI DI LEGNARO

GIANFRANCO PRETE

Laboratori Nazionali di Legnaro, viale dell'Università 2
Legnaro, 35020 Italy

A. ANDRIGHETTO[1], C. ANTONUCCI[2], M. BARBUI[1], L. BIASETTO[1,3], G. BISOFFI[1], S. CARTURAN[1], L. CELONA[4], F. CERVELLERA[1], S. CEVOLANI[2], F. CHINES[4], M.CINAUSERO[1], P. COLOMBO[3], M. COMUNIAN[1], G. CUTTONE[4], A. DAINELLI[1], P. DI BERNARDO[5], E. FAGOTTI[1], M. GIACCHINI[1], F.GRAMEGNA[1], M. LOLLO[1], G. MAGGIONI[1], M. MANZOLARO[1,3], G. MENEGHETTI[3], G.E. MESSINA[4], A. PALMIERI[1], C. PETROVICH[2], A. PISENT[1], L. PIGA[1,5], M. RE[4], V. RIZZI[1], D. RIZZO[4], M. TONEZZER[1], D. ZAFIROPOULOS[1], P. ZANONATO[5]

1) INFN Laboratori Nazionali di Legnaro, Italy,
2) ENEA, Bologna, Italy
3) Dipartimento di Ingegneria Meccanica, University of Padova, Italy
4) INFN Laboratori Nazionali del Sud, Catania, Italy
5) Dipartimento di Scienze chimiche, University of Padova, Italy

The SPES Direct Target Project at the Laboratori Nazionali di Legnaro (Italy), is devoted to the construction of a Radioactive Ion Beam (RIB) Facility within the framework of the new European RIB panorama. The SPES Project aims for the production of neutron-rich unstable nuclei by the fission of natural Uranium target induced by a primary proton beam of 40 MeV, 200 μA directly impinging on a multi-sliced Uranium Carbide (UC_x) target. The idea is an evolution of the existing HRIBF facility at Oak Ridge National Laboratory (USA) where a proton primary beam of 40 MeV is also used. The goal is to reach a high number of fission products (up to 10^{13} fission/s) still keeping a relatively low power deposition inside the target. The description of the whole facility, together with the details on the Direct Target configuration, will be illustrated. Thermo-mechanical calculations, pellets preparation and characterization, effusion-diffusion model predictions for the estimation of the release fraction of different radioactive nuclides will be discussed. A scaled (1:5) prototype of the Target system has been built and tested recently at HRIBF.

1. Introduction

SPES is an INFN project to develop a Radioactive Beam facility as an intermediate step toward EURISOL. The Laboratori Nazionali di Legnaro (LNL) was chosen as the site for the facility construction. The LNL capability to play a role in this research field is related to the presence of the superconducting linac ALPI, able to re-accelerate exotic ions at 8÷13 MeV/amu, the well consolidated know-how in linac construction, the existing detectors and the related know-how.

Moreover, the necessary real estate is available thanks to the extension of the Laboratory site (more than a factor two in area respect to actual size). Primary services and new infrastructures, like a 40 MW power station, are currently under implementation.

2. The SPES project overview

The main goal of the proposed facility is to provide an accelerator system to perform forefront research in nuclear physics by studying nuclei far from stability. The SPES project is concentrating on the production of neutron-rich radioactive nuclei with mass in the range 80-160 amu starting from a fission rate of 10^{13} fission/s induced by a 40 MeV proton beam impinging onto a direct UCx target. The emphasis to neutron-rich isotopes is justified by the fact that this vast territory has been little explored, at exceptions of some decay and in-beam spectroscopy following fission. Therefore, reactions in inverse kinematics will allow a new class of data to be obtained.

The feasibility of the SPES project has been positively analyzed both from the point of view of the production target and of the proton driver. The most critical element of the SPES project is the Direct Target. Up to day the proposed target represent an innovation in term of capability to sustain the primary beam power. The design is carefully oriented to optimise the radiative cooling taking advantage of the high operating temperature of 2000°C.

An extensive simulation of the target behaviour has been performed to characterize the thermal properties and the release process. Experimental work to bench mark the simulations was carried out at HRIBF, the Oak Ridge National Laboratory ISOL facility (USA).

The design of the proton driver will take advantage from the TRASCO project using the high current proton source TRIPS, developed at LNS and transferred at LNL where it is now in operation, and the RFQ able to deliver a proton beam of 30mA 5MeV.

The high energy part of the driver will follow the design of the Drift Tube Linac (DTL) under construction for the Linac4 project at CERN.

The low energy, high current proton beam will be used to feed a Neutron Facility devoted to the Boron Neutron Capture Therapy (BNCT) and to produce a neutron beam for cross section measurements of astrophysical interest and material irradiation (LENOS). The possibility to perform neutron cross section measurements on unstable target produced by RIB will be envisaged.

The high energy DTL will be designed with the aim to extend its capability to produce a 50Hz pulsed proton beam up to 1 mA and 100 MeV. A switching system will allow the operation of two beams with a repetition rate of 25Hz each.

The production target will be designed following the ISOLDE and EXCYT projects and special care will be devoted to the safety and radioprotection of the system. According to the estimated level of activation in the production target area of 10^{13} Bq, a special infrastructure will be designed. The use of up-to-date techniques of nuclear engineering will result in a high security level of the installation. The radiation management and the control system will be integrated and redundancies will be adopted in the design.

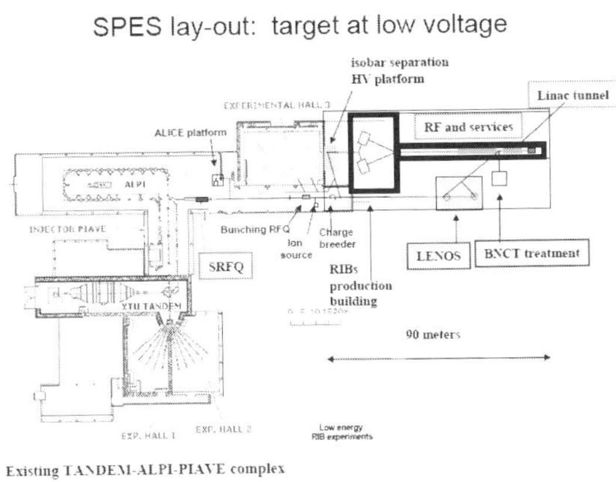

Figure 1 The SPES layout with the new facility connected to the existing Tandem-Alpi-Piave complex.

The isotopes will be extracted and ionized at +1 with a source directly connected with the production target. Several kinds of sources will be used according to the beam of interest. A laser source will be implemented in collaboration with INFN-Pavia with the aim to produce a beam as pure as possible.

The selection and the transport of the exotic beam at low energy and low intensity is a challenging task. Techniques applied for the EXCYT beam will be of reference for the beam diagnostic and an online identification station will be part of the diagnostic system.

To optimize the reacceleration, a Charge Breeder will be developed to increase the charge state to +N before to inject the exotic beam in the Bunching RFQ and in the existing PIAVE Superconductive RFQ which represents the first re-acceleration stage before the injection in ALPI.

The expected beam on experimental target will have a rate on the order of 10^8-10^9 pps for ^{132}Sn, ^{90}Kr, ^{94}Kr and 10^7-10^8 pps for ^{134}Sn, ^{95}Kr with energies of 9-13 MeV/amu. The SPES lay-out is shown in fig.1.

3. Rear beam production

The evaluation of the in-target yield for physics experiment at SPES has been determined starting from the production yield (fission fragment distribution), which was calculated mainly through a Monte Carlo simulation based on transportation model MCNPX [1]. The target is designed with the aim to reach a fission rate of about 10^{13} fission/s, considering this number a challenge.

The final radioactive beam current depends on the efficiencies of several chemical-physical processes and beam transport elements For ISOL facilities, the total efficiency is extremely case dependent and lies between 10^{-2} to 10^{-6}. To give an evaluation of the final beam the exotic species must be followed all along their path.

The first tricky parameter is the release efficiency, related to the diffusion inside the material; a complicated phenomenon which is not completely known, especially when the material is at high temperature. It strictly depends on the material structure and on the temperature at which the material is maintained. [3]

A statistical approach was used to describe the effusion of atoms inside the target powder. After the diffusion the atoms follow a random walk up to exit the container following the effusion process. Experimental data available from ISOLDE-CERN [4], ORNL [5] and Gatchina [6, 7] have been used.

Table 1. Release time and Total Release Fraction for Sn and Kr evaluated for the SPES target configuration.

element	Diff. time (s)	Nr of Coll.	Eff. Time (s)	Release Time (s)	T1/2 (s)	TRF (%)
132Sn	1	10^5	0.2	1.2	40	98
133Sn	1	10^5	0.2	1.2	1	40
Kr	10	10^5	0.1	10	1	15

Calculations were performed with RIBO [8], a CERN code, and GEANT4. The final results of our calculation indicate a release time of 1s for Sn isotopes and 10s for Kr for the chosen target configuration, mainly due to the diffusion time, as reported in table1.

The production rates of exotic beam on the experimental target were evaluated taking into account the following efficiencies as well as the total release fraction of each isotope.

The assumed +1 and +1/+N ionization efficiencies are 90% (+1) and 12% (+1/+N) respectively for Kr and Xe, but only 30% (1+) and 4% (1+/n+) respectively for Zn, Sr, Sn, I and Cd. These values are obtained by the SPIRAL2 project and are expected for an optimized extraction in which the source is specifically designed for each beam. In some case the use of an ECR source is required. The Linac ALPI transmission efficiency is considered 50%.

Using these quantities an evaluation of the beam current on target, which can be obtained with the new SPES facility, are then shown in Fig.2 for some isotopes.

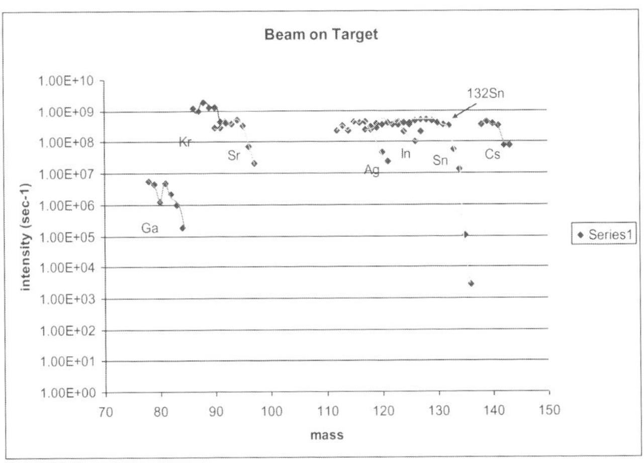

Figure 2. Beam on target: Intensities evaluated considering emission, ionization and acceleration efficiencies (see text) for different isotopes

In table 2 we report, for sake of comparison, some world-wide facilities looking to the fission rate and to the power deposited in the production target.

SPES is located to a high production rate of fission fragments similar to Spiral2. Nevertheless a comparison with Spiral2, from the point of view of driver, layout and cost, is doubtful as the driver of the Spiral2 facility is oriented to the production of high intensity stable beams and in this sense the two facilities cannot be compared at all.

Table 2 Comparison between several ISOL facilities

	Primary beam	Power on target	target	Fissions per sec or notes
ISOLDE	p 1 GeV, 2 μA	0.4 KW	Direct Convert.	$4 \cdot 10^{12}$ spallation
HRIBF	p 40 MeV, 10 μA	0.4 KW	Direct	$4 \cdot 10^{11}$
TRIUMF	p 450 MeV, 40 μA (250 MeV in target) Up-grade: 70 μA	10 KW 17 KW	Direct	Spallation No fission
EXCYT	^{13}C 45 AMeV	0.5 KW	Direct	Light ions by direct reactions $4\ 10^{8}\ ^{8}$Li ($9 \cdot 10^{6}$ extracted)
SPIRAL2	d 40 MeV, 5 mA	200 KW	Convert.	10^{13}- 10^{14}
SPES	p 40 MeV 200 μA	8 KW	Direct	10^{13}

4. Direct Target prototype

The most critical challenge for the SPES Direct Target is its capability to sustain the beam power of 8KW with safe conditions and without melting. According to the classical Stefan-Boltzmann equation the energy irradiated by a black body is proportional to the forth power of its temperature.

At 2000°C, the target operating temperature, the radiative phenomena is the main responsible for the system cooling. The target was designed as a high radiating system made by a stack of separated disks. The geometry was evaluated by ENEA [2] and Ansys calculations to optimise the radiative cooling.

The power distribution was evaluated according to the following design:

1. the incident 40 MeV proton beam has a current of 200 μA. The beam profile spans uniformly over a circle distribution, which matches the disk radii;
2. the window, necessary to separate the beam line from the target void regions, is made of one (or two) thin carbon foil of 400 μm total thickness;
3. the UCx target (ρ=2.5 g/cm3) is made of seven disks about 1.3 mm thick each;
4. the beam dump is made of three carbon disks about 0.9 mm thick each;
5. the box containing the disks is made of graphite.

It turns out that a power of 0.19 kW is deposited in the window, 4.1 kW in the seven UCx disks, 1.7 kW into the three dump disks and 2.2 kW is lost outside the disks (due to proton scattering). Thus, the average power deposition for the UCx target disks is about 4.1 kW / 7 disks = 0.58 kW/disk. With disks of 4 cm diameter a power density of 145 W/g is obtained, a reasonable value similar to the HRIBF one.

A pellet production work was started in 2006 with the Chemical Department of Padua University for the production of SiC and LaC2 pellets which are very similar to UCx from the point of view of the physical-chemical properties but much easier to handle from the radiation safety point of view. Several pellets of diameter up to 4 cm were produced.

Special care was devoted to the characterization of the samples with the aim to select the proper production procedure. The electron microscope was used as well as the x-ray diffrattometry for the internal structure study and a measurement of the emissivity was implemented as this parameter is crucial for the thermal properties of the target.

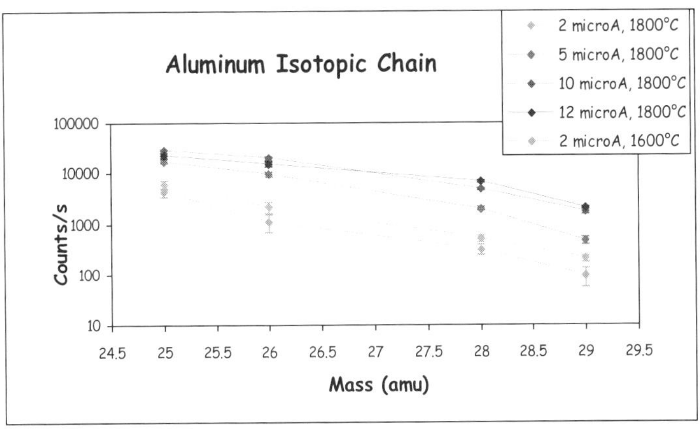

Figure 3. production rate of Aluminum isotopes at HRIBF with the SPES scaled prototype target.

An in-beam test was performed on Jan 2007at HRIBF with a SiC scaled prototype (1.3 cm pellets, to fit the HRIBF target system) to benchmark the thermal properties of the SPES target configuration. The experimental results on the temperature measurements fit very well with the FEM calculations. A more efficient heater was suggested, realized and successfully tested in beam. The thermal analysis of the SPES-like configuration predicts the capability to sustain a beam current of 20 μA (40 MeV protons) without melting. In this scaled configuration the current of 20 μA supplies the same power density of the final SPES target. On the contrary for a standard packed configuration a maximum current of 7 μA is expected to reach the melting temperature of 2300°C. During the experiment a current up to 12 μA was used with a limit imposed by the driver accelerator.

From the point of view of the beam production the experiment aims to study the production of Al isotopes. Results are reported in figure 3 and show a good production rate, considerably better than the standard production at HRIBF which is similar to the lower curve.

5. Proton driver

The proposed driver is a 40 MeV proton linac, operating at 352 MHz, pulsed at high repetition rate, composed by a room temperature radiofrequency quadrupole (RFQ) and a Drift Tube Linac (DTL) with permanent magnet quadrupole focusing. The driver will deliver a 50Hz pulsed beam with mean current value of 200 μA on the production target. The driver has a current capability of 500 μA and can be upgraded up to 100 MeV 1 mA.

The proton source is an off resonance ECR source (TRIPS). Together with the RFQ it produces a high current cw proton beam of 5 MeV and 35 mA necessary for the operation of the 150 kW beryllium target of the BNCT neutron source and the LENOS neutron facility.

Source was transferred from LNS to LNL at the end of 2005 [9]. Installation was completed in late July 2006 and beam extraction was successfully tested in September 2006 [10].

The RFQ, initially developed for the TRASCO project, has two working regimes, pulsed and cw. The operating frequency is 352.2 MHz, with the design choice of using a single 1.3 MW klystron already used at LEP. The RF power will be fed by means of eight high power loops. The achievement of the longitudinal field stabilization for the operating mode will be performed with two coupling cells in order to reduce the effect of perturbating quadrupole modes and with 24 dipole stabilizing rods in order to reduce the effect of perturbating dipole modes.

The RFQ structure consists of six modules 1.18 m long each made of OFHC copper. The vacuum ports are on the first and fourth segments and the couplers on the other four. Particularly challenging are the very tight mechanical tolerances (20 μm) necessary for the purity of the accelerating mode (as required by beam dynamics) that have to be kept in presence of a large power density. To date, the first two modules (RFQ1 and RFQ2) underwent the complete construction cycle and the remaining four modules (RFQ3, RFQ4, RFQ5 and RFQ6) were pre-braze assembled and RF characterized and are ready for brazing at CERN [11].

The last part of the proton driver is the DTL, to bust the protons up to 40 MeV. CERN is developing for Linac4 a more advanced mechanical design of a DTL; we plan to build a common LNL-CERN prototype, based on CERN mechanical design, to test these choices up to high power RF. The scope is to arrive to a common design with CERN and to produce two similar structures in the industry.

On the other hand the requirements for the proton driver accelerator could be met by a commercial proton cyclotron, but in this case no further improvement in the proton power beam is possible; nevertheless this solution will be taken into account for the final decision according to beam characteristics and cost.

6. Reacceleration

The linear accelerator ALPI, with a β range between about 0.04 and 0.2 and CW operation, represents an ideal re-accelerator for the radioactive beams. Radioactive ions can be accelerated above the Coulomb Barrier with high efficiency, and with a quasi-continuous time structure well suited for experiments. A time structure suitable for TOF measurements can be implemented by a low energy bunching system.

ALPI underwent a number of significant upgrades, in recent years, which made it a world-class facility in heavy ion stable beam accelerators and which will represent an important added value for its use as a RIB accelerator as well. Alpi has an equivalent acceleration voltage of 40 MV and may accelerates heavy ions in the region of Tin at energies between 6 and 13 AMeV according to their charge state (19+ or 40+ respectively). To allow the reacceleration with ALPI a new bunching RFQ will be developed and the PIAVE Superconductive RFQ, which represents the actual injector from the ECR heavy ion source, will be moved to accept the low energy exotic beam.

7. Conclusion

The SPES facility is expected to run the first exotic beam in 2013. Before starting the construction the R&D program will continue for key development subsystems to receive adequate answers; such items like control system final target design and proton driver analysis, will be completely mature in about one year.

At the same time the detailed design and the procedure needed for the construction authorization will be implemented.

Design and construction of the complete facility will require 4 years, with the installation and commissioning of parts of the machine beginning immediately after the completion of the buildings and related infrastructures. Critical parts as RIB target and high current RFQ are in advanced construction stage and will be ready for laboratory test before the building construction. The implementation of the ALPI re-accelerator is already scheduled starting from the current year.

The evaluated budget is in the order of 45-48 Meuro, compatible with the INFN Road Map for the Nuclear Physics development.

References

1. J.S. Hendricks et al., MCNPX vers. 2.5.e – LA-UR-04-0569 (2004)
2. A. Andrighetto, S. Cevolani, C. Petrovich – Europ. Journal of Physics A25(2005) 41-47
3. J. Crank, The Mathematics of Diffusion, Clarendon Press (1956).
4. http://www94.web.cern.ch/ISOLDE/
5. K.Carter WATD, TRIUMF (2006) http://watd.triumf.ca/program.html
6. M.Barbui at al. LNL Annual report 2006, 182
7. M.Barbui at al. LNL Annual report 2006, 184
8. M.Santana http://ribo.web.cern.ch/ribo
9. E. Fagotti et al., Annual Report 2005, INFN-LNL 210(2006), 150.
10. E. Fagotti, A. Palmieri et al., Annual Report 2006, 191
11. A. Palmieri et al LNL Annual Report 2006, 193

SECTION II

NUCLEAR STRUCTURE AND NUCLEAR FORCES

MODERN ASPECTS OF NUCLEAR STRUCTURE THEORY

JOCHEN WAMBACH,[1,2]

[1] *Institut für Kernphysik, TU-Darmstadt, Germany*
[2] *Gesellschaft für Schwerionenforschung, Darmstadt, Germany*

I review recent progress in constructing high-precision nucleon-nucleon potentials and multi-nucleon forces from Chiral Perturbation Theory. In conjunction with ab-initio many-body approaches a consistent framework is in reach to understand the structure of nuclei on the basis of Quantum Chromodynamics.

1. Introduction

One of the ultimate goals of nuclear physics is to elucidate the relation of basic nuclear properties to the underlying degrees of freedom of the strong interaction, namely quarks and gluons. One would like to know how the richness of nuclear spectra eventually depends on the basic parameters of Quantum Chromodynamics (QCD), such as quark masses, the number of flavors or colors or the strong coupling constant g_s. This question is not purely academic but has direct bearing on signatures of time-dependent changes of the constants of nature.[1] Bridging the gap between QCD and the nuclear many-body problem is mandatory in other aspects as well. Since neutrons and protons, the basic degrees of freedom in nuclei, are not point-like particles but composites of quarks and gluons their vacuum properties will be modified in the presence of other nucleons. A stringent theoretical framework is therefore required to quantify such changes, thus providing a unified picture of nucleons and nuclei. On the other hand, a detailed understanding of how basic QCD aspects translate into the effective interactions of nucleons will provide guidelines for extrapolations to unknown regions of the nuclear chart.

Much progress has made in recent years in setting up a consistent framework which is based on the pertinent symmetry principles of QCD and their breaking pattern. This leads to a consistent effective field theory for the low-energy structure of nucleons and the systematic evaluation of nuclear forces. In conjunction with ab-initio non-relativistic methods for solving the many-body Schrödinger equation one thus begins to make the connection between details of nuclear spectra and fundamental QCD constants.

2. Quantum Chromodynamics

The strong-interaction sector of the Standard Model is governed by Quantum Chromodynamics (QCD), a local gauge theory involving quarks as the matter particles and gluons as the force carriers. Both carry 'color' as fundamental quantum numbers. The Lagrange density is very similar to that of Quantum Electrodynamics (QED), except that the gauge group is color-$SU(3)$ instead of charge-U(1)[2] and reads

$$\mathcal{L}_{QCD} = -\frac{1}{4} G^a_{\mu\nu} G^{\mu\nu}_a + \bar{q}\gamma^\mu (i\partial_\mu - g_s \Lambda_a A^a_\mu) q - m_q \bar{q} q \ . \tag{1}$$

In addition to color, the quark fields carry the usual Dirac as well as flavor indices. The quark mass matrix is denoted by m_q. Because of the Lie character of the gauge group the gluonic field strength tensor contains a non-linear term in the vector potential A^a_μ involving the strong-coupling constant g_s and the $SU(3)$ structure constants:

$$G^a_{\mu\nu} = \partial_\mu A^a_\nu - \partial_\nu A^a_\mu - g_s f_{abc} A^b_\mu A^c_\nu \ . \tag{2}$$

In contrast to QED this non-Abelian term implies a self-interaction of the gauge bosons and renders the theory very rich and complicated, even at the classical level. Besides the local gauge symmetry, the Lagrangian exhibits extra symmetries. One is the so-called conformal or dilatation symmetry which holds for massless quarks and implies that $\mathcal{L}_{QCD}$ is invariant under a global scale transformation $x \to \lambda x$ of the space-time coordinates. It is a consequence of the fact that QCD (as well as QED) in the massless limit contains no dimensionfull parameters. In QCD scale invariance is broken at the quantum level by an anomaly giving rise to a fundamental scale, Λ_{QCD} which can be inferred from the 'running' of the strong coupling constant or $\alpha_s(Q) = g_s^2(Q)/4\pi$[3–5] and is of the order of 200 MeV at a renormalization scale of $\sim$ 2 GeV. While α_s becomes perturbative at large momenta, Q, it is extremely large in the infra-red regime $Q \sim \Lambda_{QCD}$. This renders the quantum theory highly non-perturbative, giving rise to a complicated structure of the proton and other hadrons as well as nuclei at this scale.

The only ab-initio method that can deal with this regime without approximations is Lattice-QCD (LQCD) by which the pertinent correlation functions are obtained numerically on a discrete Euclidean space-time lattice. For baryons and mesons such calculations demand resources at the forefront of supercomputing technology and are not yet at the stage to make reliable predictions for the structure of hadrons in the case of realistic masses for the (very) light up- and down quarks. Calculations for nuclei heavier than the deuteron are prohibitive at present. To make the connection of the QCD degrees of freedom to the structure of nuclei, one has to follow a different route.

3. Nuclear Effective Field Theory

There is another global symmetry of $\mathcal{L}_{QCD}$, which also holds in the limit of vanishing quark masses. This is chiral symmetry, which is almost exactly realized for up- and down quarks. It implies an invariance under $SU(2)_L \times SU(2)_R$ rotations in flavor space or, equivalently, vector and axial-vector transformations. The axial symmetry is spontaneously broken in the QCD vacuum as can be inferred for instance from the absence of parity doublets in the low-lying hadron spectrum. As a consequence pions appear as (nearly) Goldstone bosons which together with the nucleon, form the building blocks of a low-energy nuclear effective field theory (NEFT) that can be stringently based on QCD.[6,7] The interactions are described by an effective Lagrangian $\mathcal{L}_{eff}$ constrained by chiral symmetry and its explicit breaking by small up- and down quark masses. In principle, $\mathcal{L}_{eff}$ contains an infinite number of operators compatible with chiral symmetry, which can ordered by a chiral counting parameter

$$\Delta = d + \frac{1}{2}\nu - 2; \qquad \Delta \geq 0 \tag{3}$$

that classifies the interaction vertices. Here d denotes the number of derivatives and ν the number of nucleon field operators. For an expansion of the effective Lagrangian in orders of Δ one then has

$$\mathcal{L}_{eff}(\pi, N) = \mathcal{L}_{eff}^{(0)} + \mathcal{L}_{eff}^{(1)} + \mathcal{L}_{eff}^{(2)} + \ldots \tag{4}$$

The parameters of $\mathcal{L}_{eff}$, the so-called low-energy constants $LECs$, are determined from the pion and π/N systems as well as from nucleonic systems (C_s, C_T, D, E). These constants contain the (integrated out) short-distance physics and are not determined by chiral symmetry but have to be obtained either from experiment or LQCD.

3.1. *Chiral Nuclear Forces*

A possible strategy[9,10] is now to construct out of the effective Lagrangian a nucleon-nucleon ($2N$) potential (and possible many-nucleon forces) by identifying a set of irreducible diagrams in a systematic low-momentum expansion $V_{2N} \propto (Q/\Lambda_\chi)^n$, where $Q \sim M_\pi$ refers to a soft scale, typical for the scattering process and $\Lambda_\chi \sim 4\pi F_\pi \sim 1$ GeV denotes the hard scale, set by spontaneous chiral symmetry breaking. This procedure is known to give correct results in meson-meson scattering. The set of diagrams is identified with an 'effective potential' which defines the kernel of a Lippmann-Schwinger equation and thus can handle non-perturbative effects such as the appearance of the deuteron bound state. The power n in the momentum expansion is given by

$$n = -4 + 2N + L + \sum_i V_i \Delta_i \tag{5}$$

where N is the number of external nucleon lines, L the number of internal loops, V_i the number of vertices and Δ_i the chiral order (3). All diagrams up to order $n = 4$ (N^3LO) have been evaluated by now[11] and are displayed in Fig. 1. The leading-order (LO) contributions are purely of $2N$ type, given by one-pion exchange and two contact terms without derivatives. These are corrected in next-to-leading order (NLO) by two-pion exchange loops and contact interactions with derivatives. Three nucleon interactions ($3N$) first appear at $n = 3$ (N^2LO) while four-body interactions ($4N$) begin to emerge at $n = 4$. Assuming that all LECs are of natural size (i.e. of order unity), which seems to be the case, it is very pleasing to see that there is a clear hierarchy of forces

$$<V_{2N}> \;\gg\; <V_{3N}> \;\gg\; <V_{4N}> \tag{6}$$

which is also observed empirically and greatly simplifies the nuclear many-body problem.

3.2. *Results for the Two-Body System*

According to Weinberg's scheme[9] discussed above the chiral two-nucleon interaction V_{2N} is to be used in a Lippmann-Schwinger equation to evaluate scattering phase shifts and thus fix the LECs, not obtained from other sources such as the $\pi\pi$- and πN systems. This has been carried out to order N^3LO in,[12] regularizing V_{2N} by a cut-off function f_Λ with cut-off values around 500 MeV. Adjusting a total of 26 parameters of the resulting phaseshifts in a combined fit to the *pp* and *np* database of the Nijmegen

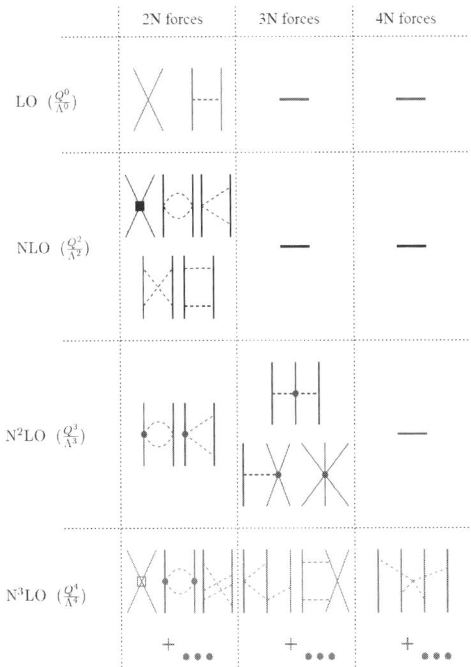

Fig. 1. The hierarchy of nuclear forces within NEFT in a low-momentum expansion $(Q/\Lambda_\chi)^n$.

analysis ($\sim$ 4500 data points) rather accurate fits can be obtained at order N^{3}LO. Based on the N^{3}LO expansion Entem et al.[13] were able to construct a high precision fit to the available scattering data with a χ^2/datum 1.1 which is comparable to other state-of-the art potentials.

With chiral potentials at hand one can begin to answer the question of how the nuclear structure depends on basic QCD parameters such as the light quark masses (or equivalently the mass of the pion, M_π). Besides the fundamental interests mentioned in the introduction there are practical applications in the extrapolation of relevant LQCD results to physical quark masses. The simplest case of the two-body system has been addressed in[11] for the deuteron binding energy and the S-wave scattering lengths in the NLO. The results for the binding energy are shown in Fig. 2. It is found that the deuteron is more strongly bound in the chiral limit $M_\pi \to 0$ with $B_D^{CL} \sim 9.6$ MeV but no further bound state is obtained in spite of the fact

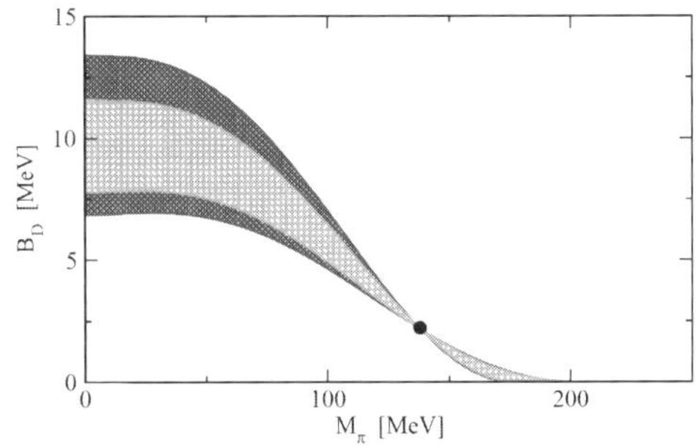

Fig. 2. The deuteron binding energy as a function of the pion mass. The shaded areas indicate the uncertainties induced by the pertinent LECS.

that the higher $S = 1$ partial waves rise linearly with momentum due to the Coulomb-like static pion propagator. The absolute values of the scattering lengths $a(^1S_0)$ and $a(^3S_1)$ are found to be smaller in the chiral limit than for the physical pion mass. The comparison with lattice simulations of these quantities[14] shows that a matching with NEFT has yet to be reached.

3.3. *Three- and Four-body Nuclei*

Given the chiral two-body potential, the role of three-body forces in the three-and four-body system can be addressed from first principles, as there exist a number of exact methods for the solution of the Schrödinger equation. Given a specific V_{2N} these methods can bench marked to give the same answer for the binding energies.[15] Such calculations have been carried out for the high-precision N^{3}LO two-body potential of Entem and Machleidt[13] in the no-core shell model (NCSM), in Fadeev and Fadeev-Yakubovski treatments as well as in the hyperspherical harmonics approach, all yielding binding energies for ^{3}H of 7.854 MeV and for ^{4}He of 25.37 MeV within less than 1 ppm accuracy.[16] Results of similar quality can also be obtained for the α-particle.[16]

It is well known that the experimental binding energies of the three- and four-body nuclei cannot be reproduced by two-body forces alone. In the past, three-body forces have been adjusted more or less ad hoc to reproduce the spectra of light nuclei evaluated in the exact Green's Function Monte Carlo method.[17] The beauty of the chiral approach is that one has a consistent scheme to evaluate them on equal footing with V_{2N}. At N^{2}LO there are three contributions to V_{3N} (see Fig. 1): one from two-pion exchange (with LECs fixed in the $\pi/\pi - N$ system), one from a one-pion exchange contact term with the LEC c_D and one from a pure contact term with c_E. These two constants can now be adjusted in ^{3}H and ^{4}He with the resulting values $c_D = 2$ and $c_E = 0.12$. Hence, both are of natural size.[16]

I should note the remarkable fact that these values seem to fix the long-standing problem of a 1^+-3^+ level inversion in ^{10}B that one always obtains with two-body interactions alone.[16] Improvements in the splitting of the $3/2^-$ ground state and $1/2^-$ first excited state of ^{11}B, which are degenerate without three-body forces, can also be observed.[16]

4. Unitary Correlation Operator Method

Exact methods for the solution of the nuclear many-body Schrödinger equation are vital for the understanding of the relationship between QCD and nuclear structure in the framework I have laid out in the previous section. At present such methods are restricted to light nuclei with $A \leq 16$. The use of realistic potentials for nuclear structure studies in heavier nuclei poses tremendous challenges. Traditional many-body methods, such as the Hartree-Fock (HF) approach or the Random Phase Approximation (RPA), cannot be used for bare nucleon-nucleon interactions. The reason is the inability of the simple model spaces underlying these approaches to describe the dominant short-range correlations, which are present in the exact many-body eigenstates. It is therefore important to devise controlled methods whose convergence properties can be stringently assessed. One such method is the 'Unititary Correlation Operator Method' (UCOM) proposed by Feldmeier and collaborators.[18–20]

4.1. *Construction of V_{UCOM}*

The two most important types of many-body correlations are those induced by the short-range repulsion and the strong tensor part of the $2N$ interaction. The basic idea of the UCOM is to include these dominant correlations in the many-body state by means of a unitary transformation. Starting from an uncorrelated wavefunction, $|\Psi\rangle$, in the simplest case a Slater determinant, a correlated wavefunction $|\tilde{\Psi}\rangle$ is defined through the application of a unitary correlation operator C ($C^\dagger C = 1$)

$$|\tilde{\Psi}\rangle = C|\Psi\rangle \,. \tag{7}$$

Alternatively, one can perform a similarity transformation of the operators of all relevant observables (e.g. the Hamiltonian, coordinate and momentum space densities, transition operators, etc.)

$$\tilde{O} = C^{\dagger} O C \ . \tag{8}$$

Due to unitarity both procedures are equivalent.

The correlation operator C is decomposed into a central correlator C_r and a tensor correlator C_Ω

$$C = C_r C_\Omega = \exp\left[-i \sum_{i<j} g_r(ij)\right] \exp\left[-i \sum_{i<j} g_\Omega(ij)\right] \ . \tag{9}$$

both defined as exponentials of Hermitian two-body generators g_r and g_Ω which encode the physics of the short-range correlations and are obtained by energy minimization in the two-body system.[21]

In practical applications the use of correlated operators is advantageous. Since these operators are defined as exponentials of two-body operators, they contain irreducible contributions for all particle numbers, which can be organized in a cluster expansion

$$\tilde{O} = C^{\dagger} O C == \tilde{O}^{[1]} + \tilde{O}^{[2]} + \tilde{O}^{[3]} + \cdots \ . \tag{10}$$

If the range of the correlators is sufficiently small compared to the average particle distance, three- and higher-order terms in the cluster expansion are small and one can restrict oneself to the two-body approximation. For the Hamiltonian this implies

$$\tilde{H}^{C2} = \tilde{H}^{[1]} + \tilde{H}^{[2]} + \tilde{V}^{[2]} = T + V_{UCOM} \ , \tag{11}$$

where $\tilde{T}^{[1]} \equiv T$ and $\tilde{T}^{[2]}$ are the one- and two-body contributions of the correlated kinetic energy, respectively, and $\tilde{V}^{[2]}$ is the two-body part of the correlated $2N$-potential. All two-body contributions are subsumed in the correlated interaction V_{UCOM}. By construction it is phase-shift equivalent to the original bare potential.

To restrict the cluster expansion to two-body terms requires special treatment of the tensor correlation functions. Since they largely originate from the one-pion exchange, the tensor force is long-ranged, and so are the tensor correlations induced in the two-body system. For $A > 2$, the long-range component of the tensor correlations between two nucleons is screened, however, due to the presence of other nucleons. Thus the range of the tensor correlator can be varied by a constraint on the volume integral of the correlation function, $I_\vartheta = \int dr\, r^2 \vartheta(r)$. The variation of the ^{3}H and ^{4}He binding energies with the triplet-even I_ϑ[21] is shown in Fig. 3. With increasing range the energy is lowered and the full Tjon-line is mapped

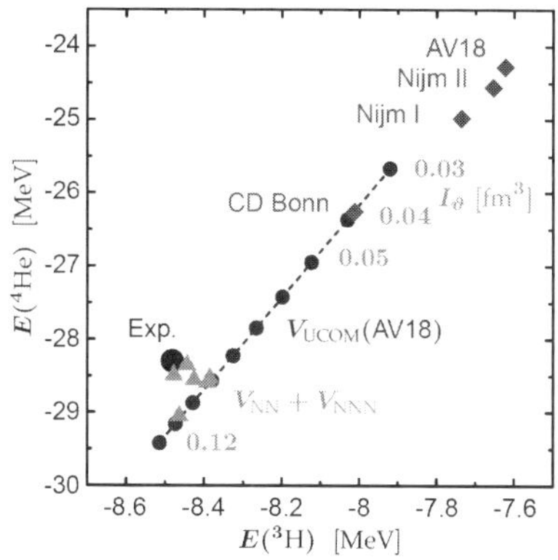

Fig. 3. Results for the Tjon-line as obtained in the NCSM using V_{UCOM} derived from the Argonne $V18$ potential.[21] The triangles denote various choices for the range in the tensor correlator.

out. This is related to the omission of three-body (and higher-order) terms in the cluster expansion. If these terms were included, the energies would be exactly the same, independent of the correlator range, because of the unitarity of the transformation. The fact that the range of the tensor correlator can be chosen such that the energies are close to experiment (e.g. for $I_\vartheta = 0.09$ fm^3) can be explained by a cancellation between genuine three-body forces and the induced three-body contributions of the cluster expansion. Thus the impact of the net three-body force on the binding energies can be minimized by a proper choice of the correlator range.

4.2. *Hartree-Fock and Many-Body Perturbation Theory*

Using V_{UCOM} adjusted within the NCSM for few-body nuclei one can perform HF calculations of nuclear ground states throughout the nuclear chart in a large oscillator basis. The results for ground-state energies of closed-shell nuclei ranging from ^{4}He to ^{208}Pb[22] are displayed in Fig. 4 for the optimal tensor correlator with $I_\vartheta = 0.09$ fm^3. Evidently, the HF binding energies are significantly smaller than the experimental ones. This is not surprising, since residual long-range correlations as they appear in the NCSM calculations cannot be described by the HF ground state.

An estimate for the impact of residual long-range correlations on the binding energies can be obtained within MBPT. The evaluation of the

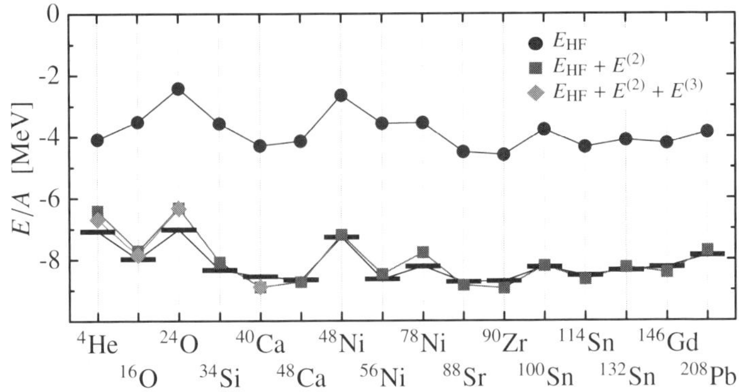

Fig. 4. The ground-state energy of various closed-shell nuclei obtained with V_{UCOM} derived from the Argonne $V18$ potential within the HF approximation (full dots) and in HF+MBPT (squares and diamonds)[22] in comparison with the experimental data.

second- and third-order perturbative contributions on top of the HF result is straightforward.[22] Fig. 4 summarizes the results for the ground-state energies including second-order correlations (for light nuclei also third order). The agreement with the experimental binding energies per nucleon is remarkably good throughout the entire mass range. The absence of any systematic deviation for larger mass numbers proves that the cancellation between genuine three-body forces and induced three-body contributions, which are observed in the NCSM for light nuclei, works throughout the nuclear chart. Furthermore, the calculations seem to establish the perturbative character of the long-range correlations.

However, the good agreement with experimental data does not hold for all observables. The charge radii obtained in HF for heavier nuclei are too small in comparison with experiment.[22] The inclusion of perturbative corrections improves the results but still leaves deviations of up to 1 fm for the heaviest nuclei. This is an indication that a net three-body force is needed to reproduce all observables, although its impact on the binding energy might be small. Similar conclusion are reached for infinite nuclear matter.[23]

5. Summary and Conclusions

In the quest for an understanding of nuclear structure in terms of the fundamental theory of the strong interaction, QCD, I have sketched a promising route that begins to take shape. It hinges on two recent developments: the emergence of stringent effective field theory methods in nuclear physics based on the spontaneous breaking of chiral symmetry in the non-perturbative vacuum of QCD and the progress in ab-initio nuclear many-body methods. When both aspects are combined, questions on how the properties of bound nuclei depend on basic QCD parameters such as quark masses can be answered.

References

1. V.V. Flambaum, *Phys. Rev.* A 73 (2006) 034101
2. Y. Nambu, In *Preludes in Theoretical Physics, in Honor of Weisskopf, V. F.*, eds. A. de-Shalit, H. Feschbach and L. van Hove, Amsterdam, North Holland, p. 133
3. D.J. Gross and F. Wilzcek, *Phys. Rev. Lett.* 30 (1973) 1334
4. H. Politzer, *Phys. Rev. Lett.* 30 (1973) 1336
5. G. 't Hooft, *Nucl. Phys.* B 254 (1985) 11
6. H. Leutwyler, *Ann. Phys. (N.Y.)* 235 (1994) 165
7. S. Weinberg, in Lecture Notes in Physics, Vol. 452 (Springer Verlag, Heidelberg 1995)
8. V. Bernard, N. Kaiser and U.-G Meissner, *Int. J. Mod. Phys.* A 4 (1995) 193
9. S. Weinberg, *Phys. Lett.* B 251 (1990) 288 ; *Nucl. Phys.* B 363 (1991) 3
10. U. van Kolck, *Phys. Rev.* C 49 (1994) 2932
11. U.-G. Meissner, *Nucl. Phys.* A 751 (2005) 149
12. E. Eppelbaum, W. Glöckle and U.-G. Meissner, *Eur. Phys. J.* A 19 (2004) 125 ; *Eur. Phys. J.* A 19 (2004) 401
13. D.R. Entem and R. Machleidt, *Phys. Lett.* B 524 (2002) 93 ; *Phys. Rev.* C 68 (2003) 041001
14. M. Fukugita et al., *Phys. Rev.* D 52 (1995) 3003
15. H. Kamada et al., *Phys. Rev.* C 64 (2001) 04401
16. P. Navratil, private communication
17. R.B. Wiringa and S.C. Pieper *Phys. Rev. Lett.* 89 (2002) 182501
18. H. Feldmeier et al., NPA 632 (1998) 61
19. T. Neff and H. Feldmeier, NPA 713 (2003) 311
20. R. Roth et al., NPA 745 (2004) 3
21. R. Roth et al., PRC 72 (2005) 034002
22. R. Roth, P. Papakonstantinou and N. Paar, PRC 73 (2006) 044312
23. R. Roth, private communication

CORRELATIONS IN NUCLEI: A REVIEW

R. SCHIAVILLA

Department of Physics, Old Dominion University, Norfolk, VA 23529, USA
and Jefferson Laboratory, Newport News, VA 23606, USA

The two preeminent features of the nucleon-nucleon interaction are its short-range repulsion and intermediate- to long-range tensor character. In the present talk, I review how these features influence a variety of nuclear properties, briefly discuss the experimental evidence in support of their presence, and finally present some very recent developments relating to their impact on two-nucleon momentum distributions in nuclei.

1. Introduction

The interaction that binds protons and neutrons together in the atomic nucleus is characterized by a strong repulsion at short ($\lesssim 0.5$ fm) distances, and a strong coupling between spatial and spin degrees of freedom in pairs of nucleons at intermediate to large ($\gtrsim 1$ fm) separations. This "tensor" character of the interaction binds the deuteron and couples S- and D-waves into a ground state with a large non-spherical component. This is in marked contrast to systems such as the hydrogen atom where the dominant $1/r^2$ Coulomb force leads to an essentially spherical ground state.

While many quantum systems share a feature of short-range repulsion, the tensor character is unique to the nuclear force. Its presence has been beautifully confirmed in electron scattering experiments on the deuteron.

In the present talk, I review how the repulsive and tensor components of the NN interaction affect the structure of nuclei, and how the presence of the correlations they induce among the nucleons is reflected in a variery of nuclear properties. In particular, I also show how tensor correlations impact strongly the momentum distributions of NN pairs in the ground state of a nucleus and lead to large differences in the np versus pp distributions at moderate values of the relative momentum in the pair. These differences should be easily observable in two-nucleon knockout processes.

2. Two-nucleon density distributions in nuclei

The two-nucleon density distributions in spin-isospin states T, SM_S are defined as [1]

$$\rho_{T,S}^{M_S}(\mathbf{r}) = \frac{1}{2J+1} \sum_{M_J} \langle JM_J | \sum_{i<j} P_{ij}^{T,SM_S}(\mathbf{r}) | JM_J \rangle \ , \tag{1}$$

where $|JM_J\rangle$ denotes the ground state of the nucleus with total angular momentum J and projection M_J, and

$$P_{ij}^{T,SM_S}(\mathbf{r}) \equiv \delta(\mathbf{r} - \mathbf{r}_{ij}) \, P_{ij}^T \, |SM_S, ij\rangle\langle SM_S, ij| \tag{2}$$

projects out the specific two-nucleon state with $\mathbf{r}_{ij} \equiv \mathbf{r}_i - \mathbf{r}_j$.

The above functions obviously reflect features of the underlying NN interaction. Of particular relevance are those in the TS=01 channel [1], illustrated in Fig. 1. Note that the expectation value of the static part of

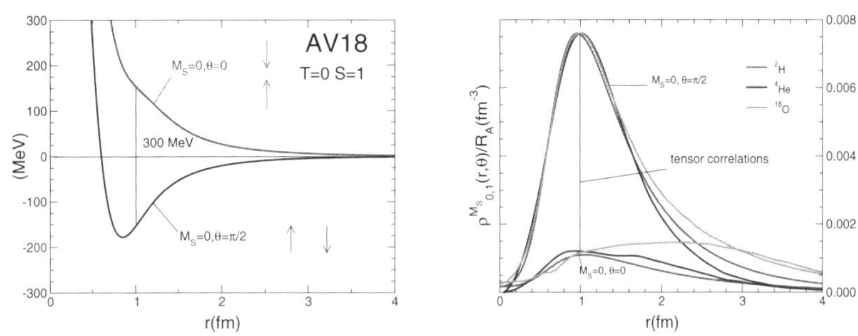

Fig. 1. Left panel: The expectation value of the static part of the NN interaction in the TS=01 channel for two different spin-spatial configurations: the NN pair in state M_S=0 located either along the z-axis or in the xy-plane. Right panel: Corresponding two-nucleon density distributions in a variey of nuclei.

the NN interaction (its dominant part) in the state with M_S=±1 and θ=0 is the same as that in the state with M_S=0 and $\theta=\pi/2$ (see left panel of Fig. 1) by symmetry, while that in the state with M_S=±1 and $\theta=\pi/2$ is half way in between those of the states with M_S=0,θ=0 and M_S=0,$\theta=\pi/2$. The interaction is very repulsive for $r \lesssim 0.5$ fm regardless of the M_S value. However, for distances $r \simeq 1$ fm, it is very attractive when the two nucleons in state M_S=0 are confined in the xy-plane, and very repulsive when they are along the z-axis. When the two nucleons are in state M_S=1, the situation is reversed: the interaction is repulsive (but not as repulsive as for M_S=0) when the two nucleons are in the xy-plane and very attractive when they are located along the z-axis. The energy difference between the two configurations θ=0 and $\pi/2$ for M_S=0 is found to be very large—a few hundreds of MeV—in all realistic NN interactions. As a result, the two-nucleon densities, displayed in the right panel of Fig. 1, are strongly anisotropic. Furthermore, the densities $\rho_{0,1}^{M_S}(\mathbf{r};A)$ in a nucleus with mass number A are proportional to those in the deuteron for internucleon separations up to $\simeq$ 2 fm: the scaling factor R_A is defined as R_A=Max $\left[\rho_{0,1}^{\pm 1}(\mathbf{r};A)\right]$/Max $\left[\rho_{0,1}^{\pm 1}(\mathbf{r};d)\right]$. That the neutron-proton relative wave function in a nucleus is similar, at small separations, to that in the deuteron had been conjectured by Levinger and Bethe [2] over fifty years ago; indeed, Fig. 1 provides a microscopic justification for that conjecture, known as the "quasi-deuteron model."

The two-nucleon density distributions in states with pair spin-isospin TS=11, 00, and 10 are drastically different from those with TS=01 [1]. They are shown for ^{4}He, ^{6}Li, and ^{16}O in Fig. 2, where all curves have been normalized to have the same peak height as for ^{16}O. The TS=11 interaction

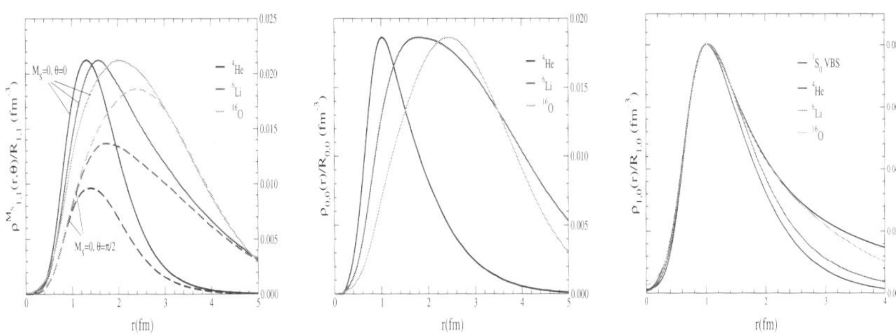

Fig. 2. The $\rho_{11}^{M_S}(r,\theta;A)/R_{1,1}$ (left panel), $\rho_{00}^{0}(r,\theta;A)/R_{0,0}$ (central panel), and $\rho_{10}^{0}(r,\theta;A)/R_{1,0}$ (right panel) density distributions for various nuclei.

has a tensor component of opposite sign with respect to that of the TS=01 interaction. As a consequence, the M_S=$\pm$1 (=0) density distributions (left panel of Fig. 2) are largest when the two nucleons are in the xy-plane (along the z-axis), namely the situation is the reverse of that illustrated in Fig. 1 for TS=01. However, the TS=11 densities are strongly A-dependent, in particular their anisotropy decreases as the number of nucleons increases.

This strong A-dependence is also a feature of the TS=00 two-nucleon densities, as seen in the central panel of Fig. 2. In contrast, the TS=10 density distributions, shown in the right panel of Fig. 2, do display, for separation distances less than 2 fm, very similar shapes. This is not surprising, since the TS=10 interaction is attractive enough to almost support a bound state.

3. Experimental evidence for tensor correlations in nuclei

The best experimental evidence for the presence of tensor correlations in the ground state of nuclei comes from measurements of the deuteron structure functions and tensor polarization by elastic electron scattering [3]. The deuteron is a special case, since for it the two-nucleon density is simply proportional to the (one-body) charge density [1], *i.e.* $\rho_{0,1}^{M=M_S}(\mathbf{r};d) \propto \rho_d^{M=M_d}(\mathbf{r}'=\mathbf{r}/2)$. In essence, the elastic (e,e') measurements have mapped out the Fourier transforms

$$F_{C,M}(q) \propto \int \rho_d^M(\mathbf{r}')e^{iqz'}\mathrm{d}^3\mathbf{r}' \,, \tag{3}$$

showing that they are very different. This is most clearly seen in the tensor polarization observable (left panel of Fig. 3), which corresponds (up to a small magnetic contribution) to the following combination

$$T_{20}(q) \simeq -\sqrt{2}\frac{F_{C,0}^2(q) - F_{C,1}^2(q)}{F_{C,0}^2(q) + 2\,F_{C,1}^2(q)} \,. \tag{4}$$

Fig. 3. Left panel: The deuteron tensor polarization observable. Also shown are the results including relativistic and two-body corrections in the nuclear charge operator. Right panel: Evolution of the ^{8}Be energy with nuclear interactions, from semirealistic (AV4$'$ with no tensor components) to fully realistic models (AV18 and AV18/IL2). Note the energy inversion between the threshold for breakup into two α's and the ^{8}Be energy occurring when tensor forces are taken into account.

Many nuclear properties are strongly influenced by tensor correlations. In addition to the two-nucleon densities discussed in the previous section, these include among others: the high-momentum and high-energy components of momentum distributions $N(k)$ and spectral functions $S(k, E)$ [4]; the strength distribution in response functions to spin-isospin probes injecting momentum and energy into the nucleus [5]; the rates of low-energy capture processes [6]; the ordering of levels in the low-energy spectra of light nuclei (up to mass number A=10) and, in particular, in the observed absence of stable A=8 nuclei [7] (see the right panel of Fig. 3). Yet, the effects of tensor correlations in nuclei with $A > 2$ are subtle and their presence is not easily isolated in the experimental data.

4. Tensor forces and two-nucleon momentum distributions

Tensor correlations also impact strongly the momentum distributions of NN pairs in the ground state of a nucleus and, in particular, lead to large differences in the np versus pp distributions at moderate values of the relative momentum in the pair [8]. These differences should be observable in two-nucleon knockout processes, such as $A(e, e'np)$ and $A(e, e'pp)$ reactions.

The probability of finding two nucleons with relative momentum $\mathbf{q}$ and total momentum $\mathbf{Q}$ in the ground state of a nucleus is proportional to the density

$$\rho^{NN}(\mathbf{q}, \mathbf{Q}) = \frac{1}{2J+1} \sum_{M_J} \langle \psi_{JM_J} \mid \sum_{i<j} P_{ij}^{NN}(\mathbf{q}, \mathbf{Q}) \mid \psi_{JM_J} \rangle , \tag{5}$$

with

$$P_{ij}^{NN}(\mathbf{q}, \mathbf{Q}) \equiv \delta(\mathbf{k}_{ij} - \mathbf{q}) \delta(\mathbf{K}_{ij} - \mathbf{Q}) \, P_{NN}(ij) , \tag{6}$$

and P_{ij}^{NN} projects on either a pp or an np state. The momentum densities defined in Eq. (5) are normalized to the total number of pp or np pairs in the nucleus, depending on whether NN=pp or np.

The np and pp momentum distributions in ^{3}He, ^{4}He, ^{6}Li, and ^{8}Be nuclei are shown in the left panel of Fig. 4 as functions of the relative momentum q at fixed total pair momentum Q=0, corresponding to nucleons moving back to back. They are calculated with the Variational Monte Carlo method, and the associated statistical errors are displayed only for the pp pairs; they are negligibly small for the np pairs. The striking features seen in all cases are: i) the momentum distribution of np pairs is much larger than that of pp pairs for relative momenta in the range 1.5–3.0 fm^{-1}, and ii) for the helium and lithium isotopes the node in the pp momentum distribution is absent in the np one, which instead exhibits a change of slope at a characteristic value of $q \simeq 1.5$ fm^{-1}. The nodal structure is much less prominent in ^{8}Be.

At small values of q the ratios of np to pp momentum distributions are closer to those of np to pp pair numbers, which in ^{3}He, ^{4}He, ^{6}Li, and ^{8}Be are respectively 2, 4, 3, and 8/3.

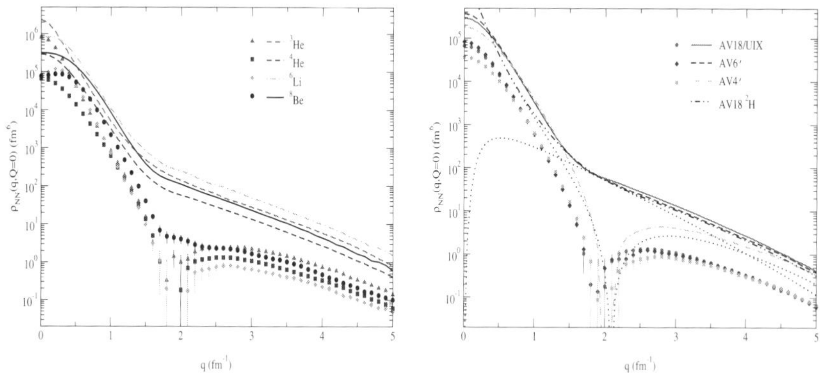

Fig. 4. Left panel: The np (lines) and pp (symbols) momentum distributions in various nuclei as functions of the relative momentum q at vanishing total pair momentum Q. Right panel: The np (lines) and pp (symbols) momentum distributions in ^{4}He obtained with different Hamiltonians. Also shown is the scaled momentum distribution for the AV18 deuteron; its separate S- and D-wave components are shown by dotted lines.

The excess strength in the np momentum distribution is due to the strong correlations induced by tensor components in the underlying NN potential. When Q=0, the pair and residual (A–2) system are in relative S-wave. Hence, in ^{3}He (^{4}He), whose spin,parity quantum numbers are respectively 1/2,+ (0,+), np pairs are in deuteron-like states, while pp pairs are in T=1 and $^1S_0$ (quasi-bound) states. In $A > 4$ nuclei, np and pp pairs are known to be *predominantly* in deuteron-like and quasi-bound states respectively [9], since these are the channels in which the NN potential is most attractive. However, the tensor force vanishes in $^1S_0$ states. Consequently, the np momentum distributions for q values larger than 1.5 fm^{-1} only differ by a scaling factor, and indeed all scale relative to the deuteron momentum distribution as shown in the right panel of Fig. 4. The right panel of Fig. 4 also shows the np and pp momentum distributions in ^{4}He obtained with a series of Hamiltonians, ranging from semirealistic (AV4′ and AV6′ without and with tensor components, respectively) to fully realistic. Note in particular the node which develops in the AV4′ np momentum distribution, due to the purely S-wave nature of the deuteron-like state in this case.

The strong isospin dependence of NN momentum distributions should lead to large differences in the cross sections for back-to-back pp- and np-pair knockout. There are strong indications from a recent analysis of a BNL experiment [10], which measured cross sections for (p, pp) and (p, ppn) processes on ^{12}C in kinematics close to two nucleons being ejected back to back, that this is indeed the case. From this analysis an upper limit to the ratio of pp to np probabilities (denoted respectively as P_{pp} and P_{np}) with (approximately) vanishing total momentum and relative momenta in the range (275–550) MeV/c has been inferred: $P_{pp}/P_{np} \lesssim 0.04^{+0.09}_{-0.04}$. Hopefully, a more precise value for this ratio will become available in the near future, when the analysis of ^{12}C$(e, e'np)$ and ^{12}C$(e, e'pp)$ data, taken at Jefferson Lab, is completed. Experimental confirmation of the smallness of this ratio would provide a direct verification of the crucial role that the tensor force plays in shaping the short-range structure of nuclei.

Acknowledgments

I dedicate this paper to the memory of Vijay R. Pandharipande, a mentor and friend, who was deeply interested in evidence for correlations in nuclei. This is also in remembrance of my colleague and friend Adelchi Fabrocini, who contributed significantly to the theoretical study of correlations in strongly interacting systems.

I wish to thank Joe Carlson, Steve Pieper, and Bob Wiringa, with whom most of the work presend in this talk was carried out. The support of the U.S. Department of Energy, Office of Nuclear Physics, under contract DE-AC05-06OR23177 is gratefully acknowledged.

References

1. J. L. Forest *et al.*, Phys. Rev. C **54**, 646 (1996).
2. J.S. Levinger and H.A. Bethe, Phys. Rev. **78**, 115.
3. A complete list of references is in R. Gilman and F. Gross, J. Phys. **G28**, R37 (2002).
4. R. Schiavilla, V. R. Pandharipande, and R. B. Wiringa, Nucl. Phys. A **449**, 219 (1986); S. C. Pieper, R. B. Wiringa, and V. R. Pandharipande, Phys. Rev. C **46**, 1741 (1992); O. Benhar, A. Fabrocini, and S. Fantoni, Nucl. Phys. A **505**, 267 (1989).
5. A. Fabrocini and S. Fantoni, Nucl. Phys. A **503**, 375 (1989); V. R. Pandharipande *et al.*, Phys. Rev. C **49**, 789 (1994).
6. A. Arriaga, V. R. Pandharipande, and R. Schiavilla, Phys. Rev. C **43**, 983 (1991); L.E. Marcucci *et al.*, Nucl. Phys. A **777**, 111 (2006).
7. R. B. Wiringa and S. C. Pieper, Phys. Rev. Lett. **89**, 182501 (2002).

8. R. Schiavilla *et al.*, Phys. Rev. Lett. **98**, 132501 (2007).
9. R.B. Wiringa, Phys. Rev. C **73**, 034317 (2006).
10. E. Piasetzky *et al.*, Phys. Rev. Lett. **97**, 162504 (2006).

CORRELATED NUCLEONS IN k AND r-SPACE

INGO SICK
Dept. für Physik und Astronomie, Universität Basel, Switzerland
E-mail: Ingo.Sick@unibas.ch

The role of correlations in nuclei, induced by the short-range repulsion of the central term of the N-N interaction and the N-N tensor force, is very important for the quantitative understanding of nuclei. The number of correlated nucleons and their distribution in k, E-space now have been measured, and found to be in good agreement with the predictions of theory.

Keywords: short-range correlations, spectral function at large k and E

1. Introduction

Mean-field (MF) ideas have played and are playing an important role in nuclear physics. When using effective interactions fitted to selected nuclear properties, many features of nuclei can be explained in this frame work. For some observables, however, MF theories badly fail. The need to employ phenomenological effective interactions also remains unsatisfactory.

In order to describe nuclei on a more fundamental level, one has to start from the nucleon-nucleon (N-N) interaction which is accurately known from N-N scattering. This interaction features a strong repulsion at small internucleon distances for the central component, and has a strong tensor component of also rather short range. The consequence of the presence of these short-range terms: pronounced *short-range correlations.* The hard, *i.e.* short-range, collisions between nucleons scatter these nucleons to states of high momentum and energy. These states are outside the model-space considered by mean-field models.

The short-range interaction thus leads to a partial occupation of orbits lying far above the Fermi energy, and correspondingly causes a partial depletion of the occupation of orbits below the Fermi edge.[1] Modern calculations that include a treatment of these correlations can be performed using Faddeev or Variational techniques, but are presently limited to mass number A≤12. For heavier nuclei the Bethe-Bruckner-Goldstone theory, self-consistent Greens function theory SCGF or the Correlated Basis Function (CBF) approach can be employed. Up to recently these were limited to infinite nuclear matter

In order to illustrate quantitatively the effect of these correlations, we quote the result of the CBF calculation for nuclear matter of Benhar *et al.*:[2]

fraction of nucleons in correlated state	20%
contribution to mean removal energy	37%
contribution to mean kinetic energy	47%

The contributions to mean removal and kinetic energies are very large, thus emphasizing that for a *quantitative* understanding of nuclei mean-field approaches cannot work unless one restricts the attention to *differential* quantities, such as excitation energies of nuclear states or spectroscopic factors, where one can hope that much of the correlation contribution cancels.

The effect of short-range correlations is best seen in the spectral function $S(k, E)$, the quantity that gives the probability to find nucleons of a given momentum k and removal energy E. As a consequence of the correlations, the spectral function shows strength in the ridge at *simultaneously* large momenta k and large removal energies E, see Fig. 1.

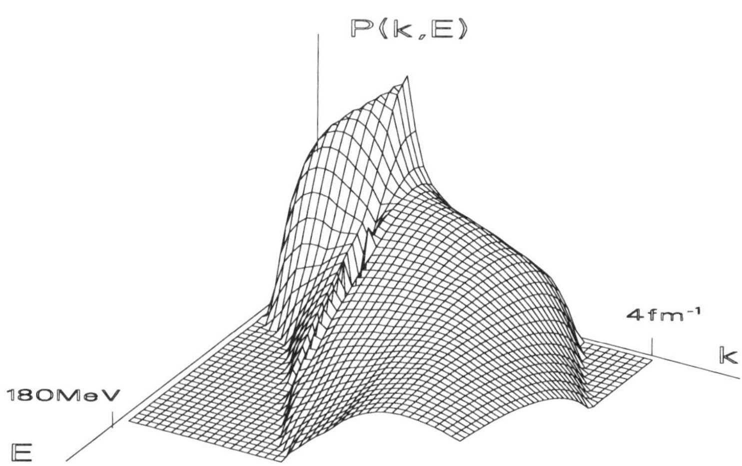

Fig. 1. Log of spectral function $S(k, E)$ of nuclear matter as given by the CBF calculation of Benhar *et al.*[2]

The analogous high-k strength together with a reduced occupation of low-k states is also found for other strongly correlated systems such as liquid helium where calculations can be more easily performed given the simpler structure of the interaction. As the repulsive core of the typical Lennard-Jones inter-atom potential is stronger than in the N-N potential, the effect of correlations is even larger than in nuclei; MF orbits are occupied to ~30% only.

The studies performed[3] for liquid ^{3}He and ^{4}He teach us a second important lesson: the correlations essentially depend on the short-range repulsion in the potential, and are in a minor way only affected by the Fermion/Boson nature of the constituents. This shows that the occurrence of MF orbits in nuclei is not, as often claimed, a consequence of the Pauli principle.

From (e,e'p) data on nuclei measured mainly at NIKHEF,[4] we have had for quite some time now convincing evidence for the depletion of the occupation of states below the Fermi edge. This is documented in Fig. 2 which shows that on average the occupation of MF orbits is ~0.7, not 1.0 as assumed in a MF description. For the most deeply bound states, the occupation is closer to the value of 0.8 as predicted for nuclear matter.

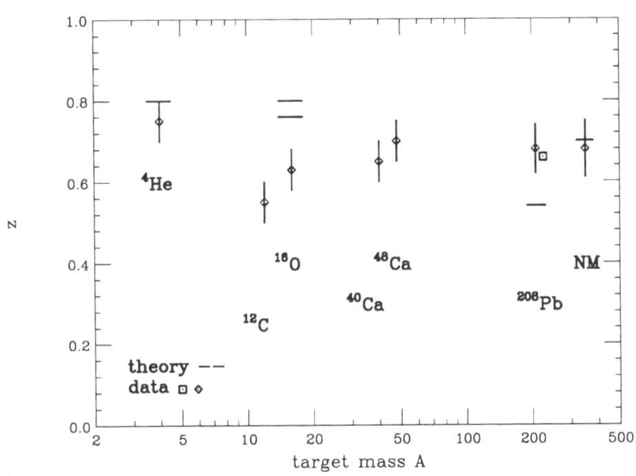

Fig. 2. Occupation of the least-bound MF orbits as function of A.

A direct observation of the strength moved to large k, E, on the other hand, has become available only very recently, with the experiment of Rohe *et al.* to be discussed below.

It is very difficult to directly measure the correlated strength, for two reasons:
– As is shown by Fig. 1 the strength is spread over a large range in E (100-200MeV) and a large range in k (600MeV/c); experiments measure the strength in a given bin of (k, E), and therefore observe a very small fraction of the strength only.
– Multi-step processes involving the recoiling proton are important for even the cleanest of experimental tools we have available, the (e,e'p) reaction. For the vast majority of experimental data published for the region of large k, E, these multi-step processes simulate strength at large k, E that covers up the correlated strength of interest.[5]

Alternative attempts to observe the correlated strength have often aimed at a measurement of the momentum distribution at large k for individual states. The high-k tail of MF momentum distributions, measurable in the standard (e,e'p) to low-lying states of the residual nucleus,[6] obviously misses completely the correlated strength (see Fig.1). Reactions like inclusive (e,e') at large momentum transfer q and low electron energy loss ω (see *e.g.* Ref.[7]) unfortunately are not sensitive either to the large-E ridge which contains most of the correlated strength. In addition, the cross section in the low-ω region depends sensitively on the final state interaction of the knocked out nucleon,[8] thus masking the effect of initial-state correlations. The same is true for reactions such as (γ,p) which are based on the idea to observe high-momentum protons liberated from nuclei using low-momentum probes such as γ's; in this case the final state interaction also masks the high-k properties of the initial state.[9]

A direct measurement of the correlated strength is important, however, if one really wants to understand quantitatively the role of correlations and gain knowledge about the distribution of the correlated strength in k and E.

2. JLab (e,e'p) experiment

Via studies of the reaction kinematics of multi-step reactions together with a systematic study of available (e,e'p) data,[5] it could be shown how multi-step processes can be suppressed to the point of becoming a manageable correction. By performing the (e,e'p) experiment at very large momentum transfer q and in *parallel* kinematics, with the momentum $\vec{k}$ of the initially bound nucleon being parallel to $\vec{q}$, the contributions of unwanted reaction mechanisms can be maximally suppressed. Unfortunately, almost all available (e,e'p) data has been measured in the experimentally much more convenient *perpendicular* kinematics. This kinematics also leads to a maximal (unwanted) contribution of meson exchange currents. This insight has been confirmed by theoretical calculations[10] to be discussed below.

The experiment of Rohe *et al.*,[11] performed in hall C at JLab in parallel kinematics, was designed to best identify this correlated strength. The high duty factor of the JLab accelerator together with the high electron energy (3GeV) allowed one to achieve both acceptable accidental/true coincidence ratios and large q, despite the unfavorable kinematics.

The HMS and SOS spectrometers were used to detect the scattered electron and the recoiling proton. Both the MF region of the spectral function and the correlated region at large k, E were covered. To better understand the role of multi-step reactions, data were taken in both perpendicular and parallel kinematics. The nuclei ^{12}C, ^{27}Al, ^{56}Fe and ^{197}Au were studied; the most extensive data set, to be discussed below, was taken for Carbon.

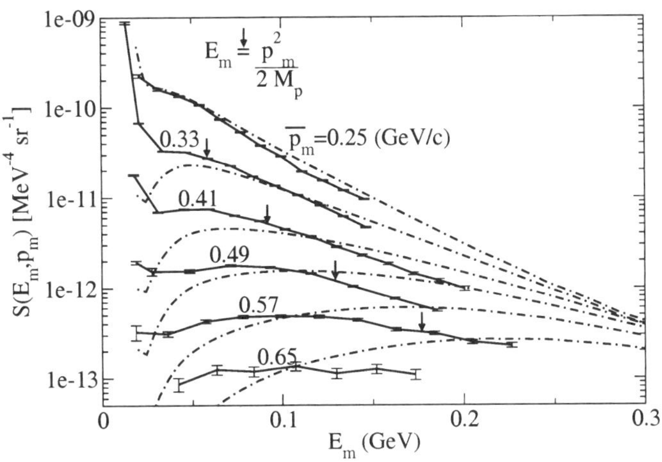

Fig. 3. Spectral function $S(k, E)$ for different bins in k for ^{12}C.

In Fig. 3 we show the experimental spectral function $S(k, E)$ for different bins of the initial proton momentum k. The data on the upper right-hand part of the figure have been removed as they correspond to Δ-excitation. The peak of the strength visible in the lower, central part of the figure represents the correlated strength, which moves, as expected from Fig. 1, to larger values of E with increasing k. Detailed comparison between data and theory indicates that the strength is peaked at somewhat lower E than expected from the naive argument $E_{max} \sim k^2/2M$, an observation that is not yet entirely understood.

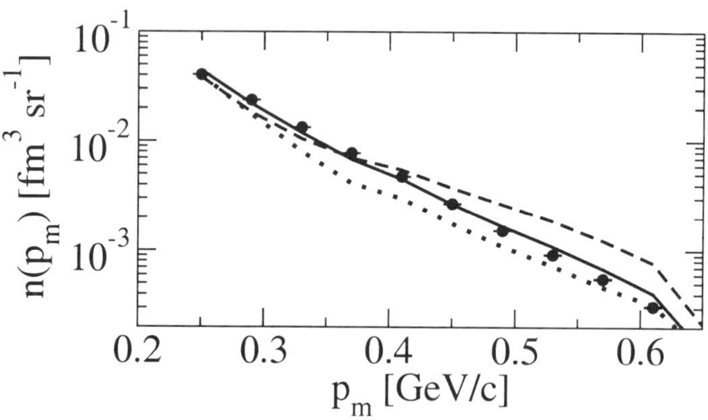

Fig. 4. Momentum distribution of ^{12}C for $k > k_F$. Experiment analyzed using CC (points) or CC1 (solid line), CBF theory (dashed), Greens function theory (dotted).

In Fig. 4 we show the spectral function integrated over E for momenta above the Fermi momentum. The resulting momentum distribution clearly shows the tail toward large k expected from short-range correlations; this tail is in quite good agreement with the predictions from theory. The most

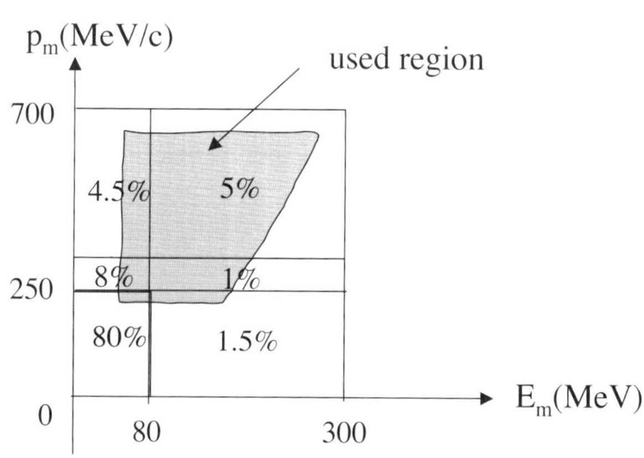

Fig. 5. Region in k and E used to determine the integrated correlated strength.

interesting quantity is perhaps the number of correlated protons, integrated over all k and E. This integral, however, can not be determined directly: both the MF strength at low k, E and the Δ-excitation strength in part of the large k, E-region cover up some of the correlated strength. One can, however, integrate over the "clean" region, displayed in Fig. 5 as "used region"; this region covers about 50% of the correlated strength. The comparison between experiment and theory,[2,12] both integrated over the same region, is shown in the table below.

correlated protons in ^{12}C	used region	total
experiment	0.59	
CBF theory	0.64	1.32
SCGF approach	0.61	1.27

We find very good agreement between data and theory for the k, E-region that can be exploited. This gives us confidence that the total number of correlated protons as predicted by theory, also shown in the table, is correct. In Carbon there are somewhat more than 20% of the protons in the correlated region.

As pointed out above, the experiment has been carried out in parallel kinematics, in order to minimize the effect of multi-step processes of the recoiling nucleon. The comparison of the data from parallel and perpendicular kinematics shows that indeed in the latter there are contributions from processes other than single nucleon knockout which dominate the strength in the region suitable for a measurement of components of large (k, E). C. Barbieri[10] has performed a Monte Carlo calculation in which the recoiling particle is followed through the nucleus and its interactions, quasi-elastic scattering and Δ-excitation, are accounted for. This calculation agrees with the experimental finding that in parallel kinematics multistep processes amount to a manageable correction, while for perpendicular kinematics — the standard one used in most previous (e,e'p) experiments — multistep-processes are so large that the data cannot be used to learn something about the contribution of correlations to the spectral function.

The (e,e'p) data taken in the MF region (dominated by the 1p and 1s shells) also provide a very useful consistency check: the strength in that region is found to amount to 80%, in agreement with the 20% deduced for the correlated strength.

The (e,e'p) experiment of Rohe *et al.* has also taken data for heavier nuclei. The correlated strength found is very similar to the one for ^{12}C, as expected. For the heaviest nucleus, ^{197}Au, the correlated strength appears, however, to be about 30% larger than predicted by a calculation based on the spectral function for (symmetric) nuclear matter calculated for different densities, and applied to nuclei using the local density approximation.[13] As the n-p interaction is known to be mainly responsible for the short-range correlations it is tempting to explain this difference by the fact that for Gold $N > Z$. It however is not yet clear whether $N > Z$ fully explains the enhancement observed for Au.[14] One also might suspect that the multistep interactions of the recoiling proton have not yet been entirely accounted for.

The various results found in this JLab experiment are discussed in more detail in the habilitation work of Daniela Rohe.[15] This work also describes several other aspects that are not touched upon here due to lack of space.

Overall, we can say that the experiment of Rohe *et al.* has for the first time given us a believable measurement of the strength of the correlations induced by the short-range repulsion and tensor force of the nucleon-nucleon interaction. The results found agree quite well with the most reliable theoretical predictions.

3. Correlated strength in r-space

In general, the correlated strength is discussed in k, E-space, where it is, to some degree, separable from the MF strength and thus can be identified. It would be interesting to alternatively look at it in radial space, for two reasons:

– The q-dependence of experimental electromagnetic form factors of single-particle transitions often disagrees with predictions based on radial wave functions fitted to the total nuclear (charge) density. This could be due to the fact that these form factors measure the properties of quasi-particle orbits, while the nuclear density contains the contribution from both quasi-particles and correlated nucleons; these do not necessarily have the same radial shape.

– Mean field calculations of densities in general explain very poorly the nuclear densities in the nuclear interior. This also could be explained if the MF and correlated densities in r-space differ substantially.

It is not obvious a priori what to expect for the distribution of the correlated strength, as one can make two opposing arguments: The presence of more high momentum components, favoring higher angular momentum states, would lead to a shift to larger r. On the other hand, the large values of E connected to correlated nucleons favors a shift to small r.

We have investigated this question for ^{12}C in the following way:[16] The quasi-particle contribution to the density can be determined from the quasiparticle (QP) momentum distributions as measured by (e,e'p) experiments. Transforming these momentum distributions to radial space and summing over all shells yields the quasi-particle density in r-space. The total density[17] can be very accurately determined from elastic electron scattering, with the proton finite size suitably unfolded. The difference between the total (point) density and the QP density then yields the correlated density $\rho_{corr}(r) = \rho_{point}(r) - \rho_{QP}(r)$. In Fig. 6 we show the resulting densities. We find that the correlated density is more concentrated toward the nuclear interior. It gives at $r = 0$ a contribution of ~30% to the total density. This shift toward small radii is due to the fact that the correlated strength appears mainly at large values of E. The important contribution of the correlated density in the nuclear interior explains the problem with QP orbitals obtained from fits of the total density as mentioned above.

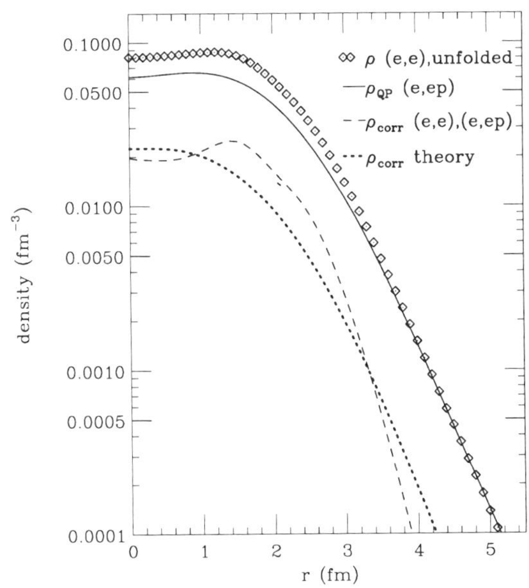

Fig. 6. Point density of ^{12}C (diamonds), experimental QP (dashed) and correlated densities (solid), and SCGF theoretical correlated density (dot-dash).[16]

At very large radii, the correlated density falls off very quickly, and the total (point) density is entirely gives by the QP-contribution. This large-r region is the region where a MF-description can be expected to work quantitatively (as long as the reduced occupation of QP orbits is properly accounted for). This is also the region where experiments employing hadronic (strongly absorbed) probes are the most sensitive.

We also observe in Fig. 6 that the experimental correlated density agrees reasonably well with the one obtained from a calculation based on self-consistent Greens function theory and the N-N interaction.

4. Conclusions

With the experiment of Rohe *et al.* the *correlated strength* has been measured for the first time. The resulting strength agrees quite well with the predictions of theories that describe nuclei starting from the N-N interaction.

The important contribution of correlated nucleons — 20% of the strength, with significantly larger contributions to the average removal and kinetic energies of the nucleons — clearly shows that for a quantitative understanding of nuclei mean-field approaches are inadequate; the short-range correlations between nucleons must be accounted for.

References

1. V. Pandharipande, I. Sick and P. deWitt Huberts, *Rev. Mod. Phys.* **69**, p. 981 (1997).
2. O. Benhar, A. Fabrocini and S. Fantoni, *Nucl. Phys. A* **505**, p. 267 (1989).
3. S. Moroni, G. Senatore and S. Fantoni, *Phys. Rev. B* **55**, p. 1040 (1997).
4. L. Lapikás, *Nucl. Phys. A* **553**, p. 297c (1993).
5. I. Sick, *World Scientific* , p. 445 (1997), Proc. Elba Conf. Electron-Nucleus Scattering, Eds. O. Benhar, A. Fabrocini.
6. I. Bobeldijk *et al.*, *Phys. Rev. Lett.* **73**, p. 2684 (1994).
7. K. Egiyan *et al.*, *Phys. Rev. C* **68**, p. 14313 (2003).
8. O. Benhar, A. Fabrocini, S. Fantoni, G. Miller, V. Pandharipande and I. Sick, *Phys. Rev. C* **44**, p. 2328 (1991).
9. R. Amado and R. Woloshyn, *Phys. Rev. Lett.* **36**, p. 1435 (1976).
10. C. Barbieri, D. Rohe, I. Sick and L. Lapikas, *Phys. Lett. B* **608**, p. 47 (2005).
11. D. Rohe *et al.*, *Phys. Rev. Lett.* **93**, p. 182501 (2004).
12. T. Frick, K. Hassaneen, D. Rohe and H. Müther, *Phys. Rev. C* **70**, p. 24309 (2004).
13. I. Sick, S. Fantoni, A. Fabrocini and O. Benhar, *Phys. Lett. B* **323**, p. 267 (1994).
14. T. Frick, H. Müther, A. Rios, A. Polls and A. Ramos, *Phys. Rev. C* **71**, p. 14313 (2005).
15. D. Rohe, *Habilitationsschrift, Universität Basel* (2004).
16. H. Müther and I. Sick, *Phys. Rev. C* **70**, p. 41301 (2004).
17. I. Sick and J. McCarthy, *Nucl. Phys. A* **150**, p. 631 (1970).

ROLES OF ALL-ORDER CORE POLARIZATIONS AND BROWN-RHO SCALING IN NUCLEAR EFFECTIVE INTERACTIONS

T.T.S. KUO*, J.W. HOLT, G.E. BROWN

Department of Physics and Astronomy, Stony Brook University,
Stony Brook, NY 11794, USA
**E-mail: kuo@tonic.physics.sunysb.edu*

J.D. HOLT

Theory Group, Physics Division, TRIUMF,
Vancover, Canada

R. MACHLEIDT

Physics Department, University of Idaho,
Moscow, Idaho 83844, USA

We carry out an all-order summation of the Kirson-Babu-Brown (KBB) core polarization diagrams for the sd-shell effective interaction. The calculation of the self-consistent induced interactions is simplified by the use of a Green's function formalism together with the energy-independent low-momentum interaction V_{low-k}. Renormalizations of the particle-hole and particle-particle vertex functions are both included. The effective interactions calculated with the all-order KBB core polarization and those with the polarization in second-order perturbation theory are found to be remarkably similar to each other, typical difference between them being less than 10%. We also study the effect of Brown-Rho scaling on the nuclear effective interactions. As the density of the nuclear medium increases, the tensor interaction is significantly weakened by the dropping of the ρ-meson mass. This behavior of the tensor force is found to be important for nuclear matter saturation.

1. Introduction

It is indeed a great pleasure for me (TTSK) to come to this beautiful conference. I am very grateful to Prof. Covello and the organizers for inviting me here. Today, I shall report on two topics: all-order core polarizations[1] and Brown-Rho scaling for in-medium mesons.[2,3] Both topics have close connections with the low-momentum NN interaction V_{low-k} which has been a widely studied topic in nuclear physics.[4–10] This interaction is derived using renormalization group methods for two nucleons in free space. When using it in nuclear many body problems, V_{low-k} should be further renormalized to take care of the medium renormalization effects. There are mainly two types of such medium effects: First the core polarization effect on the effective interaction between two valence nucleons. Next is the medium effect on mesons by which the NN interaction V_{NN} is mediated. According to Brown-Rho scaling, the in-medium meson mass, particularly the mass of the ρ meson, is significantly different from that in free space. Thus V_{NN} in medium may be importantly different from that in free space.

Before discussing the above topics in some details, let me first briefly review V_{low-k}. In nuclear physics, there are a number of high-precision V_{NN} models;[11–14] they all reproduce the low-energy NN scattering phase shifts and deuteron properties very well. When applying such V_{NN} to nuclear many body problems, there are, however, two well known difficulties: The first difficulty concerns the strong repulsive core contained in V_{NN}; to take care of this repulsive core one needs to use either the Brueckner G-matrix approach or a correlated basis wave function approach such as that of Jastrow. Both are fairly complicated. The second difficulty is about the non-uniqueness of V_{NN}.

By comparing the momentum-space matrix elements of the above potential models, it is readily seen that these potentials are vastly different from each other,[9] despite the fact that they all give nearly identical results for the NN phase shifts and deuteron. This is clearly not a desirable situation, as, in principle, we should have a unique NN potential. We are confronted with the ambiguity in deciding which of the above potential models is the "correct" one to be used in nuclear many body calculations.

In the past several years, there has been much progress in treating low energy nuclear physics using the renormalization group (RG) and effective field theory (EFT) approach,[15–17] which is the basis for constructing V_{low-k}. Briefly speaking, V_{low-k} is obtained by integrating out the high momentum components of modern realistic NN potentials V_{NN}. This "integrating out" is performed under an important requirement, namely certain low-energy physics must be preserved by this integration procedure. Specifically, we require that the deuteron binding energy and low energy phase shifts, below a cut-off scale Λ, of V_{NN} are preserved by V_{low-k}. This preservation may be satisfied by the following T-matrix equivalence approach.[4–9]

We start from the half-on-shell T-matrix

$$T(k', k, k^2) = V_{NN}(k', k) + \int_0^\infty q^2 dq V_{NN}(k', q) \frac{1}{k^2 - q^2 + i0^+} T(q, k, k^2), \tag{1}$$

where notice that the intermediate state momentum q is integrated from 0 to ∞. We then define an effective low-momentum T-matrix by

$$T_{low-k}(p', p, p^2) = V_{low-k}(p', p) + \int_0^\Lambda q^2 dq V_{low-k}(p', q) \frac{1}{p^2 - q^2 + i0^+} T_{low-k}(q, p, p^2), \tag{2}$$

noticing that the intermediate state momentum is integrated from 0 to Λ, the momentum space cut-off. We require the above T-matrices to satisfy the condition

$$T(p', p, p^2) = T_{low-k}(p', p, p^2); \ (p', p) \leq \Lambda. \tag{3}$$

The above equations define the effective low momentum interaction V_{low-k}. The iteration method of Lee-Suzuki-Andreozzi[18,19] is commonly used in calculating V_{low-k} from the above T-matrix equivalence equations.

We now discuss the choice of the cut-off momentum Λ. Realistic NN potential models all contain a number of parameters, which are adjusted to fit the NN scattering phase shifts up to $E_{lab} \approx 350$ MeV. This E_{lab} corresponds to relative momentum ~ 2.05 fm^{-1}. Clearly, beyond this momentum there is no experimental input in constraining the parameters contained in the NN potentials. This suggests that we should use a cut-off momentum of $\Lambda \sim 2$ fm^{-1}, since the high ($> \Lambda$) momentum components of the NN potential are not at all certain; they are model dependent and can not be determined by low energy scatterings. V_{low-k} is obtained by integrating out those high momentum components.

As mentioned earlier, there are large differences between the various V_{NN} models. But for $\Lambda \sim 2$ fm^{-1}, the V_{low-k} derived from them are almost identical, suggesting a nearly unique low-momentum NN interaction.[9] This result is encouraging and is not a surprise in retrospect. The differences between the various potentials come mainly from the different short range repulsions contained in them, other parts of them being very similar. The short range (high momentum) components of them are not constrained by low energy phase shifts, rather they are essentially put in "by hand". When we integrate out those high-momentum components of V_{NN}, we are basically filtering out the short range parts of V_{NN} which are not essential for low energy physics. The resulting V_{low-k}'s thus become nearly identical to each other.

2. All-order core polarization diagrams

Comparing with V_{NN}, V_{low-k} is clearly more convenient to work with. It is a smooth potential (no hard core), and can be used directly in shell model calculations without having to first calculate the G-matrix which is rather complicated; the G-matrix is both energy and Pauli operator dependent while V_{low-k} is not. In addition, V_{low-k} is essentially unique as discussed earlier.

When applying V_{low-k} to open-shell nuclei such as ^{18}O and ^{210}Po,[20] it is essential to include the medium renormalization effect due to core polarization. Here the nucleus is treated as consisting of a small number of valence nucleons outside a closed core; these valence nucleons interact through a renormalized effective interaction V_{eff}, which has mainly two components: the direct interaction between the valence nucleons and the indirect interaction mediated by core polarization (CP). The importance of the CP effect has been recognized for a long time.[21–23] For example, the spectrum of ^{18}O given by V_{low-k} alone is too compressed, while the spectrum given by (V_{low-k}+ CP) agrees remarkably well with the experimental spectra.[22] The well-known quadrupole-quadrupole (P_2) effective interaction is also well reproduced by CP.[23] It has to be noted that in most calculations reported in the literature, the CP contribution was calculated using second-order perturbation method, namely only the second-order CP diagram was included. Naturally, there was the concern about the importance of the higher-order CP diagrams.[24–28] We shall discuss that with the V_{low-k} interaction an all-order summation of the CP diagrams can be carried out using the Kirson-Babu-Brown (KBB) induced interaction approach; in fact we have found that the results of all-order and second-order CP calulations are remarkably close to each other.

The KBB induced interaction approach provides an elegant way for summing large classes of planar diagrams. It is a self consistent approach. To illustrate, let us consider the particle-hole (ph) irreducible vertex f. As displayed in Fig. 1, f has the general form $f = V + V_{CP}$, where V, usually called the driving term, is the first-order ph diagram and V_{CP} is the core polarization term representing all the other diagrams on the RHS of the equation. Note the V_{CP} itself is composed of f vertices, each of which is the same as the f on the LHS of the equation. Thus this equation is a self-consistent equation; the condition for convergence is that the f's on

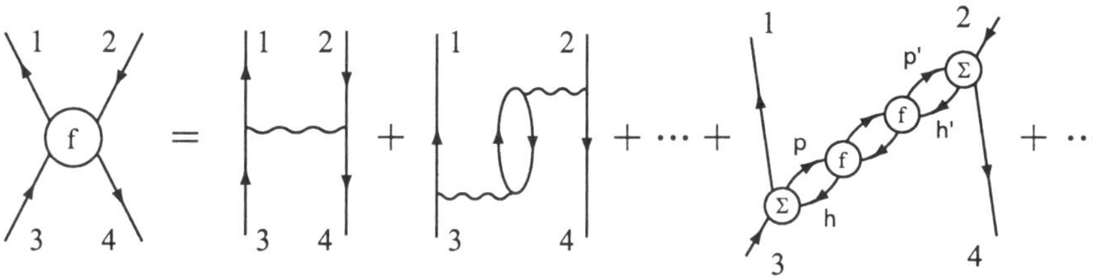

Fig. 1. Self consistent equation for the ph vertex f.

both sides of the equation become the same. The vertex function f of Fig. 1 is given by

$$f = V + \Sigma g_{ph} \Sigma + \Sigma g_{ph} f g_{ph} \Sigma + \Sigma g_{ph} f g_{ph} f g_{ph} \Sigma + \cdots , \tag{4}$$

where g_{ph} is the free ph Green function, and Σ denotes the vertex for particle-core and hole-core coupling which shall be discussed later.

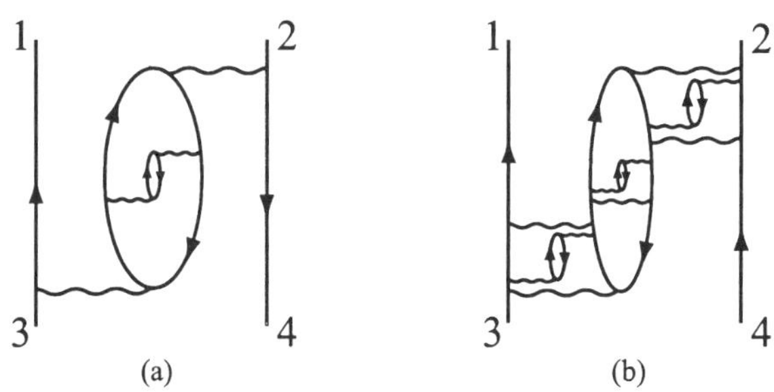

Fig. 2. Typical diagrams generated by KBB equations.

Note that Eq. (4) is not a linear equation for f; its solution is complicated and is usually carried out by iteration. Let us define the ph Green's function

$$G_{ph} = g_{ph} + g_{ph} f G_{ph}. \tag{5}$$

Then Eq(4) is simplified to

$$f = V + \Sigma G_{ph} \Sigma. \tag{6}$$

Diagrams of the types shown in Fig. 2 are all included in our calculation and are summed to all orders. Diagram (a) of the figure is a high-order diagram of the ph vertex f. Diagram (b) is a particle-particle (pp) diagram, belonging to the pp vertex Γ. We employ a Green's function approach where the true ph and pp Green's functions are the basic building blocks for the CP diagrams. The structure of Γ is similar to the ph vertex f of Fig. 1; it is given by

$$\Gamma = V + \Sigma G_{ph} \Sigma. \tag{7}$$

The particle-core and hole-core coupling vertex is given by

$$\Sigma = V + V G_{ph} V + V G_{pp} V, \tag{8}$$

where G_{pp} is the pp Green's function

$$G_{pp} = g_{pp} + g_{pp} \Gamma G_{pp}. \tag{9}$$

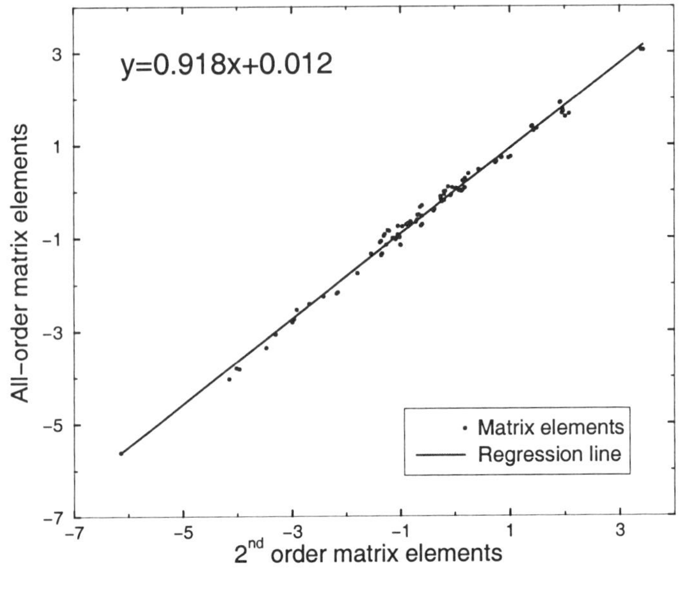

Fig. 3. Comparison of sd-shell matrix elements calculated with all-order and 2nd-order CP.

The self-consistent vertex functions are obtained from the solutions of the above set of five coupled equations. We have solved them using an iteration scheme: Starting from the nth order vertex functions $f^{(n)}$ and $\Gamma^{(n)}$, we calculate the nth order Green's function $G_{ph}^{(n)}$ and $G_{pp}^{(n)}$. Using them, we then calculate the (n+1)th order vertex functions $f^{(n+1)}$ and $\Gamma^{(n+1)}$. That V_{low-k} is energy independent has largely simplified the calculation, comparing with the case of using a G-matrix interaction which is energy dependent.

In Fig. 3 the 63 sd-shell matrix elements $\langle abJT|V_{eff}|cdJT\rangle$ calculated with the above KBB all-order CP are compared with those obtained from the second-order CP.[1] As seen, the all-order matrix elements are remarkably close to the second-order ones, the former being about 10% weaker than the latter. It is of interest that our result is in difference with early calculations[25] where the net effect of all-order CP is vanishingly small. A main difference between our calculations and those of[25] concerns the interaction employed; we use V_{low-k} while a G-matrix interaction was used in the latter. Another difference is in the treatment of the valence-core coupling vertex; it is given by Eq. (8) in our work while in[25] the *pp* ladders were not included for this vertex. Our all-order CP effect has been found to be essential for attaining good agreement between theoretical and experimental spectra.[1]

3. Brown-Rho scaling and nuclear effective interactions

Nucleon-nucleon interactions are mediated by meson exchanges, and clearly the in-medium modifications of meson masses are important for such interactions.[29] These modifications could arise from the partial restoration of chiral symmetry at finite density/temperature or from traditional many-body effects. Particularly important are the vector mesons, for which there is now strong evidence from both theory[3,30–32] and experiment[33,34] that the masses decrease on the level of $\sim 10-15\%$ at normal nuclear matter density and zero temperature. This in-medium decrease of meson masses is often referred to as the Brown-Rho scaling.[3] For densities below that of nuclear matter, it is expected[30] that the masses decrease linearly with the density n:

$$\frac{m_V^*}{m_V} = 1 - C\frac{n}{n_0}, \tag{10}$$

where m_V^* is the vector meson mass in-medium and C is a constant.

We study the consequences for nuclear many-body calculations by replacing the NN interaction in free space with a density-dependent interaction with medium-modified meson exchanges. Several versions of medium-modified NN potentials have been employed. A simple way to obtain such potentials is by modifying the meson masses and relevant parameters of the one-boson-exchange NN potentials (e.g. Bonn and Nijmegen). We have also employed the medium-dependent multi-boson potentials of Rapp et al.[29] Nuclear matter is particularly well-suited for studying the effects of

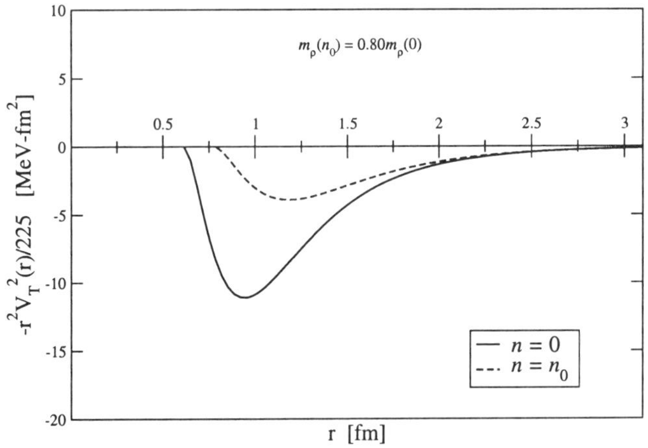

Fig. 4. The second-order tensor force from π and ρ-meson exchange at zero density and nuclear matter density, assuming that the value of C in Eq. (10) is 0.20.

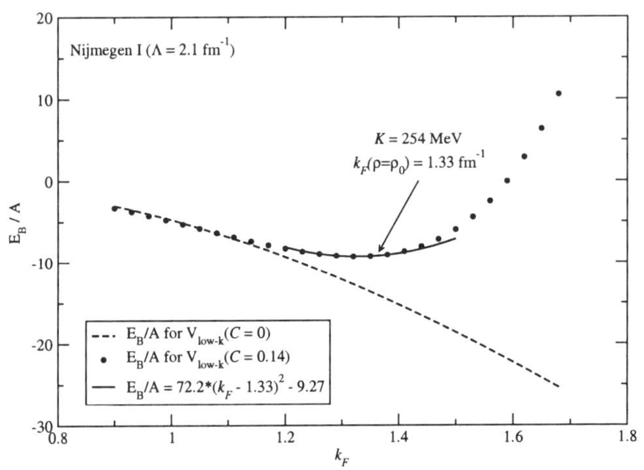

Fig. 5. Hartree-Fock calculation of the binding energy per nucleon using the medium-modified Nijmegen I potential.

dropping masses.[2,29] Here the density is uniform, and so there is no ambiguity in how to incoporate the scaling. It is well known that low momentum interactions without three-body forces saturate nuclear matter at too high a density and too high an energy. However, a large contribution to the binding energy comes from the second order tensor force. The two most important contributions to the tensor force come from π and ρ-meson exchange, which act opposite to one another:

$$V_\rho^T(r) = \frac{f_\rho^2}{4\pi} m_\rho \tau_1 \cdot \tau_2 \left(-S_{12} \left(\frac{1}{(m_\rho r)^3} + \frac{1}{(m_\rho r)^2} + \frac{1}{3 m_\rho r} \right) e^{-m_\rho r} \right) \quad (11)$$

$$V_\pi^T(r) = \frac{f_\pi^2}{4\pi} m_\pi \tau_1 \cdot \tau_2 \left(S_{12} \left(\frac{1}{(m_\pi r)^3} + \frac{1}{(m_\pi r)^2} + \frac{1}{3 m_\pi r} \right) e^{-m_\pi r} \right) \quad (12)$$

Since the ρ meson is expected to decrease in mass at finite density while the pion mass is generally protected by chiral invariance, the second-order tensor force should decrease as the density increases, as shown in Fig. 4.

In Fig. 5 we show the saturation properties of nuclear matter in Hartree-Fock approximation using an unmodified $V_{\text{low}-k}$ and a medium-modified low-momentum Nijmegen I potential in which the vector meson masses were scaled by 14% at nuclear matter density. Also, the mass of the light scalar σ particle, treated as correlated 2π exchange in which the two pions interact essentially through t-channel ρ-meson exchange, should receive a residual mass modification from a decreased ρ mass. Based on the work of Rapp *et al.* we decrease the σ mass less than the vector masses, in this case by 7% at nuclear matter density. The result of these in-medium corrections is to saturate nuclear matter at approximately the correct density but at too low a binding energy. In principle, one could also calculate corrections from three-body forces, but it has been argued[2] that the use of medium-modified meson masses implicitly accounts for 3N forces at least partially.

In summary, we have studied two medium renormalization effects for the nuclear effective interactions. For valence nucleons, the CP diagrams are important, and can be summed to all orders using the Kirson-Babu-Brown induced interaction approach. Our calculations indicate that the commonly used second-order approximation reproduces well the all-order CP result. The in-medium modification of the meson masses has significantly affected the NN tensor force and is important for nuclear saturation. The effects of this modification on other nuclear properties are being studied and will be reported in a future publication.

References

1. J.D. Holt, J.W. Holt, T.T.S. Kuo, G.E. Brown and S.K. Bogner, Phys. Rev. C72 041304(R) (2005).
2. J.W. Holt, G.E. Brown, J.D. Holt and T.T.S. Kuo, Nuc. Phys. A785 (2007) 322.
3. G.E. Brown and M. Rho, Phys. Rev. Lett. **66** (1991) 2720.
4. S. Bogner, T.T.S. Kuo and L. Coraggio, Nucl. Phys. **A684**, 432c (2001).
5. S. Bogner, T.T.S. Kuo, L. Coraggio, A. Covello and N. Itaco, Phys. Rev. **C65** , 051301(R) (2002).
6. T.T.S. Kuo, S. Bogner and L. Coraggio, Nucl. Phys. **A704**, 107c (2002).
7. L. Coraggio, A. Covello, A. Gargano, N. Itako, T.T.S. Kuo, D.R. Entem and R. Machleidt, Phys. Rev. **C66** , 021303(R) (2002).
8. A. Schwenk, G.E. Brown and B. Friman, Nucl. Phys. **A703**, 745 (2002).
9. S. K. Bogner, T. T. S. Kuo, and A. Schwenk, Phys. Rep. **386**, 1 (2003).
10. A. Schwenk, report in this proceedings.
11. R. Machleidt, Phys. Rev. **C63**, 024001 (2001).
12. V.G.J. Stoks et al., Phys. Rev. **C49**, 2950 (1994).
13. R. B. Wiringa et al., Phys. Rev. **C51**, 38 (1995).
14. D.R. Entem, R. Machleidt and H. Witala, Phys. Rev. **C65**, 064005 (2002).
15. P. Lepage, "How to Renormalize the Schroedinger Equation", (1997) [nuc-th/9706029]
16. E. Epelbaum, W. Glöckle, and Ulf-G. Meissner, Nucl. Phys. **A637**, 107 (1998).
17. P. Bedaque et. al. (eds.), Nuclear Physics with Effective Field Theory II, (1999) World Scientific Press.
18. K. Suzuki and S. Y. Lee, Prog. Theor. Phys. **64**, 2091 (1980).
19. F. Andreozzi, Phys. Rev. **C54**, 684 (1996).
20. L. Coraggio, A. Covello, A. Gargano, N. Itaco, T.T.S.Kuo, D.R. Entem and R. Machleidt, Phys. Rev. C66, 021303(R) (2002).
21. G. F. Bertsch, Nucl. Phys. **74**, 234 (1965).
22. T. T. S. Kuo and G. E. Brown, Nucl. Phys. **85**, 40 (1966).
23. G. E. Brown, *Unified Theory of Nuclear Models and Forces* (North-Holland, Amsterdam, 1971).
24. B. R. Barrett *et al.*, Nucl. Phys. **A148**, 145 (1970).
25. M. W. Kirson, Ann. Phys. **66**, 624 (1971); *ibid.* **68**, 556 (1971); *ibid.* **82**, 345 (1974).
26. J. P. Vary *et al.*, Phys. Rev. C **7**, 1776 (1973).
27. D. W. L. Sprung *et al.*, Can. J. Phys. **50**, 2768 (1972).
28. M. Hjorth-Jensen, T. T. S. Kuo, and E. Osnes, Phys. Rep. **261**, 126 (1995).
29. R. Rapp, R. Machleidt, J.W. Durso, and G.E. Brown, Phys. Rev. Lett. **82** (1999) 1827.
30. T. Hatsuda and S.H. Lee, Phys. Rev. C **46** (1992) R34.
31. M. Harada and K. Yamawaki, Phys. Rept. **381** (2003) 1.
32. S. Klimt, M. Lutz, and W. Weise, Phys. Lett. B **249** (1990) 386.
33. D. Trnka *et al.*, Phys. Rev. Lett. **94** (2005) 192303.
34. M. Naruki *et al.* Phys. Rev. Lett. **96** (2006) 092301.

AB INITIO AND *AB EXITU* NO CORE SHELL MODEL

J. P. VARY*

Department of Physics and Astronomy, Iowa State University, Ames, IA 50011, USA
**E-mail: jvary@iastate.edu*

P. NAVRÁTIL, V. G. GUEORGUIEV, W. E. ORMAND

Lawrence Livermore National Laboratory, P.O. Box 808, L-414, Livermore, CA 94551, USA

A. NOGGA

Forschungszentrum Jülich, Institut für Kernphysik (Theorie), D-52425 Jülich, Germany

P. MARIS

Department of Physics and Astronomy, Iowa State University, Ames, IA 50011, USA

A. SHIROKOV

Skobeltsyn Institute of Nuclear Physics, Moscow State University, Moscow, 119991 Russia

We outline two complementary approaches based on the no core shell model (NCSM) and present recent results. In the *ab initio* approach, nuclear properties are evaluated with two-nucleon (NN) and three-nucleon interactions (TNI) derived within effective field theory (EFT) based on chiral perturbation theory (ChPT). Fitting two available parameters of the TNI generates good descriptions of light nuclei. In a second effort, an *ab exitu* approach, results are obtained with a realistic NN interaction derived by inverse scattering theory with off-shell properties tuned to fit light nuclei. Both approaches produce good results for observables sensitive to spin-orbit properties.

Keywords: Light nuclear properties; Chiral effective field theory; Inverse scattering potentials, Many-body theory.

1. Introduction

Recent advances in microscopic many-body methods have opened new paths for investigating both the strong interaction itself as well as the many facets of nuclear phenomena evident in light nuclei. Once the *ab initio* no core shell model (NCSM) was introduced and shown to be reasonably convergent,[1] opportunities emerged for precision testing of the properties of the strong interaction in the nuclear medium. Our focus in this presentation will be on two recent and complementary efforts to determine important features of the strong interaction through the resulting properties of nuclei in the p-shell. Selected observables are especially sensitive to the three-nucleon interaction (TNI) and to the off-shell properties of the NN interaction.

In the first approach,[2] we invoke the power of chiral perturbation theory (ChPT)[3] that provides a promising bridge with the accepted relativistic quantum field theory of the strong interactions, QCD. Beginning with the pionic or the nucleon-pion system[4] one works with systems of increasing nucleon number.[5-7] One makes use of the explicit and spontaneous breaking of chiral symmetry to systematically expand the strong interaction in terms of a characteristic small momentum of a few hundred MeV/c divided by the chiral symmetry breaking scale of about 1 GeV/c. Results should be trustworthy for observables dominated by momentum scales below this characteristic small momentum and thus, we expect the derived interactions to be valid for low-energy nuclear properties.

We adopt the potentials of ChPT at the orders presently available, N3LO for the NN interaction[8] and N2LO for the TNI[9] . The ChPT expansion divides the interactions into perturbative and non-perturbative elements. The latter are represented by a finite set of constants at each order of perturbation theory that are not presently calculable from QCD but can be fixed by measured properties of nuclei provided the many-body methods are sufficiently accurate. Once the non-perturbative constants are determined, the resulting Hamiltonian predicts, in principle, all other nuclear properties, including those of heavier nuclei with no residual freedom. We refer to this first effort as an application of the *ab initio* NCSM since the NN interaction is completely fixed by properties of the two-body system. Important components of TNI and higher-body interactions are also fixed by the ingredients of the NN terms. Only residual non-perturbatve chiral TNI couplings are fixed, as necessary, by the properties of nuclei beyond $A = 2$. Eventually, independent methods such as lattice QCD should fix all these parameters and complete the *ab initio* NCSM so the need for fitting would be eliminated.

In the second approach[10] the NN interaction is taken as a finite rank separable form in an oscillator representation for each partial wave. The coefficients are determined, to the extent possible, by inverse scattering techniques[11] using the available NN data. Subsequently, one investigates off-shell freedoms with phase-shift equivalent unitary transformations to tune the interaction to fit the properties of light nuclei. By fitting the ^3He and ^{16}O binding energies and the ^6Li low-lying spectra we obtain the interaction "JISP16", which represents "J-matrix Inverse Scattering Potential tuned up to ^{16}O". We achieve soft interactions with this approach that describe all the data conventionally fit by realistic NN interactions and provide good fits to light nuclear properties.[10] We consider this as a phenomenological approach designed to explore regions of NN interactions that are not yet explored by other methods. We hope that the phase-shift equivalent transformation methods that prove successful in this *ab exitu* NCSM will be useful for minimizing higher-body forces in other approaches. This would be helpful for gaining access to heavier nuclei within the NCSM.

In many instances we find the results of both approaches to be similar and we cite the example of ^{10}B in the present work.

2. No core shell model (NCSM)

The NCSM casts the diagonalization of the infinite dimensional many-body Hamiltonian matrix as a finite matrix in a harmonic oscillator (HO) basis with an equivalent "effective Hamitonian" derived from the original Hamiltonian.[1] The finite matrix is defined by N_{max}, the maximum number of oscillator quanta shared by all nucleons above the lowest configuration. We solve for the effective Hamiltonian by approximating it as either a 3-body interaction[12] based on our chosen NN+TNI from ChPT (our *ab initio* application) or a 2-body interaction based on JISP16 (our *ab exitu* application). With these "cluster approximations", convergence is guaranteed with increasing N_{max} or with increased cluster size at fixed N_{max}.[1]

The NCSM is the only approach currently available to solve the resulting many-body Schrödinger equation for mid-p-shell nuclei while preserving all symmetries when the interactions are non-local, a feature of all the interactions employed in this work.

3. *Ab initio* NCSM with interactions from ChPT

In order to motivate the inclusion of the TNI, we begin by showing, in Fig. 1, the natural parity excitation spectra of ^{10}B with the ChPT N3LO NN

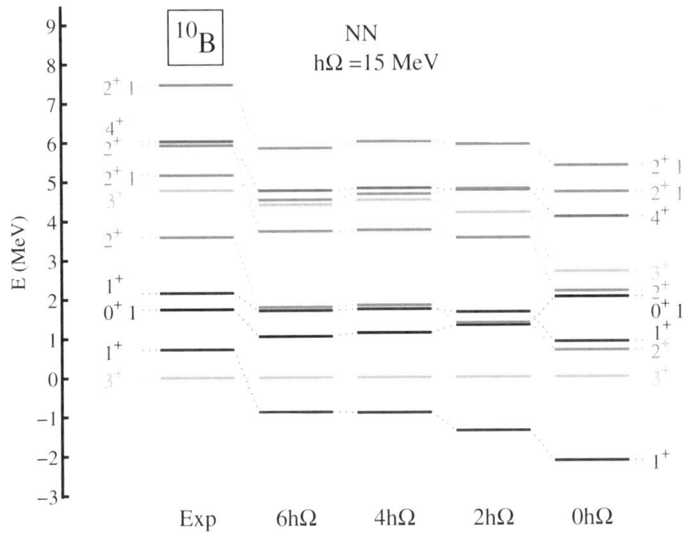

Fig. 1. Experimental and theoretical excitation spectra of ^{10}B with respect to the lowest 3^+ state. The NCSM results are obtained with the chiral N3LO potential[8] at an oscillator energy, $\hbar\Omega = 15\ MeV$ as a function of $N_{max}\hbar\Omega$, indicated at the bottom of each spectrum. Note the reasonable convergence as one proceeds up to $N_{max} = 6$ where the dependence on $\hbar\Omega$ (not shown here) is found to be weak.

interaction alone (excluding TNI), using our 3-body cluster renormalization to the finite basis specified by N_{max}. The figure displays results for basis spaces from $N_{max} = 0 - 6$. We note the now-accepted defect with conventional realistic NN interactions: theory and experiment differ by an inversion of the two lowest levels. In addition, the theory spectrum is somewhat compressed relative to experiment. Over the past few years, these deficiencies, as well as others in mid-p-shell nuclei, such as binding energies, spectral properties and electromagnetic transition rates, have been ascribed to the need for TNI's. In terms of physics, the inadequacy of the realistic NN interactions appears as insufficient spin-orbit splitting in the mean field generated by those interactions, though the mean field itself is not calculated directly. We summarize here the role of the TNI and the role of off-shell modified NN interactions in correcting this inadequacy.

We define the two non-perturbative coupling constants of the TNI, not fixed by 2-body data, as C_D (C_E), the strength of the $N - \pi - NN$ (TNI)

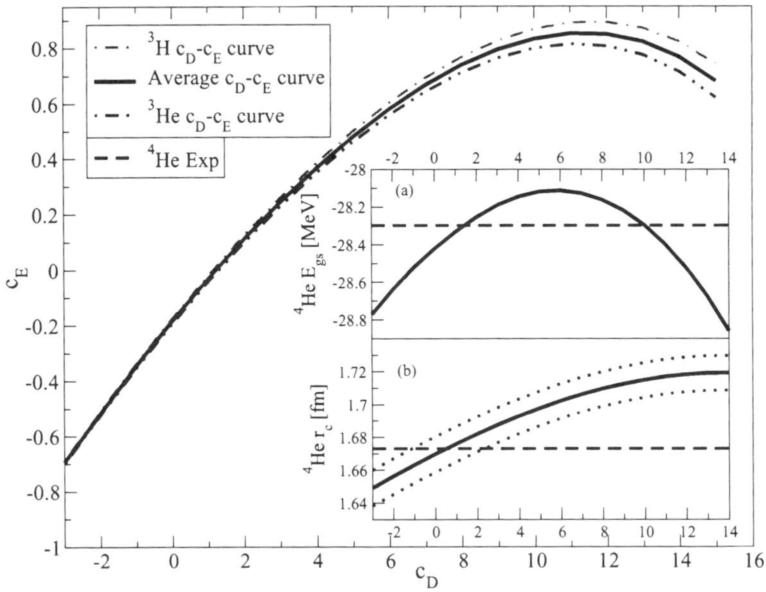

Fig. 2. Relations between C_D and C_E for which the binding energy of ^{3}H (8.482 MeV) and ^{3}He (7.718 MeV) are reproduced. (a) ^{4}He binding energy along the averaged curve. The experimental ^{4}He binding energy (28.296 MeV) defines two points of intersection using the averaged $A = 3$ $C_D - C_E$ curve. (b) ^{4}He charge radius. Dotted lines represent the spread in r_c due to uncertainties in the proton charge radius.

contact term. Fig. 2 shows the trajectories of these two parameters as determined from fitting the binding energies of the $A = 3$ & 4 systems as well as the average of the two trajectories. Our approach is similar to the one used in a detailed investigation[13] of ^7Li. The ^4He results use the average of the $A = 3$ fits and the inset shows two crossing points where the ^4He binding is reproduced. Note the expanded scale. The second inset (b) depicts the corresponding rms charge radius of ^4He.

Our results on the radii of the $A = 3$ systems are in good agreement with experiment as well. While the uncertainties in the ^3H and ^3He charge radii obscure the differences between the intersection points, the ^4He charge radius (inset (b) of Fig. 2) indicates a preference for $C_D \sim 0$ with a broad span of reasonable results around it. This led us to investigate observables in the mass 10-13 range where we find good results for $C_D \sim -1$.[2]

An example of the improvement obtained with the TNI of ChPT ($C_D = -1$, $C_E = -0.33$) is shown in Fig. 3 as compared/contrasted with Fig.

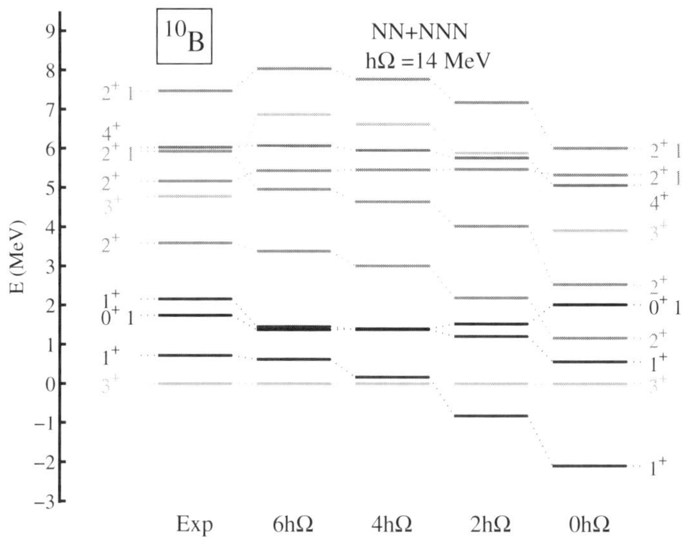

Fig. 3. Experimental and theoretical excitation spectra of ^{10}B with respect to the lowest 3^+ state. The NCSM results are obtained with the chiral N3LO potential[8] at an oscillator energy, $\hbar\Omega = 14\ MeV$ as a function of N_{max} indicated at the bottom of each spectrum. Note the reasonable convergence again as one proceeds up to $N_{max} = 6$

1. The correct level ordering is now obtained and the spectrum is more spread out in closer agreement with experiment. In addition, the binding energy shifts towards agreement with experiment as seen below in Table 1. Refinements in our NCSM techniques will soon allow us to obtain the spectrum at $N_{max} = 8$ to extend the convergence trends for Figs. 1 and 3.

4. *Ab exitu* NCSM with interaction from inverse scattering

Turning to a sample of results with the *ab exitu* NCSM we present in Fig. 4 the binding energies of stable p-shell nuclei relative to experiment using the JISP16 interaction. Both bare and effective interaction results are presented as well as some initial extrapolations to the infinite basis limit.[14] The bare interaction results are strict upper bounds to the exact ground state energy so the bare curves will drop as the basis space is increased (direction of increased binding in the theory). The effective interaction results do not follow a variational principle. The results show a tendency to underbind nuclei in mid-p-shell and to overbind at the upper end. Results in larger

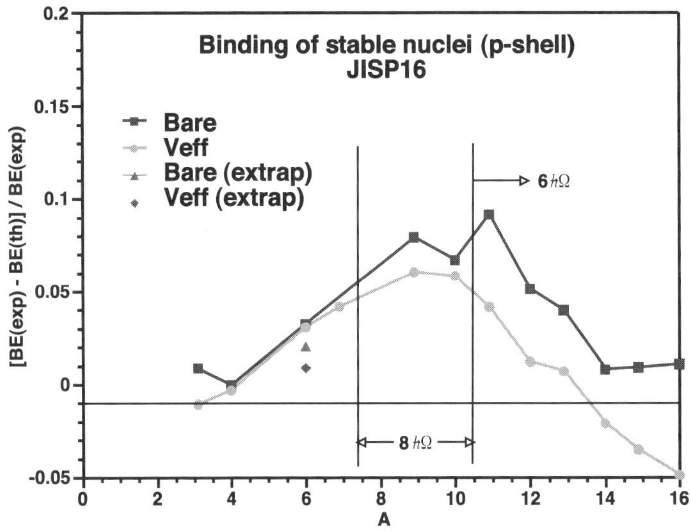

Fig. 4. Fractional difference between theory and experiment for the binding energies of stable p-shell nuclei. The results are quoted with the specified N_{max} values and with both the bare and effective JISP16 interactions. The effective interactions are evaluated at the 2-body cluster level. The oscillator energy, $\hbar\Omega$ is taken, in each case, to be the value at which an extremum in the binding energy occurs.

basis spaces will help clarify these trends.

Towards this goal, we present in Fig. 5 initial results for the ground state energy of ^{12}C in a larger $N_{max} = 8$ basis space using the bare JISP16 interaction and compare with the results obtained in smaller basis spaces. While the convergence trend is encouraging, we note that JISP16 seems on a path to produce modest overbinding. More analyses are in progress to obtain an extropolated ground state energy and its uncertainty.[14]

5. Concluding remarks

Table 1 contains selected experimental and theoretical results for ^{10}B. The binding energy and rms deviation between the experimental and theoretical excitation energies improve substantially with the inclusion of TNI. The JISP16 results lie intermediate to the N3LO and N3LO+TNI interaction results. Other observables are in reasonable accord with experiment considering that (1) we use bare electromagnetic operators, and (2) moments and transition rates are expected to be more sensitive to enlarging the basis

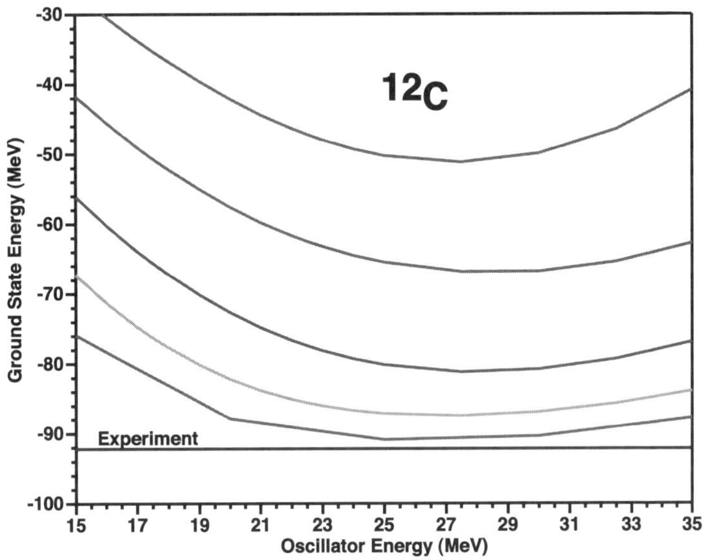

Fig. 5. Ground state energy for ^{12}C as a function of the oscillator energy, $\hbar\Omega$, for $N_{max} = 0 - 8$ for the bare JISP16 interaction. The $N_{max} = 8$ curve is closest to experiment and each curve above it corresponds to decrements by two units in N_{max}.

spaces as we plan to do. The JISP16 results employ partial waves, $J \leq 4$. If we retain only $J \leq 3$ partial waves in JISP16, the excitation energies change by less than 10 keV and the binding energy decreases by 35 keV.

These results required substantial computer resources. The $N_{max} = 6$ spectrum shown in Fig. 3 and a set of additional experimental observables, takes an hour on 3500 processors on the LLNL-Thunder machine. Our largest run that is reported here, the ^{12}C with JISP16 in the $N_{max} = 8$ basis ($dimension = 6 \times 10^8$) took 2.3 hours on 15,400 processors (33,350 cpu hours) at the ORNL Jaguar facility. All runs produce the lowest 15 converged eigenvectors and a suite of observables (rms radii, electromagnetic moments and transitition rates, electroweak transition rates, etc.).

We demonstrated here that TNI's make substantial contributions to improving the spectra and other observables. In addition, phase-equivalent transformations of an interaction obtained from inverse scattering, JISP16, produces appealing fits to light nuclear properties. However, there is considerable room for further improvement in both approaches. Our leading

Table 1. Properties of ^{10}B from experiment and theory. E2 transitions are in e^2 fm^4 and M1 transitions are in μ_N^2. The rms deviations of excited state energies are quoted for the lowest 9 states whose spin–parity assignments are well established and that are known to be dominated by $0\hbar\Omega$ configurations. Results are obtained in the basis spaces $N_{\max} = 6(8)$ with $\hbar\Omega = 14$ MeV for the ChPT (JISP16) interaction. In the $N3LO + TNI$ column we show selected sensitivity to changing C_D by ±1. "N/A" indicates a result yet to be calculated. The experimental values are from Ref.[15,16]

Nucleus/property	Exp	$N3LO$ $+TNI$	$N3LO$	JISP16
^{10}B : $\lvert E(3^+,0)\rvert$ [MeV]	64.751	64.78	56.11	59.751
r_p [fm]	2.30(12)	2.197	2.256	2.210
$Q(3_1^+,0)$ [e fm^2]	+8.472(56)	+6.327	+6.803	6.716
$\mu(3_1^+0)$ [μ_N]	+1.801	+1.837	+1.853	N/A
$E_x(3_1^+0)$ [MeV]	0.0	0.0	0.0	0.0
$E_x(1_1^+0)$ [MeV]	0.718	0.523	-1.128	0.193
$E_x(0_1^+1)$ [MeV]	1.740	1.279	0.913	1.034
$E_x(1_2^+0)$ [MeV]	2.154	1.432	1.643	2.221
$E_x(2_1^+0)$ [MeV]	3.587	3.178	1.643	3.430
$E_x(3_2^+0)$ [MeV]	4.774	6.729	4.193	5.792
$E_x(2_1^+1)$ [MeV]	5.164	5.315	4.419	4.861
$E_x(2_2^+0)$ [MeV]	5.92	4.835	3.555	5.201
$E_x(4_1^+0)$ [MeV]	6.025	5.960	4.790	5.685
$E_x(2_2^+1)$ [MeV]	7.478	7.823	5.565	7.320
$rms(Exp - Th)$ [MeV]	-	0.823	1.482	0.536
B(E2;$1_1^+0 \to 3_1^+0$)	4.13(6)	3.05(62)	4.380	3.736
B(E2;$1_2^+0 \to 3_1^+0$)	1.71(0.26)	0.50(50)	0.082	0.578
B(M1;$2_1^+0 \to 3_1^+0$)	0.0015(3)	0.0000	N/A	0.0012
B(M1;$2_1^+1 \to 3_1^+0$)	0.041(4)	0.216	N/A	0.125
B(M1;$2_2^+0 \to 3_1^+0$)	0.050(12)	0.053	N/A	0.056
B(M1;$4_1^+0 \to 3_1^+0$)	0.043(7)	0.002	N/A	0.003
B(M1;$2_2^+1 \to 3_1^+0$)	-	4.020	N/A	4.148
B(GT;$3_1^+0 \to 2_1^+1$)	0.083(3)	0.07(1)	0.102	0.040
B(GT;$3_1^+0 \to 2_2^+1$)	0.95(13)	1.22(2)	1.487	1.241

suggestions include: (1) extend the TNI's to the order consistent with the NN interaction, N3LO; (2) extend the basis spaces to higher N_{max} values to further improve convergence; (3) examine sensitivity of TNI's to the choice of regulator; and (4) include four-nucleon interactions at a consistent order of ChPT. In addition, further exploration of the phase-shift equivalent transformations appears warranted.

Our overall conclusion is that these results support a full program of deriving the NN interaction and its mulit-nucleon partners in the consistent approach provided by chiral effective field theory. It is straightforward, but challenging, to extend this research thrust in the directions indicated. However, the favorable results to date and the need for addressing fundamental symmetries of strongly interacting systems with enhanced predictive power firmly motivate this path.

This work was supported in part by the U. S. Department of Energy Grants DE-FG02-87ER40371 and DE-FC02-07ER41457. This work was also partly performed under the auspices of the U. S. Department of Energy by the University of California, Lawrence Livermore National Laboratory under contract No. W-7405-Eng-48. Support from the LDRD contract No. 04–ERD–058 and from U.S. DOE/SC/NP (Work Proposal Number SCW0498) is acknowledged. This work was also supported in part by the Russian Foundation of Basic Research, Grant 05-02-17429

References

1. P. Navrátil, J. P. Vary and B. R. Barrett, Phys. Rev. Lett. **84**, 5728 (2000); Phys. Rev. C **62**, 054311 (2000).
2. P. Navrátil, V.G. Gueorguiev, J. P. Vary, W. E. Ormand and A. Nogga, Phys. Rev. Lett. **99**, 042501 (2007).
3. S. Weinberg, Physica **96A**, 327 (1979); Phys. Lett. B **251**, 288 (1990); Nucl. Phys. **B363**, 3 (1991).
4. V. Bernard, *et al.*, Int. J. Mod. Phys. **E4**, 193 (1995).
5. C. Ordonez, L. Ray, and U. van Kolck, Phys. Rev. Lett. **72**, 1982 (1994); Phys. Rev. C **53**, 2086 (1996).
6. U. van Kolck, Prog. Part. Nucl. Phys. **43**, 337 (1999).
7. P. Bedaque and U. van Kolck, Ann. Rev. Nucl. Part. Sci. **52**, 339 (2002).
8. D. R. Entem and R. Machleidt, Phys. Rev. C **68**, 041001(R) (2003).
9. E. Epelbaum, W. Glöckle, and Ulf-G. Meissner, Nucl. Phys. **A637**, 107 (1998); **A671**, 295 (2000).
10. A. M. Shirokov, J. P. Vary, A. I. Mazur and T. A. Weber, Phys. Letts. B **644**, 33 (2007).
11. A. M. Shirokov, A. I. Mazur, S. A. Zaytsev, J. P. Vary and T. A. Weber, Phys. Rev. C **70**, 044005 (2004); A. M. Shirokov, J. P. Vary, A. I. Mazur, S. A. Zaytsev and T. A. Weber, Phys. Letts. B **621**, 96 (2005).
12. P. Navrátil and W. E. Ormand, Phys. Rev. Lett. **88**, 152502 (2002).
13. A. Nogga, P. Navrátil, B. R. Barrett and J. P. Vary Phys. Rev. C **73**, 064002 (2006).
14. P. Maris, J. P. Vary and A. Shirokov, to be published.
15. F. Ajzenberg-Selove, Nucl. Phys. **A490**, 1 (1988).
16. A. Ozawa, I. Tanihata, T.Kobayashi, Y. Sugihara, O. Yamakawa, K. Omata, K. Sugimoto, D. Olson, W. Christie, and H. Wieman, Nucl. Phys. **A608**, 63 (1996); H. De Vries, C. W. De Jager, and C. De Vries, At. Data Nucl. Data Tables **36**, 495 (1987).

AB-INITIO COUPLED CLUSTER THEORY FOR OPEN QUANTUM SYSTEMS

G. HAGEN[!*], D. J. DEAN[!] and T. PAPENBROCK[!@]

[!] *Physics Division, Oak Ridge National Laboratory, P.O. Box 2008, Oak Ridge, TN 37831, USA*
[@] *Department of Physics and Astronomy, University of Tennessee, Knoxville, TN 37996, USA*
[*] *E-mail: hageng@ornl.gov*

M. HJORTH-JENSEN

Department of Physics and Center of Mathematics for Applications, University of Oslo, N-0316 Oslo, Norway
Centre of Mathematics for Applications, University of Oslo, N-0316 Oslo, Norway

We discuss peculiarities of open-quantum systems, as compared to closed quantum systems. We emphasize the importance of taking continuum degrees of freedom into account when dealing with systems with a tendency to decay through emission of fragments. In this context we introduce the Coupled-Cluster theory, and argue that this method allows for an accurate description of such systems starting from nucleon-nucleon degrees of freedom. We present *Ab-initio* Coupled Cluster calculations with singles and doubles excitations (CCSD) for the ground states of the helium isotopes $^{3-10}$He. The calculated masses and decay widths are in semi-quantitative agreement with experiment. The discrepancy with experiment is suspected to be attributed to the three-nucleon force (3NF) which is not included at this point.

Keywords: Style file; LaTeX; Proceedings; World Scientific Publishing.

1. Introduction

Recently, there has been a growing effort in the experimental and theoretical understanding of nuclei located far away from the valley of beta-stability. Moving away from the stability line, one approaches the very limit of existence of matter. At this limit, the number of protons/neutrons which can be added to a given nucleus is exhausted, and the protons/neutrons literally start to drip from the nucleus, i.e. these borders of the nuclear chart have been labeled the proton and neutron drip line. As one moves away from the stability line, exotic features, which are not seen in the well-bound and stable nuclei, start to emerge, such as extreme matter clusterizations, melting and reorganizing of shell structure, ground states embedded in the continuum, and extreme dilute and extended matter densities. The physics of such nuclei was triggered by Tanihata's discovery (1985) [1] of vastly spatially extended nuclei (^{6,8}He; ^{11}Li; ^{11}Be) at the neutron drip line. These nuclei exhibit an extreme clusterization into parts of normal nuclear density together with a *halo* of vastly extended and dilute matter. Some of these nuclei exhibit another peculiar feature, in that the one-neutron decay threshold is above the two-neutron decay threshold. Some of these nuclei, like ^{6}He and the cardinal case of ^{11}Li, have been labeled *Borromean* nuclei. Their name originates from the three Borromean rings. The heraldic symbol of the Italian Borromeo family are interlocked in such a way that if any of them were removed, the other two would also fall apart. The Borromean property of different nuclei is clearly exhibited by the hadronic stability of various isotope chains. In the helium chain we see that while ^{5}He is unbound, ^{6}He is bound; ^{7}He is unbound but ^{8}He is again bound. Another exotic feature, which appears at the limits of the nuclear chart, is the emergence of new magic numbers and reorganizing of normal shell structure. One such example is seen along the oxygen isotope chain. There is experimental evidence that N=14/16 are new magic numbers in extreme neutron-rich nuclei [2,3]. In standard shell model theory, one would expect ^{28}O to be a stable nucleus, since it is a doubly magic nucleus, but rather it is observed that ^{24}O is the last stable nucleus [3], and therefore defining the drip line in the oxygen chain. These rare isotopes, at the drip lines, play also a crucial role in the synthesis of elements. The famous Hoyle resonance in ^{12}C has been labeled the doorway to our universe. The processes leading to this doorway proceed from ^{4}He via *Borromean* systems: (i) in Red Giants via the triple α-process; (ii) in Supernovae via $^9_4$Be$_5$ also Borromean; and (iii) a less probable side-route to ^{9}Be via ^{6}He. Recall that neither $(\alpha\alpha)$, (αn), nor (nn) can form bound binary systems, only resonances.

The properties of nuclei located far away from the stability line suggests that one should depart from standard shell model approaches. In standard shell model approaches, one usually builds the wave function starting from an oscillator basis. By doing this, one considers the system as closed from the scattering environment, since nucleons move in an infinite oscillator well. Such a description cannot account for vastly extended nuclei, and of nuclei with a tendency to decay by emission of fragments. Rather, one should treat such nuclei as open quantum systems, so that nucleons can correlate and interact through continuum degrees of freedom. The natural starting point in an open-quantum system formulation is to use a single-particle basis which treats bound-, resonant-, and continuum states on equal footing. Such a basis is called the Berggren basis [4]. It is a generalization of the normal completeness defined along the real energy axis, so that resonances are also taken into account. This basis has been successfully applied to nuclei with open-quantum system characteristics in the recently developed Gamow shell model [5–7] and in Coupled-Cluster approaches [8]. In these Proceedings we revisit our Coupled-Cluster results in Ref. [8]. The Coupled-Cluster method is a microscopic theory so that all nucleon degrees of freedom are taken into account. The Coupled-Cluster method is an ideal compromise between accuracy on the one hand and computational cost on the other hand. The soft scaling with system size allows one to treat enormous basis sets which is necessary when continuum effects are taken into account. Due to the discretization of the continuum for each partial wave, the total size of the single-particle basis can be orders of magnitude larger than the size of the oscillator basis, where only bound states enter. Coupled-Cluster with singles and doubles excitations scales as $n_o^2 n_u^4$, where n_o is the number of occupied orbits and n_u is the number of unoccupied orbits. This clearly has advantages when compared to other methods such as direct diagonalization methods used in shell-model and no-core shell model approaches, which scales combinatorially with the system size. This compromise between computational cost and accuracy motivates our Coupled-Cluster approach to loosely bound and unbound nuclei. In Ref. [8] we started from a Gamow-Hartree-Fock basis generated from a low-momentum interaction, $V_{\text{low}-k}$ [9], derived from the N^3LO interaction model [10]. We calculated the ground states of the helium chain within Coupled-Cluster theory within the singles and doubles approximation (CCSD). Using a Berggren basis, this allowed us to calculate the lifetimes of these nuclei starting from nucleon degrees of freedom. The outline of these Proceedings is the following. In Sec. 2 we outline the Coupled-Cluster theory and the various approximations we use. In Sec. 3 we discuss the concept of open-quantum systems, and in Sec. 3.1 we present our Coupled-Cluster results for the helium chain. Finally, in Sec. 4 we conclude.

2. Coupled Cluster Theory

Coupled-cluster theory originated in nuclear physics in the late fifties, with Coester and Kümmel being the pioneers [11,12]. The theory was further developed and applied in nuclear theory in the 1970s, and the status of the field was summarized by the Bochum group [13] in 1978. After that, Coupled-Cluster theory only saw sporadic applications in nuclear theory. On the other hand, Coupled-Cluster theory had enormous growth and popularity in quantum chemistry, and has become the state-of-the-art many-body theory, see [14] and references therein. Recently, Coupled-Cluster theory has been revived in nuclear theory in the spirit of the quantum chemistry approach. (See Refs. [8,15,16] and references therein.)

In Coupled-Cluster theory the exact many-body wave function is written as

$$|\psi\rangle = e^{\hat{T}}|\phi\rangle \,, \tag{1}$$

where $|\phi\rangle = \prod_{i=1}^{A} \hat{a}_i^\dagger|0\rangle$ is a given uncorrelated reference state, built by filling oscillator orbitals or Hartree-Fock orbitals up to the Fermi level. The correlation are built on the reference state by acting with the exponentiated correlation operator T,

$$\hat{T} = \hat{T}_1 + \hat{T}_2 + \ldots + \hat{T}_A, \tag{2}$$

which is a linear combination of particle-hole excitation operators,

$$\hat{T}_k = \frac{1}{(k!)^2} \sum_{i_1,\ldots,i_k;a_1,\ldots,a_k} t_{i_1\ldots i_k}^{a_1\ldots a_k} \hat{a}_{a_1}^\dagger \ldots \hat{a}_{a_k}^\dagger \hat{a}_{i_k} \ldots \hat{a}_{i_1} \,. \tag{3}$$

Here, and in the following, $i, j, k, \ldots$ label occupied single-particle orbitals, while $a, b, c, \ldots$ refer to unoccupied orbitals. The only approximation which appears, in applications of Coupled-Cluster theory is the truncation of the correlation operator T at a given particle-hole excitation level. The most popular approximation is the Coupled-Cluster method with Singles and Doubles excitations (CCSD) which means one truncates T at the two-particle-two-hole level, $T = T_1 + T_2$.

In order to obtain the ground-state wave function, one has to solve for the particle-hole excitation amplitudes given in Eq. 3. In the CCSD approximation, the coupled-cluster equations are given by

$$E = \langle\phi|\overline{H}|\phi\rangle \,, \;\; 0 = \langle\phi_i^a|\overline{H}|\phi\rangle \,, \;\; 0 = \langle\phi_{ij}^{ab}|\overline{H}|\phi\rangle \,. \tag{4}$$

Here $|\phi_{i_1\ldots i_n}^{a_1\ldots a_n}\rangle = \hat{a}_{a_n}^\dagger \ldots \hat{a}_{a_1}^\dagger \hat{a}_{i_1} \ldots \hat{a}_{i_n}|\phi\rangle$ is a $np - nh$ excitation of the reference state $|\phi\rangle$, and

$$\overline{H} = e^{-\hat{T}} \hat{H} e^{\hat{T}} = \left(\hat{H} e^{\hat{T}}\right)_c \tag{5}$$

is the similarity-transformed Hamiltonian (note that $\overline{H}$ is non-Hermitian). The last expression on the right-hand side of Eq. (5) indicates that only fully connected diagrams contribute to the construction. This means that only terms where the excitation operators T_1 and T_2 connect with the Hamiltonian enter the final expression. This property makes the Coupled-Cluster wave function fully linked, and ensures that the theory is size-extensive and the energy of the system scales correctly with the number of particles (see Ref. [14] for more details on size-extensivity).

We already mentioned that Coupled-Cluster theory is an ideal compromise between accuracy on the one hand and computational cost on the other. CCSD is the most common approximation, and already accounts for $\sim 90\%$ of the full correlation energy [14]. CCSD scales with the number of occupied particles (n_o) and unoccupied particles (n_u) as $O(n_o^2 n_u^4)$. Including full triples, we get to the next level of approximation which is CCSDT, and which accounts for $\sim 99\%$ of the full correlation energy, and CCSDT scales with the system size as $O(n_o^3 n_u^5)$. In these Proceedings we will present Coupled-Cluster results with up to three-particle-three-hole excitations in the correlation operator T (CCSDT). There are also various approximations to the full CCSDT equations, and the most popular of these schemes is the CCSD(T) approach [17]. The CCSD(T) approximation is relatively inexpensive compared to CCSDT; no storage of triples amplitudes is required and the computational cost is a non-iterative $O(n_o^3 n_u^4)$ step. There is also a family of iterative triples correction schemes known as CCSDT-n [18].

3. *Ab-initio* description of open quantum systems

Before we discuss the Coupled-Cluster results for the helium chain using Coupled-Cluster theory, we would like to point out some essential features of open-quantum systems as compared to closed-quantum systems. As already mentioned in the Introduction, the closed-quantum system formulation assumes that the particles are totally isolated from the external scattering continuum. Such a system could be particles trapped in a well with infinite walls, such as the oscillator potential (see the right illustration in fig.1). Such a description is not suitable if one aims at describing structure and reaction properties of loosely bound and unbound nuclei near the nuclear drip lines. In order to theoretically account for properties of these loosely bound and unbound nuclei, one should rather take an approach which treats the bound- and scattering continuum on equal footing. In order to achieve this, one should start by defining the single-particle basis from a finite-well potential, such as, for example, the Woods-Saxon potential. This kind of

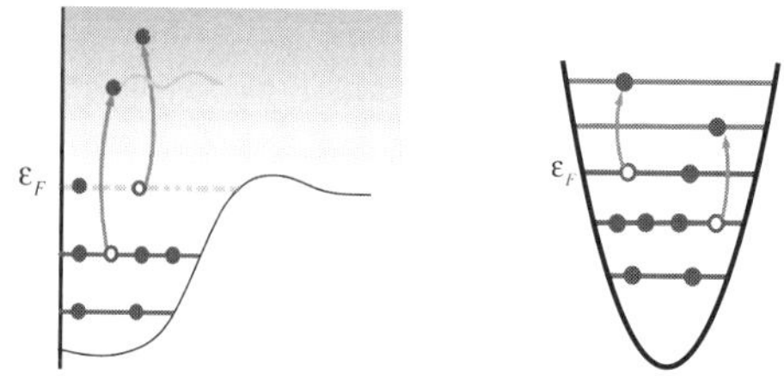

Fig. 1. The left figure gives an illustration of an open-quantum formulation where particle can excite and correlate through continuum degrees of freedom. The right-hand side of the figure gives an illustration of a closed-quantum system, where particles are trapped in an infinite well and never interact with the external environment.

potential has a much richer spectrum than the oscillator potential. Any finite well potential supports bound-, resonant, and a continuum of scattering states. Constructing a basis from such a potential, and building up the many-body states from this basis, allows for a correct description of reaction and structure properties. Such an approach also unifies structure and reaction theory. In our Coupled-Cluster approach to these nuclei, we start with a Woods-Saxon basis in the complex energy plane. We generate the self-consistent Hartree-Fock basis from the low-momentum interaction generated from the N^3LO interaction model [10]. Our reference state is then constructed by filling the lowest energy states up to the Fermi-level in the Hartree-Fock potential. See Refs. [7,8] for details regarding the construction of the basis and the interaction used in the calculations. It should be noted that the Fermi-level might even be unbound, which is the case in ^{5}He (see the right illustration in fig.1). Hereafter we obtain the correlated many-body wave function by employing the Coupled-Cluster method. In this way, nucleons are scattered and correlated through continuum degrees of freedom, and the particles might even decay, resulting in an unstable many-body correlated wave function. This allows for the calculation of lifetimes and decay widths.

3.1. *CC results for Helium chain*

In this section we summarize the results for the helium chain using CCSD and a Berggren basis. For details regarding basis size and convergence, see Ref. [8]. In order to keep the basis size managable we used an effective interaction of the low-momentum type, $V_{\text{low}-k}$, with cutoff $\Lambda = 1.9\text{fm}^{-1}$. $V_{\text{low}-k}$ preserves all two-body scattering observables by construction. However, since high-momentum components are removed from the model, V_{low-k} induces many-body forces in the A−body system;

$$\hat{V} = V_{\rm NN}(\Lambda) + V_{\rm 3N}(\Lambda) + \dots , \tag{6}$$

where the momentum cutoff Λ denotes the cutoff or resolution scale. A fully renormalized theory will be independent of the resolution scale (Λ). In these Proceedings we only include up two two-body forces, so naturally there will be a strong dependence on the cutoff (Λ). The hope is that three-nucleon forces are sufficient to approximately renormalize the model, i.e. there is a cutoff regime where observables vary very slowly with Λ. These topics will be pursued in the near future.

Before we turn to the helium results using a Berggren basis, we would like to investigate how well the single-reference Coupled-Cluster method works for open-shell nuclei. This is relevant since many of the helium isotopes have open-shell character. ^{6}He is the most difficult case in a single reference formulation since it has two neutrons in the $p_{3/2}$ shell, and there is no way to construct a single-reference state with good J. In order to see how well single-reference Coupled-Cluster works for these nuclei, we compare exact diagonalization for the various Coupled-Cluster approximations given in Sec.2, i.e. we include up to full triples in our Coupled-Cluster calculations (CCSDT). In Table 1 we present the Coupled-Cluster results for the helium

Method	^{3}He	^{4}He	^{5}He	^{6}He	$\langle J\rangle$, ^{6}He
CCSD	-6.21	-26.19	-21.53	-20.96	0.61
CCSD(T)	-6.40	-26.27	-21.88	-22.60	0.65
CCSDT-1	-6.41	-28.27	-21.89	-22.85	0.29
CCSDT-2	-6.41	-28.26	-21.89	-22.78	0.25
CCSDT-3	-6.42	-26.27	-21.92	-22.90	0.26
CCSDT	-6.45	-26.28	-22.01	-22.52	0.04
Exact	-6.45	-26.3	-22.1	-22.7	0.00

isotopes $^{3-6}$He and compare with exact diagonalization. Here we started with an oscillator basis (see Ref. [8]), since we are not computing lifetimes but investigating how well single-reference Coupled-Cluster theory works for open-shell nuclei. It is seen that for $^{3-5}$He, CCSD performs well, with an error being at most $\sim$ 0.5MeV for ^{5}He. For ^{6}He, CCSD does not perform that well; the CCSD energy is about 1.5MeV away from exact results. This is due to the truly open-shell character of ^{6}He. The various triples approximations all improve on the Coupled-Cluster results. The Coupled-Cluster

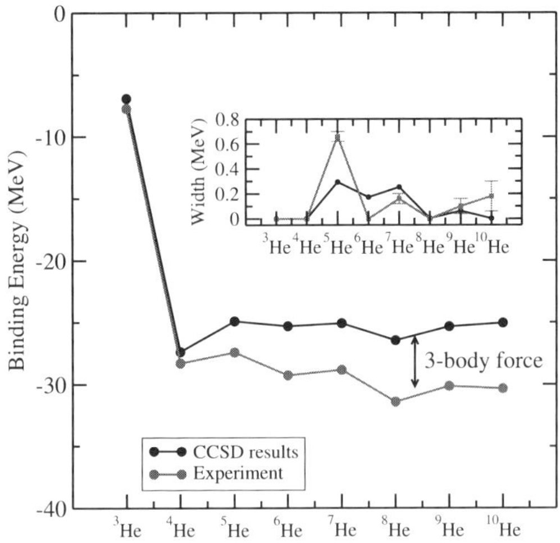

Fig. 2. CCSD results (black dotted line) and experimental values (red dotted line) for the ground state of the helium chain $^{3-10}$He using a Gamow-HF basis and a low-momentum interaction generated from the N^{3}LO interaction model.

results with full triples included (CCSDT) is the most computationally expensive method considered here, but on the other hand, it provides us with very accurate results, even for the case of ^{6}He. In order to investigate how accurate the various Coupled-Cluster approximations perform for the open-shell nucleus ^{6}He, we calculate the expectation value of J. It is clearly seen that the expectation value J improves when including more correlations in the Coupled-Cluster wave function. In the full CCSDT calculation of ^{6}He, the expectation value of J came down to $J \sim 0.04$. For the self-consistent and iterative triples approximations CCSDT-n ($n = 1, 2, 3$) [18], the expectation value of J has a value of ~ 0.3, which is clearly an improvement to the calculated value of J using CCSD and CCSD(T). From this exercise, one can conclude that CCSDT is an accurate approximation to the exact wave function of truly open-shell nuclei, like ^{6}He.

Next, we turn to the calculation of the ground states of the helium isotopes $^{3-10}$He using a Gamow-Hartree-Fock basis and Coupled-Cluster theory with up to singles and doubles excitations (CCSD). Fig.2 summarizes the results presented in Ref. [8]. These results are obtained in the largest model space considered, which comprised more than 1000 single-particle orbitals. For more details on convergence and model-space, see Refs. [7,8]. The black dotted line gives our calculated masses, while the red dotted line gives the experimental mass values. The inset gives our calculated widths of the helium isotopes compared with experimental values. To summarize, we see that our results are in semi-quantitative agreement with experiment. With this interaction, all helium isotopes lack binding as compared to experiment. However, the even/odd mass pattern is reproduced fairly well. We see that ^{5}He is unstable with respect to one-neutron emission, while ^{6}He is stable towards one-neutron emission. However, ^{6}He is not stable towards two-neutron emission. This is mainly due to missing three-body forces and inclusion of full triples in our calculation. ^{8}He is stable towards one-, two-, and three-neutron emission but not stable against emission of four neutrons to the continuum and ^{4}He. We believe that the growing discrepancy between theory and experimental mass values as we move along the helium chain, is due to the lack of three-nucleon forces. But for larger systems, triples corrections should play a more important role as well. By combining both of these missing ingredients, we believe that our results should be closer the experimental values.

4. Conclusions

We have discussed the peculiarities of open-quantum systems, and what differentiates such a description from a standard closed formulation. We have pointed out that in order to properly account for properties of nuclei away from the valley of beta-stability, the coupling with continuum degrees of freedom cannot be neglected. An open-quantum system formulation allows for an accurate description of particle decay and emission, and properly accounts for clusterization phenomena taking place near the scattering thresholds. We applied this open-quantum system formalism to the *ab-initio* calculation of masses and lifetimes of the rare helium isotopes. Our results are in semi-quantitative agreement with experiment. The discrepancy with experiment is believed to be attributed to missing three-nucleon forces which has not been included at this point. Nevertheless, we reproduce the even/odd mass pattern of the helium chain. In the near future, we will repeat these calculations with properly renormalized three-nucleon forces. Further, we will include three-particle-three-hole excitations in the Coupled-Cluster wave function in order to properly describe truly open-shell nuclei like the Borromean nuclei ^{6}He.

Acknowledgments

This research was supported in part by the Laboratory Directed Research and Development program of Oak Ridge National Laboratory (ORNL), by the U.S. Department of Energy under Grant Nos. DE-FG02-96ER40963 (University of Tennessee) and DE-AC05-00OR22725 with UT-Battelle, LLC (ORNL), and by the Natural Sciences and Engineering Research Council of Canada (NSERC). Computational resources were provided by the National Center for Computational Sciences at Oak Ridge and the National Energy Research Scientific Computing Facility.

References

1. I.Tanihata *et al*, Phys. Rev. Lett. **55**, 2676 (1985); Phys. Lett. **B160**, 380 (1985); I. Tanihata, J. Phys., **G22**,157 (1996).
2. A. Ozawa *et al*, Phys. Rev. Lett. **84**, 5493 (2000).
3. M. Thoennessen, Rep. Prog. Phys. **67**, 1187 (2004).
4. T. Berggren, Nucl. Phys. A **109**, 265 (1968).
5. N. Michel, W. Nazarewicz, M. Płoszajczak, and K. Bennaceur, Phys. Rev. Lett. **89**, 042502 (2002).
6. R. Id Betan, R. J. Liotta, N. Sandulescu, and T. Vertse, Phys. Rev. Lett. **89**, 042501 (2002).
7. G. Hagen, M. Hjorth-Jensen, and N. Michel, Phys. Rev. C **73**, 064307 (2006).
8. G. Hagen, D.J. Dean, M. Hjorth-Jensen, and T. Papenbrock, nucl-th/0610072.
9. S.K. Bogner, T.T.S. Kuo, and A. Schwenk, Phys. Rept. **386**, 1 (2003), nucl-th/0305035.
10. D. R. Entem and R. Machleidt, Phys. Lett. B **524**, 93 (2002).
11. F. Coester, Nucl. Phys. **7**, 421 (1958).
12. F. Coester and H. Kümmel, Nucl. Phys. **17**, 477 (1960).
13. H. Kümmel, K.H. Lührmann, and J.G. Zabolitzky, Phys. Rep. **36**, 1 (1978).
14. R.J. Bartlett and M. Musiał, Rev. Mod. Phys. **79**, 291 (2007).
15. G. Hagen, T. Papenbrock, D.J. Dean, A. Schwenk, A. Nogga, P. Piecuch, and M. Włoch, Phys. Rev. C **76**, 034302 (2007).
16. G. Hagen, D.J. Dean, M. Hjorth-Jensen, A. Schwenk, and T. Papenbrock, Phys. Rev. C **76**, 044305 (2007).
17. K. Raghavachari *et al.*, Chem. Phys. Lett. **157**, 479 (1989).
18. J. Noga and R. J. Bartlett, Chem. Phys. Lett. **134**, 126 (1987).

SYMPLECTIC NO-CORE SHELL MODEL

JERRY P. DRAAYER, TOMÁŠ DYTRYCH, KRISTINA D. SVIRATCHEVA, AND CHAIRUL BAHRI

Department of Physics and Astronomy, Louisiana State University, Baton Rouge, LA 70803, USA

JAMES P. VARY

Department of Physics and Astronomy, Iowa State University, Ames, IA 50011, USA

Results from a full $6\hbar\omega$ No-Core Shell Model (NCSM) calculation for low-lying states in ^{12}C and ^{16}O using a realistic nucleon-nucleon interaction are found to project at approximately the 90% level onto a few of the leading 0p-0h and 2p-2h symplectic representations. The latter span a symplectic space that is typically only a very small fraction (under 1%) of the NCSM model space, and grows slowly with increasing $\hbar\Omega$ oscillator strength parameter. The results are nearly independent of $\hbar\Omega$ and whether bare or renormalized effective interactions are used in the analysis.

Keywords: ab-initio no core shell model, symplectic Sp(3,R) shell model, model space truncation scheme, ^{12}C, ^{12}O.

1. Introduction

A unified microscopic description of low-lying states in light nuclei requires a theory that can incorporate diverse degrees of freedom, ranging from single-particle effects, pairing correlations and a few particle-hole excitations, to α-clustering phenomena and enhanced shape deformations. Furthermore, if such a theory is designed to employ realistic interactions, it could relate to the fundamental strong interaction as well as quark constituents and be applicable in regions of the nuclear chart where experimental data necessary to adjust effective interactions might not be available.

We propose the Symplectic No-Core Shell Model (SpNCSM)[1] as the first *ab initio* approach capable of meeting these criteria. It combines the NCSM,[2] which uses modern realistic interactions and yields a good description of the low-lying states in few-nucleon systems[3] as well as in more complex nuclei like ^{12}C,[2,4] with the symplectic $\mathrm{Sp}(3,\mathbb{R})$ model[5,6] that can yield a dramatic reduction of the NCSM space and hence serve as a powerful basis space reduction (truncation) scheme. This approach allows one to advance *ab initio* calculations to heavier nuclei and to account for even higher $\hbar\Omega$ configurations required to realize experimentally measured B(E2) values without an effective charge, and to accommodate highly deformed spatial configurations including α-cluster structures,[7] which may be essential for modeling, e.g., the second 0^+ state in ^{12}C and ^{16}O.

We recently showed in a 'proof-of-principle' study[1] that realistic eigenstates for low-lying states determined in NCSM calculations for light nuclei with the JISP16 realistic interaction,[8] project predominantly onto a few of the most deformed $\mathrm{Sp}(3,\mathbb{R})$-symmetric basis states that are free of spurious center-of-mass motion. This suggests the presence of an underlying symplectic structure, which is not *a priori* imposed on the interaction and furthermore that is found to remain essentially unaltered after a Lee-Suzuki similarity transformation is introduced to accommodate the truncation of the infinite Hilbert space through a renormalization of the bare interaction. This in turn provides insight into the physics of a nucleon system and its geometry since collective states shaped by monopole/quadrupole-vibrational and rotational modes are described naturally by irreducible representations (irreps) of $\mathrm{Sp}(3,\mathbb{R})$.

We focus on the 0^+_{gs} ground state and the lowest 2^+_1 and 4^+_1 states in the deformed ^{12}C nucleus as well as the 0^+_{gs} in the 'closed-shell' ^{16}O nucleus. The corresponding NCSM eigenstates are reasonably well converged in the $N_{max} = 6$ (or $6\hbar\Omega$) model space with an effective interaction based on the JISP16 realistic interaction. In addition, calculated binding energies as well as other observables for ^{12}C such as $\mathrm{B}(\mathrm{E2};2^+_1 \rightarrow 0^+_{gs})$, $\mathrm{B}(\mathrm{M1};1^+_1 \rightarrow 0^+_{gs})$, ground-state proton rms radii and the 2^+_1 quadrupole moment all lie reasonably close to the measured values. While symplectic algebraic approaches achieve a good reproduction of low-lying energies and B(E2) values in light nuclei[9–13] using phenomenological or semi-microscopic interactions, here, for the first time, we establish the dominance of the symplectic symmetry in light nuclei, and hence their propensity towards the development of collective motion as unveiled through *ab initio* NCSM calculations.

2. Symplectic $\mathrm{Sp}(3,\mathbb{R})$ Basis

The symplectic shell model,[5,6] based on the noncompact symplectic $\mathfrak{sp}(3,\mathbb{R})$ algebra[a], is known to underpin the successful Bohr-Mottelson collective

---

[a]We use lowercase (capital) letters for algebras (groups).

model and has also been shown to be a multiple oscillator shell generalization of Elliott's SU(3) model. The significance of the symplectic symmetry for a microscopic description of a quantum many-body system of interacting particles emerges from the physical relevance of its 21 generators. For A nucleons the bilinear products of the particle momentum ($p_{s\alpha}$) and coordinate ($x_{s\beta}$) operators, $T_{\alpha\beta} = \sum_s p_{s\alpha}p_{s\beta}$, $L_{\alpha\beta} = \sum_s(x_{s\alpha}p_{s\beta} - x_{s\beta}p_{s\alpha})$, $S_{\alpha\beta} = \sum_s(x_{s\alpha}p_{s\beta} + p_{s\alpha}x_{s\beta})$, and $Q_{\alpha\beta} = \sum_s x_{s\alpha}x_{s\beta}$ with $\alpha, \beta = 1, 2, 3$ for the 3 spatial directions and $s = 1, \ldots, A$, realize the symplectic $\mathfrak{sp}(3, \mathbb{R})$ algebra. Hence, the many-particle kinetic energy, the mass quadrupole moment operator, and the angular momentum are all elements of the $\mathfrak{sp}(3, \mathbb{R}) \supset \mathfrak{su}(3) \supset \mathfrak{so}(3)$ algebraic structure. It also includes monopole and quadrupole collective vibrations reaching beyond a single shell to higher-lying and core configurations, as well as vorticity degrees of freedom for a description of the continuum from irrotational to rigid rotor flows. Alternatively, the elements of the $\mathfrak{sp}(3, \mathbb{R})$ algebra can be represented as bilinear products in harmonic oscillator (HO) raising and lowering operators, which means the basis states of a $\mathrm{Sp}(3, \mathbb{R})$ irrep can be expanded in a 3-D HO (m-scheme) basis which is the same basis used in the NCSM, thereby facilitating calculations and symmetry identification.

The basis states within a $\mathrm{Sp}(3, \mathbb{R})$ irrep are built by applying multiples of the symplectic raising operators to a np-nh (n-particle-n-hole, $n =$ 0, 2, 4, ...) lowest-weight $\mathrm{Sp}(3, \mathbb{R})$ state (symplectic bandhead), which by definition is annihilated by the symplectic lowering operator. The raising operator induces a $2\hbar\Omega$ 1p-1h monopole or quadrupole excitation (one particle raised by two shells) together with a smaller $2\hbar\Omega$ 2p-2h correction for eliminating the spurious center-of-mass motion. If one were to include all possible lowest-weight np-nh starting state configurations ($n \leq N_{max}$), and allowed all multiples thereof, one would span the full NCSM space.

3. Results and Discussions

3.1. *Reproduction of NCSM results in a* $\mathrm{Sp}(3, \mathbb{R})$ *subspace*

The lowest-lying eigenstates of ^{12}C and ^{16}O were calculated using the NCSM[b] as implemented through the Many Fermion Dynamics (MFD) code[14] with an effective interaction derived from the realistic JISP16 *NN* potential[8] for different $\hbar\Omega$ oscillator strengths. For both nuclei we con-

[b]In order to speed up the calculations, we retained only the largest amplitudes of the NCSM states, those sufficient to account for at least 98% of the total norm.

structed all of the 0p-0h and $2\hbar\Omega$ 2p-2h symplectic bandheads and generated their Sp$(3,\mathbb{R})$ irreps up to $N_{max} = 6$ ($6\hbar\Omega$ model space). Analysis of overlaps of the symplectic states with the NCSM eigenstates in the 0, 2, 4 and $6\hbar\Omega$ subspaces reveals the dominance of the 0p-0h Sp$(3,\mathbb{R})$ irreps [Fig. 1(a,b,c)]. For the 0^+_{gs} and the lowest 2^+ and 4^+ states in ^{12}C there are nonnegligible overlaps for only 3 of the 13 0p-0h Sp$(3,\mathbb{R})$ irreps, namely, the leading (most deformed) representation specified by the shape deformation of its symplectic bandhead, (0 4), and spin $S = 0$ together with two (1 2) $S = 1$ irreps with different bandhead constructions for protons and neutrons. For the ground state of ^{16}O there is only one possible 0p-0h Sp$(3,\mathbb{R})$ irrep, (0 0) $S = 0$ (Fig. 1d). In addition, among the $2\hbar\Omega$ 2p-2h Sp$(3,\mathbb{R})$ irreps only a small fraction of the most deformed configurations contributes significantly to the overlaps accounting for 5% (10%) in ^{12}C (^{16}O) (Fig. 2). The results reveal that approximately 85-90% of the NCSM eigenstates fall within a subspace spanned by the few most significant 0p-0h and $2\hbar\Omega$ 2p-2h Sp$(3,\mathbb{R})$ irreps.

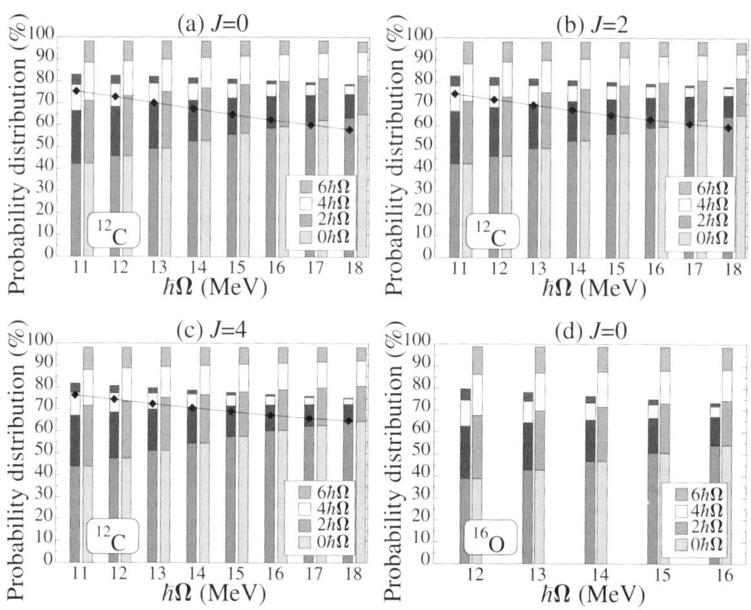

Fig. 1. Probability distribution for the (a) 0^+_{gs}, (b) 2^+_1 and (c) 4^+_1 states in ^{12}C and (d) 0^+_{gs} in ^{16}O over $0\hbar\Omega$ (blue, lowest) to $6\hbar\Omega$ (green, highest) subspaces for the 3 0p-0h Sp$(3,\mathbb{R})$ irrep case (left) and NCSM (right) together with the (0 4) irrep contribution (black diamonds) in ^{12}C as a function of the $\hbar\Omega$ oscillator strength in MeV for $N_{max} = 6$.

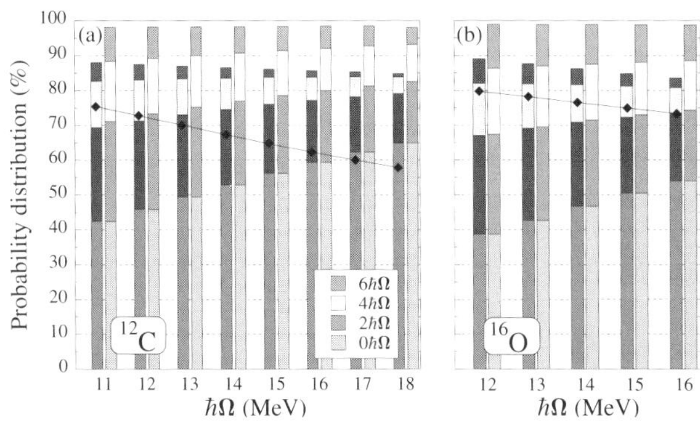

Fig. 2. Ground 0^+ state probability distribution over $0\hbar\Omega$ (blue, lowest) to $6\hbar\Omega$ (green, highest) subspaces for the most dominant 0p-0h + $2\hbar\Omega$ 2p-2h Sp$(3,\mathbb{R})$ irrep case (left) and NCSM (right) together with the leading irrep contribution (black diamonds), (0 4) for ^{12}C (a) and (0 0) for ^{16}O (b), as a function of the $\hbar\Omega$ oscillator strength, $N_{max} = 6$.

The largest contribution comes from the leading Sp$(3,\mathbb{R})$ irrep (Fig. 2, black diamonds), growing to 80% of the NCSM wavefunctions for the lowest $\hbar\Omega$. The results can be also interpreted as a strong confirmation of Elliott's SU(3) model since the projection of the NCSM states onto the $0\hbar\Omega$ space [Fig. 2, blue (lowest) bars] is a projection of the NCSM results onto the SU(3) shell model. Clearly, the simplest of Elliott's collective states can be regarded as a good first-order approximation in the presence of realistic interactions, whether the latter is restricted to a $0\hbar\Omega$ model space or richer multi-$\hbar\Omega$ NCSM model spaces.

The 0^+_{gs} and 2^+_1 states in ^{12}C, constructed in terms of the three Sp$(3,\mathbb{R})$ irreps with probability amplitudes defined by the overlaps with the NCSM wavefunctions for $N_{max} = 6$ case, were also used to determine $B(E2 : 2^+_1 \to 0^+_{gs})$ transition rates. The Sp$(3,\mathbb{R})$ $B(E2 : 2^+_1 \to 0^+_{gs})$ values reproduce almost exactly ($\sim 100\%$) the NCSM results.

3.2. *Large reduction of model space dimension*

As N_{max} is increased the dimension of the $J = 0, 2$, and 4 symplectic space built on the 0p-0h Sp$(3,\mathbb{R})$ irreps for ^{12}C grows very slowly compared to the NCSM space dimension (Fig. 3a). The dominance of only three irreps additionally reduces the dimensionality of the symplectic model space, which remains a small fraction of the NCSM basis space even when the most dominant $2\hbar\Omega$ 2p-2h Sp$(3,\mathbb{R})$ irreps are included. The space reduction is even more dramatic in the case of ^{16}O (Fig. 3b). This means that a space

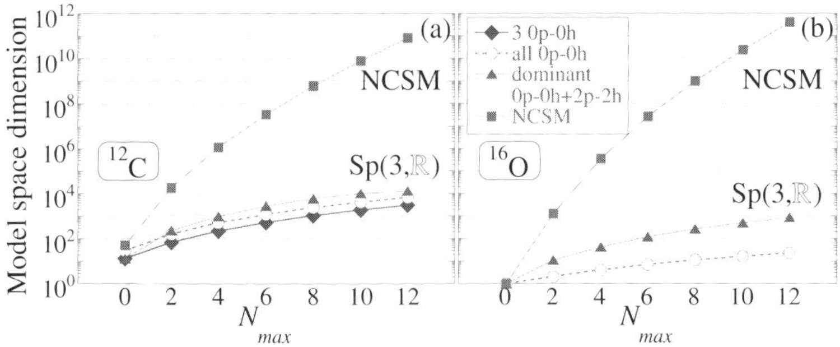

Fig. 3. NCSM space dimension vs. the maximum $\hbar\Omega$ excitations, N_{max}, compared to that of the Sp(3, $\mathbb{R}$) subspace: (a) $J = 0, 2,$ and 4 for ^{12}C, and (b) $J = 0$ for ^{16}O.

spanned by a set of symplectic basis states is computationally manageable even when high-$\hbar\Omega$ configurations are included.

In short, the symplectic subspace for the low-lying states in ^{12}C and ^{16}O that achieves large overlaps with the realistic NCSM eigenstates and reproduction of the NCSM estimates for the B(E2) transition rates comprises only a small fraction of the full NCSM model space.

3.3. Sp(3, $\mathbb{R}$) *invariance within NCSM spin components*

Another striking property of the low-lying eigenstates is revealed when the spin projections of the converged NCSM states are examined. Specifically, as shown in Fig. 4, their Sp(3, $\mathbb{R}$) symmetry and hence the geometry of the nucleon system being described is nearly independent of the $\hbar\Omega$ oscillator strength. The symplectic symmetry is present with equal strength in the spin parts of the NCSM wavefunctions for ^{12}C as well as ^{16}O regardless of whether the bare or the effective interactions are used. This suggests that the Lee-Suzuki transformation, which effectively compensates for the finite space truncation by renormalization of the bare interaction, does not affect the Sp(3, $\mathbb{R}$) symmetry structure of the spatial wavefunctions. Hence, the symplectic structure detected in the present analysis for $6\hbar\Omega$ model space is what would emerge in NSCM evaluations with a sufficiently large model space to justify use of the bare interaction.

4. Conclusions

We have shown that *ab initio* NCSM calculations with the JISP16 nucleon-nucleon interaction display a very clear symplectic structure, which is unaltered whether the bare or effective interactions for various $\hbar\Omega$ strengths

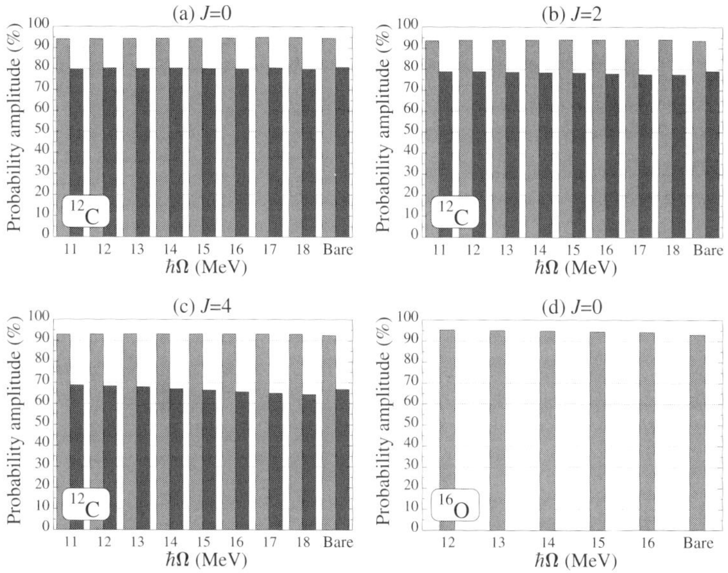

Fig. 4. Projection of the $S = 0$ (blue, left) [and $S = 1$ (red, right)] $\mathrm{Sp}(3,\mathbb{R})$ irreps onto the corresponding significant spin components of the NSCM wavefunctions for (a) 0^+_{gs}, (b) 2^+_1, and (c) 4^+_1 in ^{12}C and (d) 0^+_{gs} in ^{16}O, for effective interaction for different $\hbar\Omega$ oscillator strengths and bare interaction.

are used. The NCSM results are reproduced remarkably well by only a few 0p-0h and $2\hbar\Omega$ 2p-2h spurious center-of-mass free symplectic irreps with dimensionality that is only $\approx 10^{-3}\%$ that of the NCSM model space. Specifically, these symplectic states account for 85-90% of the NCSM wavefunctions for the lowest 0^+_{gs}, 2^+_1 and 4^+_1 states in ^{12}C and the ground state in ^{16}O and they closely reproduce the NCSM $B(E2)$ estimates.

The comparisons with NCSM results open the path for consequent studies as the results suggest either the effective nucleon-nucleon interaction possesses a heretofore unappreciated symmetry, namely Sp(3,R) and the complementary (spin-isospin) supermultiplet symmetry, or the nuclear many-body system acts as a filter that allows the symplectic symmetry to propagate in a coherent way into the many-body dynamics while reducing the effects of symplectic symmetry breaking terms.

In short, the results confirm for the first time the validity of the $\mathrm{Sp}(3,\mathbb{R})$ approach when realistic interactions are invoked in a NCSM space. This demonstrates the importance of the $\mathrm{Sp}(3,\mathbb{R})$ symmetry in light nuclei while reaffirming the value of the simpler Elliott SU(3) model upon which it is based. The results further suggest that a Sp-NCSM extension of the *ab-initio* NCSM may be a practical scheme for reaching heavier nuclei and achieving convergence to measured $B(E2)$ values without the need for introducing an effective charge.

Acknowledgments

Discussions with many colleague, but especially Bruce R. Barrett, are gratefully acknowledged. This work was supported by the US National Science Foundation, Grant Nos 0140300 & 0500291, and the Southeastern Universities Research Association, as well as, in part, by the US Department of Energy Grant No. DE-FG02-87ER40371. TD acknowledges supplemental support from the Graduate School of Louisiana State University.

References

1. T. Dytrych, K. D. Sviratcheva, C. Bahri, J. P. Draayer and J. P. Vary, Phys. Rev. Lett. **98**, 162503 (2007).
2. P. Navrátil, J. P. Vary and B. R. Barrett, Phys. Rev. Lett. **84**, 5728 (2000); Phys. Rev. C **62**, 054311 (2000).
3. P. Navrátil and B. R. Barrett, Phys. Rev. C **57**, 562 (1998); **59**, 1906 (1999); P. Navrátil, G. P. Kamuntavičius, and B. R. Barrett, Phys. Rev. C **61**, 044001 (2000).
4. P. Navrátil and W. E. Ormand, Phys. Rev. C **68**, 034305 (2003).
5. G. Rosensteel and D. J. Rowe, Phys. Rev. Lett. **38**, 10 (1977).
6. D. J. Rowe, Reports on Progr. in Phys. **48**, 1419 (1985).
7. K. T. Hecht and D. Braunschweig, Nucl. Phys. A**295**, 34 (1978); Y. Suzuki, Nucl. Phys. A**448**, 395 (1986).
8. A. M. Shirokov, A. I. Mazur, S. A. Zaytsev, J. P. Vary and T. A. Weber, Phys. Rev. C **70**, 044005 (2004); A. M. Shirokov, J. P. Vary, A. I. Mazur, S. A. Zaytsev and T. A. Weber, Phys. Letts. B **621**, 96(2005); A. M. Shirokov, J. P. Vary, A. I. Mazur, and T. A. Weber, Phys. Letts. B **644**, 33 (2007).
9. G. Rosensteel and D. J. Rowe, Ann. Phys. N.Y. **126**, 343 (1980).
10. J. P. Draayer, K. J. Weeks and G. Rosensteel, Nucl. Phys. A**413**, 215 (1984).
11. J. Escher and A. Leviatan, Phys. Rev. C **65**, 054309 (2002).
12. F. Arickx, J. Broeckhove and E. Deumens, Nucl. Phys. A**377**, 121 (1982).
13. S. S. Avancini and E. J. V. de Passos, J. Phys. G **19**, 125 (1993).
14. J. P. Vary, "The Many-Fermion-Dynamics Shell-Model Code," Iowa State University, 1992 (unpublished); J. P. Vary and D. C. Zheng, *ibid* 1994 (unpublished).

ROLE OF DEFORMED SYMPLECTIC CONFIGURATIONS in *AB INITIO* NO-CORE SHELL MODEL RESULTS

T. DYTRYCH[1], K. D. SVIRATCHEVA[1], C. BAHRI[1], J. P. DAAYER[1], AND J. P. VARY[2]

[1]*Department of Physics and Astronomy, Louisiana State University, Baton Rouge, Louisiana 70803, USA*
[2]*Department of Physics and Astronomy, Iowa State University, Ames, IA 50011, USA*

A set of methods for the construction of an arbitrary n-particle-n-hole (np-nh) translationally invariant symplectic irreducible representation in a spherical harmonic oscillator basis is presented. The methods are used to construct all possible 2p-2h as well as the most deformed 4p-4h irreducible representation of $\mathrm{Sp}(3,\mathbb{R})$ in ^{12}C and in ^{16}O. We use the results to demonstrate the significant role these 2p-2h $\mathrm{Sp}(3,\mathbb{R})$ states plays in a description of the low-lying states calculated within the framework of the no-core shell model using the realistic JISP16 nucleon-nucleon interaction.

1. Introduction

The *ab initio* no-core shell model (NCSM)[1] approach has emerged as a prominent method for modeling properties of light nuclei at the microscopic level. Utilizing modern realistic internucleon interactions,[2–4] the NCSM has achieved favorable agreement between theory and experiment for various low-lying states of light nuclei.

Despite its success, the NCSM with the largest model spaces currently attainable, i.e. $N_{\mathrm{max}} = 6$ for the upper p-shell nuclei, is still not capable of reproducing the observed excitation energies of low-lying states dominated by multiple-particle-multiple-hole modes.[1,5] Examples of the latter are the 0_2^+ states in ^{12}C and ^{16}O, both highly deformed with dominating multiple-particle-multiple-hole configurations[6] and a pronounced α-cluster structure.[7,8] The intrinsic structure of these states essentially requires yet larger NCSM model spaces[1] in order to accurately reproduce their excitation energies. However, the dimensionalities of such model spaces are well-beyond present computing capabilities. Clearly, physically relevant schemes that bring about a strong model space reduction are needed. The well-developed microscopic cluster method,[7–9] which reduces the model space by exploiting the cluster nature of investigated states, demonstrates the feasibility and effectivness of such an approach.

This work focuses on the symplectic shell-model[10] approach. The scheme utilizes irreducible representations (irreps) of the symplectic Sp(3, $\mathbb{R}$) group, which is the dynamical symmetry underlying a microscopic description of the nuclear collective phenomena. Recently, the preponderance of only few 0p-0h symplectic irreps in the NCSM eigenstates in ^{12}C and ^{16}O was reported,[11] which confirms the validity of this scheme*. Studies of the relationships of symplectic and cluster model wave functions[12] indicate that the restriction of the symplectic model space to the dominant 0p-0h irreps is not sufficient to desribe highly deformed states with a pronounced α-cluster structure. For a more microscopic theory of deformed states in light nuclei, a symplectic model space with highly deformed $2\hbar\Omega$ 2p-2h and $4\hbar\Omega$ 4p-4h Sp(3, $\mathbb{R}$) irreps must be included. It is the purpose of this contribution to discuss the role of the latter irreps within the ground-state rotational band in ^{12}C, and the ground and 0_2^+ state in ^{16}O, as determined within the framework of the NCSM with the JISP16 realistic interaction.

2. Construction of a translationally invariant symplectic basis space

For A nucleons, the 21 distinct bilinear products of the particle momentum ($p_{s\alpha}$) and coordinate ($x_{s\beta}$) operators, $T_{\alpha\beta} = \sum_s p_{s\alpha}p_{s\beta}$, $L_{\alpha\beta} = \sum_s(x_{s\alpha}p_{s\beta} - x_{s\beta}p_{s\alpha})$, $S_{\alpha\beta} = \sum_s(x_{s\alpha}p_{s\beta} + p_{s\alpha}x_{s\beta})$, and $Q_{\alpha\beta} = \sum_s x_{s\alpha}x_{s\beta}$ with $\alpha, \beta = 1, 2, 3$ for the 3 spatial directions and $s = 1, \ldots, A$, yield a realization of the symplectic $\mathfrak{sp}(3, \mathbb{R})$ algebra.[10] It follows from this that the elements of the $\mathfrak{sp}(3, \mathbb{R})$ algebra include important observables such as the many-particle kinetic energy, the mass quadrupole moment and angular momentum operators, together with multi-shell vibrations and vorticity degrees of freedom for a description of rotational dynamics in a continuous range from irrotational to rigid rotor flows.

The symplectic generators as described above are not translationally invariant. This problem can be overcome by constructing the Sp(3, $\mathbb{R}$) generators in terms of coordinates defined with respect to the center-of-mass

---

*In accordance with the classification of symplectic irreps by their starting-state configuration, we denote an irrep constructed over a $N\hbar\Omega$ n-particle-n-hole configuration as a $N\hbar\Omega$ np-nh irrep.

(c.m.) momentum and position, i.e. $x'_{s\alpha} = x_{s\alpha} - X_\alpha$, $p'_{s\alpha} = p_{s\alpha} - P_\alpha$. The c.m. momentum and position operators are defined as $P_{s\alpha} = \sum_s p_{s\alpha}$ and $X_\alpha = \frac{1}{A}\sum_s x_{s\alpha}$, respectively. The translationally invariant (intrinsic) $\mathrm{Sp}(3,\mathbb{R})$ generators, written in SU(3)-coupled form, are given as

$$\begin{aligned}
A^{(2\,0)}_{\mathfrak{L}M} &= \tfrac{1}{\sqrt{2}}\sum_i \left[b_i^\dagger \times b_i^\dagger\right]^{(2\,0)}_{\mathfrak{L}M} - \tfrac{1}{\sqrt{2}A}\sum_{s,t}\left[b_s^\dagger \times b_t^\dagger\right]^{(2\,0)}_{\mathfrak{L}M} \\
B^{(0\,2)}_{\mathfrak{L}M} &= \tfrac{1}{\sqrt{2}}\sum_i \left[b_i \times b_i\right]^{(0\,2)}_{\mathfrak{L}M} - \tfrac{1}{\sqrt{2}A}\sum_{s,t}\left[b_s \times b_t\right]^{(0\,2)}_{\mathfrak{L}M} \\
C^{(1\,1)}_{\mathfrak{L}M} &= \sqrt{2}\sum_i \left[b_i^\dagger \times b_i\right]^{(1\,1)}_{\mathfrak{L}M} - \tfrac{\sqrt{2}}{A}\sum_{s,t}\left[b_s^\dagger \times b_t\right]^{(1\,1)}_{\mathfrak{L}M},
\end{aligned} \tag{1}$$

together with $H_0^{(00)} = \sqrt{3}\sum_i \left[b_i^\dagger \times b_i\right]^{(00)} + \frac{3}{2}(A-1)$, where the sums are over all A particles of the system. Clearly, the intrinsic $\mathrm{Sp}(3,\mathbb{R})$ generators (1) can be obtained from the translationally non-invariant $\mathrm{Sp}(3,\mathbb{R})$ generators [first term in (1)] after the subtraction of the c.m. two-body operators [last term in (1)], $A^{(20)\mathrm{cm}}_{\mathfrak{L}M} = \frac{1}{\sqrt{2}}\left[\mathfrak{B}^\dagger \times \mathfrak{B}^\dagger\right]^{(20)}_{\mathfrak{L}M}$, $B^{(02)\mathrm{cm}}_{\mathfrak{L}M} = \frac{1}{\sqrt{2}}\left[\mathfrak{B} \times \mathfrak{B}\right]^{(02)}_{\mathfrak{L}M}$, and $C^{(11)\mathrm{cm}}_{\mathfrak{L}M} = \sqrt{2}\left[\mathfrak{B}^\dagger \times \mathfrak{B}\right]^{(11)}_{\mathfrak{L}M}$. These c.m. operators together with the c.m. excitation number operator,

$$\hat{N}^{\mathrm{cm}} = \mathfrak{B}^\dagger \cdot \mathfrak{B}, \tag{2}$$

are expressed by means of the c.m. harmonic oscillator ladder operators,

$$\mathfrak{B}^\dagger_\alpha = \sqrt{\tfrac{Am\Omega}{2\hbar}}\left(X_\alpha - \tfrac{i}{Am\Omega}P_\alpha\right) = \tfrac{1}{\sqrt{A}}\sum_{i=1}^A b^\dagger_{i\alpha} = (\mathfrak{B}_\alpha)^\dagger \tag{3}$$

The c.m. operators generate the $\mathrm{Sp}(3,\mathbb{R})$ group with representations equivalent to the $\mathrm{Sp}(3,\mathbb{R})$ representation of a single-particle space, namely $(k\,0)$ for a k-$\hbar\Omega$ c.m. excitation.

The symplectic basis states (labeled in standard notation[10]) are constructed by acting with polynomials $\mathcal{P}$ in the symplectic raising operator, $A^{(2\,0)}$ (1), on a set of basis states of the symplectic bandhead, $|\sigma\rangle$, which is a $\mathrm{Sp}(3,\mathbb{R})$ lowest-weight state,

$$|\sigma n\rho\omega\kappa(LS)JM_J\rangle = \left[\mathcal{P}^n(A^{(2\,0)}) \times |\sigma\rangle\right]^{\rho\omega}_{\kappa(LS)JM_J}, \tag{4}$$

where $\sigma \equiv N_\sigma\,(\lambda_\sigma\,\mu_\sigma)$ labels $\mathrm{Sp}(3,\mathbb{R})$ irreps with $(\lambda_\sigma\,\mu_\sigma)$ denoting a SU(3) lowest-weight state, $n \equiv N_n\,(\lambda_n\,\mu_n)$, and $\omega \equiv N_\omega\,(\lambda_\omega\,\mu_\omega)$. The quantum number $N_\omega = N_\sigma + N_n$ is the total number of oscillator quanta related to the eigenvalue, $N_\omega\hbar\Omega$, of a harmonic oscillator (HO) Hamiltonian that is free of spurious modes. The $(\lambda_n\,\mu_n)$ set gives the overall SU(3) symmetry of $\frac{N_n}{2}$ coupled raising operators in $\mathcal{P}$ and $(\lambda_\omega\,\mu_\omega)$ specifies the SU(3) symmetry of the symplectic state. The symplectic basis states, which are generated by the translationally invariant form of the $A^{(2\,0)}$ symplectic raising operators (1), are free of c.m. spurious excitations provided that the $|\sigma\rangle$ symplectic bandhead is also free of such spurious excitations.

The symplectic bandhead is a SU(3) $\times$ SU(2) irrep that contains a Sp(3, $\mathbb{R}$) lowest-weight state. Consequently, the basis states of the symplectic bandhead, $|\sigma\kappa(L_0 S_0)J_0 M_0\rangle$, are annihilated upon action of the $B^{(02)}_{\mathfrak{L}M}$ symplectic lowering operators,

$$B^{(02)}_{\mathfrak{L}M}\,|\sigma\kappa(L_0 S_0)J_0 M_0\rangle = 0. \tag{5}$$

By construction, the symplectic bandhead, and hence the basis states of the corresponding Sp(3, $\mathbb{R}$) irrep (4), can be expanded in a m-scheme basis. In our study, the symplectic bandheads are constructed in a proton-neutron formalism as $\mathrm{SU(3)}_{(\lambda_\sigma\,\mu_\sigma)}\times\mathrm{SU(2)}_{S_0}$-symmetric linear combinations of m-scheme configurations of a given N_σ. The construction formula is given as,

$$|(\gamma)\sigma\kappa(L_0 S_0)J_0 M_0\rangle = \left[\mathcal{P}^{(\lambda_\pi\,\mu_\pi)}_{S_\pi}(a^\dagger_\pi)\times\mathcal{P}^{(\lambda_\nu\,\mu_\nu)}_{S_\nu}(a^\dagger_\nu)\right]^{\rho_\sigma(\lambda_\sigma\,\mu_\sigma)}_{\kappa(L_0 S_0)J_0 M_0}|0\rangle\,, \tag{6}$$

where γ schematically denotes the additional quantum numbers included to distinguish between different bandhead constructions for protons and neutrons. In (6), $\mathcal{P}^{(\lambda_\pi\,\mu_\pi)}_{S_\pi}$ and $\mathcal{P}^{(\lambda_\nu\,\mu_\nu)}_{S_\nu}$ are polynomials of proton ($a^\dagger_\pi$) and neutron ($a^\dagger_\nu$) creation operators coupled to good SU(3)$\times$SU(2) symmetry, that is to definite $(\lambda\,\mu)$ and spin S values.

The symplectic bandheads generated by the procedure described above are not translationally invariant with the exception of those constructed within the $0\hbar\Omega$ model space. We utilize U(3) symmetry preserving c.m. projecting operators[13] to eliminate spurious c.m. excitations in the symplectic bandheads. The projection technique is based on the fact that a general SU(3) symmetry adapted A-particle state of $n_{\max}\hbar\Omega$ excitations above the lowest energy configuration can be written in a SU(3)-coupled form as

$$|(\lambda\,\mu)\alpha\rangle=\sum_{n=0}^{n_{\max}}\sum_{(\lambda\,\mu)_{\mathrm{intr}}} c_n^{(\lambda\,\mu)_{\mathrm{intr}}}\,|(n\,0)\otimes(\lambda\,\mu)_{\mathrm{intr}};(\lambda\,\mu)\alpha\rangle\,, \tag{7}$$

where $\alpha=\{\kappa LM\}$ and $(c_n^{(\lambda\,\mu)_{\mathrm{intr}}})^2$ yield the probability amplitudes. The SU(3) quantum numbers, $(\lambda\,\mu)_{\mathrm{intr}}$, label the intrinsic wavefunctions of $(n_{\max}-n)\hbar\Omega$ excitations that are coupled with the c.m. SU(3) irreps $(n\,0)$ of $n\hbar\Omega$ excitations into the final SU(3) symmetry $(\lambda\,\mu)$.

In order to eliminate the c.m. spuriosity from a given A-particle SU(3)-symmetric state, one needs to project out all the $n\geq 1$ terms on the right hand side of the expansion (7) as they describe excited (spurious) c.m. motion. This is done by employing the c.m. number operator $\hat{N}^{\rm cm}$ (2), in a U(3) symmetry preserving projecting operator,

$$\hat{P}(n_{\max}) = \prod_{k_0=1}^{n_{\max}} \left(1 - \frac{\hat{N}^{\rm cm}}{k_0}\right). \tag{8}$$

3. Results and Discussions

The lowest-lying eigenstates of the deformed ^{12}C nucleus and the "closed-shell" ^{16}O nucleus were calculated using the NCSM as implemented through the Many Fermion Dynamics (MFD) code[14] with an effective interaction derived from the realistic JISP16 NN potential[2] for different $\hbar\Omega$ oscillator strengths and with the bare interaction. We analyze the symplectic symmetry structure of the $J=0^+_{gs}$ and the lowest $J=2^+(\equiv 2^+_1)$ and $J=4^+(\equiv 4^+_1)$ states in ^{12}C and the 0^+_{gs} ground state in ^{16}O, which appear to be reasonably well converged in the $N_{max} = 6$ NCSM basis space.

The methods described above were used to construct all translationally invariant $2\hbar\Omega$ 2p-2h Sp(3, $\mathbb{R}$) bandheads in ^{12}C and ^{16}O. For the latter nucleus, we also generated the most deformed $4\hbar\Omega$ 4p-4h, $(\lambda_\sigma\, \mu_\sigma) = (8\,4)$, symplectic bandhead. The projection of the symplectic bandheads onto the NCSM eigenstates yields 20 most dominant $2\hbar\Omega$ 2p-2h symplectic bandheads, which are used to generate the corresponding Sp(3, $\mathbb{R}$) translationally invariant irreps up to $N_{max} = 6$ ($6\hbar\Omega$ model space).

The symplectic model space remains a small fraction of the NCSM basis space even when the most dominant $2\hbar\Omega$ 2p-2h Sp(3, $\mathbb{R}$) irreps are included (Fig. 1). The space reduction is even more dramatic in the case of ^{16}O, where only the $J = 0$ symplectic space can be taken into account for the 0^+ states under consideration (Fig.1b). This means that a space spanned by a set of symplectic basis states is computationally manageable even when high-$\hbar\Omega$ configurations are included.

The symplectic model space expansion by means of the most important $2\hbar\Omega$ 2p-2h Sp(3, $\mathbb{R}$) irreps improves the overlaps between the NCSM eigenstates and the Sp(3, $\mathbb{R}$) symmetric basis by about 5% (Fig. 2). Overall, approximately 85% of the NCSM eigenstates fall within a subspace spanned by the three most significant 0p-0h and 20 $2\hbar\Omega$ 2p-2h Sp(3, $\mathbb{R}$) irreps.

A much more interesting scenario is observed for the lower-lying 0^+ states in ^{16}O. The 0p-0h symplectic model space analysis has revealed the dominance of the $(0\,0)S = 0$ symplectic irrep, namely 80−75% of the NCSM realistic wavefunction for values of the oscillator strength $\hbar\Omega = 12$MeV to 16MeV.[11] The $0\hbar\Omega$ projection of the $(0\,0)S = 0$ Sp(3, $\mathbb{R}$) irrep reflects

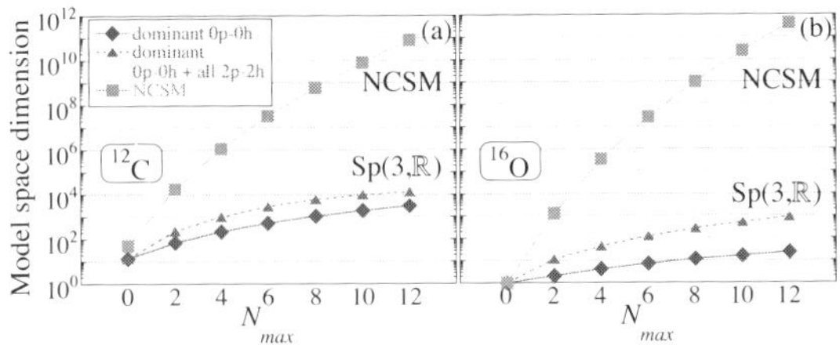

Fig. 1. Dimension of the NCSM model space as a function of maximum allowed $\hbar\Omega$ excitations, N_{max}, compared to Sp$(3,\mathbb{R})$ space with (a) $J = 0, 2$, and 4 for ^{12}C, and (b) $J = 0$ for ^{16}O.

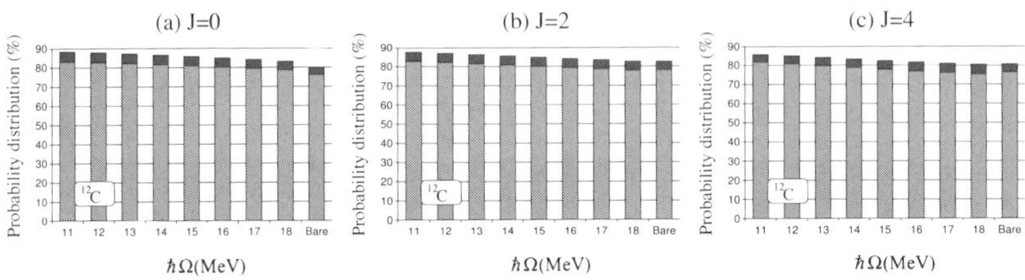

Fig. 2. Projection of the most dominant 0p-0h (orange) and $2\hbar\Omega$ 2p-2h (blue) Sp$(3,\mathbb{R})$ irreps onto (a) the ground, (b) 2_1^+, and 4_1^+ NCSM wavefunctions in ^{12}C as a function of the $\hbar\Omega$ oscillator strength and when the bare interaction is used.

the spherical shape preponderance in the ground state of ^{16}O, specifically around $40-55\%$ for the same $\hbar\Omega$ range. In addition, a relatively significant mixture, 13–25%, of slightly prolate deformed shape is observed. This shape is described by the $2\hbar\Omega$ $(2\,0)$ 1p-1h and weaker 2p-2h Sp$(3,\mathbb{R})$-symmetric excitations built over the $(0\,0)S = 0$ symplectic bandhead. Orthogonal to these excitations, the $(2\,0)S = 0$ 2p-2h symplectic bandhead constructed at the $2\hbar\Omega$ level are found the most dominant among the $2\hbar\Omega$ 2p-2h Sp$(3,\mathbb{R})$ bandheads. This means that the dominance of the $2\hbar\Omega$ 2p-2h Sp$(3,\mathbb{R})$ bandheads in the ground state of ^{16}O is governed in such a way to preserve the shape coherence of all the significant $2\hbar\Omega$ excitations.

The inclusion of the most important $2\hbar\Omega$ 2p-2h Sp$(3,\mathbb{R})$ irreps improves the overlaps with the ^{16}O ground state by 10%. Overall, the ground state in ^{16}O projects at 85%-90% level onto the $J = 0$ symplectic symmetry adapted basis (Fig. 3a) with a total dimensionality of only $\approx 0.001\%$ of the NCSM space.

The correct description of the 0_2^+ state in ^{16}O has been a long standing and extensively studied problem. Nevertheless, even up-to-date methods

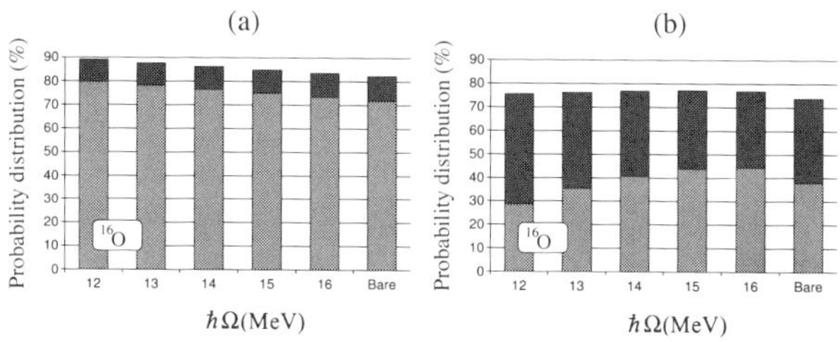

Fig. 3. Projection of the 0p-0h (orange) Sp(3, ℝ) irrep and the most dominant $2\hbar\Omega$ 2p-2h (blue) Sp(3, ℝ) irreps onto (a) the ground, and (b) 0_2^+ NCSM wavefunctions in ^{16}O as a function of the $\hbar\Omega$ oscillator strength and when the bare interaction is used.

such as *ab initio* NCSM with $N_{\max} = 6$ model space[5] or coupled-cluster calculations up to 3p-3h[15] fail to reproduce the observed excitation energy of this highly deformed state. The symplectic no-core shell-model (Sp-NCSM) emphasizes the fact that this state can be described naturally in terms of a particular set of Sp(3, ℝ) irreps.

We focus our investigation on determining whether the highly deformed multiple-particles-multiple-hole symplectic irreps are realized within the 0_2^+ NCSM eigenstate. It is important to note that this state is not fully converged. The results show that the 0_2^+ NCSM eigenstate is dominated by the $2\hbar\Omega$ configurations composing from 45% up to 55% of the wave function. The projection of this NSCM wavefunction onto the symplectic basis reveals a large contribution of the $2\hbar\Omega$ 2p-2h Sp(3, ℝ) irreps (Fig. 3b). This contribution tends to decrease with increasing $\hbar\Omega$ oscillator strength. It reaches 47% for $\hbar\Omega = 12$ MeV and declines down to 33% for $\hbar\Omega = 16$ MeV, with a clear dominance of the leading, $(4\,2)$, $2\hbar\Omega$ 2p-2h irrep. The leading most deformed $4\hbar\Omega$ 4p-4h symplectic irrep, $(\lambda_\sigma\,\mu_\sigma) = (8\,4)$, contribute rather insignificantly (0.31%-0.16%) to the NCSM 0_2^+ state. Overall, the symplectic basis projects at the 80% level onto the first excited 0^+ state.

4. Conclusion

Our study reveals for the first time the important role of the most deformed $2\hbar\Omega$ 2p-2h symplectic irreps for the description of NCSM eigenstates determined by a realistic interaction. Specifically, NCSM wavefunctions for the lowest 0_{gs}^+, 2_1^+ and 4_1^+ states in ^{12}C and the ground state in ^{16}O project at the 85-90% level onto very few 0p-0h and $2\hbar\Omega$ 2p-2h spurious center-of-mass free symplectic irreps. The addition of the $2\hbar\Omega$ 2p-2h symplectic configurations improved the overlaps by 5% (for ^{12}C) and 10% (for ^{16}O). The outcome also shows that the most deformed $2\hbar\Omega$ 2p-2h Sp(3, ℝ) irreps, predominantly (4 2), are strongly represented in the 0_2^+ state in ^{16}O. The total dimensionality of the symplectic basis remains only a fraction ($\approx 10^{-3}\%$) of the NCSM space, even when the most dominant 2p-2h irreps are included. The results also suggest that a Sp-NCSM may be a practical scheme for modeling cluster-like phenomena as these modes can be accommodated within the general framework of the Sp(3, $\mathbb{R}$) model with 0p-0h, $2\hbar\Omega$ 2p-2h, $4\hbar\Omega$ 4p-4h, and maybe higher $n\hbar\Omega$ np-nh starting state configurations if extended to large model spaces (high N_{max}).

Acknowledgments

Discussions with many colleague, but especially Bruce R. Barrett, are gratefully acknowledged. This work was supported by the US National Science Foundation, Grant Nos 0140300 & 0500291, and the Southeastern Universities Research Association, as well as, in part, by the US Department of Energy Grant No. DE-FG02-87ER40371. Tomáš Dytrych acknowledges supplemental support from the Graduate School of Louisiana State University.

References

1. P. Navrátil, J. P. Vary and B. R. Barrett, Phys. Rev. Lett. **84**, 5728 (2000);
2. A. M. Shirokov, J. P. Vary, A. I. Mazur and T. A. Weber, Phys. Letts. B **644**, 33 (2007).
3. R. Machleidt, Phys. Rev. C **63**, 024001 (2001).
4. D. R. Entem and R. Machleidt, Phys. Rev. C **68**, 041001 (2003).
5. J. P. Vary, "How Effective are Strong Interactions," in Blueprints for the Nucleus, C. Johnson, Editor, Int. J. Mod. Phys. E **14**, 1 (2005).
6. H. Morinaga, Phys. Rev. **101**, 254 (1956).
7. M. Chernykh, H. Feldmeier, T. Neff, P. von Neumann-Cosel and A. Richter, Phys. Rev. Lett. **98**, 032501 (2007).
8. I. N. Filikhin and S. L. Yakovlev, Physics of Atomic Nuclei **63**, 343 (2000).
9. R. F. Bishop, M. F. Flynn, M. C. Boscá, E. Buenda and R. Guardiola, Phys. Rev. C **42**, 1341 (1990).
10. G. Rosensteel and D. J. Rowe, Phys. Rev. Lett. **38**, 10 (1977);
11. T. Dytrych, K. D. Sviratcheva, C. Bahri, J. P. Draayer and J. P. Vary, to be published in Phys. Rev. C.
12. Y. Suzuki, Nucl. Phys. A**448**, 395 (1986);
13. K. T. Hecht, Nucl. Phys. A **170**, 34 (1971).
14. J. P. Vary, "The Many-Fermion-Dynamics Shell-Model Code," Iowa State University, 1992 (unpublished); J. P. Vary and D. C. Zheng, *ibid* 1994 (unpublished).
15. M. Wloch, D. J. Dean, J. R. Gour, M. Hjorth-Jensen, K. Kowalski, T. Papenbrock and P. Piecuch, Phys. Rev. Lett. **94**, 212501 (2005).

NUCLEAR STRUCTURE, DOUBLE BETA DECAY AND NEUTRINO MASS

AMAND FAESSLER

Institute for Theoretical Physics, University of Tuebingen,
Auf der Morgenstelle 14, Tuebingen, Germany
E-mail: amand.faessler@uni-tuebingen.de

The neutrinoless double beta decay is not allowed in the Standard Model (SM) but it is allowed in most Grand Unified Theories (GUT's). The neutrino must be a Majorana particle (identical with its antiparticle) and must have a mass to allow the neutrinoless double beta decay. Apart of one claim that the neutrinoless double beta decay in ^{76}Ge is measured, one has only upper limits for this transition probability. But even the upper limits allow to give upper limits for the electron Majorana neutrino mass and upper limits for parameters of GUT's and the minimal R-parity violating supersymmetric model. One further can give lower limits for the vector boson mediating mainly the right-handed weak interaction and the heavy mainly right-handed Majorana neutrino in left-right symmetric GUT's. For that one has to assume that the specific mechanism is the leading one for the neutrinoless double beta decay and one has to be able to calculate reliably the corresponding nuclear matrix elements. In the present contribution, one discusses the accuracy of the present status of calculating the nuclear matrix elements and the corresponding limits of GUT's and supersymmetric parameters.

Keywords: double beta decay, Majorana neutrino, nuclear matrix elements

1. Physics beyond the Standard Model and the Double Beta Decay

In the standard model the neutrinoless double beta decay is forbidden but it is allowed in most Grand Unified Theories (GUT's) where the neutrino is a Majorana particle (identical with its antiparticle) and where the neutrinos have a mass.

In GUT's, each beta decay vertex in Fig. 1 can occur in eight different ways (see Fig. 2).

The hadronic current changing a neutron into a proton can be left- or right-handed, the vector boson exchanged can be the light one W_1 or the

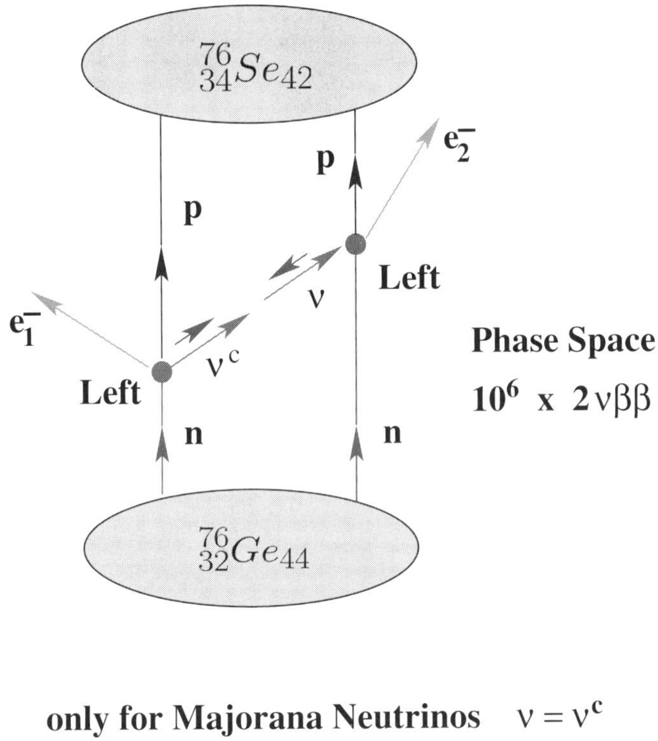

Fig. 1. Neutrinoless double beta decay of ^{76}Ge through ^{76}As to the final nucleus ^{76}Se. The neutrino must be a Majorana particle that means identical with its antiparticle and must have a mass to allow this decay.

heavy orthogonal combination W_2 mediating mainly a right-handed weak interaction and two different leptonic currents changing a neutrino into an electron. So the simple vertex for the beta decay can occur in eight different ways.

With two such vertices and the exchange of a light or a heavy mainly right-handed Majorana neutrino one already has 128 different matrix elements describing the neutrinoless double beta decay[1] (see Fig. 3).

At the R-parity violating vertex (see Fig. 4) of the up and the down quarks and the SUSY electron $\tilde{e}$, one has the R-parity violating coupling constant λ'_{111} which is new compared to the standard model.[2] The formation of a pion by a down and a up quark produces a long range neutrinoless double beta decay transition operator. Due to the short range Brueckner repulsive correlations between two nucleons, this is increasing the transition

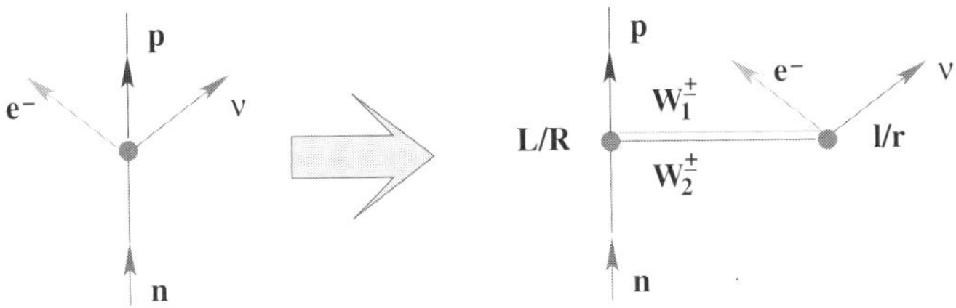

$$l/r = \bar{e}\,\gamma^{\mu}\left(1 \mp \gamma^{5}\right)\nu$$

$$L/R = p^{+}\left(g_V \mp g_A\,\vec{\sigma}\right)n$$

$$g_V = 1 \ ; \ g_A = 1.25$$

Fig. 2. The single beta decay changes a neutron into a proton by the emission of an electron and (in the standard model) an antineutrino. The microscopic version is shown on the right side of the figure.

$$\hat{H}^{\mathrm{GUT}}_{\mathrm{Weak}} = \frac{G_F \cos\vartheta_c}{\sqrt{2}}\left[1 \cdot l \cdot L + t\,g\,\vartheta\; r \cdot L + t\,g\,\vartheta\; R \cdot l + \frac{M_1^2}{M_2^2}\, r \cdot R\right]$$

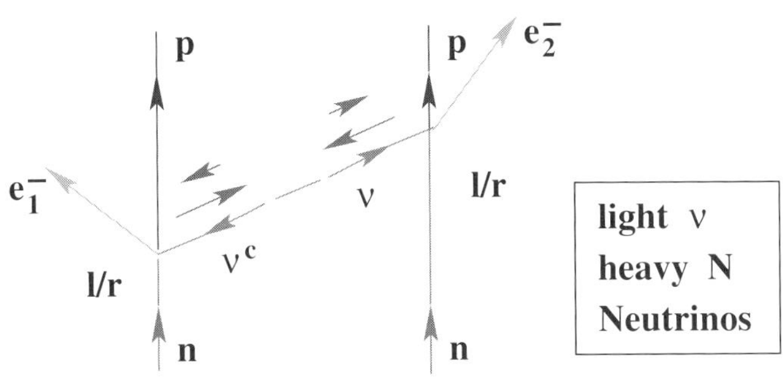

Fig. 3. Diagram of the double beta decay

probability by a factor 10 000. The upper limit derived from upper limits of the neutrinoless transition probability for λ'_{111} is therefore more stringent and one can derive from the neutrinoless transition probability in ^{76}Ge an upper limit for $|\lambda'_{111}|\langle 10^{-4}$.

To calculate the neutrinoless double beta decay transition probability, we use Fermi's Golden Rule in second order.

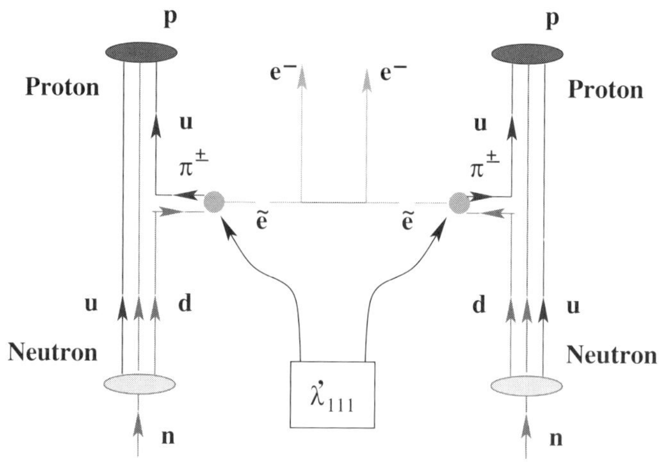

Fig. 4. Matrix element for the neutrinoless double beta decay within R-parity violating supersymmetry

$$
\begin{aligned}
T &= \sum_k \frac{< f|\hat{H}_W|k >< k|\hat{H}_W|i >}{E_i - E_k}, \\
E_i &= E_{0+}(^{76}Ge), \\
E_k &= E_k(e^-) + E_k(\nu) + E_{\text{nucleus}}(^{76}As; k), \\
T &= M_m < m_\nu > + M_\theta < tg\vartheta > \\
&\quad + M_{WR} < \left(\frac{M_2}{M_R}\right)^2 > \\
&\quad + M_{SUSY}\lambda'^2_{111} + M_{VR} < \frac{m_p}{M_{VR}} >, \\
w &= \frac{2\pi}{\hbar}|T|^2 p_f \leq 4.4 \cdot 10^{-33}[sec^{-1}],
\end{aligned}
\tag{1}
$$

Here $|k\rangle$ are the intermediate nuclear states in ^{76}As with a Majorana neutrino and one electron. The sum over k includes also an integration over neutrino energies. E_{0+} (^{76}Ge) is the ground state energy of ^{76}Ge.

To calculate the nuclear matrix elements, the most reliable method has turned out to be the quasiparticle random phase approximation (QRPA) to calculate at the example of ^{76}Ge the wave function of the initial ^{76}Ge, the excited states of the intermediate nucleus ^{76}As and the ground state of the final nucleus ^{76}Se.[3]

The QRPA approach describes the intermediate excited nuclear states

$|m\rangle$ for example in ^{76}As as a coherent superposition of two quasiparticle excitations and two quasiparticle annihilations relative to the initial ground state (in our example) of ^{76}Ge.

$$\begin{aligned} a_i^+ &= u_i c_i^+ - v_i c_{\bar{i}}, \\ A_\alpha^+ &= [a_i^+ a_k^+]_{J\alpha}, \\ Q_m^+ &= \sum_\alpha [X_\alpha^m A_\alpha^+ - Y_\alpha^m A_\alpha], \\ |m> &= Q_m^+ |g>, \end{aligned} \tag{2}$$

The QRPA equation for the determination of the intermediate states by X_α^m and Y_α^m are derived from the many-body Schroedinger equation by using quasi-boson commutation relations for the quasiparticle pairs $A_\alpha^\dagger$. One uses therefore for deriving the QRPA equations that two quasiparticle states behave like bosons. This is a usual approximation one often is using in physics. For example, one describes a pair of quarks and antiquarks as a meson and treats it as a boson, although it consists out of a fermion pair.

To test the quality of this approximation for the two neutrino and the neutrinoless double beta decay, we include the exact commutation relations of the fermion pairs at least as ground state expectation values. This is called renormalized QRPA (R-QRPA) applied for in[4] to the two neutrino double beta decay and in[5] for the first time to the interesting neutrinoless double beta decay.

The R-QRPA includes the Pauli principle into the QRPA and reduces by that the number of quasiparticles in the ground state. RQRPA stabilizes the intermediate nuclear wave functions compared to the QRPA approach. One is relatively close to a phase transition to stable proton-neutron Gamow-Teller correlations. Since they are not included in the basis, the QRPA solution is collapsing if the 1^+ proton-neutron nucleon-nucleon two-body matrix elements are increased. This one studies usually by multiplying all nucleon-nucleon particle-particle matrix elements of the Brueckner reaction matrix of the Bonn (or the Nijmegen or Argonne) potential with a factor g_{pp} (typical: $g_{pp} \approx 0.9$). With an increasing g_{pp} the QRPA solution collapses like a spherical solution is collapsing if a nucleus gets deformed by increasing the quadrupole force. The inclusion of the Pauli principle in the RQRPA approach moves this instability to much larger nucleon-nucleon matrix elements (larger factors g_{pp}) and the agreement for the two neutrino double beta decay with the experimental value is more in the range of $g_{pp} = 1$.

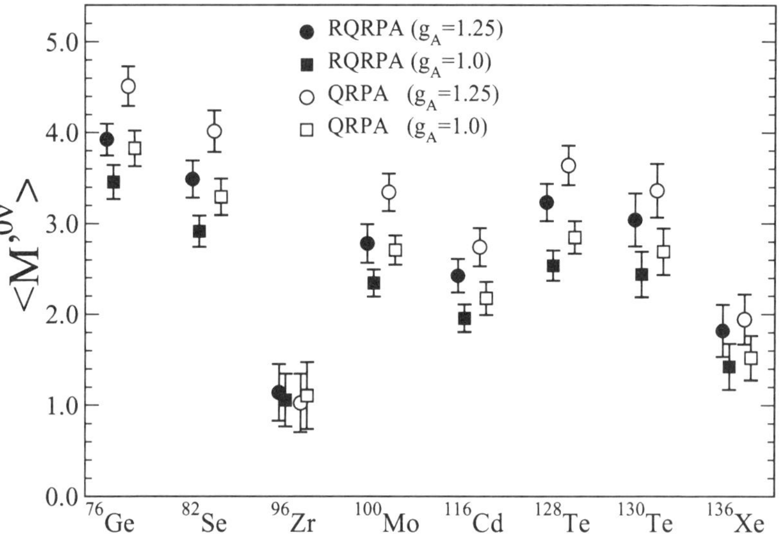

Fig. 5. Neutrinoless double beta decay matrix elements at the Majorana neutrino mass calculated for different initial double beta decay nuclei. (for more details see text)

Fig. 5 shows the neutrinoless double beta decay matrix elements calculated in 36 different ways for each initial nucleus indicated at the figure.[6] The approach includes three different forces (Bonn, Nijmegen, Argonne) and three different basis sets (about two oscillator shells, about three oscillator shells and about five to six oscillator shells) and four different approaches QRPA and the renormalized QRPA (RQRPA) with two different axial vector coupling constants $g_A = 1.25$ and with the quenched value $g_A = 1.0$. In each of the 36 calculations for each nucleus, the factor g_{pp} in front of the two nucleon-nucleon matrix element is adjusted to reproduce the experimental value of the two neutrino double beta decay. The error bars given in the figure include the 1σ error of the theory plus the experimental error of the two neutrino double beta decay transition probability. The error is less than $\pm$ 30 %. The main part of the error indicated in Fig. 5 are the experimental error of the two neutrino double beta decay probability. So when this measurements are improved, one is also automatically improving the prediction of the neutrinoless double beta decay matrix elements.

Table I shows the results for the life-time with our matrix elements and assumed neutrino mass $\langle m \rangle = 50$ meV.

Table 1. Averaged $0\nu\beta\beta$ nuclear matrix elements $\langle M'^{0\nu}\rangle$ and their variance σ (in parentheses) evaluated in the RQRPA and QRPA. M_{GT}^{exp} and g_A denote the $2\nu\beta\beta$-decay nuclear matrix element deduced from $T_{1/2}^{2\nu-exp}$ and axial-vector coupling constant, respectively. In column 6 the $0\nu\beta\beta$ half-lives evaluated with the RQRPA average nuclear matrix element and for assumed $\langle m_{\beta\beta}\rangle = 50$ meV are shown. For ^{136}Xe there are four entries; the upper two use the upper limit of the 2ν matrix element while the lower two use the ultimate limit, vanishing 2ν matrix element. ^{150}Nd is included for illustration. It is treated as a spherical nucleus; deformation will undoubtedly modify its 0ν matrix element.

Nuclear transition	g_A	M_{GT}^{exp} [MeV^{-1}]	$\langle M'^{0\nu}\rangle$ RQRPA	$\langle M'^{0\nu}\rangle$ QRPA	$T_{1/2}^{0\nu}$ ($\langle m_{\beta\beta}\rangle = 50$ meV) [yrs]
$^{76}Ge \rightarrow {}^{76}Se$	1.25	0.15 ± 0.006	3.92(0.12)	4.51(0.17)	$0.86^{+0.08}_{-0.07}\ 10^{27}$
	1.00	0.23 ± 0.01	3.46(0.13)	3.83(0.14)	$1.10^{+0.13}_{-0.11}\ 10^{27}$
$^{82}Se \rightarrow {}^{82}Kr$	1.25	0.10 ± 0.009	3.49(0.13)	4.02(0.15)	$2.44^{+0.32}_{-0.26}\ 10^{26}$
	1.00	0.16 ± 0.008	2.91(0.09)	3.29(0.12)	$3.50^{+0.46}_{-0.38}\ 10^{26}$
$^{96}Zr \rightarrow {}^{96}Mo$	1.25	$0.11^{+0.03}_{-0.06}$	1.20(0.14)	1.12(0.03)	$0.98^{+1.1}_{-0.31}\ 10^{27}$
	1.00	$0.17^{+0.05}_{-0.1}$	1.12(0.11)	1.21(0.07)	$1.12^{+1.3}_{-0.35}\ 10^{27}$
$^{100}Mo \rightarrow {}^{100}Ru$	1.25	0.22 ± 0.01	2.78(0.19)	3.34(0.19)	$2.37^{+0.41}_{-0.32}\ 10^{26}$
	1.00	0.34 ± 0.015	2.34(0.12)	2.71(0.14)	$3.33^{+0.47}_{-0.39}\ 10^{26}$
$^{116}Cd \rightarrow {}^{116}Sn$	1.25	0.12 ± 0.006	2.42(0.16)	2.74(0.19)	$2.86^{+0.50}_{-0.39}\ 10^{26}$
	1.00	0.19 ± 0.009	1.96(0.13)	2.18(0.16)	$4.39^{+0.77}_{-0.61}\ 10^{26}$
$^{128}Te \rightarrow {}^{128}Xe$	1.25	0.034 ± 0.012	3.23(0.12)	3.64(0.13)	$4.53^{+0.64}_{-0.53}\ 10^{27}$
	1.00	0.053 ± 0.02	2.54(0.08)	2.85(0.08)	$7.35^{+1.1}_{-0.88}\ 10^{27}$
$^{130}Te \rightarrow {}^{130}Xe$	1.25	$0.036^{+0.03}_{-0.009}$	2.95(0.12)	3.26(0.12)	$2.16^{+0.33}_{-0.46}\ 10^{26}$
	1.00	$0.056^{+0.05}_{-0.15}$	2.34(0.07)	2.59(0.06)	$3.42^{+0.51}_{-0.83}\ 10^{26}$
$^{136}Xe \rightarrow {}^{136}Ba$	1.25	0.030	1.97(0.13)	2.11(0.11)	$4.55^{+0.68}_{-0.56}\ 10^{26}$
	1.00	0.045	1.59 (0.09)	1.70 (0.07)	$6.38^{+1.12}_{-0.91}\ 10^{26}$
	1.25	0	1.67(0.13)	1.78(0.11)	$7.00^{+0.84}_{-0.71}\ 10^{26}$
	1.00	0	1.26 (0.09)	1.35 (0.07)	$1.11^{+0.17}_{-0.14}\ 10^{27}$
$^{150}Nd \rightarrow {}^{150}Sm$	1.25	$0.07^{+0.009}_{-0.03}$	4.16(0.16)	4.74(0.20)	$2.23^{+0.41}_{-0.21}\ 10^{25}$
	1.00	$0.11^{+0.014}_{-0.05}$	3.30(0.16)	3.72(0.20)	$3.55^{+0.87}_{-0.42}\ 10^{25}$

The present calculation includes the higher order currents according to Towner and Hardy[8] and short range correlations in the neutrinoless double beta decay transition probability with a Jastrow factor.

Table 2. A lower limit of the experimental lifetimes for the 0ν double beta decay of ^{76}Ge is indicated in the second row with the experimental reference in the third row. The fourth row indicates the upper limit of the Majorana neutrino mass derived with the help of our RQRPA matrix element for $g_A = 1.25$.

Nucleus	^{76}Ge
$\tau^{1/2}$ (exp) [y.]	$> 1.9\ 10^{25}$
Klapdor et al.	hep-ph/0512263
$\langle m\rangle$ [eV]	< 0.34

2. Summary

In this contribution we investigated the accuracy of the matrix elements for the neutrinoless double beta decay. This accuracy determines the accuracy of the Majorana neutrino mass extracted from the neutrinoless double beta decay. We used the Quasiparticle Random Phase Approach (QRPA) and the renormalized QRPA (RQRPA) approximation which includes the Pauli principle and reduces the ground state correlations and stabilises the wave function of the excited states of the intermediate nuclei. To get a feeling for the uncertainty we calculated the neutrinoless double beta decay for three different nucleon-nucleon forces (Bonn, Nijmegen, Argonne) and three different basis sets (small with about two oscillator shells, intermediate with about three oscillator shells and large with about five oscillator shells). In addition we performed the calculations with QRPA, with RQRPA (including the Pauli principle) and for two different axial vector coupling constants $g_A = 1.25$ and $g_A = 1.00$. In each of the 36 calculations for every nucleus we adjust a factor g_{pp} multiplying the nucleon-nucleon two-body matrix elements of the order between 0.85 and 1.1 to fit the experimental two neutrino transition probabilities. We give the theoretical error of the neutrinoless transition probability as the 1σ error of the nine calculations for each method plus the experimental error of the two neutrino transition probability. In many cases the theoretical error of the neutrinoless nuclear matrix element is less than $\pm$ 30 %. But the main part of the error as indicated in Fig. 5 comes from the experimental error of the two neutrino transition probability. An improvement of the measurement of the two neutrino transition probability would even further reduce the uncertainty and by that further reduce the error of the Majorana neutrino mass which can be determined if the neutrinoless double beta decay is measured.

2.1. *Problems to be solved:*

One open problem is connected with the fact that two-quasi particle excitations of the initial and the final nucleus to the intermediate states yield a slightly different Hilbert space and thus also slightly different intermediate states. This is treated by including an overlap matrix between these states. We are working on a better treatment and methods to see which error is induced by this effect.

A second problem is connected with the proton and neutron number non-conservation in the BCS treatment of pairing. Particle number projection before variation and the use of the Lipkin-Nogami method did show that this is only a problem if one of the nuclei involved has a closed shell.[9]

A third problem are nuclear deformations for example for ^{150}Nd.[10]

Acknowledgement: I would like to thank Dr. F. Simkovic, Dr. V. Rodin and Prof. P. Vogel for important contributions to the results presented here.

References

1. T. Tomoda, A. Faessler, Phys. Lett. **B199**, (1987) 475
2. A. Faessler, S. Kovalenko, F. Simkovic, J. Schwieger, PRL **78**, (1997) 183
3. O. Civitarese, A. Faessler, T. Tomoda, Phys. Lett. **B194**, (1987) 11
4. J. Toivanen, J. Suhonen, Phys. Lett. **75**, (1995) 410 and Phys. Rev. **C55**, (1997) 2314
5. J. Schwieger, F. Simkovic, A. Faessler, Nucl. Phys. **A600**, (1996) 179
   F. Simkovic, J. Schwieger, M. Veselsky, G. Pantis, A. Faessler, Phys. Lett. **B393**, (1997) 267
6. V.A. Rodin, A. Faessler, F. Simkovic, P. Vogel, Phys. Rev. **C68**, (2003) 044302; Nucl. Phys. **A766**, (2006) 107, nucl-th/0503063 and Erratum to be published in Nucl. Phys. **A**, (2007)
7. J. Suhonen, S.B. Khadkikar, A. Faessler, Nucl. Phys. **A529**, (1991) 727
8. I.S. Towner, J.C. Hardy, preprint nucl-th/9504015 v1, (12 Apr 1995)
9. Ch.C. Moustakidis, T.S. Kosmas, F. Simkovic, Amand Faessler, nucl-th/0511013
10. R. Alvarez-Rodriguez, P. Sarriguren, E. Moya de Guerra, L. Pacearescu, Amand Faessler, F. Simkovic, Phys. Rev. **C70**, (2004) 064309, nucl-th/0411039

DOUBLE BETA DECAY TO THE FIRST 2^+ STATE

A. A. RADUTA[a),b)] and C. M. RADUTA [b)]

[a)] *Department of Theoretical Physics and Mathematics, Bucharest University, POBox MG11, Romania*

[b)] *Department of Theoretical Physics, Institute of Physics and Nuclear Engineering, Bucharest, POBox MG6, Romania*

The Gamow-Teller transition operator is written as a polynomial in the QRPA boson operators by using the prescription of the boson expansion technique. The $2\nu\beta\beta$ process ending on the first 2^+ state in the daughter nucleus takes place via one, two and three boson states describing the odd-odd intermediate nucleus. The approach uses a projected spherical single particle basis. The GT transition amplitude as well as the half lives were calculated for eighteen nuclei. Results are compared with the available data as well as with the predictions obtained with other methods.

Keywords: particle-hole interaction, particle-particle interaction, pn correlations, boson expansion, half life, deformed states

1. Introduction

One of the most exciting subjects of nuclear physics is that of double beta decay. The interest is generated by the fact that in order to describe quantitatively the decay rate one has to treat consistently the neutrino properties as well as the nuclear structure features. The process may take place in two distinct ways:

• by a $2\nu\beta\beta$ decay where the initial nuclear system, the mother nucleus, is transformed in the final stable nuclear system, usually called the daughter nucleus, two electrons and two anti-neutrinos;

• by the $0\nu\beta\beta$ process where the final state does not involve any neutrino and therefore the lepton number is not conserved. Such a process is forbidden in the standard gauge theories like $SU5, SU(2)\times U(1)$.

The $2\nu\beta\beta$ decay is a very rare process while for $0\nu\beta\beta$, no experimental evidence is yet available. The latter decay mode is especially interesting since one hopes that its discovery might provide a definite answer to the question whether the neutrino is a Majorana or a Dirac particle. In order to make predictions for $0\nu\beta\beta$ decay one needs reliable nuclear matrix elements. However, no trustful test is yet available for the necessary matrix elements. Fortunately, the same matrix elements are involved also in the $2\nu\beta\beta$ decay amplitude. Due to this circumstance one has attempted to use for $0\nu\beta\beta$ decay those m.e. which describe realistically the $2\nu\beta\beta$ process.

The $2\nu\beta\beta$ decay is by itself an interesting phenomenon. Moreover, due to the feature mentioned above this phenomenon deserves a special attention. The contributions over several decades have been reviewed by many authors.[1–6]

Although none of the double beta emitters is a spherical nucleus most formalisms use a single particle spherical basis. In the middle of 90's we treated the $2\nu\beta\beta$ process in a pnQRPA formalism using a projected spherical single particle basis which resulted in having a unified description of the process for spherical and deformed nuclei.[7,8] Recently the single particle basis[9,10] has been improved by accounting for the volume conservation while the mean field is deformed.[11,12] The improved basis has been used for describing quantitatively the ground state to ground state double beta decay rates as well as the corresponding half lives.[13,14] The manners in which the physical observable is influenced by the nuclear deformations of mother and daughter nuclei are in detail commented. One expects that for the transition $0^+ \rightarrow 2^+$, the nuclear deformation is even more important. This can be understood even at the first glance since the larger the deformation of the daughter nucleus the lower the energy of the first 2^+ state. Consequently, the Q value is expected to be larger. By contrast to the ground to ground transition, the transition $0^+ \rightarrow 2^+$ is forbidden within the pnQRPA approach. Therefore a higher RPA description is necessary.

The aim of this talk is to review the recent results on the $2\nu\beta\beta$ decay $0^+ \rightarrow 2^+$, where 0^+ is the ground state of the emitter, while 2^+ is a single quadrupole phonon state describing the daughter nucleus. The adopted procedure is the boson expansion (BE) method as formulated in our previous paper,[15] but using a projected spherical single particle basis. Although the BE approach has been widely used for two alike fermion operators, the procedure has been extended for proton-neutron operators only in the beginning of 90's ([15]) for spherical single particle basis and recently for a deformed mean field.[16]

The formalism employed for describing the transition $0^+ \rightarrow 2^+$ involves two basic ingredients :

- the boson expansion approach for the Gamow-Teller transition operator. In this way the $2\nu\beta\beta$ decay $0^+ \to 2^+$, forbidden in the framework of the pnQRPA formalism, becomes an allowed process;

• a projected spherical single particle basis. Using such a basis one may unitarily treat the transitions of spherical and deformed nuclei. Moreover, situations when the mother and daughter nuclei have different nuclear deformations could be accounted for.

In order to have a self sustained description of the results, here we devote a separate section to each of the mentioned ingredients.

The formalism and the results will be described as follows. In Section II the projected single particle basis to be used, is presented. The model Hamiltonian, written in second quantization for the projected spherical basis, is defined in Section II. The states involved in the process are are described within a QRPA formalism. Section IV deals with the boson expansion of the Gamow-Teller double beta transition operator while the GT amplitude is presented in Section V. Numerical applications to eighteen nuclei are commented in Section VI. The main results and conclusions are summarized in Section VII.

2. A projected spherical single particle basis

The single particle mean field is determined by a particle-core Hamiltonian:

$$\tilde{H} = H_{sm} + H_{core} - M\omega_0^2 r^2 \sum_{\lambda=0,2} \sum_{-\lambda\leq\mu\leq\lambda} \alpha^*_{\lambda\mu} Y_{\lambda\mu}, \tag{2.1}$$

where H_{sm} denotes the spherical shell model Hamiltonian while H_{core} is a harmonic quadrupole boson (b^+_μ) Hamiltonian associated to a phenomenological core. The interaction of the two subsystems is accounted for by the third term of the above equation, written in terms of the shape coordinates $\alpha_{00}, \alpha_{2\mu}$. The quadrupole shape coordinates and the corresponding momenta are related to the quadrupole boson operators by the canonical transformation:

$$\alpha_{2\mu} = \frac{1}{k\sqrt{2}}(b^\dagger_{2\mu} + (-)^\mu b_{2,-\mu}),\ \pi_{2\mu} = \frac{ik}{\sqrt{2}}((-)^\mu b^\dagger_{2,-\mu} - b_{2\mu}), \tag{2.2}$$

where k is an arbitrary C number. The monopole shape coordinate is determined from the volume conservation condition. In the quantized form, the result is:

$$\alpha_{00} = \frac{1}{2k^2\sqrt{\pi}}\left[5 + \sum_\mu (2b^\dagger_\mu b_\mu + (b^\dagger_\mu b^\dagger_{-\mu} + b_{-\mu}b_\mu)(-)^\mu)\right]. \tag{2.3}$$

Averaging $\tilde{H}$ on the eigenstates of H_{sm}, hereafter denoted by $|nljm\rangle$, one obtains a deformed boson Hamiltonian whose ground state is, in the harmonic limit, described by a coherent state

$$\Psi_g = exp[d(b^+_{20} - b_{20})]|0\rangle_b, \tag{2.4}$$

with $|0\rangle_b$ standing for the vacuum state of the boson operators and d a real parameter which simulates the nuclear deformation. On the other hand, the average of $\tilde{H}$ on Ψ_g is similar to the Nilsson Hamiltonian.[18]

Due to these properties, it is expected that the best trial functions to generate a spherical basis are:

$$\Psi^{pc}_{nlj} = |nljm\rangle\Psi_g. \tag{2.5}$$

The projected states are obtained by acting on these deformed states with the projection operator P^I_{MK}. The subset of projected states :

$$\Phi^{IM}_{nlj}(d) = \mathcal{N}^I_{nlj}P^I_{MI}[|nljI\rangle\Psi_g] \equiv \mathcal{N}^I_{nlj}\Psi^{IM}_{nlj}(d), \tag{2.6}$$

are orthogonal with the normalization factor denoted by $\mathcal{N}^I_{nlj}$.

Although the projected states are associated to the particle-core system, they can be used as a single particle basis. Indeed, when a matrix element of a particle like operator is calculated, the integration on the core collective coordinates is performed first, which results in obtaining a final factorized expression: one factor carries the dependence on deformation and one is a spherical shell model matrix element. The single particle energies are approximated by the average of the particle-core Hamiltonian $H' = \tilde{H} - H_{core}$ on the projected spherical states defined by Eq.(2.6):

$$\epsilon^I_{nlj} = \langle\Phi^{IM}_{nlj}(d)|H'|\Phi^{IM}_{nlj}(d)\rangle. \tag{2.7}$$

The off-diagonal matrix elements of H' is ignored at this level. Their contribution is however considered when the residual interaction is studied. As shown in Ref.,[9] the dependence of the new single particle energies on deformation is similar to that shown by the Nilsson model.[18] The properties of this basis and how to use it for treating one and two-body interactions has been reviewed in several places.[12–14]

3. Description of the states involved in the process

The states, involved in the $2\nu\beta\beta$ process are described by the following many body Hamiltonian:

$$H = \sum \frac{2}{2I+1}(\epsilon_{\tau\alpha I} - \lambda_{\tau\alpha})c^{\dagger}_{\tau\alpha IM}c_{\tau\alpha IM} - \sum \frac{G_{\tau}}{4}P^{\dagger}_{\tau\alpha I}P_{\tau\alpha I'}$$
$$+ 2\chi \sum \beta^{-}_{\mu}(pn)\beta^{+}_{-\mu}(p'n')(-)^{\mu} - 2\chi_1 \sum P^{-}_{1\mu}(pn)P^{+}_{1,-\mu}(p'n')(-)^{\mu}$$
$$- \sum_{\tau,\tau'=p,n} X_{\tau,\tau'}Q_{\tau}Q^{\dagger}_{\tau'}. \tag{3.1}$$

The operator $c^{\dagger}_{\tau\alpha IM}(c_{\tau\alpha IM})$ creates (annihilates) a particle of type τ (=p,n) in the state Φ^{IM}_{α}, when acting on the vacuum state $|0\rangle$. In order to simplify the notations, hereafter the set of quantum numbers $\alpha(= nlj)$ will be omitted. The two body interaction consists of four terms, the pairing, the dipole-dipole particle-hole (ph) and particle-particle (pp), and the QQ interaction. The corresponding strengths are denoted by $G_{\tau}, \chi, \chi_1, X_{\tau\tau'}$, respectively. All of them are separable interactions, with the factors defined by the following expressions:

$$P^{\dagger}_{\tau I} = \sum_{M} \frac{2}{2I+1} c^{\dagger}_{\tau IM} c^{\dagger}_{\widetilde{\tau IM}}, \quad \beta^{-}_{\mu}(pn) = \sum_{M,M'} \frac{\sqrt{2}}{\hat{I}} \langle pIM|\sigma_{\mu}|nI'M'\rangle \frac{\sqrt{2}}{\hat{I}'} c^{\dagger}_{pIM} c_{nI'M'},$$
$$P^{-}_{1\mu}(pn) = \sum_{M,M'} \frac{\sqrt{2}}{\hat{I}} \langle pIM|\sigma_{\mu}|nI'M'\rangle \frac{\sqrt{2}}{\hat{I}'} c^{\dagger}_{pIM} c^{\dagger}_{\widetilde{nI'M'}}, \quad \hat{I} = \sqrt{2I+1}$$
$$Q^{(\tau)}_{2\mu} = \sum_{i,k} q^{(\tau)}_{ik} \left(c^{\dagger}_{i} c_{k}\right)_{2\mu}, \quad q^{(\tau)}_{ik} = \frac{\sqrt{2}}{\hat{I}_k} \langle I_i||r^2 Y_2||I_k\rangle. \tag{3.2}$$

The remaining operators from Eq.(3.1) can be obtained from the above operators, by hermitian conjugation. The one body term and the pairing interaction terms are treated first through the standard BCS formalism and consequently replaced by the quasiparticle one body term $\sum_{\tau IM} E_{\tau} a^{\dagger}_{\tau IM} a_{\tau IM}$. The Hamiltonian terms describing the quasiparticle correlations becomes a quadratic expression in the dipole, $A^{\dagger}_{1\mu}(pn)$, and quadrupole, $A^{\dagger}_{2\mu}(\tau\tau')$, two quasiparticles and quasiparticle density operators ,$B^{\dagger}_{1\mu}(pn), B^{\dagger}_{2\mu}(\tau\tau')$. Quasiparticle pn correlations are treated within the pnQRPA formalism. Since the final state in the daughter nucleus is the collective 2^+, we have also to describe the quadrupole charge conserving phonons. The two kinds of phonons are defined by:

$$\Gamma^{\dagger}_{n\mu} = \sum_{k}[X_n(k)A^{\dagger}_{n\mu}(k) - Y_n(k)A_{n,-\mu}(k)(-)^{n-\mu}], n = 1,2. \quad (3.3)$$

with the amplitudes X and Y determined by the $QRPA$ equations. In order to distinguish between the phonon operators acting in the RPA space associated to the mother and daughter nuclei respectively, one needs an additional index. Also, an index labeling the solutions of the RPA equations is necessary. Thus, the two kinds of bosons will be denoted by:

$$_j\Gamma^{\dagger}_{1\mu}(k),\ j = i, f;\ k = 1, 2, ...N_s^{(1)};_j\Gamma^{\dagger}_{2\mu}(k),\ j = i, f;\ k = 1, 2, ...N_s^{(2)}. \quad (3.4)$$

Acting with $_i\Gamma^{\dagger}_{1\mu}(k)$ and $_f\Gamma^{\dagger}_{1\mu}(k)$ on the vacuum states $|0\rangle_i$ and $|0\rangle_f$ respectively, one obtains two sets of non-orthogonal states describing the intermediate odd-odd nucleus. By contrast, the states $_i\Gamma^{\dagger}_2(k)|0\rangle_i$ and $_f\Gamma^{\dagger}_2(k)|0\rangle_f$ describe different nuclei, namely the initial and final ones, participating in the process of $2\nu\beta\beta$ decay. The mentioned indices are however omitted whenever their presence is not necessary

4. The Boson Expansion (BE) procedure

In the BE formalism, the operators $A^{\dagger}_{1\mu}(p,n)$,$A_{1\mu}$,$B^{\dagger}_{1\mu}(p,n)$, $B_{1\mu}$ are written as polynomials in the QRPA boson operators with the expansion coefficients determined such that their mutual commutation relations are preserved in each order of approximation.[19] Based on this criterion the boson expansions of the quadrupole two quasiparticle and quadrupole quasiparticle density charge conserving operators have been obtained by Belyaev and Zelevinsky in Ref.[19] For charge non-conserving two quasiparticle and quasiparticle density dipole operators the expansion has been derived by one of us (A.A.R, in collaboration) in Ref.([15]). The latter expansion has the peculiarity that the commutator algebra cannot be satisfied restricting the expansion to the proton-neutron dipole bosons. However, this goal can be touched if the boson operators space is enlarged by adding the charge conserving quadrupole two quasiparticle bosons. The last step consists in expressing the quasiboson operators $\overset{0}{A}{}^{\dagger}_{1\mu}(pn)$, $\overset{0}{A}_{1\mu}(pn)$, $\overset{0}{A}{}^{\dagger}_{2\mu}(pp)$, $\overset{0}{A}{}^{\dagger}_{2\mu}(pp)$, $\overset{0}{A}{}^{\dagger}_{2\mu}(nn)$, $\overset{0}{A}_{2\mu}(nn)$ (these are, in fact the operators denoted by the same symbol but without the index "0", with the commutators approximated to be of boson type) as linear combination of the QRPA bosons. In this way the basic operators mentioned above are written as polynomials of pn and $pp + nn$ QRPA bosons. The expansions involve not only the collective but also non-collective QRPA bosons. The expansion coefficients may be found in Ref.[15]

The boson expansions associated to the two quasiparticle and quasiparticle density proton-neutron operators have the property that preserve the matrix elements in a boson basis. Actually, this can be used as a criterion to determine the expansion coefficients. For example, the coefficients multiplying the linear and quadratic terms in bosons are determined as:

$$\mathcal{A}^{(1,0)}_{k_1}(j_pj_n) = \langle 0| \left[\Gamma_{1\mu}(k_1), A^{\dagger}_{1\mu}(j_p,j_n)\right] |0\rangle, \tag{4.1}$$

$$\mathcal{B}^{(2,0)}_{k_1k_2}(j_pj_n) = \sum_{\mu_1,\mu_2} C^{1\ \ 2\ \ 1}_{\mu_1\ \mu_2\ \mu}\langle 0| \left[\Gamma_{1\mu_1}(k_1), \left[\Gamma_{1\mu_2}(k_2), B^{\dagger}_{1\mu}(j_p,j_n)\right]\right] |0\rangle.$$

It is worth mentioning that the matrix element of the double commutator involved in Eq.4.1 does not depend on the order in which the two commutators are calculated. However, the commutation order is important when one determines the expansion coefficients of the third order monomial in bosons. The ordering in the mentioned commutators is chosen such that the mutual commutator equations of the basic operators $A^{\dagger}_{1\mu}, B^{\dagger}_{1\ mu}$ are satisfied in each order of approximation. The comparison of the boson expansion formulated in Ref.[15] and other approaches is given in Ref.[24]

5. The Gamow-Teller transition amplitude

Here we are interested to describe the Gamow-Teller two neutrino double beta decay of an even-even deformed nucleus. In our treatment the Fermi transitions, contributing with about 20% to the total rate, and the "forbidden" transitions are ignored, which is a reasonable approximation for medium and heavy nuclei. The $2\nu\beta\beta$ process is conceived as two successive single β^- virtual transitions. The first transition connects the ground state of the mother nucleus to a magnetic dipole state 1^+ of the intermediate odd-odd nucleus which, subsequently decays to the first state 2^+ of the daughter nucleus. The second leg of the transition is forbidden within the pnQRPA approach but non-vanishing within a higher pnQRPA approach,[15]

If the energy carried by leptons in the intermediate state is approximated by the sum of the rest energy of the emitted electron and half the Q-value of the double beta decay process from the ground state of the mother nucleus to the first excited state 2^+ of the daughter nucleus,

$$\Delta E = m_ec^2 + \frac{1}{2}Q^{(0\rightarrow 2)}_{\beta\beta}, \tag{5.1}$$

the reciprocal value of the $2\nu\beta\beta$ half life can be factorized as:

$$T^{2\nu}_{1/2}(0^+_i \rightarrow 2^+_f)^{-1} = G_{02}|M^{(02)}_{GT}|^2, \tag{5.2}$$

where G_{02} is the Fermi integral which characterizes the phase space of the process while the second factor is the GT transition amplitude which, in the second order of perturbation theory, has the expression:

$$M_{GT}^{(02)} = \sqrt{3}\sum_{k,m} \frac{{}_i\langle 0||\beta^+||k,m\rangle_i \; {}_i\langle k,m|k',m'\rangle_f \; {}_f\langle k',m'||\beta^+||2_1^+\rangle_f}{(E_{k,m}+\Delta E_2)^3}. \quad (5.3)$$

Here $\Delta E_2 = \Delta E + E_{1^+}$, with E_{1^+} standing for the experimental energy for the first state 1^+. The intermediate states $|k,m\rangle$ are k-boson states with $k = 1, 2, 3$ labeled by the index m, specifying the spin and the ordering label of the RPA roots. Note that by contrast to the case of ground to ground transition here the denominator has a cubic power which results in obtaining a suppression of the corresponding GT amplitude. Inserting the boson expansion of the operators $A_{1\mu}^\dagger, B_{1\mu}^\dagger$ into the expression of the β^+ transition operator one can check that several non-vanishing factors, show up at the numerator. The non-vanishing matrix elements contributing to the double beta transition $0^+ \to 2^+$ are represented pictorially in Fig.1.

6. Numerical application

Calculations were performed for eighteen nuclei which have been previously considered in Refs.[13,14] for studying the double beta ground to ground decay. Among these, eleven are proved to be, indeed, double beta ground to ground emitters, while the remaining ones are suspected to have this property due to the corresponding positive Q-value. Since the excitation energies for the states 2^+ in the daughter nuclei are not large, the Q-values characterizing the double beta transition $0^+ \to 2^+$ are also positive. For some of the selected nuclei, experimental data either for the half life of the process or for the low bounds of the half lives are available.

The single particle space, the pairing interaction treatment and the pnQRPA description of the dipole states describing the intermediate odd-odd nuclei used here, are identical with those from Refs.[13,14] for ground to ground transition. The parameters determining the single particle energies and the strengths of the pairing and dipole interactions are taken from the quoted works . However, the microscopic Hamiltonian used here involves in addition the QQ interaction between alike nucleons. As we already mentioned this interaction is needed in order to define the charge conserving quadrupole phonon operators used by the boson expansion procedure. Moreover, this interaction describes the final state, i.e. 2^+, in the daughter nucleus. The strength of the QQ interaction was fixed by requiring that the first root of the QRPA equation for the quadrupole charge

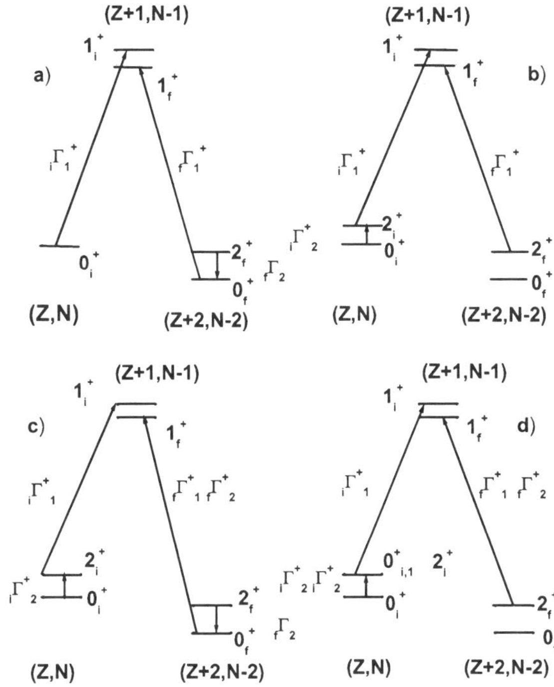

Fig. 1. One illustrates various GT transitions $0^+ \to 2^+$ via one (a)) two (b)) and (c)) and three (d)) phonon states.

conserving boson is close to the experimental energy of the first 2^+ state.

Having the QRPA states defined, the GT amplitude has been calculated by means of Eq.(5.3), while the half life with Eq.(5.2). The Fermi integral for the transition $0^+ \to 2^+$, denoted by G_{02}, was computed by using the analytical result given in Ref.[4] The final results are collected in Table I. One notices that the half life is influenced by both the phase space integral (through the Q-value) and the single particle properties which determine the transition amplitude. Indeed, for ^{128}Te and ^{134}Xe the small Q-value causes a very large half life, while in ^{48}Ca the opposite situation is met. By contrary, the Q value of ^{110}Pd is about the same as for ^{76}Ge but, due to the specific single particle and pairing properties of the orbits participating coherently to the process, the half life for the former case is more than three orders of magnitude less than in the latter situation. The transition matrix elements reported in Ref.[20] are larger than those given here, despite the fact that the higher pnQRPA approaches in the two descriptions are similar.[24] The reason is that in that case a spherical single particle basis is used whereas here we use a deformed basis. The same effect of deformation on the GT matrix elements was pointed out by Zamick and Auerbach in Ref.[26] Indeed, they calculated the GT transition matrix elements for the neutrino capture $\nu_\mu + ^{12}C \rightarrow ^{12}N + \mu^-$ using different structures for the ground states of ^{12}C and ^{12}N: a) spherical ground states; b) asymptotic limits of the wave functions and 3) deformed states with an intermediate deformation of $\delta = -0.3$. The results for the transition rate were $\frac{16}{3}$, 0 and 0.987, respectively. Similar results are obtained also for the spin M1 transitions in ^{12}C. The ratio between the transition rates obtained with spherical and deformed basis explains the factor of 5 overestimate in the calculations of Ref.[27] where a spherical basis is used. It is worth mentioning the good agreement between our prediction for ^{100}Mo and that of Ref.[23] obtained with a deformed SU(3) single particle basis. To have a reference value for the matrix elements associated to the transition $0^+ \rightarrow 2^+$, the M_{GT} values for the transitions $0^+ \rightarrow 0^+$ are also listed.[14] The ratio of the transition $0^+ \rightarrow 0^+$ and $0^+ \rightarrow 2^+$ matrix elements is quite large for ^{76}Ge (398), ^{100}Mo (224) and ^{96}Zr (136) but small for ^{110}Pd (5.26), ^{134}Xe (6.3) and ^{150}Na (20.46). However, these ratios are not directly reflected in the half lives, since the phase space factors for the two transitions are very different from each other and, moreover, the differences depend on the atomic mass of the emitter. The composing terms of the transition amplitude are suggestively represented in Fig.1. The term corresponding to Fig.1 d) has a negligible contribution and, therefore, has been ignored. The terms corresponding to the panels a), b) and c) of Fig. 1 are denoted by $M^{(1)}_{GT}$, $M^{(2)}_{GT}$, $M^{(3)}_{GT}$ respectively. Our calculations indicate that for some nuclei the term $M^{(1)}_{GT}$ prevails while for other $M^{(2)}_{GT}$ is the dominant partial amplitude. None of the cases has $M^{(3)}_{GT}$ as a dominant term. The contributions coming from $M^{(1)}_{GT}$ and $M^{(2)}_{GT}$ have opposite sign. There are two exceptions, ^{104}Ru and ^{154}Sm, where the two contributions add coherently.

Table I. The GT transition amplitudes and the half lives of the double beta decay $0^+ \rightarrow 2^+$ are given. For comparison, we give also the available experimental results as well as some theoretical predictions obtained with other formalisms: [a),b)]Ref.,[23c)] Ref.,[21d),e)23] for different nuclear deformation, $\beta = 0.28$ and $\beta = 0.19$, respectively. The M_{GT} values for the ground to ground transitions are also listed. For ^{100}Mo we mention the result of Ref.[23] obtained with an SU(3) deformed single particle basis [a)] and with a spherical basis [b)].

Nucleus	$\lvert M_{GT}^{(0\to 0)}\rvert$	$\lvert M_{GT}^{(0\to 2)}\rvert$	$T_{1/2}^{(0\to 2)}[yr]$		
	[MeV^{-1}]	[MeV^{-3}]	present	Exp.	Ref.[20]
^{48}Ca	0.043	$0.901\cdot 10^{-3}$	$1.72\cdot 10^{24}$		
^{76}Ge	0.222	$0.558\cdot 10^{-3}$	$5.75\cdot 10^{28}$	$>1.1\cdot 10^{21}$	$1.0\cdot 10^{26}$
^{82}Se	0.096	0.259×10^{-3}	1.7×10^{27}	$5.8\cdot 10^{23}$	$^{c)}$ $3.3\cdot 10^{26}$
^{96}Zr	0.113	$0.834\cdot 10^{-3}$	$2.27\cdot 10^{25}$	$> 7.9\cdot 10^{19}$	$4.8\cdot 10^{21}$
^{100}Mo	0.305	$0.136\cdot 10^{-2}$	$1.21\cdot 10^{25}$	$>1.6\cdot 10^{21}$	$3.9\cdot 10^{24}$ $^{a)}2.5\cdot 10^{25}$ $^{b)}1.2\cdot 10^{26}$
^{104}Ru	0.781	0.028	$6.2\cdot 10^{28}$		
^{110}Pd	0.263	0.050	$1.48\cdot 10^{25}$		
^{116}Cd	0.116	$0.507\cdot 10^{-2}$	$3.4\cdot 10^{26}$	$>2.3\cdot 10^{21}$	$1.1\cdot 10^{24}$
^{128}Te	0.090	$0.229\cdot 10^{-2}$	$4.7\cdot 10^{33}$	$>4.7\cdot 10^{21}$	$1.6\cdot 10^{30}$
^{130}Te	0.055	$0.620\cdot 10^{-3}$	$6.94\cdot 10^{26}$	$>4.5\cdot 10^{21}$	$2.7\cdot 10^{23}$
^{134}Xe	0.039	$0.621\cdot 10^{-2}$	$5.29\cdot 10^{35}$		
^{136}Xe	0.039	$0.249\cdot 10^{-2}$	$3.88\cdot 10^{26}$		$2.0\cdot 10^{24}$
^{148}Nd	0.559	$1.408\cdot 10^{-3}$	$9.97\cdot 10^{27}$		
^{150}Nd	0.546	$2.668\cdot 10^{-3}$	$1.5\cdot 10^{23}$	$> 8\cdot 10^{18}$	$^{d)}7.2\cdot 10^{24}$ $^{e)}1.2\cdot 10^{25}$
^{154}Sm	0.311	$1.238\cdot 10^{-3}$	$1.41\cdot 10^{29}$		
^{160}Gd	0.624	$6.799\cdot 10^{-3}$	$4.56\cdot 10^{25}$		
^{232}Th	0.262	$2.142\cdot 10^{-3}$	$1.399\cdot 10^{30}$		
^{238}U	0.152	$0.692\cdot 10^{-3}$	$2.84\cdot 10^{29}$		

7. Conclusions

The main results presented in this lecture can be summarized as follows. The GT transition rate has been calculated within a boson expansion formalism. The single particle basis is generated through an angular momentum projection procedure, from a deformed set of states. In some publications we treated various situations when the mother and daughter had different deformations, in the context of the ground to ground double beta transition and pointed out the role of deformation on the transition rate. Here we showed that the transition $0^+ \to 2^+$ is even more affected by the nuclear deformation.

Concerning the quantitative description, the results reveal the following features. There are five nuclei whose half lives fall in the range accessible to experiment. These are: ^{48}Ca, ^{96}Zr, ^{100}Mo, ^{110}Pd, ^{150}Nd, ^{160}Gd. Comparing with the results obtained in Refs.,[20,22] the half lives presented here are larger. The reason is that, by contrast, we use a deformed single particle basis. The agreement we obtain for ^{100}Mo with the calculations from Ref.,[23] where a deformed SU(3) basis is used, supports the above statement. Moreover our results agree with the earlier calculations of Zamick and Auerbach for ^{12}C, showing that the nuclear deformation suppresses the GT matrix elements.

References

1. H. Primakof and S. Rosen, Rep. Prog. Phys. 22 (1959) 125.
2. W. C. Haxton, G. J. Stephenson, Jr. Prog. Part. Nucl. Phys. 12 (1984) 409.
3. A. Faessler, Prog. Part. Nucl. Phys. **21** (1988) 183.
4. J. Suhonen and O. Civitarese, Phys.Rep. **300** (1998) 123.
5. J.D. Vergados, Phys. Rep. **361** (2001) 1.
6. A. A. Raduta, Prog. Part. Nucl. Phys. **48**(2002) 233.
7. A. A. Raduta, A. Faessler and D.S. Delion, Nucl. Phys. **A 564** (1993) 185.
8. A. A. Raduta, D. S. Delion and A. Faessler, Nucl. Phys. **A 617** (1997) 176.
9. A. A. Raduta, D. S. Delion and N. Lo Iudice, Nucl. Phys. **A 564** (1993) 185.
10. A. A. Raduta, N. Lo Iudice and I. I. Ursu, Nucl. Phys. **A584** (1995) 84.
11. A. A. Raduta *et al*, Phys. Rev. **B 59** (1999) 8209.
12. A. A. Raduta, A. Escuderos, E. Moya de Guerra, Phys. Rev.C65 2002)024312.
13. A.A. Raduta, *et al.*,Phys. Rev. **C 69**, (2004) 064321.
14. A. A. Raduta, C. M. Raduta, A. Escuderos, Phys. Rev. **C71**,(2005)024307.
15. A.A.Raduta, A.Faessler and S.Stoica, Nucl. Phys. **A534** (1991) 149.
16. A.A. Raduta, C. M. Raduta, Phys. Lett. 647 B (2007)265.
17. M.J. Hornish *et al.*, Phys. Rev. **C 74** (2006) 044314.
18. S.G.Nilsson,Mat.Fys.Medd. K. Dan. Vid.Selsk.**29** no. 16 (1955).
19. S.T.Belyaev and G. Zelevinski, Nucl. Phys. **39** (1962) 582.
20. J. Toivanen, J. Suhonen, Phys. Rev. C **55** (1997) 2314.
21. M. Aunola and J. Suhonen, Nucl. Phys. A **602** (1996) 133.
22. O. Civitarese, J. Suhonen, Nucl. Phys. A **575** (1994) 62.
23. J. G. Hirsch,*et al.*, Phys. Rev. **51** (1995) 2252.
24. A. A. Raduta, J. Suhonen, Phys. Rev. C **53** (1996) 176.
25. A. A. Raduta, D. Ionescu and A. Faessler, Phys. RFev. **C 65** (2002) 233.
26. L. Zamick and N. Auerbach, Nucl. Phys. **A 658** (1999) 285.
27. E. Kolbe, *et al*, Phys. Rev. **C 52** 1995) 3437.

PION-INDUCED INTERACTION AND SINGLE-PARTICLE SPECTRA IN RELATIVISTIC HARTREE-FOCK

NGUYEN VAN GIAI*

Institut de Physique Nucléaire, IN2P3 and Université Paris-Sud, 91405 Orsay, France
**E-mail: nguyen@ipno.in2p3.fr*

W.H. LONG, J. MENG

School of Physics, Peking University, Beijing 100871, China

H. SAGAWA

Center for Mathematical Sciences, University of Aizu, Aizu-Wakamatsu, 965-8580 Fukushima, Japan

The relativistic mean field theory cannot describe pion-induced effects because exchange terms are dropped. The relativistic Hartree-Fock (RHF) approach allows for the treatment of such effects. A density-dependent RHF model is presented and it is applied to the study of single-particle spectra along the chains of N=82 isotones and Z=50 isotopes. The effects of the pion-induced tensor interaction on these spectra are discussed.

Keywords: Relativistic Hartree-Fock; tensor coupling of pion; single-particle spectra

1. Introduction

A large part of the present understanding of nuclear structure is based on self-consistent mean field descriptions making use of effective interactions directly parametrized so as to reproduce selected nuclear properties. During the past decade, a great success has been achieved by the relativistic mean field (RMF) theory, not only in stable nuclei but also in exotic regions[1–9]. Of special interest is the fact that the RMF model provides a natural mechanism for explaining the spin-orbit splittings of single-particle levels. This feature becomes even more of central importance with the experimental observation that nuclei near drip lines undergo modifications of their shell structure, where the spin-orbit potential must play an essential role.

One of the basic open problems is the role of one-pion exchange process, which is known to play a fundamental role in the meson-exchange interaction. However, the RMF model is not the appropriate framework to study pion-related processes because it is essentially a Hartree approximation where the Fock (exchange) contributions are altogether dropped, while the Hartree (direct) contributions of pions are zero due to the parity conservation in spherical and axially deformed nuclei. Recent progress in the relativistic Hartree-Fock description of nuclear structure, namely the density-dependent relativistic Hartree-Fock (DDRHF) approach[10] has brought a new insight to consider this problem. Within the DDRHF theory, the effective meson-nucleon coupling strengths including the one-pion exchange are determined in a similar way to the RMF model and a quantitative description of the nuclear structure properties can be successfully achieved[11,12] . Thus, DDRHF opens the possibility to investigate the role of one-pion exchange processes in nuclear structure problems within the framework of a relativistic mean field theory.

In this work, we study the role of pion-exchange processes on single-particle spectra by concentrating on some specific cases.

2. Density-dependent effective Hamiltonians

We start from an effective Lagrangian density $\mathcal{L}$ constructed with the degrees of freedom associated with the nucleon field(ψ), two isoscalar meson fields (σ and ω), two isovector meson fields (π and ρ) and the photon field (A). The parameters of the model are the effective meson masses and meson-nucleon couplings. These couplings are considered to be functions of the baryonic density ρ_b.

The determination of the parameters of the model is explained in Refs.10,11. They are obtained by a least square fit of bulk properties of nuclear matter plus the binding energies of a set of selected nuclei. Several combinations of the coupling strengths (g_σ and g_ω for the two isoscalar mesons, f_π for the pseudo-vector pion, g_ρ and f_ρ for the vector and tensor couplings of the ρ meson) are thus obtained. In Fig.1 we show the density dependence of the coupling strengths for two of our RHF models: the model PKO1 contains the scalar σ-, vector ω- and ρ-, pseudo-vector π-couplings, but as a simplification the tensor ρ-coupling has been dropped. Recently, a new version called PKA1 which contains the additional tensor ρ-coupling f_ρ has been constructed.[13] In Fig.1 we show the density dependence of the couplings of the two models, PKO1 and PKA1. For comparison, the coupling strengths of the density-dependent RMF model DD-ME2 of Ref.6 are also shown. In the rest of this talk we shall concentrate on the results calculated with the model PKO1, i.e., on the effects of the pion coupling on the single-particle spectra. The effects of the ρ-tensor coupling will be discussed elsewhere.[13]

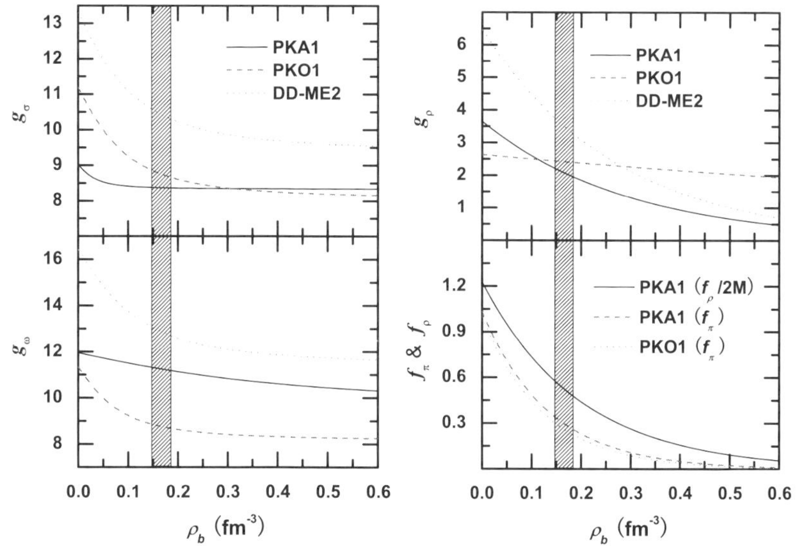

Fig. 1. (color online) The density-dependent meson-nucleon couplings in the isoscalar (left panel: g_σ and g_ω) and isovector (right panel: g_ρ, f_ρ and f_π) channels as functions of density for the DDRHF effective interactions PKA1 and PKO1, and the RMF model DD-ME2. The shadowed area denotes the empirical saturation density region.

3. Single-particle spectra

In the recent paper of Schiffer et al.,[14] it was shown that a set of states outside the proton core $Z = 50$ and the neutron core $N = 82$ may provide a unique information to determine the evolution of the nuclear shell structure. This is why we choose here to discuss the pion effects taking these nuclei as an illustrative example.

The one-pion exchange process with pseudo-vector coupling contains two types of contributions, the central coupling and the non-central tensor coupling.[15] Recently, the tensor type force was shown to have a distinct effect on the evolution of the nuclear shell structure in non-relativistic

Hartree-Fock models.[16,17] Here, we will study the behavior of single-particle energies of the states $\{\nu 1h_{9/2}, \nu 1i_{13/2}\}$ (ν denotes neutron states) in the $N = 82$ isotones and the states $\{\pi 1g_{7/9}, \pi 1h_{11/2}\}$ (π denotes the proton states) in the $Z = 50$ isotopes, especially the influence of the tensor component of the π-coupling.

The origin of the tensor nature of the nucleon-nucleon interaction induced by one-pion exchange is easily seen by making a non-relativistic reduction. Then, the one-pion exchange potential can be divided into a central and a tensor component, $V_\pi^c(\boldsymbol{q})$ and $V_\pi^T(\boldsymbol{q})$:

$$V_\pi^c(\boldsymbol{q}) = \frac{-1}{3}\left[\frac{f_\pi}{m_\pi}\right]^2 \frac{\boldsymbol{\sigma}_1 \cdot \boldsymbol{\sigma}_2 \boldsymbol{q}^2}{m_\pi^2 + \boldsymbol{q}^2} \vec{\tau}_1 \cdot \vec{\tau}_2, \tag{1a}$$

$$V_\pi^T(\boldsymbol{q}) = \frac{-1}{3}\left[\frac{f_\pi}{m_\pi}\right]^2 \frac{3\boldsymbol{\sigma}_1 \cdot \boldsymbol{q}\boldsymbol{\sigma}_2 \cdot \boldsymbol{q} - \boldsymbol{\sigma}_1 \cdot \boldsymbol{\sigma}_2 \boldsymbol{q}^2}{m_\pi^2 + \boldsymbol{q}^2} \vec{\tau}_1 \cdot \vec{\tau}_2. \tag{1b}$$

In a relativistic framework, the central part of the π-induced interaction is given by

$$V_\pi^c = \frac{1}{2} \sum_{\alpha\beta\gamma\delta} c_\alpha^\dagger c_\beta^\dagger c_\gamma c_\delta \int d\boldsymbol{r}_1 d\boldsymbol{r}_2 \bar{f}_\alpha(\boldsymbol{r}_1) \bar{f}_\beta(\boldsymbol{r}_2) \times [\Gamma_\pi^c v_\pi]_{(1,2)} f_\gamma(\boldsymbol{r}_2) f_\delta(\boldsymbol{r}_1), \tag{2}$$

where c_i and $c_i^\dagger$ correspond to nucleon annihilation and creation operators, and the interaction vertex including the propagator is

$$[\Gamma_\pi^c v_\pi]_{(1,2)} = \vec{\tau}_1 \cdot \vec{\tau}_2 \, [f_\pi \gamma\gamma_5]_1 \cdot [f_\pi \gamma\gamma_5]_2 \frac{e^{-m_\pi |\boldsymbol{r}_1 - \boldsymbol{r}_2|}}{12\pi |\boldsymbol{r}_1 - \boldsymbol{r}_2|}. \tag{3}$$

The tensor part of the π-induced interaction can be obtained by subtracting out the central component from the total contribution.

The contributions of the π-N coupling to the energy differences $\Delta E(Z) = E_{\nu 1i_{13/2}} - E_{\nu 1h_{9/2}}$ along the isotonic chain $N = 82$ are shown in the left panel of Fig.2 . They are separated into the central and tensor contributions. A similar comparison is made for the energy differences $\Delta E(N) = E_{\pi 1h_{11/2}} - E_{\pi 1g_{7/2}}$ in $Z = 50$ isotopes, in the right panel of Fig. 2. The main conclusion of this figure is that, there is a strong isospin dependence in the energy differences $\Delta E(Z)$ and $\Delta E(N)$ coming from the π-N coupling, and practically all the effect is due to the tensor component of the π-induced interaction.

The effect of a tensor force on single-particle energies can be understood as follows. First, the force acts only between nucleons of different

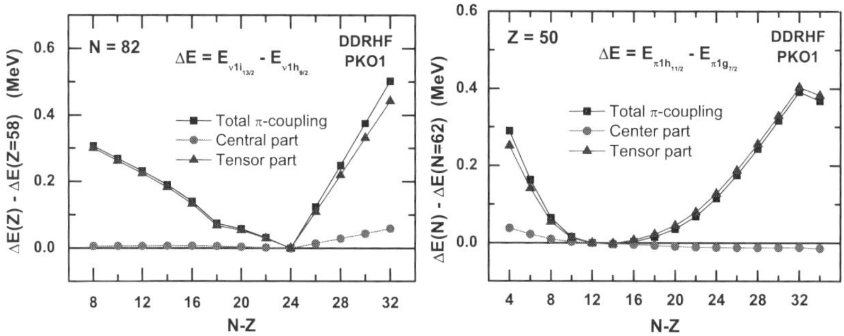

Fig. 2. (color online) Central and tensor pion contributions to the energy differences $\Delta E(Z) = E_{\nu 1i_{13/2}} - E_{\nu 1h_{9/2}}$ in $N = 82$ isotones (left panel), and $\Delta E(N) = E_{\pi 1h_{11/2}} - E_{\pi 1g_{7/2}}$ in $Z = 50$ isotopes (right panel), as a function of neutron excess. The results are calculated with PKO1.

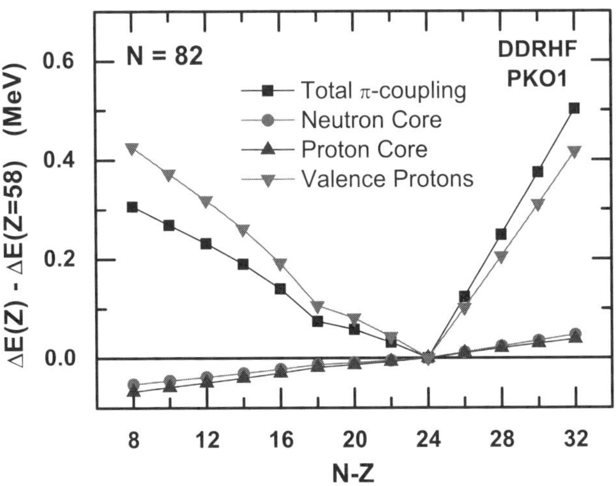

Fig. 3. (color online) The contributions of π-coupling to the energy difference $\Delta E = E_{\nu 1i_{13/2}} - E_{\nu 1h_{9/2}}$ as a function of $N - Z$ along the isotonic chain $N = 82$. They are separated into contributions from the neutron core ($N = 82$), proton core ($Z = 50$) and proton valence orbitals. The results correspond to PKO1.

isospin, i.e., it is effective only in the proton-neutron channel. Let us consider for example the single-particle energy $E(i_\nu)$ of a neutron state i_ν, its tensor part will result from interactions with all occupied proton states k_π. Among these states we can distinguish those belonging to the proton core (in the present case the core is Z=50), to the proton valence orbitals

$(1g7/2, 2d5/2, 2d3/2, 1h11/2)$ and to the neutron core. Usually, the core is such that both $j_>$ and $j_<$ orbitals are occupied and therefore the corresponding tensor contributions largely compensate each other.[16] Only in the valence subspace such compensation does not occur and therefore, one expects that most of the effect of a tensor force on single-particle energies comes from valence orbitals of the other kind of nucleons. In Fig.3 we show the quantities $\Delta E(Z) = E_{\nu 1 i_{13/2}} - E_{\nu 1 h_{9/2}}$ with the separate contributions from the neutron core, the proton core and the valence proton orbitals. The expected behavior is indeed clearly seen.

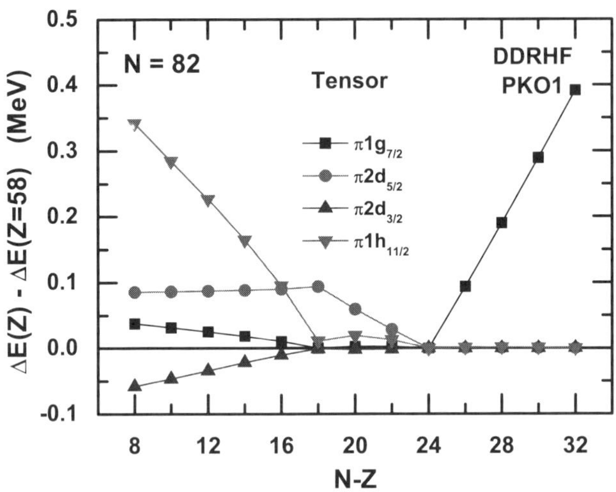

Fig. 4. (color online) The contributions of the tensor π-coupling from the 4 valence proton orbitals to the energy difference $\Delta E = E_{\nu 1 i_{13/2}} - E_{\nu 1 h_{9/2}}$. The results correspond to PKO1.

One can understand further the microscopic mechanism governing the effect of the tensor interaction on single-particle energies by analyzing the separate contribution of each valence orbital. This is done on the example of the energy $\Delta E(Z) = E_{\nu 1 i_{13/2}} - E_{\nu 1 h_{9/2}}$ in the isotonic chain N=82, which is shown in Fig.4. The general rule is that the tensor contribution of an orbital nl is minimal and close to zero when both $j_> = l + 1/2$ and $j_< = l - 1/2$ are filled whereas it is maximal when $j_> = l + 1/2$ is filled while $j_< = l - 1/2$ is empty. When N-Z decreases from 32 to 8 the number of protons is increasing from Z=50 (no valence protons) to Z=74 and one is filling successively the $1g7/2$, $2d5/2$, $2d3/2$ and $1h11/2$ orbitals. In the proton core the $1g9/2$ orbital is filled but the $2d$ and $1h$ orbitals are empty and therefore, one can see in Fig.4 that the above rule is well obeyed.

4. Conclusion

In this paper we have briefly examined the effects of the pion-induced tensor interaction on the evolution of single-particle spectra along a chain of isotopes or isotones, in the framework of the RHF approach. The tensor character of the nucleon-nucleon interaction is contained in the part of the force mediated by the π and the tensor-ρ couplings. It is essential to keep the exchange (Fock) terms in the mean field in order to have non vanishing pionic and tensor-ρ effects. Here, we have concentrated the discussion on the pionic effects. The π-N pseudo-vector coupling generates a tensor component in the N-N effective interaction, and this tensor force plays an important role in non-closed shell nuclei. Because of its isovector character it acts mostly between neutrons and protons, and it influences strongly the single-particle spectra in nuclei with a sizable neutron excess. One important feature of the RHF approach is that the π-N coupling is determined on the same footing as that of the other effective mesons (σ, ω, vector-ρ), i.e., by the fit to bulk properties of nuclear matter and nuclei, and not by an *a posteriori* fit of single-particle spectra.

Another important contribution to the effective N-N tensor interaction comes from the Lorentz tensor coupling of the ρ meson. Because of the limitations of the present talk we have not discussed this aspect here. In Ref.13 this tensor-ρ coupling has been considered in the effective Lagrangian PKA1, and it is shown that the combination of both pseudovector-π and tensor-ρ couplings can improve considerably the shell pattern in the regions N $\sim$ 50-82 and Z $\sim$ 82-126.

Acknowledgments

This work is partly supported by the National Natural Science Foundation of China under Grant No. 10435010, and 10221003, and the Japanese Ministry of Education, Culture, Sports, Science and Technology by Grant-in-Aid for Scientific Research under the program number (C(2)) 16540259, and the European Community project Asia-Europe Link in Nuclear Physics and Astrophysics CN/Asia-Link 008(94791).

References

1. P.-G. Reinhard, Rep. Prog. Phys. **52**, 439 (1989).

2. P. Ring, Prog. Part. Nucl. Phys. **37**, 193 (1996).
3. G.A. Lalazissis and P. Ring, Phys. Rev. **C 55**, 540 (1997).
4. B.D. Serot and J.D. Walecka, Int. J. Mod. Phys. **E 6**, 515 (1997).
5. S. Typel and H.H. Wolter, Nucl. Phys. **A 656**, 331 (1999).
6. T. Nikšić, D. Vretenar, P. Finelli and P. Ring, Phys. Rev. **C 66**, 024306 (2002).
7. M. Bender, P.-H. Heenen and P.-G. Reinhard, Revs. Mod. Phys. **75**, 121 (2003).
8. W.H. Long, J. Meng, N. Van Giai and S.G. Zhou, Phys. Rev. **C 69**, 034319(2004).
9. J. Meng, H. Toki, S.G. Zhou, S.Q. Zhang, W.H. Long and L.S. Geng, Prog. Part. Nucl. Phys. **57**, 470 (2006).
10. W.H. Long, N. Van Giai and J. Meng, Phys. Lett. **B 640**, 150 (2006).
11. W.H. Long, N. Van Giai and J. Meng, arXiv:nucl-th/0608009; W.H. Long, Ph.D thesis, Université Paris-Sud (2005), unpublished.
12. W.H. Long, H. Sagawa, J. Meng and N. Van Giai, Phys. Lett. **B 639**, 242(2006).
13. W.H. Long, H. Sagawa, N. Van Giai and J. Meng, Phys. Rev. **C 76**, 034314 (2007).
14. J.P. Schiffer et al., Phys. Rev. Lett. **92**, 162501 (2004).
15. A. Bouyssy, J.-F. Mathiot, N. Van Giai and S. Marcos, Phys. Rev. **36**, 380 (1987).
16. T. Otsuka, T. Suzuki, R. Fujimoto, H. Grawe and Y. Akaishi, Phys. Rev. Lett. **95**, 232502 (2005).
17. G. Colò, H. Sagawa, S. Fracasso and P.F. Bortignon, Phys. Lett. **B 646**, 227 (2007).

SECTION III

SHELL MODEL AND NUCLEAR STRUCTURE

STRUCTURE OF $A=14$ NUCLEI: THE NCSM AND THE SHELL MODEL

I. TALMI*

Department of Particle Physics, The Weizmann Institute of Science, Rehovot 76100, Israel

**E-mail: igal.talmi@weizmann.ac.il*

No-core shell model (NCSM) calculations of energy levels and Gamow-Teller (GT) transitions in $A=14$ nuclei were recently published. The results are in good agreement with some experimental data and in disagreement with other ones. The simple shell model, in which states of the p-shell and some intruders from the s,d-shell are calculated from two nucleon two-body effective interactions, presents a good description of $A=14$ states. Looking at the data it seems that the NCSM reproduces rather well the levels which, according to the simple shell model, belong to the p-shell. It seems to miss those levels which are intruder states from the next s,d-shell. A possible reason for this fact is discussed. It is indeed a challenge to the NCSM to understand the reason for this feature and find the way to correct it.

Keywords: $A=14$ levels, shell model, no-core calculations

1. Introduction

In a recent paper, "Gamow-Teller Strengths in the $A=14$ Multiplet: A Challenge to the Shell Model",[1] some Gamow-Teller (GT) matrix elements between states of $A=14$ nuclei were considered. New experimental data were compared with theoretical predictions. Some nice agreements between calculations and experiment were found but also some disagreements were noticed. The theoretical predictions presented there were obtained by no-core shell model calculations (NCSM). Hence, it is actually these no-core calculations which face the challenge. The NCSM is a very interesting and promising approach to *ab initio* calculations of wave functions and energies of nuclear states. Starting from available interactions between free nucleons, an effective interaction is derived.[2] Harmonic oscillator wave functions are then used as a convenient set of states to obtain an approximate solution of the nuclear many body problem. To obtain accurate solutions, combinations of states in several oscillator shells are used, their number limited only by computational problems. The results reported in Ref.[1] were obtained by including excitations up to $6\hbar\omega$. The NCSM is an ambitious program which is still being developed. Therefore, it is interesting to see how well it agrees with experiment and in particular, where it seems to need improvements.

The term "shell model" is usually referred to a far less ambitious model. It starts from the opposite end of the nuclear structure problem, from the regular features which are experimentally observed in nuclei. These regularities have been well described by the simple shell model in which nucleons occupy orbits in definite shells. In A=14 nuclei, nucleons occupy the p-shell which includes the $1p_{3/2}$ and $1p_{1/2}$ orbits with possible "intruder" states from the next s,d shell. States and energies are then determined by introducing an effective two-body interaction between nucleons.

The aim of this note is to show how, in the case of A=14 nuclei, the simple shell model can meet the challenge posed in Ref.[1]. Some results of the simple model are mentioned in that paper. Still, the impression is that these "$0\hbar\omega$ calculations" are less reliable than the more correct ones in which excitations into higher and higher shells are taken into account. This, however, is not the case since there is no exact theory for the interaction between free nucleons. The simple ("$0\hbar\ \omega$ ") shell model is not a crude approximation to the NCSM although n equals 0 and not 6 in the number of shells included in the calculation. It is a well defined model, different from the NCSM. There should be no conflict between these two approaches, they may complement one another. One is easier to apply and to obtain results whereas the other is more fundamental and hence much more difficult to use.

The basis of the simple shell model is the assumption that an exact theory of nuclear structure will justify its use as a good approximation and will also lead to a two body effective interaction between valence nucleons. For the time being, order and spacings of single nucleon orbits as well as matrix elements of the two-body effective interaction are determined from experiment.[3,4] The merit of this approach, as well as its justification, is the good agreement of its predictions with experiment. This way, a consistent description was obtained of a large body of experimental energies and wave functions of nuclear states

As reported in Ref. [1], there are two kinds of disagreements between results of the NCSM calculations and experimental data (Fig.1). The authors explain that "neither a reproduction of the total experimental B(GT) strength nor a detailed description of fragmentation into three final 2^+

states is possible with NCSM, even using very large ($6\hbar\omega$) model spaces". Similarly, the elaborate NCSM calculations do not reproduce the position of the first excited 0^+ state. These calculations do not reproduce also the next excited 0^+ state at 9.75 MeV excitation.

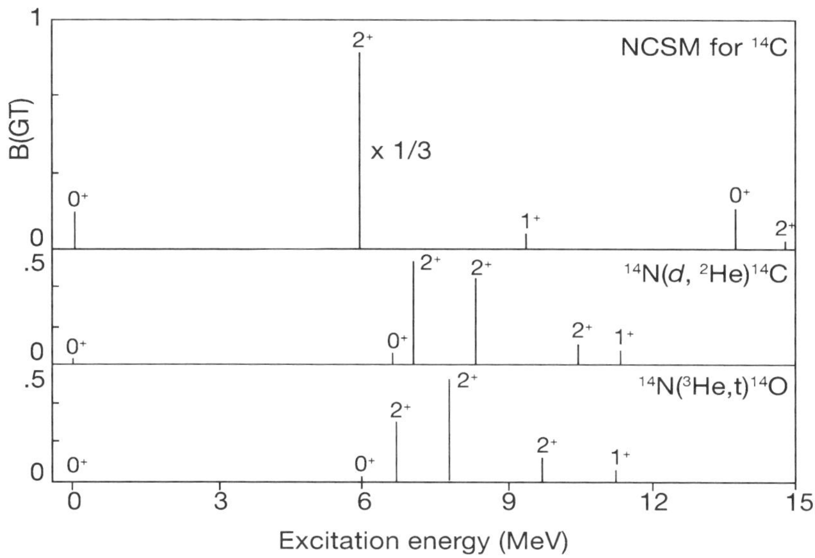

Fig. 1. Energies and Gamow-Teller transition strengths of A=14 Nuclei

2. Intruder states in p-shell nuclei

The situation in the simple shell model is rather different, as mentioned several times in Ref.[1] The first excited 0^+ state as well as the two extra 2^+ states are due to excitations of two p-nucleons into the next $2s,1d$ shell.[5] The 2^+ state which is due to excitations within the p-shell is mixed with the "intruder" 2^+ states leading to fragmentation of the GT strength. The authors state that it "is interesting to note that simplified calculationsconsidering the lowest 2 particle - 4 hole configurations in ^{14}C made of a $^{12}\text{C} \bigotimes 2n$ configuration with the two neutrons in the $1d_{5/2}$ and $2s_{1/2}$ orbitals seem to reproduce the experimental finding of a further 0^+ state and three 2^+ states in the excitation region of interest". Yet they seem to dismiss "these early studies" which "were very phenomenological in scope" and hence, "contrary to the shell model calculations of" the NCSM approach. As explained above, the simple shell model *is phenomenological.* This is its strength and its limitation. In the following, a brief discussion of the intruder states will be given. No detailed calculations will be presented but simple energy considerations are sufficient to give semi-quantitative predictions of those states. Some matrix elements between states of s, d and p-nucleons needed for a detailed calculation were determined by Millener and Kurath,[6] but they will not be used here.

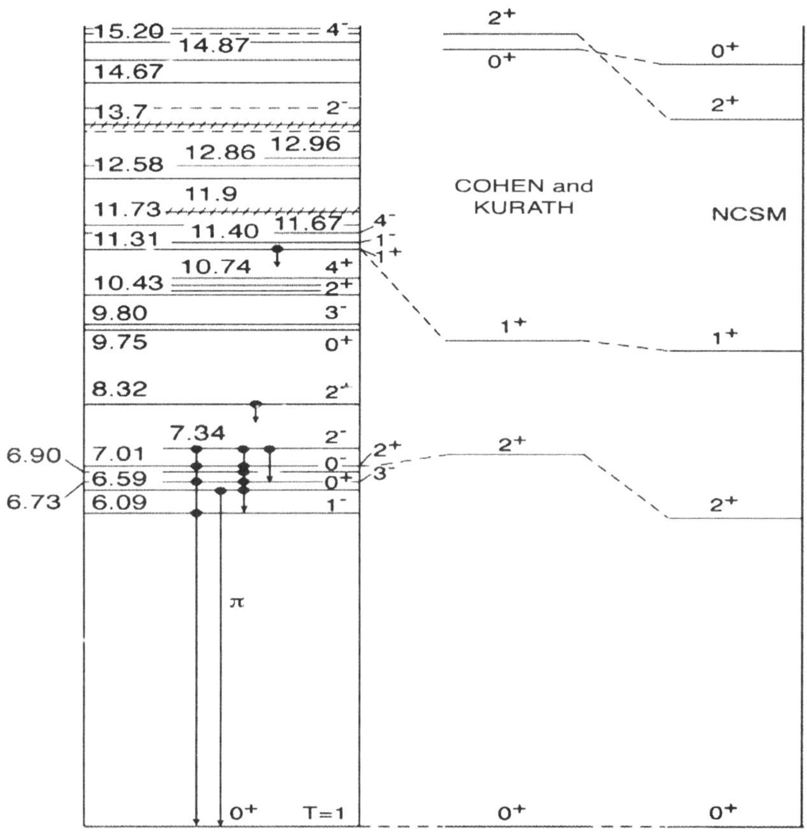

Fig. 2. Experimental and calculated levels of ^{14}C

Energy levels of p-shell configurations were calculated by Cohen and Kurath[7] who determined matrix elements of the effective interaction from some experimental energies. The agreement with experiment which they obtained is satisfactory. For the ^{14}C nucleus they obtained the first excited 0^+ state to be higher than the ground state by 13.75 MeV. The first excited 2^+ state was calculated to lie at 7.21 MeV above the ground state in good agreement with the measured one at 7.01 MeV. From their interaction the next higher 2^+ state is predicted to be at 15.27 MeV. It is interesting and probably significant that the positions of levels calculated by NCSM are in fair agreement with some experimental levels. These are all levels, including the 1^+ level, which are due to p-nucleons and were calculated by Cohen and Kurath[7] (Fig.2).

The situation described above implies that the experimental 0^+ level at 6.59 MeV and the 2^+ levels at 8.32 and 10.43 MeV, also shown in Fig.2, are due to two nucleon excitations from the p-shell into the s,d-shell. The three observed 2^+ states and the first excited 0^+ state contain considerable components of $s.d$ states. This may be seen also from the rather large Coulomb energy differences between corresponding levels of ^{14}C and ^{14}O. These are due to the radial wave functions of $1d$-neutrons and specially of 2s-neutrons, which extend to larger distances than the tightly bound p-nucleons. This behavior should be contrasted with the Coulomb energy difference between the 1^+ states which are due to p-nucleons. That difference is practically equal to that of the ground states in spite of the rather high excitation energy of of the 1^+ state, 11.31 MeV in ^{14}C (Fig. 1).

The appearance of such intruder states at low excitations should not be surprising. In ^{13}C, $1/2^+$ and $5/2^+$ levels, presumably due to excitations of a p-neutron into the $2s_{1/2}$ and $1d_{5/2}$ orbits were found at rather low excitations, 3.09 and 3.85 MeV respectively. This fact may pose some difficulty for a schematic approach where energy separations between major shells are taken to be much higher. In the phenomenological approach, the experimental data *determine* those spacings. In fact, a simple shell model calculation starting from ^{13}C levels, predicted that the ground state of ^{11}Be should be $1/2^+$ [8] as later verified by experiment (Fig.3).

In ^{14}C there are odd parity levels with spins 1(6.09 MeV), 0(6.90), 3(6.73) and 2(7.34) (Fig.2). They may be interpreted as due mainly to a $1p_{1/2}$ neutron coupled to $2s_{1/2}$ and $1d_{5/2}$ neutrons respectively.[9] Indeed, the single particle excitation energy of the $p_{1/2}s_{1/2}$ configuration is simply calculated to be 6.31 MeV which is in very good agreement with the center of mass excitation energy of the 0^- and 1^- levels, 6.29 MeV. This is consistent with the observation that the average interaction energy (monopole part) of two neutrons in different orbits is rather weak and may be slightly

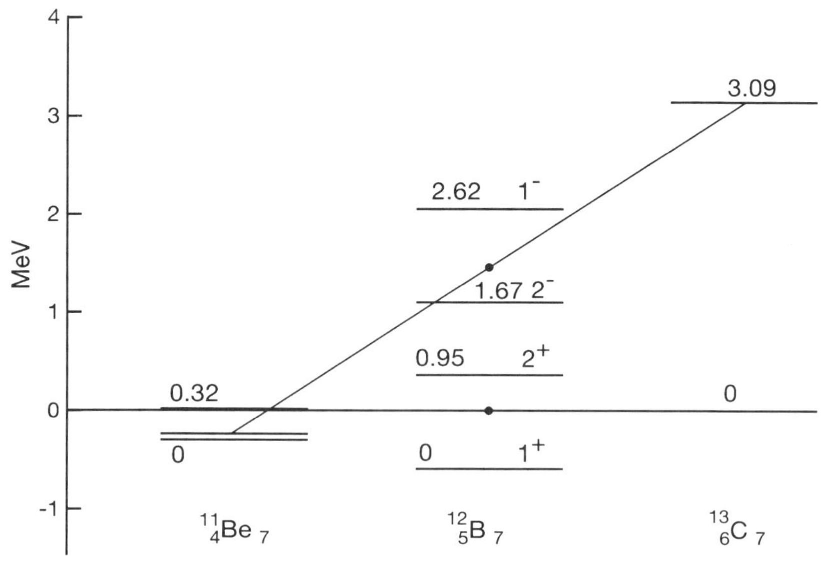

Fig. 3. Experimental and calculated levels of ^{11}Be

repulsive.[9,10] The analogous predicted excitation of the $p_{1/2}s_{5/2}$ configuration is 7.17 MeV in fair agreement with the center of mass of the 2^- and 3^- levels which is 6.98 MeV. Also the Coulomb energy differences between these states in ^{14}C and ^{14}N are larger than the difference between the ^{14}C ground state and the T=1, J=0 state of ^{14}N. This is true in particular, for the states in which s-nucleons are present.

The positions of 0^+ states, due to excitation of two p-nucleons into the s,d-shell, can be simply obtained. Neutrons in $s_{1/2}$ and $d_{5/2}$ orbits are bound to ^{12}C by 1.86 and 1.10 MeV respectively. Hence, the single nucleon energy of the $(s_{1/2})^2$ configuration relative to the ^{14}C ground state is given by

$$B.E.(^{14}C) - B.E.(^{12}C \bigotimes s^2_{1/2}) = \qquad (1)$$
$$[B.E.(^{12}C) + 13.12] - [B.E.(^{12}C) + 3.72] = 9.40MeV$$

This number will be reduced if the interaction between the s-neutrons in the $s^2_{1/2}$ J=0 state will be included. It could bring this energy down to 6.59 MeV which is the experimental excitation energy of the first excited 0^+state. In fact, the situation is similar to that in ^{16}C (Fig.4). The ground state of ^{15}C is $1/2^+$ and the interaction energy between the two neutrons outside the p-shell in ^{16}C is 3.04 MeV.

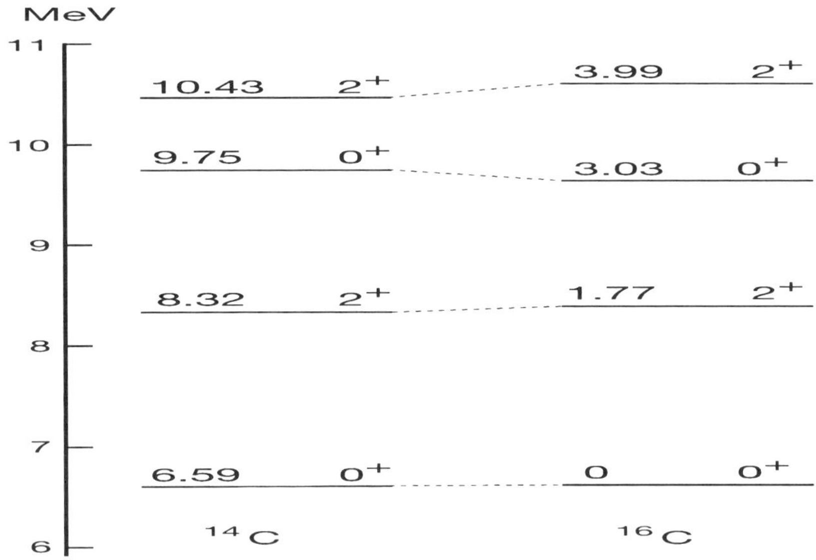

Fig. 4. Intruder states in ^{14}C and corresponding levels of ^{16}C

The next intruder 0^+ state should belong to the $d^2_{5/2}$ configuration. A similar calculation yields for its excitation energy due to single nucleon energies the value of 11 MeV. Adding possible values of the interaction between d-nucleons could lower the excitation energy of this state to the observed position at 9.75 MeV. This assignment seems correct since, as mentioned above, the calculated position of the excited 0^+ state of the p^{10} configuration should lie at 13.75 MeV above the ground state.[7] The exact positions of these four 0^+ states in the ^{14}C spectrum are determined by the possible mixing between them and thus, by the non-diagonal matrix elements of the effective interaction. The analogy with ^{16}C applies also to this state. In ^{15}C the first excited state is a $5/2^+$ state and its position above the $1/2^+$ ground state, 0.740 MeV, is close to the spacing between the corresponding states in ^{13}C. Also in ^{16}C the first excited 0^+ state lies about 3 MeV above the ground state which is fairly equal to the difference 9.75-6.59=3.16 MeV in ^{14}C. The striking similarity between positions of intruder states in ^{14}C and levels in ^{16}C is presented and discussed by Fortune *et al.*[5]

Two 2^+ intruder states belong to the $2s_{1/2}$,$1d_{5/2}$ and $2d^2_{5/2}$ configurations in J=2 states coupled to the 0^+ ground state of ^{12}C. The interaction energy between intruder neutrons in states with J=2 should be weaker than that of the $2s_{1/2}^2$ 0^+ state. The single nucleon energy of the $2s_{1/2}$,$1d_{5/2}$ 2^+ state is higher by 1.86-1.1=.76 MeV from that of the $2s_{1/2}^2$ 0^+ state. Hence, this state could be identified with the observed 8.32 MeV J=2 level. In ^{16}C there is a 2^+ state which lies 1.77 MeV above the ground state. The analogous state in ^{14}C should thus lie 1.77 MeV above the 0^+ state at 6.59 MeV, *i.e.* at 8.36 MeV. The $2d_{5/2}^2$ 2^+ state should have a weaker interaction than the $2d_{5/2}^2$ 0^+ state and hence, expected to lie a couple of MeV above the latter. It could be identified with the 10.43 MeV level with J=2. From the similarity with ^{16}C this level should lie at 10.58 MeV. Still, in that region of excitation energy, other 2^+ states may be present.

There should be two other 2^+ levels due to intruder $2s_{1/2}^2$ and $1d_{5/2}^2$ states with J=0 coupled to the lowest J=2 state of ^{12}C. The excitation energy of the latter is 4.44 MeV so that these two levels should lie about 4 MeV above the intruder 0^+ states. The one based on the $2s_{1/2}^2$ 0^+ state is also a possible candidate for the 2^+ level at 10.43 MeV. It is clear, however, that without a detailed calculation, the exact positions of all 2^+ levels in ^{14}C cannot be predicted. Their final positions should be determined by the matrix elements of the interaction between them as well as by the matrix elements between them and the two 2^+ states of the p-shell. One of the latter is the 7.01 MeV level and the other is predicted by Cohen and Kurath[7] to be at 15.27 MeV excitation.

3. Gamow-Teller transitions

Let us now look at the Gamow-Teller matrix elements between states of A=14 nuclei. In Ref.[1] matrix elements between the $J=1^+$ ground state of ^{14}N and states with $J=2^+$, $J=1^+$ in ^{14}C are considered. As mentioned above, there is only one 2^+ state emerging from the NCSM calculation in the energy range reached by experiment. The calculated Gamow-Teller strength to that state is rather large. The simple shell model calculation provides additional 2^+ states which, by mixing with the p-shell 2^+ state, share that strength. Still, the summed strength to the three lowest 2^+ levels is less than half of the calculated one. This difficulty arises also in the simple shell model. Using it, the calculated strength is roughly the same as in Ref.[1], B(GT)$\sim$2.5. The reduction in Gamow-Teller matrix elements is observed in almost all cases considered in the simple shell model. Perhaps an effective Gamow-Teller operator with a reduced coupling constant should be used with the simple shell model. This problem has attracted a big effort both in theory and experiment but it will not be further discussed here.

The other Gamow-Teller strength of interest is to the 1^+ state which lies 11.31 MeV above the ^{14}C ground state. Its excitation energy is about 2 MeV higher than the one calculated within the p-shell by Cohen and Kurath[?] and by NCSM. It could be safely identified with the only 1^+, T=1 state of a two hole p^{-2} configuration, $p_{3/2}^{-1}p_{1/2}^{-1}$ which is identical with the $^3P_1$ state of LS-coupling. Hence, the only non-vanishing Gamow-Teller matrix element between this state and the ^{14}N $J=1^+$, $T=0$ ground state is due to the $^1P_1$ component of the latter. The strength is thus given by 2×1.251 multiplied by the square of the amplitude of the $^1P_1$ component. It was mentioned in Ref.[1] that the ^{14}N ground state is almost a pure D state "supported e.g., by a study of the ^{12}C(^{3}He,p) reaction, where the angular distributions to the ground state and the 3.95 MeV state of ^{14}N were characterized by ΔL=2 and 0 respectively".[11] Indeed, in the state $p_{1/2}^2$, J=1, the coefficient of the D-state component is equal to $(20/27)^{1/2}$ (roughly 0.86). If reasonable matrix elements of the effective interaction are adopted,[12] the coefficient of the D-state in the g.s. of ^{14}N is calculated to be higher, .967 (and that of $^3S_1$ in the 3.95 MeV state to be .928) but even this value does not yield the observed attenuation of the GT decay of the ^{14}C ground state. The 1^+ ground state of ^{14}N cannot be a *pure* D-state.

The non-diagonal matrix element between the $^3D_1$ and the $^1P_1$ states in the ^{14}N g.s. is proportional to the spin-orbit splitting between the $1p_{3/2}$ and the $1p_{1/2}$ orbits. Its effect may be diminished if there is a large difference between the diagonal elements of the interaction in D and P states. Putting the $^1P_1$ amplitude equal to zero, or close to it, however, amounts to vanishing spin-orbit splitting which destroys the basis the observed closed shells and magic numbers. The very large attenuation of the Gamow-Teller matrix element of the ^{14}C ground state beta decay to the ground state of ^{14}N may be explained in the simple shell model by using only p-shell wave functions.

Many years ago it was shown that if tensor forces are present in the two-body interaction, an "accidental" cancellation may arise between the Gamow-Teller matrix elements from the $^3P_0$ to the $^1P_1$ component and from the $^1S_0$ to the $^3S_1$ component.[12,13] In this calculation a strong spin orbit interaction is present and its magnitude is taken from experiment. The Gamow-Teller operator was assumed to be the usual one or proportional to it. The condition of accidental cancellation does not determine the size of the P amplitude in the ^{14}N ground state. It is related to the S amplitude but both could be rather small. In Ref.[1] it is stated that the authors "try to shed light on this long-standing problem by studying GT transitions from the ^{14}N g.s. to excited states of ^{14}C and ^{14}O" but a few lines below they explain that "such experiments cannot provide new insight into the suppressed g.s. transitions". Some NCSM calculations yield the strong attenuation of the ^{14}C Gamow-Teller transition strength. It would be interesting to see if this attenuation is due to a mechanism similar to the one in the simple shell model.

Using the matrix elements of the effective interaction estimated in Ref.[12], the amplitude of $^1P_1$ in the ^{14}N ground state turns out to be about 0.2. Its square is 0.04 which means that the strength of the Gamow-Teller transition to the 1^+ state in ^{14}C is only 4% of the total strength to the 2^+ states. The negligible GT strength to the 1^+ state compared to the strong GT transition to the 2^+ state(s) was predicted by the NCSM calculations and was verified by experiment.[1] As shown above, it is in fair agreement with the simple shell model.

4. Summary

It seems that the NCSM calculations for A=14 nuclei reproduce rather well, energies and GT transitions of states which in the simple shell model belong to the p-shell. The difficulties arise in the case of intruder states and the challenge for NCSM is to overcome them. Perhaps this is a good case to find out if there is some relation between the NCSM wave functions of the states reported in Ref.[1] and those of the simple shell model in the p-shell. In some schematic models in which a central potential well is assumed, states of the next shell may lie rather high and have no chance to be intruder states and certainly not to become ground states. As its name suggests, in the no-core model no potential well is assumed. In NCSM calculations, $h\omega$ seems to determine only the set of oscillator wave functions used in the calculations, and is not simply related to energy differences between shells. Still, using an expansion in harmonic oscillator wave functions, single nucleon states in higher shells have higher kinetic energies. Thus, such states have effectively single nucleon energies higher by $nh\omega\ /2$ than in the lower, n=0 shell. The oscillator parameter used in the calculations in Ref.[2] is $h\omega$ =14 MeV. Hence, it is rather difficult to bring states with s,d nucleons down to their observed positions. This is a challenge facing the no-core calculations which involves not only A=14 states with positive parity, but also the negative parity states in ^{14}C and positive parity ones in ^{13}C and in ^{11}Be.

References

1. A. Negret et al., Phys. Rev. Lett. **97**, 062502 (2006).
2. S. Aroua et al., Nucl. Phys. **A720**, 71 (2003).
3. S. Goldstein and I. Talmi, Phys. Rev. **102**, 589 (1956).
4. Examples may be found in I. Talmi, Simple Models of Complex Nuclei, Harwood, 1993.
5. H. T. Fortune et al, Phys. Rev. Lett. **40**, 1236 (1978).
6. D. J. Millener and D.Kurath, Nucl. Phys. **A255**, 315 (1975).
7. S. Cohen and D. Kurath, Nucl. Phys. **73**, 1 (1965).
8. I. Talmi and I. Unna, Phys. Rev. Lett. **4**, 469 (1960).
9. I. Talmi and I. Unna, in Annual Review of Nuclear Science **10**, 353 (1960).
10. I. Talmi, Rev. Mod. Phys. **34**, 704 (1962).
11. C. H. Holbrow, R. Middleton and W. Focht, Phys. Rev. **183**, 880 (1969).
12. I. Talmi, in: Proceedings of the International Symposium on Electromagnetic Properties of Atomic Nuclei, H. Horie and H. Ohnuma (Eds.), Tokyo Institute of Technology, Tokyo, 1984.
13. B. Jancovici and I. Talmi, Phys. Rev. **95**, 289 (1954).

CI AND EDF APPLICATIONS IN LIGHT NUCLEI TOWARDS THE DRIP LINE

B. A. BROWN

Department of Physics and Astronomy and National Superconducting Cyclotron Laboratory, Michigan State University, East Lansing, Michigan 48824-1321, USA

W. RICHTER

Department of Physics, University of the Western Cape, Private Bag X17, Bellville 7535, South Africa

New sd-shell Hamiltonians, USDA and USDB, are discussed. The application to sd-shell energies is reviewed and results are shown for M1 transition matrix elements and magnetic moment data. The predictions for the properties of the oxygen isotopes near the drip-line nucleus ^{24}O are shown and compared with recent data. The empirical values for the proton and neutron single-particle energies in ^{24}O are compared with Skyrme EDF calculations. The systematic properties of the single-particle energies from Skyrme EDF for ^{24}O, ^{34}Si and ^{42}Si are shown and discussed.

Keywords: Shell Model, Hartree-Fock, neutron-unbound states, sd-shell

1. CI Hamiltonians for the sd-Shell

The new USDA/B Hamiltonians[1] for the sd-shell nuclei provide an improved level of precision for nuclear structure properties in the A=16-40 mass region. In this section we will summarize some of the background for these new interactions and then focus on some new results for the properties of the oxygen isotopes out to and beyond the drip line nucleus ^{24}O. In the next section we will discuss some results from Skyrme energy-density functional (EDF) calculations for ^{24}O, ^{34}Si and ^{42}Si.

Shell-model configuration interaction (CI) methods are used for the $(0d_{5/2}, 0d_{3/2}, 1s_{1/2})$ set of orbitals. The USDA/B Hamiltonians were obtained from a least-squares fit of three single-particle energies and 63 two-body matrix elements (TBME) to binding energy and energy level data for 608 states in 77 nuclei. For the older USD Hamiltonian[2] (based on the computer codes and data available in 1980) many states in the middle of the shell shell were not considered in the fit due to the prohibitive computational time. Now it is easy to included all states. In addition, since 1980 many new experimental data for binding enegies and energy levels have been obtained for neutron-rich nuclei. When the fit error matrix is diagonalized, the largest eigenvalues give the linear combinations of parameters that are most well determined. N linear combinations of TBME were allowed to vary, and the remaining $63 - N$ linear combinations of TBME were constrained to the values obtained from the renormalized G matrix (RGSD). The RGSD calculations were based on the Bonn-A NN potential and include diagrams up to third-order as well as folded diagrams for an assumed ^{16}O core (table 20 of Ref[3]). Details of the procedure are given in.[1] As was assumed for USD, the single-particle energies (SPE) are taken to be constant (mass independent). One could add some mass-dependence to the SPE, but it has little effect on the rms since it can be compensated by a change in the TBME (in particular in the monopole TBME combinations). "Effective SPE" for a given configuration in the sd-shell are obtained by the addition of the monopole linear-combinations of the TBME to the original SPE. A mass dependence of the two-body matrix elements of the form $(18/A)^{0.3}$ was assumed. This accounts qualitatively for the mass dependence expected from the evaluation of a medium-range interaction with harmonic-oscillator radial wavefunctions that change in size as $A^{1/6}$.

Fig. (1) shows the rms deviation between experiment and theory as a function of N (solid line). It also shows the rms deviation of the TBME between the RGSD and fitted values. USDA corresponds to 30 linear combinations with an rms of 170 keV (where a plateau in the solid line starts). USDB corresponds to 56 linear combinations with a rms of 130 keV. The fitted and input RGSD (G matrix) TBME are shown in Fig. (2). The USDA TBME are strongly correlated with the RGSD values with an rms deviation between the TBME of 290 keV [see Fig. (1)]. Compared to the maximal magnitude of the TBME (6 MeV), this represents only a 5% rms difference. However, this specific 5% change is critical for obtaining accurate binding energies and spectra. The USDB Hamiltonian, where 56 parameters were varied, is compared with RGSD in middle of Fig. (2); the result is a 375 keV rms difference for the TBME (6% of the largest). There are several reasons why the RGSD TBME may differ from those actually required by the sd-shell data: (i) the perturbation expansion may not be converged;[4] (ii) the oscillator basis used for matrix elements and energy denominators of the perturbation diagrams is an approximation; (iii) the RGSD should be taken as an average or interpolation of results for the ^{16}O and ^{40}Ca cores (and not just the ^{16}O core) and (iii) real three-body forces are required as observed

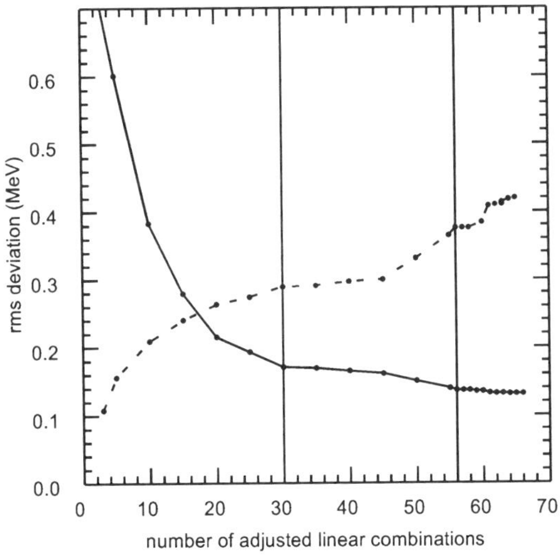

Fig. 1. Rms deviations as a function of the number of fitted linear combinations (see text for details).

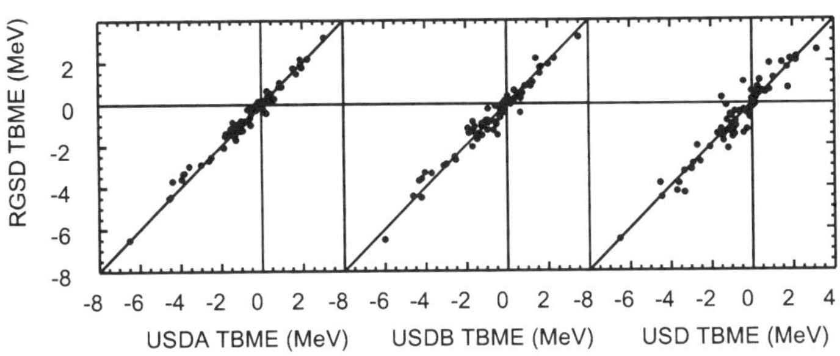

Fig. 2. Comparison of the fitted and RGSD (renormalized G matrix) TBME.

for ab-initio calculations of light nuclei.[5] Three-body forces contribute to effective one- and two-body matrix elements when they are averaged over the nucleons in the closed core[6] (for the entire sd-shell we should take an average of the three-body effective matrix elements for the ^{16}O and ^{40}Ca cores).

Thus we have two new Hamiltonians, a conservative one, USDA, closest to RGSD that gives a good but not the best fit to data, and another, USDB, that differs more from RGSD but gives the best fit to data. Calculation of other observables such as spectroscopic factors, moments, gamma decay and beta decay rates will further test the properties of USDA and USDB and work is underway on this. As an example we show in Figs. 3 and 4 the results for the $M1$ matrix elements [defined by $B(M1) = M^2(M1)/(2J_i+1)$] and magnetic moments. The panels on the left show theory with the free-nucleonic g-factors and panels on the right show the theory with common set of effective g-factors obtained from a fit to all of the data; for protons $g_s = 5.127$ and $g_l = 1.147$ and for neutrons $g_s = -3.543$ and $g_l = -0.090$. The overall agreement with experiment is impressive, in particular with the effective g-factors. The results for USDA are similar to those of USDB.

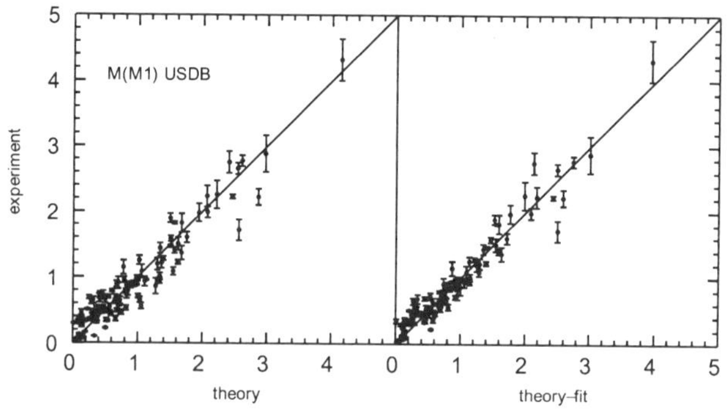

Fig. 3. Comparison of experimental and USDB matrix elements for $M1$ gamma decay. The result are obtained with free-nucleon g-factors (theory on the left-hand side) and with a fit to effective g-factors (theory-fit on the right-hand side).

Spectra were calculated for 87 sd-shell nuclei for both USDA and USDB and compared to experiment. The compete set of comparisons can be found on the web.[7] As an example, we show the results for ^{26}Al and ^{26}Mg in Fig. (5). Experiment is shown on the right-hand side with lines that indicate the J value for the known positive-parity states. The lowest J value is labeled. Levels with an unknown J^π assignment are shown by the shortest lines, with the first such level for ^{26}Al lying at about 4 MeV. The theoretical levels are shown on the right-hand side, also indicated by lines of different length for the J value. The experimental and theoretical levels joined by

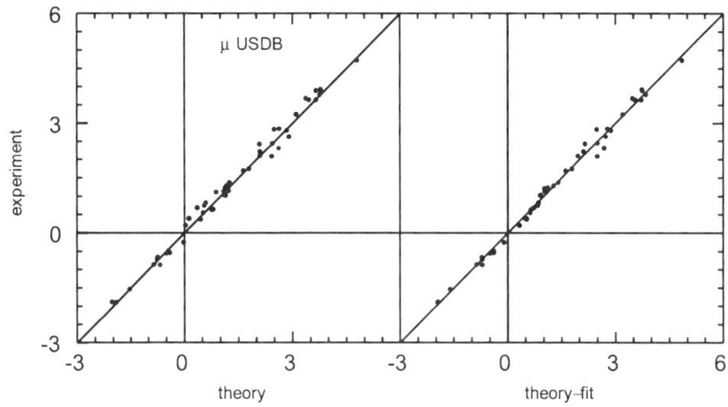

Fig. 4. Comparison of experimental and USDB magnetic moments. The result are obtained with free-nucleon g-factors (theory on the left-hand side) and with a fit to effective g-factors (theory-fit on the right-hand side).

a line in the middle are those included in the fit. The slope of these lines shows the difference between experiment and theory for the excitation energy. The ground-state binding energies are shown at the bottom. Starting at about 4 MeV we cannot make a level to level assignment between theory and experiment due to the incomplete experimental information. From 4 to 6 MeV the experimental and theoretical level densities are similar, but there are too many levels of unknown spin-parity to make definitive associations between experiment and theory. Above 6 MeV the experimental level density becomes higher, indicating the onset of intruder states.

In comparison to recent data, the original USD interaction overbinds the most neutron-rich fluorine by up to 1.5 MeV and it predicts ^{26}O to be bound. The oxygen isotopes beyond N=16 are known to be unbound.[8] Both of these problems are corrected with USDA and USDB. The improved USDA and USDB results for the ground-state binding energies of the neutron-rich fluorine and oxygen isotopes with $N \geq 16$ are related to the increased energy for the effective $d_{3/2}$ SPE. We note that this particular problem with USD was corrected when it was applied to the $sd-pf$ model space.[9] The nuclei with Z=10-12, N=20 are under-bound by USDA/B by up to 2.5 MeV confirming the intruder state (island-of-inversion) interpretation for these nuclei.[10]

The binding energies of the most neutron-rich oxygen isotopes are shown in Fig. 6. The arrows show the expected neutron-decay transitions. The filled circles indicate the location of data. Three of these were included in

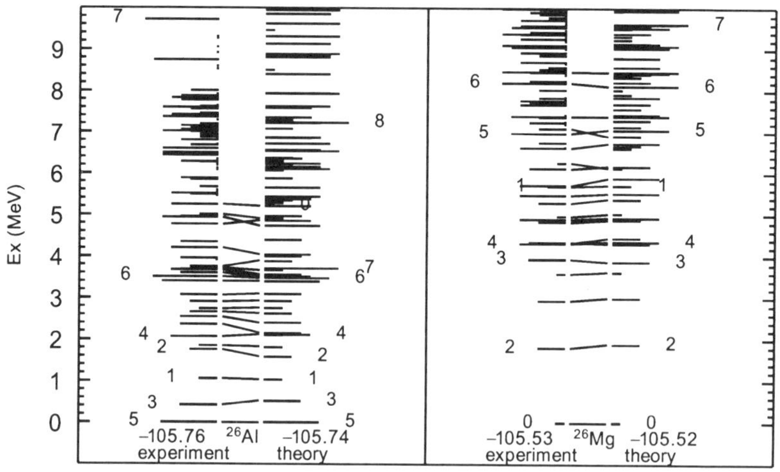

Fig. 5. Comparison of experimental and USDB theoretical levels for ^{26}Al. See the text for details.

the fit (0^+ and 2^+ for ^{22}O and $1/2^+$ for ^{23}O). The $3/2^+$ state in ^{23}O was recently observed in an inverse kinematic (d,p) reaction on ^{22}O which showed the $d_{3/2}$ single-particle nature of that state.[11] This is in good agreement with theory. The $5/2^+$ state in ^{23}O has also recently been observed by its low-energy neutron decay to the ^{22}O ground state.[12] The wavefunction for the $5/2^+$ state is dominated by the configuration $[d^5_{5/2}, s^2_{1/2}]$ and the wavefunction for the ^{22}O ground state is dominated by $[d^6_{5/2}]$ component. The spectroscopic factor between these is zero. Configuration mixing leads to a small spectroscopic factor of 0.059 and a width of 90 eV (based on its observed neutron decay energy of 45 keV). The energy and decay properties of this state are in good agreement with experiment. The ground state of ^{26}O is bound to one-neutron decay by 0.80 (0.95) MeV with USDA (USDB), but it is unbound to two-neutron decay by 0.50 (0.35) MeV. The possibility for di-neutron decay will be theoretical challenge for three-body decay models and would be extremely interesting to observe experimentally.

2. EDF Results for Nuclei Near the Neutron Drip Line

We noted that a major change between the old USD and the new USDA/B Hamiltonians can be qualitatively understood as a 1.5 MeV upward shift

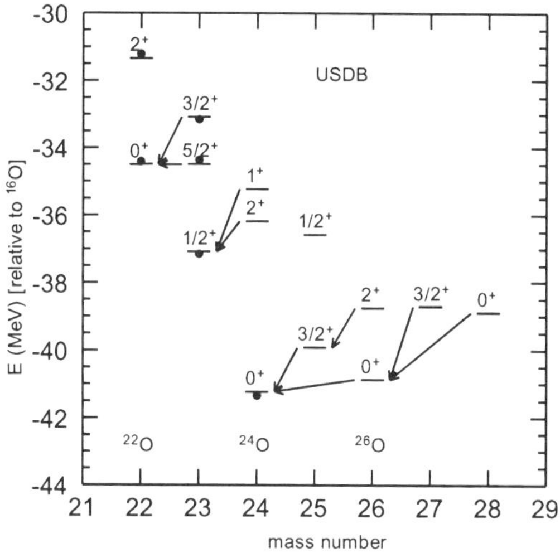

Fig. 6. Comparison of experimental and USDB binding energies for neutron-rich oxygen isotopes. The filled circles indicate the location of data (see text).

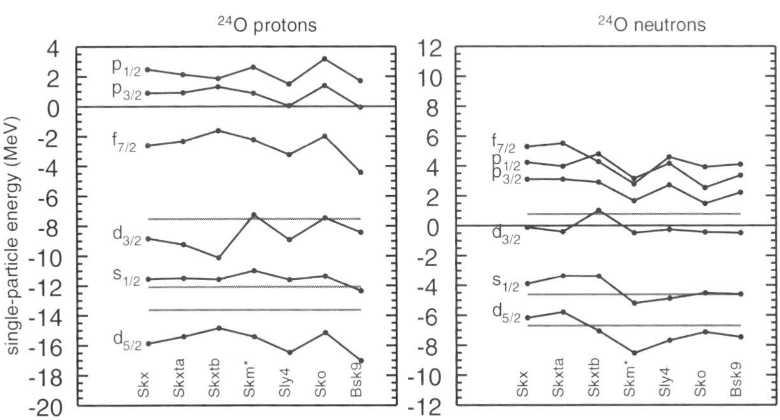

Fig. 7. USDB single-particle energies for ^{24}O (long lines) compared to the results from several Skyrme-type EDF calculations.

of the neutron $d_{3/2}$ SPE in the neutron oxygen and fluorine isotopes. This type of uncertainty in the extrapolation of SPE in CI-type calculations

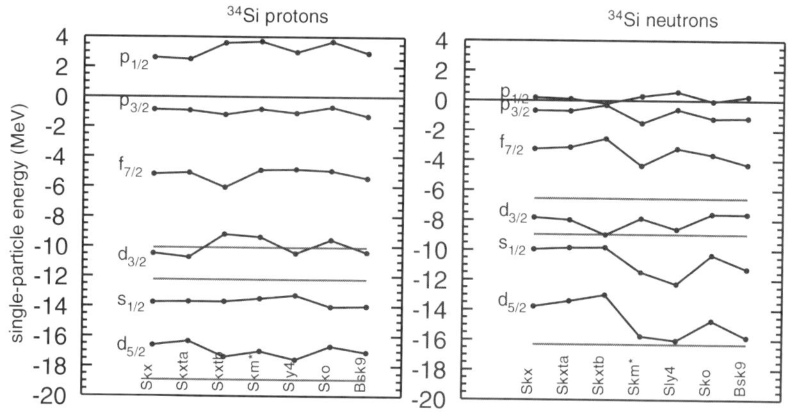

Fig. 8. USDB single-particle energies for ^{34}Si (long lines) compared to the results from several Skyrme-type EDF calculations.

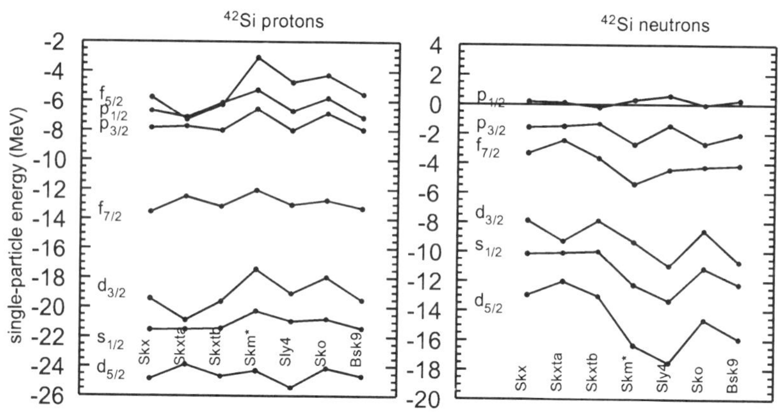

Fig. 9. Single-particle energies from Skyrme EDF calculations for ^{42}Si.

with fitted Hamiltonians (related to the fact that data for levels related to a specific orbital are missing) is a significant limitation to this method. EDF models provide an alternative method for extrapolating single-particle properties. The Skyrme functional is still based on fitted parameters, but they are global parameters obtained from the entire chart of nuclear properties.[13] The region of ^{24}O provides a test of EDF models at the nuclear drip line. CI and EDF Hamiltonians can be compared on the same footing when the same configuration is used for both. For example for ^{24}O we can compare the effective single-particle energies for the $[(\nu d_{5/2})^6]$ configuration with the single-particle levels in the potential well obtained from an EDF calculation with the same configuration (for EDF it would be more precise to calculate the binding energies for all neighboring nuclei, e.g. ^{23}O and ^{25}O, and then take the energy differences, but in practice these differences are close to those obtained for the SPE in the potential well). We compare with results from a variety of Skyrme interactions in Fig. 7 (Skx,[13] Skxta/b,[14] Skm*,[15] Sly4,[16] Sko[17] and Bsk-9[18]). The ^{24}O comparison is shown in Fig. 7. The importance of the shell gaps obtained for the SPE for deformation and shape coexistence in EDF models was shown by Reinhard et al.[19] Our emphasis is on the connection to "experimental" values of the SPE that can be derived from empirical CI Hamiltonians such as USDA/B The SPE for unbound states were estimated in the following way.

When the state is unbound, the overall potential depth is increased until it becomes bound by 0.2 MeV. This single-particle wavefunction $\mid i >$ is then used to calculate $< i \mid T + V \mid i >$ where V is the unmodified Skyrme potential. The energies obtained in this way should be close to the centroid of the resonance obtained from a nucleon scattering phase shift calculation, at least when the barrier (Coulomb plus centrifugal) is not too small.

The Fermi-energy for protons is below the $d_{5/2}$ SPE and is very large (about -18 MeV) in contrast to the near zero Fermi-energy for neutrons (just above the $d_{3/2}$ SPE). Related to this, we see neutron SPE spectrum is compressed compared to protons due to the small binding energy of neutrons. The difference between the Skyrme results come from the various assumptions and procedures used for fitting the Skyrme parameters. For this set of Skyrme only Skxta/b contain the new tensor interaction. Comparison with Skx (no tensor) shows the relative changes in SPE that come from the EDF tensor interaction. None of the Skyrme interactions can reproduce the all of the data. But it is notable that Skxtb is the only one that gets a positive energy for the $d_{3/2}$ neutron SPE. For ^{34}Si none of the Skyrme models can reproduce the $s_{1/2} - d_{5/2}$ proton gap recently discussed by Cottle.[20] ^{42}Si has recently been observed to have a low energy 2^+ state.[21] This may be related to the small $p_{3/2} - f_{7/2}$ neutron shell gap (about 2 MeV) found for all Skyrme interactions (Fig. 9). We conclude that EDF models are necessary for the extrapolation into new regions of nuclei near the drip lines and that new shell structure can emerge near the drip line due to the weakly-bound SPE. But the EDF models need to be improved in terms of reproducing known data on SPE in order to make more reliable extrapolations. Progress in nuclear structure theory will involve confrontation between all the theoretical methods we have developed - ab-initio, CI, EDF and symmetry based models - that will lead to a convergence in our quantitatively understanding of nuclear properties.

Acknowledgments

Support for this work was provided from US National Science Foundation grant number PHY-0555366

References

1. B. A. Brown and W. A. Richter, Phys. Rev. C **74**, 034315 (2006).
2. B. H. Wildenthal, Prog. Part. Nucl. Phys. **11**, 5 (1984); B. A. Brown and B. H. Wildenthal, Ann. Rev. of Nucl. Part. Sci. **38**, 29 (1988).
3. M. Hjorth-Jensen, T. T. S. Kuo and E. Osnes, Phys. Rep. **261**, 125 (1995).
4. B. R. Barrett and M. W. Kirson, Nucl. Phys. **A148**, 145 (1970).
5. S. C. Pieper, V. R. Pandharipande, R. B. Wiringa and J. Carlson, Phys. Rev. C **64**, 014001 (2001).
6. A. P. Zuker, Phys. Rev. Lett. **90**, 042502 (2003).
7. www.nscl.msu.edu/~brown/resources/resources.html
8. D. Guillemaud-Mueller et al., Phys. Rev. C **41**, 937 (1990); O. Tarasov et al., Phys. Lett. B **409**, 64 (1997); H. Sakurai et al., Phys. Lett. B **448**, 180 (1999); M. Fauerbach et al., Phys. Rev. C **53**,7647 (1996) and M. Thoennessen et al., Phys. Rev. C **68**, 044318 (2003).
9. T. Utsuno, T. Otsuka, T. Mizusaki and M. Honma, Phys. Rev. C **60**, 054315 (1999).
10. E. K. Warburton, J. A. Becker, and B. A. Brown, Phys. Rev. C **41**, 1147 (1990).
11. Z. Elekes et al., Phys. Lett. B **599**, 17 (2004).
12. A. Schiller et al., Phys. Rev. Lett. **99**, 112501 (2007).
13. B. A. Brown, Phys. Rev. C **58**, 220 (1998).
14. B. A. Brown, T. Duguet, T. Otsuka, D. Abe and T. Suzuki, Phys. Rev. C **74**, 061303(R) (2006).
15. J. Bartel, P. Quentin, M. Brack, C.Guet, and M. B. Hakansson, Nucl. Phys. **A386**, 79 (1982).
16. E. Chabanat, P. Bonche, P. Haensel, J. Meyer and T. Schaeffer, Nucl. Phys. A **635**, 231 (1998).
17. P. G. Reinhard, D. J. Dean, W. Nazarewicz, J. Dobaczewski, J. A. Maruhn and M. R. Strayer, Phys. Rev. C **60**, 014316 (1999).
18. S. Goriely, M. Samyn, J. M. Pearson and M. Onsi, Nucl. Phys. A750, 425 (2005).
19. P. G. Reinhard, D. J. Dean, W. Nazarewicz, J. Dobaczewski, J. A. Maruhn and M. R. Strayer, Phys. Rev. C **60**, 014316 (1999).
20. P. D. Cottle, Phys. Rev. C **76**, 027301 (2007).
21. B. Bastin et al., Phys. Rev. Lett. **99**, 022503 (2007).

REALISTIC SHELL-MODEL CALCULATIONS FOR EXOTIC NUCLEI AROUND CLOSED SHELLS

A. COVELLO[1,2*], L. CORAGGIO[2], A. GARGANO[2] and N. ITACO[1,2]

[1] *Dipartimento di Scienze Fisiche, Università di Napoli Federico II, Complesso Universitario di Monte S. Angelo, Via Cintia, I-80126, Napoli, Italy*

[2] *Istituto Nazionale di Fisica Nucleare, Sezione di Napoli, Complesso Universitario di Monte S. Angelo, Via Cintia, I-80126 Napoli, Italy*

** E-mail: covello@na.infn.it*

We report on a study of neutron-rich nuclei around doubly magic ^{132}Sn in terms of the shell model employing a realistic effective interaction derived from the CD-Bonn nucleon-nucleon potential. The short-range repulsion of the bare potential is renormalized by constructing a low-momentum potential, $V_{\rm low-k}$, that is used directly as input for the calculation of the effective interaction. We present results for the four nuclei beyond the $N = 82$ shell closure ^{134}Sn, ^{134}Sb, ^{136}Sb, and ^{136}Te. Comparison shows that our results are in very good agreement with the experimental data presently available for these exotic nuclei. We also present our predictions of the hitherto unknown spectrum of ^{136}Sn.

1. Introduction

The study of neutron-rich nuclei around doubly magic ^{132}Sn is a subject of special interest, as it offers the opportunity for testing the basic ingredients of shell-model calculations, in particular the two-body effective interaction, when moving toward the neutron drip line. In this context, great attention is currently focused on nuclei beyond the $N = 82$ shell closure. The experimental study of these nuclei is very difficult, but in recent years some of them have become accessible to spectroscopic studies. It is therefore challenging to perform shell-model calculations to try to explain the available data as well as to make predictions that may be verified in a not too distant future.

On these grounds, we have recently studied[1–5] several nuclei beyond ^{132}Sn within the framework of the shell model employing realistic effective interactions derived from the CD-Bonn nucleon-nucleon (NN) potential.[6] A main difficulty encountered in this kind of calculations is the strong short-range repulsion contained in the bare NN potential V_{NN}, which prevents its direct use in the derivation of the shell-model effective interaction $V_{\rm eff}$. As is well known, the traditional way to overcome this difficulty is the Brueckner G-matrix method. Instead, in the calculations mentioned above we have made use of a new approach[7] which consists in deriving from V_{NN} a low-momentum potential, $V_{\rm low-k}$, that preserves the deuteron binding energy and scattering phase shifts of V_{NN} up to a certain cutoff momentum Λ. This is a smooth potential which can be used directly to derive $V_{\rm eff}$, and it has been shown[7,8] that it provides an advantageous alternative to the use of the G matrix.

In this paper, we focus attention on ^{134}Sn, on the two odd-odd Sb isotopes ^{134}Sb and ^{136}Sb, and on ^{136}Te. The first and third nucleus with an N/Z ratio of 1.68 and 1.67, respectively, are at present the most exotic nuclei beyond ^{132}Sn for which information exists on excited states. We compare our results with the available experimental data and also report our predictions for low-energy states which have not been seen to date. We hope that this may stimulate, and be helpful to, future experiments on these nuclei. With the same motivation, we also present our predictions for the spectrum of the hitherto unknown ^{136}Sn with four neutrons outside ^{132}Sn, which makes an N/Z ratio of 1.72.

Following this introduction, in Sec. 2 we give a brief description of the theoretical framework in which our shell-model calculations are performed while in Sec. 3 we present results and predictions for the nuclei mentioned above. Some concluding remarks are given in Sec. 4.

2. Outline of theoretical framework

The starting point of any realistic shell-model calculation is the free NN potential. There are, however, several high-quality potentials, such as Nijmegen I and Nijmegen II,[9] Argonne V_{18},[10] and CD-Bonn,[6] which fit equally well (χ^2/datum ≈ 1) the NN scattering data up to the inelastic threshold. This means that their on-shell properties are essentially identical, namely they are phase-shift equivalent. In our shell-model calculations we have derived the effective interaction from the CD-Bonn potential. This may raise the question of how much our results may depend on this choice of the NN potential. We shall comment on this point later in connection with the $V_{\rm low-k}$ approach to the renormalization of the bare NN potential.

The shell-model effective interaction $V_{\rm eff}$ is defined, as usual, in the following way. In principle, one should solve a nuclear many-body Schrödinger equation of the form

$$H\Psi_i = E_i\Psi_i, \tag{1}$$

with $H = T + V_{NN}$, where T denotes the kinetic energy. This full-space many-body problem is reduced to a smaller model-space problem of the form

$$PH_{\text{eff}}P\Psi_i = P(H_0 + V_{\text{eff}})P\Psi_i = E_iP\Psi_i. \tag{2}$$

Here $H_0 = T + U$ is the unperturbed Hamiltonian, U being an auxiliary potential introduced to define a convenient single-particle basis, and P denotes the projection operator onto the chosen model space.

As pointed out in the Introduction, we "smooth out" the strong repulsive core contained in the bare NN potential V_{NN} by constructing a low-momentum potential $V_{\text{low-k}}$. This is achieved by integrating out the high-momentum modes of V_{NN} down to a cutoff momentum Λ. The integration is carried out with the requirement that the deuteron binding energy and phase shifts of V_{NN} up to Λ are preserved by $V_{\text{low-k}}$. A detailed description of the derivation of $V_{\text{low-k}}$ from V_{NN} as well as a discussion of its main features can be found in Refs. 7 and 11. However, we should mention here that shell-model effective interactions derived from different phase-shift equivalent NN potentials through the $V_{\text{low-k}}$ approach lead to very similar results.[4] In other words, $V_{\text{low-k}}$ gives an approximately unique representation of the NN potential.

Once the $V_{\text{low-k}}$ is obtained, we use it, plus the Coulomb force for protons, as input interaction for the calculation of the matrix elements of the shell-model effective interaction. The latter is derived by employing a folded-diagram method, which was previously applied to many nuclei using the G matrix.[12] Since $V_{\text{low-k}}$ is already a smooth potential, it is no longer necessary to calculate the G matrix. We therefore perform shell-model calculations following the same procedure as described for instance in Refs. 13 and 14, except that the G matrix used there is replaced by $V_{\text{low-k}}$. More precisely, we first calculate the so-called $\hat{Q}$-box[15] including diagrams up to second order in the two-body interaction. The shell-model effective interaction is then obtained by summing up the $\hat{Q}$-box folded diagram series using the Lee-Suzuki iteration method.[16]

3. Calculations and results

In our calculations we assume that ^{132}Sn is a closed core and let the valence protons occupy the five levels $0g_{7/2}$, $1d_{5/2}$, $1d_{3/2}$, $2s_{1/2}$, and $0h_{11/2}$ of the

50-82 shell, while for the valence neutrons the model space includes the six levels $0h_{9/2}$, $1f_{7/2}$, $1f_{5/2}$, $2p_{3/2}$, $2p_{1/2}$, and $0i_{13/2}$ of the 82-126 shell.

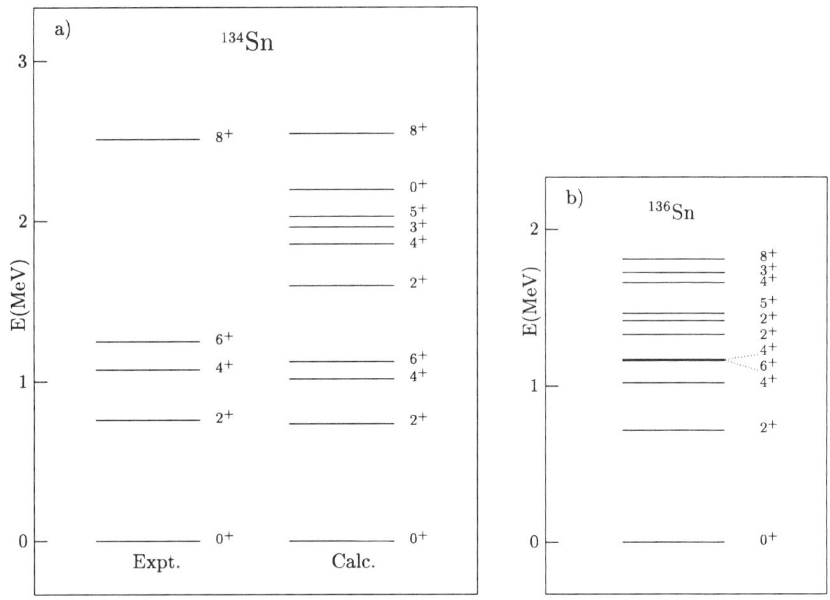

Fig. 1. a) Experimental and calculated spectrum of ^{134}Sn. b) Predicted spectrum of ^{136}Sn.

As mentioned in the previous section, the two-body matrix elements of the effective interaction are derived from the CD-Bonn NN potential renormalized through the $V_{\text{low-k}}$ procedure with a cutoff momentum $\Lambda = 2.2\ \text{fm}^{-1}$. The computation of the diagrams included in the $\hat{Q}$-box is performed within the harmonic-oscillator basis using intermediate states composed of all possible hole states and particle states restricted to the five shells above the Fermi surface. The oscillator parameter used is $\hbar\omega = 7.88$ MeV. As regards the single-particle energies, they have been taken from experiment. All the adopted values are reported in Ref. 1.

In Fig. 1a) we compare the calculated spectrum of ^{134}Sn with the available experimental data while in Fig. 1b) we show the calculated spectrum of the hitherto unknown ^{136}Sn up to about 1.8 MeV. From Fig. 1a) we see that while the theory reproduces very well all the observed levels, it also predicts, in between the 6^+ and 8^+ states, the existence of five states with spin ≤ 5. Clearly, the latter could not be seen from the γ-decay of the 8^+ state populated in the spontaneous fission experiment of Ref. 17.

Recently, the $B(E2; 0^+ \to 2_1^+)$ value in ^{134}Sn has been measured[18] using Coulomb excitation of neutron-rich radioactive ion beams. We have calculated this $B(E2)$ with an effective neutron charge of $0.70\,e$. We obtain $B(E2; 0^+ \to 2_1^+) = 0.033\ e^2\mathrm{b}^2$, in excellent agreement with the experimental value $0.029(4)\ e^2\mathrm{b}^2$.

A comparison between figures 1a) and 1b) shows that the three lowest calculated states, 2^+, 4^+, and 6^+, lie at practically the same energy in ^{134}Sn and ^{136}Sn. In the latter nucleus, however, above the 6^+ level there are seven states in an energy interval of about 650 keV. This pattern is quite different from that predicted for the spectrum of ^{134}Sn, where a rather pronounced gap (about 0.5 MeV wide) exists between the 6^+ state and the next excited state with $J^\pi = 2^+$.

Let us now come to ^{134}Sb. In Fig. 2 we show the energies of the first eight calculated states, which are the members of the $\pi g_{7/2}\nu f_{7/2}$ multiplet, and compare them with the eight lowest-lying experimental states. The wave functions of these states are characterized by very little configuration mixing, the percentage of the leading component having a minimum value of 88% for the $J^\pi = 2^-$ state while ranging from 94% to 100% for all other states.

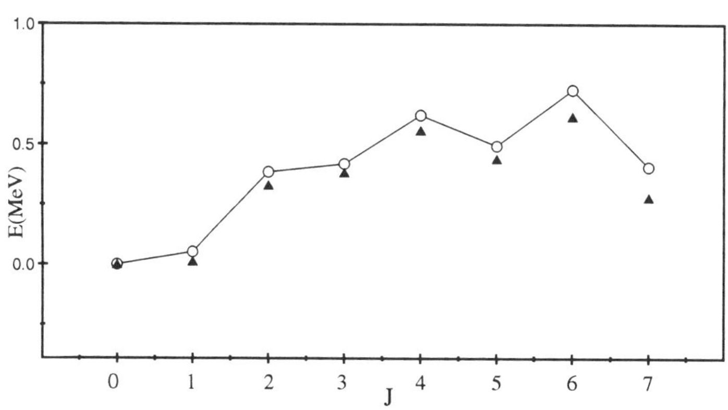

Fig. 2. Proton-neutron $\pi g_{7/2}\nu f_{7/2}$ multiplet in ^{134}Sb. The theoretical results are represented by open circles and the experimental data by solid triangles.

We see that the agreement between theory and experiment is very good, the discrepancies being in the order of a few tens of keV for most of the states. The largest discrepancy occurs for the 7^- state, which lies at about 130 keV above its experimental counterpart. It is an important outcome of our calculation that we predict almost the right spacing between the 0^- ground state and first excited 1^- state. In fact, the latter has been observed at 13 keV excitation energy, our value being 53 keV.

An analysis of the various terms which contribute to our effective neutron-proton interaction has evidenced the crucial role of core-polarization effects, in particular those arising from $1p-1h$ excitations, in reproducing the correct behavior of the multiplet. A detailed discussion of this point can be found in Ref. 2.

As regards ^{136}Sb, with two more valence neutrons, its ground state was identified as 1^- in the early β-decay study of Ref. 19 while the spectroscopic study of Ref. 20 led to the observation of a μs isomeric state, which was tentatively assigned a spin and parity of 6^-. Very recently,[5] new experimental information has been obtained on the μs isomeric cascade, leading to the identification of two more excited states.

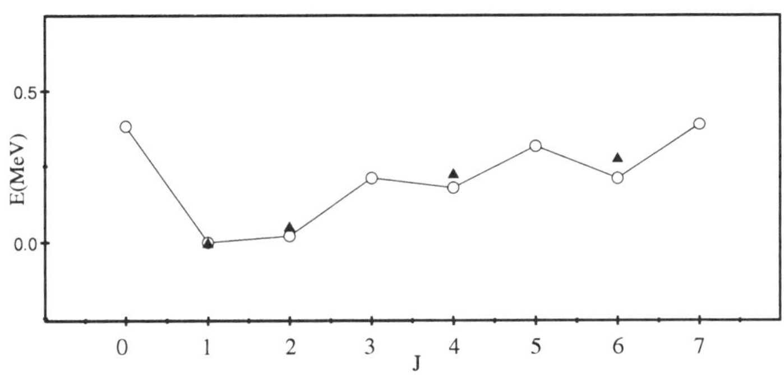

Fig. 3. Low-lying levels in ^{136}Sb. The theoretical results are represented by open circles and the experimental data by solid triangles.

This achievement is at the origin of our realistic shell-model calculation for this nucleus,[5] whose results we are now going to present. In Fig. 3, we show the four observed levels together with the calculated yrast states having angular momentum from 0^- to 7^-, which all arise from the $\pi g_{7/2}\nu(f_{7/2})^3$ configuration. These states may be viewed as the evolution of the $\pi g_{7/2}\nu f_{7/2}$ multiplet in ^{134}Sb. From Fig. 3 we see that the agreement between theory and experiment is very good, the largest discrepancy being about 70 keV.

An important piece of information is provided by the measured half life of the 6^- state, from which a $B(E2; 6^- \to 4^-)$ value of $170(40)\ e^2\ \mathrm{fm}^4$ is extracted. Using effective proton and neutron charges of $1.55\ e$ and $0.70\ e$, respectively, we obtain the value $131\ e^2\ \mathrm{fm}^4$, which compares very well with experiment. It is worth mentioning that these values of the effective charges have been consistently used in our previous calculations for nuclei in the ^{132}Sn region.[4]

Finally, we turn to the two-proton, two-neutron nucleus ^{136}Te, which in recent years has been the subject of great experimental and theoretical interest.[21–24] Particular attention has been focused on the first 2^+ state, which shows a significant drop in energy as compared to ^{134}Te, and on the $B(E2; 0^+ \to 2^+)$ value, recently measured using Coulomb excitation of radioactive ion beams.[21]

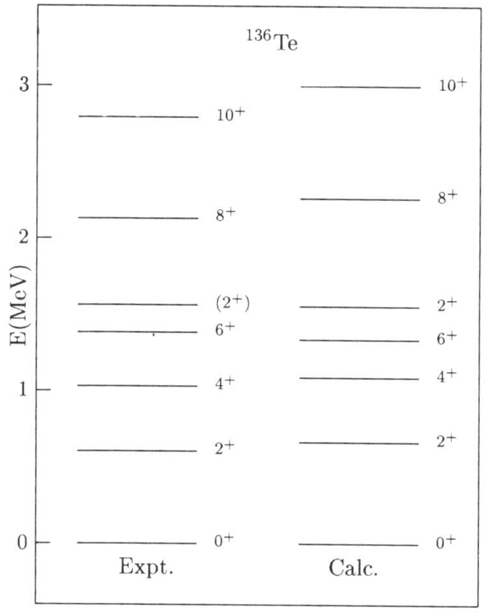

Fig. 4. Experimental and calculated spectrum of ^{136}Te.

From Fig. 4 we see that also in this case the calculated spectrum reproduces very well the experimental one.

As regards the $B(E2; 0^+ \to 2^+)$, the experimental value reported in Ref. 21 is $0.103(15)\ e^2\ \mathrm{b}^2$, to be compared with our value $0.180\ e^2\ \mathrm{b}^2$. The latter has been obtained with the same effective charges used for ^{136}Sb. We see that the calculated value overestimates the experimental one by a factor of 1.7. However, two new measurements have been recently performed which yield values about 20%[27] and 50%[28] higher than that of Ref. 21. Both values are closer to our prediction.

4. Concluding remarks

We have presented here the results of a shell-model study of neutron-rich nuclei, focusing attention on the four nuclei ^{134}Sn, ^{134}Sb, ^{136}Sb, and ^{136}Te, with an N/Z ratio of 1.68, 1.63, 1.67, and 1.61, respectively. We have also presented our predictions for the spectrum of the hitherto unknown ^{136}Sn with four neutrons outside ^{132}Sn, which makes an N/Z ratio of 1.72. In this connection, it may be recalled that the N/Z ratio of the last stable Sn isotope, ^{124}Sn, is 1.48.

In our study we have made use of a realistic two-body effective interaction derived from the CD-Bonn NN potential, the short-range repulsion of the latter being renormalized by use of the low-momentum potential $V_{\mathrm{low-k}}$. It is worth emphasizing that no adjustable parameter appears in our calculation of the effective interaction.

We have shown that all the experimental data presently available for these nuclei are well reproduced by our calculations. These data, however, are still rather scanty and we hope that our study may stimulate further experimental efforts to gain more spectroscopic information on exotic nuclei around ^{132}Sn.

Acknowledgments

This work was supported in part by the Italian Ministero dell'Istruzione, dell'Università e della Ricerca (MIUR).

References

1. L. Coraggio, A. Covello, A. Gargano, and N. Itaco, *Phys. Rev. C* **72**, 057302 (2005).
2. L. Coraggio, A. Covello, A. Gargano, and N. Itaco, *Phys. Rev. C* **73**, 031302(R) (2006).
3. A. Covello, L. Coraggio, A. Gargano, and N. Itaco, *Eur. Phys. J. ST* **150**, 93 (2007).
4. A. Covello, L. Coraggio, A. Gargano, and N. Itaco, *Prog. Part. Nucl. Phys.* **59**, 401 (2007).

5. G. S. Simpson, J. C. Angelique, J. Genevey, J. A. Pinston, A. Covello, A. Gargano, U. Köster, R. Orlandi, and A. Scherillo, *Phys. Rev. C* **76**, 041303(R) (2007).
6. R. Machleidt, *Phys. Rev. C* **63**, 024001 (2001).
7. S. Bogner, T. T. S. Kuo, L. Coraggio, A. Covello, and N. Itaco, *Phys. Rev. C* **65**, 051301(R) (2002).
8. A. Covello, in *Proc. Int. School of Physics "E. Fermi", Course CLIII*, eds. A. Molinari, L. Riccati, W. M. Alberico, and M. Morando (IOS Press, Amsterdam, 2003), p. 79.
9. V. G. J. Stoks, R. A. M. Klomp, C. P. F. Terheggen, and J. J. de Swart, *Phys. Rev. C* **49**, 2950 (1994).
10. R. B.Wiringa, V. G. J. Stoks, and R. Schiavilla, *Phys. Rev. C* **51**, 38 (1995).
11. A. Covello, L. Coraggio, A. Gargano, and N. Itaco, *Journal of Physics: Conference Series* **20**, 137 (2005).
12. A. Covello, L. Coraggio, A. Gargano, and N. Itaco, *Acta Phys. Pol. B* **32**, 871 (2001), and references therein.
13. M. F. Jiang, R. Machleidt, D. B. Stout, and T. T. S. Kuo, *Phys. Rev. C* **46**, 910 (1992).
14. A. Covello, F. Andreozzi, L. Coraggio, A. Gargano , T. T. S. Kuo, and A. Porrino, *Prog. Part. Nucl. Phys.* **38**, 165 (1997).
15. T. T. S. Kuo and E. Osnes, *Lecture Notes in Physics*, Vol. 364, (Springer-Verlag, Berlin, 1990).
16. K. Suzuki and S. Y. Lee, *Prog. Theor. Phys.* **64**, 2091 (1980).
17. A. Korgul *et al.*, *Eur. Phys. J. A* **7**, 167 (2000).
18. J. R. Beene *et al.*, *Nucl. Phys. A* **746**, 471c (2004).
19. P. Hoff, J. P. Omtved, B Fogelberg, H. Mach, and M. Hellström, *Phys. Rev. C* **56**, 2865 (1997).
20. M. N. Mineva *et al.*, *Eur. Phys. J. A* **11**, 9 (2001).
21. D. C. Radford *et al.*, *Phys. Rev. Lett.* **88**, 222501 (2002).
22. J. Terasaki, J. Engel, W. Nazarewicz, and M. Stoitsov, *Phys. Rev. C* **66**, 054313 (2002).
23. N. Shimizu, T. Otsuka, T. Mizusaki, and M. Honma, *Phys. Rev. C* **70**, 054313 (2004).
24. B. A. Brown, N. J. Stone, R. J. Stone, I. S. Towner, and M. Hjorth-Jensen, *Phys. Rev. C* **71**, 044317 (2005).
25. F. Andreozzi, L. Coraggio, A. Covello, A. Gargano, T. T. S. Kuo, and A. Porrino, *Phys. Rev. C* **56**, R16 (1997).
26. L. Coraggio, A. Covello, A. Gargano, and N. Itaco, *Phys. Rev. C* **65**, 051306(R) (2002).
27. H. Mach *et al.*, contribution to these Proceedings.
28. C. Baktash (private communication).

SURPRISING FEATURES OF SIMPLE NUCLEAR SYSTEMS JUST ABOVE ^{132}Sn

H. MACH[1,2], R. NAVARRO-PÉREZ[3,4], L.M. FRAILE[5,6], U. KÖSTER[5,7], B.A. BROWN[8], A. COVELLO[9], A. GARGANO[9], O. ARNDT[10], A. BLAZHEV[11], N. BOELAERT[11,12], M.J.G. BORGE[13], R. BOUTAMI[13], H. BRADLEY[1,14], N. BRAUN[11], Z. DLOUHY[15], C. FRANSEN[11], H.O.U. FYNBO[16], CH. HINKE[17], P. HOFF[18], A. JOINET[5,19], A. JOKINEN[20,21], J. JOLIE[11], A. KORGUL[22], K.-L. KRATZ[23,24], T. KRÖLL[17], W. KURCEWICZ[22], J. NYBERG[1], E-M. REILLO[13], E. RUCHOWSKA[25], W. SCHWERDTFEGER[26], G.S. SIMPSON[27], B. SINGH[28], M. STANOIU[29,30], O. TENGBLAD[13], P.G. THIROLF[26], V. UGRYUMOV[15], AND W.B. WALTERS[31].

[1] *Department of Nuclear and Particle Physics, Uppsala University, Sweden*
[2] *Inst. for Structure and Nuclear Astrophysics, University of Notre Dame, USA*
[3] *Department of Physics, University of Notre Dame, USA*
[4] *Departamento de Actuaría, Física y Matemáticas, Universidad de las Américas Puebla, San Andrés Cholula, Puebla 72810, México*
[5] *ISOLDE, PH Department, CERN, CH-1211 Geneva 23, Switzerland*
[6] *Facultad de CC. Físicas, Universidad Complutense, E-28006 Madrid, SPAIN*
[7] *Institut Laue-Langevin, B.P. 156, F-38042 Grenoble Cedex, France*
[8] *Dept. of Physics and Astronomy and NSCL, Michigan State University, USA*
[9] *Dipartimento di Scienze Fisiche, Università di Napoli Federico II and Istituto Nazionale di Fisica Nucleare, Napoli, Italy*
[10] *Institut für Kernchemie, Universität Mainz, D-55128 Mainz, Germany*
[11] *Institut für Kernphysik, Universität zu Köln, D-50937 Köln, Germany*
[12] *Dept. of Subatomic and Radiation Physics, Ghent University, Belgium*
[13] *Instituto de Estructura de la Materia, CSIC, E-28006 Madrid, Spain*
[14] *School of Physics, The University of Sydney, Sydney, Australia*
[15] *Nuclear Physics Institute, AS CR, CZ 25068, Rez, Czech Republic*
[16] *Institut for Fysik og Astronomi, Aarhus Universitet, Aarhus, Denmark*
[17] *Physics Department, Technical University Munich, 85748 Garching, Germany*
[18] *Department of Chemistry, University of Oslo, Blindern, Oslo, Norway*
[19] *Centre d'Etude Spatiale des Rayonnements, 9 Avenue du Colonel Roche, 31028 Toulouse Cedex 4, France*
[20] *University of Jyväskylä, Department of Physics, Jyväskylä, Finland*
[21] *Helsinki Institute of Physics, P.O.Box 64, FIN-00014 Helsinki, Finland*
[22] *Institute of Experimental Physics, University of Warsaw, Warsaw, Poland*
[23] *Max-Planck-Institut für Chemie, Otto-Hahn-Institute, Mainz, Germany*

[24] *HGF Virtuelles Institut für Struktur der Kerne und Nukleare Astrophysik, (VISTARS) Mainz, Germany*
[25] *The Andrzej Sołtan Institute for Nuclear Studies, PL 05-400 Świerk, Poland*
[26] *Department für Physik, LMU München, Garching, Germany*
[27] *LPSC, Université Joseph Fourier Grenoble 1, CNRS/IN2P3, Institut National Polytechnique de Grenoble, F-38026 Grenoble Cedex, France*
[28] *Dept. of Physics and Astronomy, McMaster Univ., Hamilton, Ont., Canada*
[29] *GSI, Postfach 110552, D-64200 Darmstadt, Germany*
[30] *Nat. Inst. of Physics and Nuclear Engineering, Bucharest-Magurele, Romania*
[31] *Dept. of Chemistry and Biochemistry, Univ. of Maryland, College Park, USA*
E-mail: hmach@nd.edu, Henryk.Mach@tsl.uu.se

The Advanced Time-Delayed Method was used to measure level lifetimes in ^{134}Sb, ^{135}Sb and ^{136}Te populated in the β decay of ^{134}Sn, βn decay of ^{136}Sn and via a chain of β decays from ^{136}Sn$\rightarrow$^{136}Sb$\rightarrow$^{136}Te, respectively. High purity Sn beams were extracted at the ISOLDE separator at CERN using a novel production technique utilizing the molecular SnS$^+$ ions to isolate Sn from a strong isobaric contamination coming from less exotic fission products. In ^{134}Sb we have measured (all results reported below are preliminary) the lifetime of the 383 keV state as $T_{1/2}$=26(5) ps and branchings for weak γ-rays de-exciting the 383 and 330 keV states. The latter were found much smaller than previously reported. Our value of B(M1;$3^-\rightarrow2^-$) = 2.0(4) μ_N^2 is one of the fastest known M1 rates at low excitation energy in all nuclei. In ^{135}Sb we have identified the missing $1/2^+$ state at the energy of 523 keV. From its lifetime, $T_{1/2}$=1.24(6) ns, we have determined a very collective B(E2;$1/2^+\rightarrow5/2^+$) = 13.0(11) W.u. We have measured lifetime for the 2_1^+ state in ^{136}Te, $T_{1/2}$ = 41.6(84) ps, which implies a B(E2) value higher by ~20% than the one previously reported. Results of shell model calculations are presented and compared with experimental data.

1. Introduction

There is a considerable interest in the spectroscopy of the exotic nuclei close to the doubly magic ^{132}Sn. The aim is to understand better the evolution of shell orbits in nuclei with large N/Z ratios and to critically test theoretical models. The information on the nuclear structure and decay properties of n-rich nuclei in this region provides key input to calculations of the astrophysical r-process. The present work is a continuation of our systematic studies in this region and in particular of ^{135}Sb, which has only a pair of neutrons and a proton outside the core ^{132}Sn. Our previous results on ^{135}Sb were recently summarized by Korgul *et al.*[1]

Our attention was largely focused on the properties of the first excited state in ^{135}Sb which has an excitation energy of only 282 keV, a much lower value than expected from the systematics of the lowest-lying $5/2^+$ state in the odd Sb isotopes with N≤82. This suggested[2] the possibility of a local shift of the proton single particle $d_{5/2}$ and $g_{7/2}$ states due to the excess of neutrons. The measurement of the l-forbidden transition, linking the proton $d_{5/2}$ and $g_{7/2}$ states, revealed[1] a very retarded M1 and non-collective E2 transitions. In Ref. [1] we have presented the results of two shell-model calculations, one which explicitly assumed a shift of 300 keV between the proton $d_{5/2}$ and $g_{7/2}$ orbits, and the other one which did not. The two calculations also employ different two-body effective interactions, which, however, are both derived from a nucleon-nucleon potential. The analyses in terms of the shell model of the data on ^{135}Sb and selected neighboring nuclei, including the new results presented below, do not provide a clear evidence for the shift of the proton $d_{5/2}$ orbit. Both models provide an overall close agreement with the experimental results, although the former approach, understandably, reproduces more accurately the energy of the first excited state in ^{135}Sb.

This presentation is focused on new experimental results obtained by the IS441 collaboration at ISOLDE. The exotic ^{134,135,136}Sn nuclei have been mass separated by a novel production technique using molecular SnS^+ ions to isolate Sn from a strong contamination of isobaric impurities.

2. Production of molecular SnS^+ ions

Until recently, the best method to produce relatively pure beams of neutron-rich tin isotopes at ISOLDE has been the selective ionization of tin with the resonance ionization laser ion source (RILIS)[3,4]. However, it required the ion source cavity to be kept hot, which caused significant surface ionization of those isobaric elements that have low ionization potentials[5], mainly Cs but also Ba. As a consequence, the β-spectroscopy experiments experienced huge background activities particularly at masses 135, 136 and 137.

In the present study we have applied a different way to produce pure ISOL beams, namely by the separation of an abundantly populated molecular sideband. The HRIBF group observed accidentally rather pure tin beams in the molecular sideband SnS^+ which was produced by a sulfur impurity in the target material[6]. The amount of possible SbS^+ and TeS^+ contaminants was not detectable. Detailed studies were performed at GSI-ISOL on the dependence of the SnS^+/Sn^+ ratio and the suppression of isobaric contaminants by leaking in a well-controlled way vapors of sulfur into the ion source[7]. Natural sulfur contains 95% ^{32}S, but also 0.75% of ^{33}S, 4.2% of ^{34}S and 0.02% of ^{36}S. Thus for neutron-rich tin isotopes, an unwanted mixture of different molecular sidebands would occur at the same mass. Since the production cross-sections drop from ^{132}Sn towards the neutron-rich side by about one order of magnitude per additional neutron, the ^{136}Sn^{32}S$^+$ beam would suffer from contaminations with ^{134}Sn^{34}S$^+$ and ^{132}Sn^{36}S$^+$ that are stronger than the wanted beam. To avoid these ambiguities we used sulfur isotopically enriched to $> 99.9\%$ ^{34}S. During the run the sulfur was continuously leaked into a standard ISOLDE UC$_x$/graphite target connected to a MK5 ion source[8]. Throughout the entire run the molecular Sn^{34}S$^+$ beams were of comparable magnitude to the respective Sn$^+$ beams, but by far purer. Isobaric contaminations in the sulfide sideband were generally negligible, except for barium sulfide. More details will be included in the upcoming paper[9] (see also[10]).

3. Fast timing measurements

We have used the Advanced Time-Delayed method described in more detail in Ref.[11]. The experimental setup included five detectors positioned in a close geometry around the Al stopper where the mass separated beam of SnS$^+$ ions was continuously deposited creating a saturated source. Time-delayed information was provided by the β- (START) and fast γ-detectors (STOP). For the latter we have used a large BaF$_2$ scintillator of Studsvik design, and separately, a smaller cylindrical 2.5 cm in diameter and 2.5 cm in length LaBr$_3$(Ce) (provided by Saint Gobain) with an effective Ce doping of 5%. The present experiment represented one of the first applications of LaBr$_3$ in fast timing. The LaBr$_3$ crystal has a similar time resolution than BaF$_2$ but about 3 times better energy resolution. In addition there were two Ge detectors with a relative efficiency of 100% each. Triple coincidence $\beta\gamma\gamma$(t) events were collected using the β-Ge-Ge, β-Ge-BaF$_2$ and β-Ge-LaBr$_3$ detectors. The first data set allowed to verify the decay schemes, identify new transitions and determine weak γ-ray branching ratios. The second and the third data sets were analysed separately and allowed for level lifetime measurements in the low picosecond to nanosecond range.

4. Results for ^{134}Sb

Our interest in the M1 transition rates in ^{134}Sb was inspired by the shell model calculations by Vadim Isakov. These calculations, which were recently published,[12] predict the B(M1) values for the lowest multiplet to be exceptionally fast of the order of $\sim$1.5 μ_n^2. Additional motivation was provided by the need to establish B(M1) values in the vicinity of ^{135}Sb

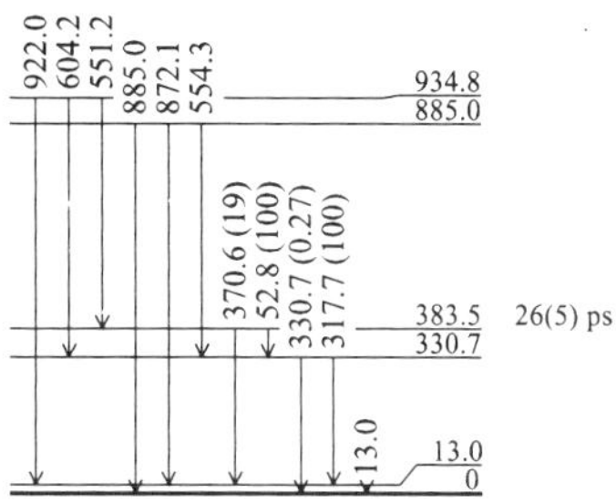

Figure 1. Partial level scheme for the decay of ^{134}Sn to ^{134}Sb determined in this work. The spin/parity for the lowest levels are (starting from the bottom): 0^-, 1^-, 2^- and 3^-, respectively.

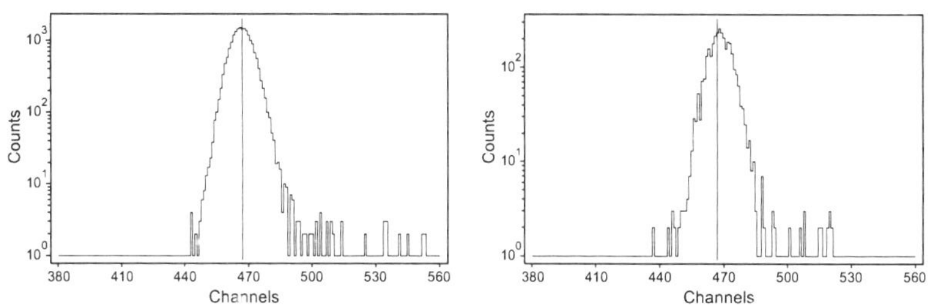

Figure 2. Time-delayed spectra started by β events and stopped by the 317-keV γ rays detected in the BaF_2 crystal. In the LEFT spectrum (reference semi-prompt spectrum) the second coincidence γ ray detected in Ge was the 554-keV transition, while in the shifted RIGHT spectrum the gate was set on the 551-keV transition. The vertical line identical in both spectra, shows the mean position of the LEFT spectrum. The displacement of the centroid of the RIGHT spectrum from the vertical line represents the meanlife of the 383-keV level in ^{135}Sb. It gives the level half-life of $T_{1/2} = 26(5)$ ps.

in order to understand better the properties of the latter nucleus. Note, however, that similar predictions for ultra-fast M1 transitions for an equivalent multiplet in ^{210}Bi were not confirmed experimentally. For the $2_1^- \rightarrow 1_1^-$ transition the predicted values for B(M1) ranged between 1.6–3.1 μ_N^2, and yet the measured[14] value is only 0.16 μ_N^2. The reason for the discrepancy is not clear. One suggestion[14] was that the measurement could have been distorted by time delayed components coming from side feeding.

Using the β-Ge-Ge data we have verified the previously proposed[15] level scheme for ^{134}Sb, and determined the intensities of weak γ rays de-exciting the the 383 and 330 keV states, see Fig. 1. Moreover, we have measured the

Table 1. Comparison of the experimental $B(M1)$ and $B(E2)$ values and the shell model calculations by A. Brown (labelled AB), A. Covello and A. Gargano (CG) and V.I. Isakov (IS). The B(M1) and B(E2) values are in the units of μ_N^2 and e^2fm^4, respectively.

Nucleus	$J_i \rightarrow J_f$	$X\lambda$	$B(X\lambda)_{exp}$ [a]	$B(X\lambda)_{th}^{AB}$	$B(X\lambda)_{th}^{CG}$	$B(X\lambda)_{th}^{IS}$
^{134}Sb	$3_1^- \rightarrow 2_1^-$	M1	2.0(4)	1.60	1.39	1.81
	$3_1^- \rightarrow 1_1^-$	E2	118(26)	84	115	116
	$2_1^- \rightarrow 0_1^-$	E2	429(238)	90	123	104
^{135}Sb	$1/2_1^+ \rightarrow 5/2_1^+$	E2	527(26)	678	566	
	$5/2_1^+ \rightarrow 7/2_1^+$	E2	$\leq$54	23	32	
	$5/2_1^+ \rightarrow 7/2_1^+$	M1	$\leq$0.00030	0.0022	0.0040	
^{136}Te	$2_1^+ \rightarrow 0_1^+$	E2	206(30) [b]			
	$2_1^+ \rightarrow 0_1^+$	E2	245(50)	452	360	

a: From this work unless noted otherwise.
b: From [13] obtained in Coulomb excitation in inverse kinematics.

lifetime of the 383 keV state as $T_{1/2} = 26(5)$ ps. Using these results, which are summarized in Fig. 2, we have obtained the experimental B(M1) and B(E2) values presented in Table 1. The measured B(M1;$3^- \rightarrow 2^-$) value of 2.0(4) μ_N^2 is one of the fastest M1 rates for any transition at low excitation energy in all known nuclei. Such a high value for B(M1) is very well reproduced by the shell model calculations by B.A. Brown[16] (labelled AB), A. Covello and A. Gargano[17] (labelled CG), and Isakov *et al.*,[12] (labelled IS). Note that the AB and CG calculation are performed in the same way as described in Ref. [1]. It should be mentioned that the lifetime of the 2^- 330-keV level has not been measured. In order to estimate the B(E2) for the 330 keV transition, we have assumed that the B(M1;$2^- \rightarrow 1^-$) is 1.8 μ_N^2.

Unlike in the case of ^{210}Bi, the ultra-fast M1 rate is confirmed in ^{134}Sb. This represents a unique transition rate. We have made a compilation of the known M1 rates for medium-heavy nuclei (A$\geq$30) and for levels below 3 MeV in the excitation energy. There is a small number of cases where the B(M1) values of ~ 1 μ_N^2 are listed, however, many of them are not reliable due to the poorly determined level lifetime or the very weak γ-ray branching ratio. Thus the number of reliable cases, like ^{134}Sb, is indeed very small. A more detailed discussion of will be presented in Ref. [18].

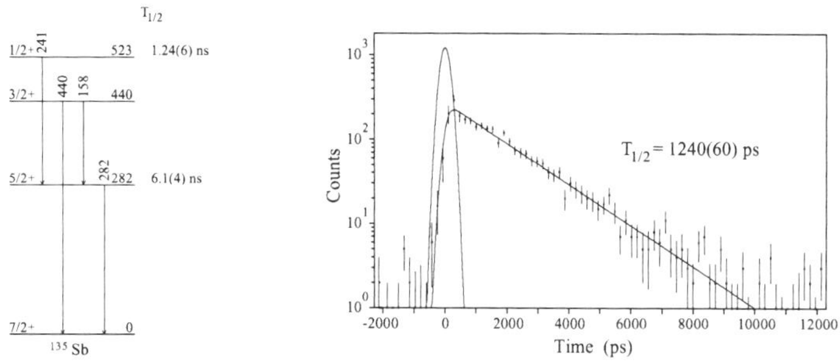

Figure 3. LEFT: partial level scheme for the βn decay of ^{136}Sn to ^{135}Sb. RIGHT: time delayed spectrum started by β events and stopped by the 241-keV γ rays detected in the BaF_2 crystal when the 282-keV transition is selected in the Ge detector. The slope is due to the lifetime of the 523-keV level in ^{135}Sb of $T_{1/2}$ = 1.24(6) ns.

5. Results for ^{135}Sb

A number of low-lying states have been observed[19] in ^{135}Sb, but not the $1/2^+$ state predicted[19] at the excitation energy between 527 and 735 keV. Such a state was not expected to be significantly populated in the β decay of the $7/2^-$ ground state in ^{135}Sn. We have succeeded in the identification of this state via the β-delayed neutron decay of the 0^+ ground state in ^{136}Sn. In this decay we have observed two γ-rays in coincidence with the 282 keV ground state transition in ^{135}Sb, see Fig. 3. The 158 keV line de-excites the known[19] $3/2^+$ state at 440 keV, while the 241 keV transition de-excites a new state at 523 keV. This state must have the spin and parity of $1/2^+$, since states of spin $3/2^+$ and higher were already observed[19] in the β decay of ^{135}Sn. There are only two low-spin states predicted[19] below 1.2 MeV in the excitation energy: the $1/2^+$ and $3/2^+$ states. Both of them are observed in the βn decay of ^{136}Sn, with the $1/2^+$ state being populated significantly stronger. Note that the predicted value of 527 keV by the AB calculations[19] (which included an explicit shift of 300 keV between the $d_{5/2}$ and $g_{7/2}$ orbits) comes very close to the experimental energy of 523 keV.

There is one more argument supporting the $1/2^+$ spin assignment for the 523-keV state, namely this state feeds the $5/2^+$ level at 282 keV via an E2 transition, necessitating the lifetime for the upper state of the order of a few ns. Indeed, the new state has a half-life of 1.24(6) ns, see Fig. 3, yielding a surprisingly collective B(E2) shown in Table 1. The shell model calculations reproduce this collective E2 rate quite well.

6. Results for ^{136}Te

The structure of ^{136}Te is also very simple with two valence neutrons and two valence protons beyond the doubly magic ^{132}Sn. Recently a very low B(E2;$0^+ \rightarrow 2^+$) value was reported[13] for ^{136}Te almost equal to the one for ^{134}Te (at N=82), which is at variance with the behavior of the Xe and Ba isotopes. This value is also lower than the predictions of shell-model calculations including those shown in Table 1. Using the Advanced Time-Delayed method, and thus utilizing the fully calibrated response of the timing γ detectors to within ~1-2 ps, we have measured via the centroid shift technique, the half-life of the 606-keV 2_1^+ state in ^{136}Te as $T_{1/2}$ = 41.6(8.4) ps (preliminary value) yielding B(E2; $2_1^+ \rightarrow 0_1^+$) = 245(50) $e^2 fm^2$. The new result implies a B(E2) value ~20% higher than that measured before in better agreement with the model predictions, see Table 1.

This study was supported in part by the NSF PHY04-57120, NSF PHY-0555366, Swedish Research Council, BMBF grant 06KY205I, the Alexander von Humboldt Foundation (WBW), Fundation for Polish Science (AK), the European Union Sixth Framework through RII3-EURONS (contract no. 506065) and the EU-RTD project TARGISOL (HPRI-CT-2001-50033) and was part of Undergraduate Research (RNP) at the Physics Department University of Notre Dame. Fast timing detectors and electronics were provided by the Fast Timing Pool of Electronics.

References

1. A. Korgul *et al.*, Eur. Phys. J. **A32**, 25 (2007), and references therein.
2. J. Shergur *et al.*, Phys. Rev. **C65**, 034313 (2002).
3. V.N. Fedoseyev *et al.*, Hyp. Int. **127**, 409 (2000).
4. U. Köster *et al.*, Spectrochimica Acta B **58**, 1047 (2003).
5. U. Köster, Nucl. Phys. **A701**, 441c (2002).
6. D.W. Stracener, Nucl. Instr. Meth. **B204**, 42 (2003).
7. R. Kirchner, Nucl. Instr. Meth. **B204**, 179 (2003).
8. S. Sundell, H. Ravn and the ISOLDE Collaboration, Nucl. Instr. Meth. **B70**, 160 (1992).
9. Oliver Arndt, Angélique Joinet *et al.*, to be published.
10. U. Köster *et al.*, Nucl. Instr. Meth. **B**, Proc. of EMIS-15, submitted.
11. H. Mach *et al.*, Nucl. Phys. **A523**, 197 (1991), and references therein.
12. V.I. Isakov *et al.*, Phys. Atom. Nucl. **70**, 818 (2007).
13. D.C. Radford *et al.*, Phys. Rev. Lett. **88**, 222501 (2002).
14. D.J. Donahue *et al.*, Phys. Rev. **C12**, 1547 (1975).
15. J. Shergur *et al.*, Phys. Rev. **C71**, 064321 (2005).
16. B. A. Brown, N. J. Stone, J. R. Stone, I. S. Towner and M. Hjorth-Jensen,

Phys. Rev. **C71**, 044317 (2005); erratum, Phys. Rev. **C72**, 029901 (2005).
17. A. Covello, L. Coraggio, A.Gargano, and N. Itaco, contribution to this Conference.
18. R. Navarro-Pérez *et al.*, to be published.
19. J. Shergur *et al.*, Phys. Rev. **C72**, 024305 (2005).

SHELL-MODEL CALCULATIONS WITH LOW-MOMENTUM NUCLEON-NUCLEON INTERACTIONS BASED UPON CHIRAL PERTURBATION THEORY

N. ITACO, L. CORAGGIO, A. COVELLO, and A. GARGANO

Dipartimento di Scienze Fisiche, Università di Napoli Federico II, and Istituto Nazionale di Fisica Nucleare, Complesso Universitario di Monte S. Angelo, Via Cintia - I-80126 Napoli, Italy

D. R. ENTEM

Grupo de Física Nuclear, IUFFyM, Universidad de Salamanca, E-37008 Salamanca, Spain

T. T. S. KUO

Department of Physics, SUNY, Stony Brook, New York 11794

R. MACHLEIDT

Department of Physics, University of Idaho, Moscow, Idaho 83844

Recently a new low-momentum nucleon-nucleon potential (N^3LOW) has been derived from chiral perturbation theory at next-to-next-to-next-to-leading order with a sharp low-momentum cutoff at 2.1 fm^{-1}. In this work we compare its perturbative properties with those of a $V_{\text{low-k}}$ potential constructed from a realistic NN potential with high-momentum components. We have performed shell-model calculations for ^{18}O using effective hamiltonians derived from both types of low-momentum potential. The results show that the N^3LOW potential is suitable to be applied perturbatively in microscopic nuclear structure calculations yielding results quite close to those obtained from $V_{\text{low-k}}$.

1. Introduction

A well-established framework to derive a shell-model effective Hamiltonian H_{eff} starting from a free-space nucleon-nucleon (NN) potential V_{NN} is the time-dependent degenerate linked-diagram perturbation theory as formulated by Kuo, Lee and Ratcliff [1,2]. However, any realistic V_{NN} is characterized by the the presence of a built-in strong short-range repulsion which dominates the high-momentum behavior of the potential. This makes the matrix elements of V_{NN} in a shell-model basis generally very large, thus preventing an order-by-order perturbative calculation of H_{eff}.

At the beginning of the 1990s Weinberg introduced into nuclear physics the method of effective field theory (EFT) to study the S-matrix for a process involving arbitrary numbers of low-momentum pions and nucleons [3,4]. This work paved the way to a number of studies on this subject (see for instance [5–10]) which eventually have lead to the construction of low-momentum V_{NN}'s based on chiral perturbation theory that are able to reproduce accurately the NN data [11,12]. However, these potentials also cannot be used directly in a perturbative nuclear structure calculation.

Inspired by EFT a new approach has been recently introduced to derive a low-momentum NN interaction $V_{\text{low-k}}$ [13–15]. The starting point in the construction of $V_{\text{low-k}}$ is a realistic model for V_{NN}. A cutoff momentum Λ is then introduced and from the original V_{NN} an effective potential, satisfying a decoupling condition between the low-and high-momentum spaces, is derived by integrating out the high-momentum components. The main result is that $V_{\text{low-k}}$ preserves exactly the onshell properties of the original V_{NN} and is a smooth potential suitable to be used directly in nuclear structure calculations. In the past few years, $V_{\text{low-k}}$ has been fruitfully employed in microscopic calculations within different perturbative frameworks such as the realistic shell model [16–21], the Goldstone expansion for doubly closed-shell nuclei [22–24], and the Hartree-Fock theory for nuclear matter calculations [25–27].

The success of $V_{\text{low-k}}$ suggests that there may be a way to construct a low-momentum potential from scratch using chiral perturbation theory, instead of taking the detour through a NN potential with high-momentum components. We have constructed such a potential at next-to-next-to-next-to-leading order (N^3LO) using a sharp cutoff at 2.1 fm^{-1} [28] (from now on we shall refer to this potential as N^3LOW). This potential reproduces the NN phase shifts up to 200 MeV laboratory energy and the deuteron binding energy.

In order to study the perturbative behavior of this new chiral low-momentum potential, we have performed shell-model calculations for ^{18}O. We employ two different effective Hamiltonians: one is based on the $V_{\text{low-k}}$ derived numerically from a 'hard' N^3LO potential [11] and the other on N^3LOW.

The paper is organized as follows. In Sec. 2 we give a brief description of the derivation of $V_{\text{low-k}}$ and an outline of the construction of the N^3LOW potential. Sec. 3 begins with a summary of the derivation of the shell-model

$H_{\rm eff}$ and is then devoted to the presentation and discussion of our results. Some concluding remarks are given in Sec. 4.

2. Low-momentum nucleon-nucleon potentials

2.1. *Potential model* $V_{\rm low-k}$

The repulsive core contained in V_{NN} is smoothed by integrating out the high-momentum modes of V_{NN} down to a cutoff momentum Λ with the requirement that the deuteron binding energy and low-energy phase shifts of V_{NN} are preserved by $V_{\rm low-k}$. This is achieved by the so-called T-matrix equivalence approach which we briefly describe in the following. We start from the half-on-shell T matrix for V_{NN}

$$T(k',k,k^2) = V_{NN}(k',k) + \mathcal{P}\int_0^{\infty} q^2 dq V_{NN}(k',q)\frac{1}{k^2-q^2}T(q,k,k^2) \quad , \tag{1}$$

where $\mathcal{P}$ denotes the principal value and k, k', and q stand for the relative momenta. The effective low-momentum T matrix, denoted by $\hat{T}$, is then defined by

$$\hat{T}(p',p,p^2) = V_{\rm low-k}(p',p) + \mathcal{P}\int_0^{\Lambda} q^2 dq V_{\rm low-k}(p',q)\frac{1}{p^2-q^2}\hat{T}(q,p,p^2) \quad , \tag{2}$$

where the intermediate state momentum q is integrated from 0 to the momentum space cutoff Λ and $(p',p)\leq\Lambda$. The above T matrices are required to satisfy the condition

$$T(p',p,p^2) = \hat{T}(p',p,p^2)\,; \quad (p',p)\leq\Lambda\,. \tag{3}$$

The above equations define the effective low-momentum interaction $V_{\rm low-k}$, and it has been shown [14] that they are satisfied by the solution

$$V_{\rm low-k} = \hat{Q} - \hat{Q}'\int\hat{Q} + \hat{Q}'\int\hat{Q}\int\hat{Q} - \hat{Q}'\int\hat{Q}\int\hat{Q}\int\hat{Q} + \ \dots \quad , \tag{4}$$

which is the well known Kuo-Lee-Ratcliff (KLR) folded-diagram expansion [1,2], originally designed for constructing shell-model effective interactions. In Eq. (4) $\hat{Q}$ is an irreducible vertex function whose intermediate states are all beyond Λ and $\hat{Q}'$ is obtained by removing from $\hat{Q}$ its terms first order in the interaction V_{NN}. In addition to the preservation of the half-on-shell T matrix, which implies preservation of the phase shifts, this $V_{\rm low-k}$ preserves the deuteron binding energy. For any value of Λ, the low-momentum potential of Eq. (4) can be calculated very accurately using iteration methods. Our calculation and hermitization of $V_{\rm low-k}$ is performed by employing an iteration method based on the Lee-Suzuki similarity transformation [29] which has been proposed by Andreozzi [30].

2.2. *Potential model* N^3LOW

As already mentioned in the Introduction, the approach to nuclear physics based upon a low-energy EFT was started by Weinberg some fifteen years ago. This nuclear EFT is characterized by the symmetries of low-energy QCD, in particular spontaneously broken chiral symmetry, and the degrees of freedom relevant for nuclear physics, nucleons and pions. The expansion based upon this EFT has become known as chiral perturbation theory (χPT), and to have this expansion convergent at a proper rate, we need that $Q \ll M_{\rm QCD}$, where Q denotes the magnitude of a nucleon three-momentum or a pion four-momentum, and $M_{\rm QCD}$ is the QCD energy scale. To enforce this, chiral NN potentials are multiplied by a regulator function that suppresses the potential for nucleon momenta $Q > \Lambda$ with $\Lambda \ll M_{\rm QCD}$. Present chiral NN potentials [11,12] typically have Λ around 2.5 fm^{-1}. Stimulated by the success of $V_{\rm low-k}$, we have constructed a NN potential at N^3LO of chiral perturbation theory that carries a sharp momentum cutoff at 2.1 fm^{-1} [28]. We have dubbed this potential N^3LOW.

The potential N^3LOW reproduces accurately the empirical deuteron binding energy, the experimental low-energy scattering parameters, and the empirical phase-shifts of NN scattering up to at least 200 MeV laboratory energy [28]. Unlike $V_{\rm low-k}$, which can be represented only numerically, this potential is given in analytic form. More precisely its analytic expression is the same as for the "hard" N^3LO potential by Entem and Machleidt of 2003 [11].

It is well known that when dealing with a low-momentum potential the role of three-body forces (3NF) may be crucial. For example, nuclear matter does not saturate without a 3NF when the two-nucleon force is represented by a low-momentum potential [27]. In this regard, one great advantage of χPT is that it generates nuclear two- and many-body forces on an equal footing (see for instance Ref. [31]). Moreover most interaction vertices that appear in the 3NF and in the four-nucleon force (4NF) also occur in the two-nucleon force (2NF), so that the corresponding parameters are already fixed in the construction of the chiral 2NF. If the 2NF is analytic, these parameters are known, and there is no problem with their consistent proliferation to the many-body force terms. This is not the case for $V_{\rm low-k}$. The values of these parameters are not explicitly known, so the parameters to be used in the 3NF, 4NF, ... must be based upon 'educated guesses', thus loosing a firm consistency between two- and many-body forces.

In the next section we shall compare the perturbative properties of N^3LOW when employed in microscopic nuclear structure calculations with those of a $V_{\rm low-k}$ potential.

3. Outline of calculations and Results

We have performed shell-model calculations for the two valence neutron nucleus ^{18}O. The shell-model effective Hamiltonian $H_{\rm eff}$ has been derived by way of the time-dependent perturbation theory [1,2]. More precisely, $H_{\rm eff}$ is expressed through the KLR folded-diagram expansion in terms of the vertex function $\hat{Q}$-box, which is taken to be composed of one- and two-body irreducible valence-linked Goldstone diagrams through third order in the interaction [32]. Finally, the series of the folded diagrams is summed up to all orders using the Lee-Suzuki iteration method [29].

The Hamiltonian $H_{\rm eff}$ contains one-body contributions, which represent the effective SP energies. In realistic shell-model calculations these are usually removed [33] and replaced by the SP energies taken from the experimental data. In this work however, in order to make a consistent study of the perturbative properties of the input potential V ($V_{\rm low-k}$ or N^3LOW) in the shell-model approach, we have employed the theoretical SP energies obtained from $H_{\rm eff}$.

We have used two $H_{\rm eff}$'s, one obtained from a $V_{\rm low-k}$ with a cutoff momentum $\Lambda = 2.1$ fm^{-1} derived from the 'hard' N^3LO chiral NN potential [11], the other from the new N^3LOW potential described in Sec. 2.2. The Coulomb force between proton-proton intermediate states has been explicitly taken into account. As regards the auxiliary potential U we only mention that we have used the harmonic oscillator potential and included first-order U-insertion diagrams. A detailed discussion about these points may be found in [28].

In order to study the convergence properties of the two above $H_{\rm eff}$'s, we have compared the convergence rate of their diagrammatic series which has to deal with the convergence of the order-by-order perturbative expansion and the sum over the intermediate states in the Goldstone diagrams

In Table 1 we show the $V_{\rm low-k}$ g.s. energies of ^{18}O relative to ^{16}O obtained at second- and third-order perturbative expansion of $H_{\rm eff}$. The energies are reported as a function of the maximum allowed excitation energy of the intermediate states expressed in terms of the oscillator quanta $N_{\rm max}$. We see that the g.s. energy practically converges at $N_{\rm max} = 12$, while the order-by-order convergence may be considered quite satisfactory, the difference between second- and third-order results being around 4 % for $N_{\max} = 16$.

Table 1. Ground state energies (in MeV) of ^{18}O relative to ^{16}O calculated with $H_{\rm eff}$ derived from the $V_{\rm low-k}$ of the "hard" N^{3}LO potential as a function of the maximum number $N_{\max}$ of the HO quanta (see text for details). Results obtained at second and third order in perturbation theory are reported.

$N_{\max}$	4	6	8	10	12	14	16
2nd	-8.191	-10.615	-12.748	-14.318	-15.037	-15.142	-15.168
3rd	-6.617	-8.987	-11.344	-13.277	-14.294	-14.487	-14.523

From inspection of Table 2 it can be seen that we obtain similar convergence features using the N^{3}LOW potential. In this case for $N_{\max} = 16$ the difference between second- and third-order results is less than 1 %.

Table 2. Same as Table 1, but with $H_{\rm eff}$ derived from N^{3}LOW.

$N_{\max}$	4	6	8	10	12	14	16
2nd	-4.806	-6.789	-9.085	-11.759	-13.671	-14.108	-14.162
3rd	-3.547	-5.366	-7.683	-10.789	-13.339	-13.992	-14.049

From Tables 1 and 2 we also see that at third order, with a sufficiently large number of intermediate states, the two interactions give very close results for the relative binding energy. Note that, from now on we shall refer to calculations with $N_{\max} = 16$ and $\hat{Q}$-box diagrams up to third order in perturbation theory.

In Fig. 1 we show the experimental relative spectrum of ^{18}O and compare it with those obtained from the two different $H_{\rm eff}$'s. As can be seen, the two theoretical spectra are very similar and in good agreement with the experimental one.

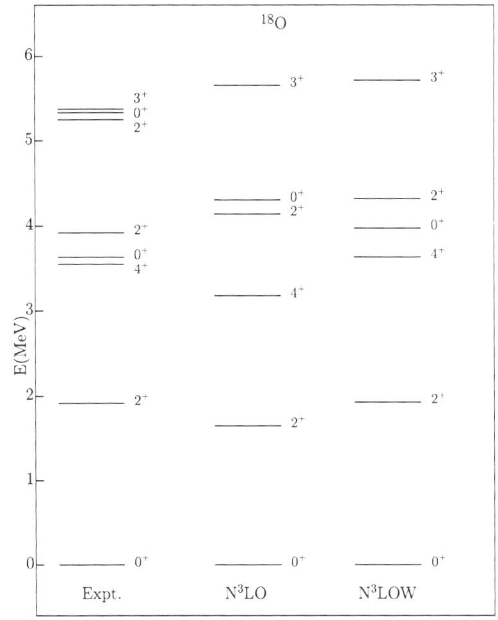

Fig. 1. Experimental and theoretical spectra of ^{18}O.

The close similarity between the two $H_{\rm eff}$'s is confirmed by the inspection of Fig. 2 and Table 3, where the corresponfing two-body matrix elements (TBME) and SP energies are compared.

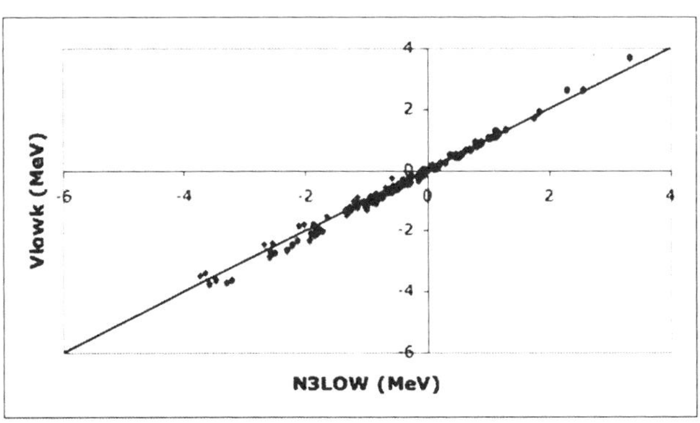

Fig. 2. Correlation plot between the TBME of $H_{\rm eff}$ obtained from the $V_{\rm low-k}$ of the "hard" N^{3}LO potential and from N^{3}LOW.

Table 3. Calculated SP relative energies of H_{eff} (in MeV) obtained from the $V_{\text{low-k}}$ of the "hard" N^3LO potential and from N^3LOW. The values in parenthesis are the absolute SP energies.

orbital	$V_{\text{low-k}}$	N^3LOW	Expt
$\nu 0d5/2$	0.0 (-5.425)	0.0 (-4.909)	0.0 (-4.144)
$\nu 0d3/2$	7.323	7.117	5.085
$\nu 1s1/2$	1.257	0.818	0.871

4. Conclusions

We have studied the convergence properties of the new low-momentum potential N^3LOW, which is derived in the framework of chiral perturbation theory at next-to-next-to-next-to-leading order with a sharp cutoff at 2.1 fm^{-1}. To this end, we have performed shell-model calculations employing an effective Hamiltonian obtained from this potential and have compared the results with those obtained using the $V_{\text{low-k}}$ derived from the "hard" N^3LO potential of Entem and Machleidt [11]. It has turned out that the two low-momentum potentials show the same perturbative behavior. On these grounds, we can conclude that the potential N^3LOW is suitable to be applied perturbatively in microscopic nuclear structure calculations yielding results which come quite close to those obtained from the $V_{\text{low-k}}$ derived from the "hard" N^3LO potential.

Acknowledgments

This work was supported in part by the Italian Ministero dell'Istruzione, dell'Università e della Ricerca (MIUR), by the U.S. DOE Grant No. DE-FG02-88ER40388, by the U.S. NSF Grant No. PHY-0099444, by the Ministerio de Ciencia y Tecnonología under Contract No. FPA2004-05616, and by the Junta de Castilla y León under Contract No. SA-104/04.

References

1. T. T. S. Kuo, S. Y. Lee and K. F. Ratcliff, *Nucl. Phys. A* **176**, p. 65 (1971).
2. T. T. S. Kuo and E. Osnes, *Lecture Notes in Physics, vol. 364* (Springer-Verlag, Berlin, 1990).
3. S. Weinberg, *Phys. Lett. B* **251**, p. 288 (1990).
4. S. Weinberg, *Nucl. Phys. B* **363**, p. 3 (1991).
5. G. P. Lepage, in *Nuclear Physics: Proceedings of the VIII Jorge Andre' Swieca Summer School*, eds. C. A. Bertulani, M. E. Bracco, B. V. Carlson and M. Nielsen (World Scientific, Singapore, 1997).

6. D. B. Kaplan, M. J. Savage and M. B. Wise, *Phys. Lett. B* **424**, p. 390 (1998).
7. E. Epelbaoum, W. Glöckle and U.-G. Meissner, *Nucl. Phys. A* **637**, p. 107 (1998).
8. P. F. Bedaque, M. J. Savage, R. Seki and U. van Kolck (eds.), *Nuclear Physics with Effective Field Theory II*, Proceedings from Institute for Nuclear Theory Vol. 9, (World Scientific, Singapore, 1999).
9. U. van Kolck, *Prog. Part. Nucl. Phys.* **43**, p. 337 (1999).
10. W. C. Haxton and C. L. Song, *Phys. Rev. Lett.* **84**, p. 5484 (2000).
11. D. R. Entem and R. Machleidt, *Phys. Rev. C* **68**, p. 041001(R) (2003).
12. E. Epelbaum, W. Glöckle and U.-G. Meissner, *Nucl. Phys. A* **747**, p. 362 (2005).
13. S. Bogner, T. T. S. Kuo and L. Coraggio, *Nucl. Phys. A* **684**, p. 432c (2001).
14. S. Bogner, T. T. S. Kuo, L. Coraggio, A. Covello and N. Itaco, *Phys. Rev. C* **65**, p. 051301(R) (2002).
15. S. Bogner, T. T. S. Kuo and A. Schwenk, *Phys. Rep.* **386**, p. 1 (2003).
16. L. Coraggio, A. Covello, A. Gargano and N. Itaco, *Phys. Rev. C* **66**, p. 064311 (2002).
17. L. Coraggio, A. Covello, A. Gargano, N. Itaco, T. T. S. Kuo, D. R. Entem and R. Machleidt, *Phys. Rev. C* **66**, p. 021303(R) (2002).
18. L. Coraggio, A. Covello, A. Gargano and N. Itaco, *Phys. Rev. C* **70**, p. 034310 (2004).
19. L. Coraggio, A. Covello, A. Gargano and N. Itaco, *Phys. Rev. C* **72**, p. 057302 (2005).
20. L. Coraggio and N. Itaco, *Phys. Lett. B* **616**, p. 43 (2005).
21. L. Coraggio, A. Covello, A. Gargano and N. Itaco, *Phys. Rev. C* **73**, p. 031302(R) (2006).
22. L. Coraggio, N. Itaco, A. Covello, A. Gargano and T. T. S. Kuo, *Phys. Rev. C* **68**, p. 034320 (2003).
23. L. Coraggio, A. Covello, A. Gargano, N. Itaco, T. T. S. Kuo and R. Machleidt, *Phys. Rev. C* **71**, p. 014307 (2005).
24. L. Coraggio, A. Covello, A. Gargano, N. Itaco and T. T. S. Kuo, *Phys. Rev. C* **73**, p. 014304 (2006).
25. J. Kuckei, F. Montani, H. Müther and A. Sedrakian, *Nucl. Phys. A* **723**, p. 32 (2003).
26. A. Sedrakian, T. T. S. Kuo, H. Müther and P. Schuck, *Phys. Lett. B* **576**, p. 68 (2003).
27. S. K. Bogner, A. Schwenk, R. J. Furnstahl and A. Nogga, *Nucl. Phys. A* **763**, p. 59 (2005).
28. L. Coraggio, A. Covello, A. Gargano, N. Itaco, D. R. Entem, T. T. S. Kuo and R. Machleidt, *Phys. Rev. C* **75**, p. 024311 (2007).
29. K. Suzuki and S. Y. Lee, *Prog. Theor. Phys.* **64**, p. 2091 (1980).
30. F. Andreozzi, *Phys. Rev. C* **54**, p. 684 (1996).
31. R. Machleidt and D. R. Entem, *J. Phys. G* **31**, p. S1235 (2005).
32. T. T. S. Kuo, J. Shurpin, K. C. Tam, E. Osnes and P. J. Ellis, *Ann. Phys. (NY)* **132**, p. 237 (1981).
33. J. Shurpin, T. T. S. Kuo and D. Strottman, *Nucl. Phys. A* **408**, p. 310 (1983).

SHELL MODEL STATES IN NEUTRON-RICH Ca AND Ar NUCLEI

B. FORNAL, R. BRODA, W. KRÓLAS, T. PAWŁAT, J. WRZESIŃSKI

Institute of Nuclear Physics, Polish Academy of Sciences, PL-31342 Cracow, Poland

R. V. F. JANSSENS, M. P. CARPENTER, T. LAURITSEN, D. SEWERYNIAK, S. ZHU

Physics Division, Argonne National Laboratory, Argonne, IL 60439, USA

N. MARGINEAN, L. CORRADI, G. DE ANGELIS, F. DELLA VEDOVA, E. FIORETTO, A. GADEA, B. GUIOT, D. R. NAPOLI, A. M. STEFANINI, J. J. VALIENTE-DOBÓN

INFN, Laboratori Nazionali di Legnaro, I-35020 Legnaro, Italy

S. LUNARDI, S. BEGHINI, E. FARNEA, P. MASON, G. MONTAGNOLI, F. SCARLASSARA, C. A. UR

Dipartimento di Fisica dell'Università di Padova, and INFN Sezione di Padova, I-35131 Padova, Italy

M. HONMA

Center for Mathematical Sciences, University of Aizu, Tsuruga, Ikki-machi, Aizu-Wakamatsu, Fukushima 965-8580, Japan

P. F. MANTICA

National Superconducting Cyclotron Laboratory, Michigan State University, East Lansing, MI 48824, USA

T. OTSUKA

Department of Physics, University of Tokyo, Hongo, Tokyo 113-0033, Japan

G. POLLAROLO

Dipartimento di Fisica Teorica, Università di Torino, and Istituto Nazionale di Fisica Nucleare, I-10125 Torino, Italy

S. SZILNER

Bošković Institute, HR-10001 Zagreb, Croatia

M. TROTTA

INFN, Sezione di Napoli, I-80126 Napoli, Italy

The low-lying yrast states in the ^{51}Ca and ^{46}Ar nuclei have been investigated. In ^{51}Ca, level structures associated with neutron excitations into the $p_{1/2}$ and $f_{5/2}$ neutron single-particle orbitals and with proton excitations across the $Z = 20$ shell gap have been identified. Shell model calculations with the GXPF1A interaction account reasonably well for the fp shell states. In the very neutron-rich argon istope, ^{48}Ar, new gamma rays have been observed.

Keywords: nuclear structure, neutron-rich nuclei, single particle states

1. Introduction

Neutron-rich nuclei around doubly-magic ^{48}Ca have recently attracted much attention due to the presence of an $N = 32$ subshell closure that was demonstrated on the basis of a series of experimental findings in the isotones ^{52}Ca, ^{54}Ti and ^{56}Cr.[1–7] The existence of this energy gap at $N = 32$ around $Z \sim 20$ arises from the sizeable energy spacing between the neutron $p_{3/2}$ orbital and the higher lying $p_{1/2}$ and $f_{5/2}$ states. The first indications that such spacing exists in neutron-rich nuclei came from studies of neutron single particle energies in ^{49}Ca performed already in the sixties.[8] It was shown that the energy separation between the $p_{3/2}$ and $p_{1/2}$ orbitals in ^{49}Ca is of the order of 2 MeV and is, in fact, very similar to the corresponding separation between the $f_{7/2}$ and $p_{3/2}$ states, which is responsible for the shell closure at $N = 28$. Apparently, with the addition of neutrons, the shell gap is preserved and manifests itself in the structural properties of the neutron-rich $N = 32$ isotones.

The energy gap above the $p_{3/2}$ neutron orbital disappears when going towards the stability line: the structures of the $N = 32$ ^{58}Fe and ^{60}Ni nuclei no longer exhibit features of a subshell closure. This can be explained by the migration in energy of the $f_{5/2}$ single particle state due to the strong proton $\pi f_{7/2}$ - neutron $\nu f_{5/2}$ monopole interaction which is primarily governed by the tensor force. This interaction causes a decrease in energy of the $\nu f_{5/2}$ single particle orbital with respect to the $\nu p_{3/2}$ and $\nu p_{1/2}$ levels as protons are added to the $\pi f_{7/2}$ shell.[9]

The studies of the location of single particle states in ^{49}Ca mentioned above also showed the existence of an approx. 2 MeV energy spacing between the two higher lying neutron orbitals $p_{1/2}$ and $f_{5/2}$. Such a finding could possibly have indicated the presence of another subshell closure associated with the filling of the $p_{1/2}$ orbital, i.e. at $N = 34$. Recently, this picture received strong support from shell model calculations with the newly developed GXPF1 interaction[10] that clearly indicated that an $N = 34$ energy gap should appear in nuclei with $Z \sim 22$, i.e. already in the Ti isotopes. In spite of those predictions, first data on the yrast structure of ^{56}Ti were found inconsistent with the presence of subshell closure in that nucleus.[11–13] It became clear that the effect of adding two protons to calcium might be underestimated by the GXPF1 interaction and that, in reality, the two $f_{7/2}$ protons cause a significant lowering of the $\nu f_{5/2}$ orbital, bringing it rather close in energy to the $\nu p_{1/2}$ state. In Ca nuclei, however, the ($\nu p_{3/2}$, $\nu p_{1/2}$) - $\nu f_{5/2}$ splitting may still be sufficient to produce a subshell closure at $N = 34$. To verify this hypothesis, the magnitude of energy separation between the $f_{5/2}$ and $p_{1/2}$ orbitals in neutron-rich Ca isotopes needs to be derived from experimental data. Obviously, this task represents a significant challenge as the states involving the $f_{5/2}$ neutron in isotopes such as ^{51,53}Ca are rather difficult to reach.

In earlier work by a part of the present collaboration, deep-inelastic reactions occurring during collisions of a ^{48}Ca beam with ^{208}Pb and ^{238}U targets, at energies approx. 20% above the Coulomb barrier, were successfully used to identify yrast structures in $^{53-56}$Ti.[3,13,14] From the observed production yields for the $^{48-56}$Ti isotopes it became clear that, particularly in reactions on the ^{238}U target, Ca isotopes with masses up to 51 or higher had to be populated as well with cross sections sufficient to display the gamma coincidence relationships within those nuclei. Firm identification of gamma rays in the ^{51}Ca case remained, however, a problem because the difficulty associated with a rather low production rate was augmented by limited information on the location of the yrast states provided by the ^{51,52}K beta-decay measurements.[6] In the present work, we have succeeded in establishing a detailed level scheme of ^{51}Ca, by supplementing the high-fold coincidence data collected at Gammasphere[15] with the ^{48}Ca+^{238}U reaction on a thick target with the results of a second, independent experiment. In the latter the same reaction was used, but with a thin target, and gamma rays were detected in coincidence with reaction products identified in a magnetic spectrometer. Using the same method, we also extended experimental information on the yrast structure of some nuclei below ^{48}Ca, including ^{46}Ar and ^{48}Ar.

2. Experimental procedure and results

The first experiment was performed at Argonne National Laboratory employing a 330-MeV ^{48}Ca beam from the ATLAS accelerator and the Gammasphere array,[15] which consisted of 101 Compton-suppressed Ge detectors. The beam was focused on a 50 mg/cm^2-thick ^{238}U target. Gamma-ray coincidence data were collected with a trigger requiring three or more Compton-suppressed gamma rays to be present in prompt coincidence. Energy and timing information of all Ge detectors firing within 800 ns of the triggering signal was stored. The beam, coming in bursts with ~0.3 ns time width, was pulsed with a 412 ns repetition rate that provided clean separation between prompt and isomeric decays.

In the second measurement the same projectile-target combination, ^{48}Ca+^{238}U, was investigated at the Laboratori Nazionali di Legnaro using the ALPI accelerator and the CLARA+PRISMA detection setup.[16–18] In this case, the 330-MeV ^{48}Ca beam was impinging on a ^{238}U target of 600 μg/cm^2 thickness, placed in the center of the CLARA germanium detector array consisting of 24 Compton-suppressed clover detectors. The PRISMA magnetic spectrometer, used to identify product nuclei, was positioned at 53 degrees with respect to beam direction; i.e., in the vicinity of the grazing angle. The spectrometer was set up for the detection of nuclei close in mass to the projectile and the event trigger required the detection in coincidence of a single gamma ray in CLARA and an ejectile at the PRISMA focal plane.

2.1. ^{51}*Ca*

Prior to the present investigations, the only information available on the ^{51}Ca nucleus came from the beta decay study of neutron-rich ^{51}K and ^{52}K by Perrot et al.[6] Excited states at 1718, 2377, 2934, 3460, 3500 and 4493 keV were proposed, but only for the 3460-keV level spin and parity $7/2^-$ was suggested. The analysis reported here started from examining the gamma-ray spectrum measured by the CLARA array in coincidence with ^{51}Ca products which is shown in Figure 1a. Among the gamma lines found in the spectrum, three transitions with respective energies of 2378, 2934 and 3462 keV had been observed earlier in beta decay. Other weak lines in the spectrum of Fig. 1a must also belong to ^{51}Ca, but their ordering and mutual

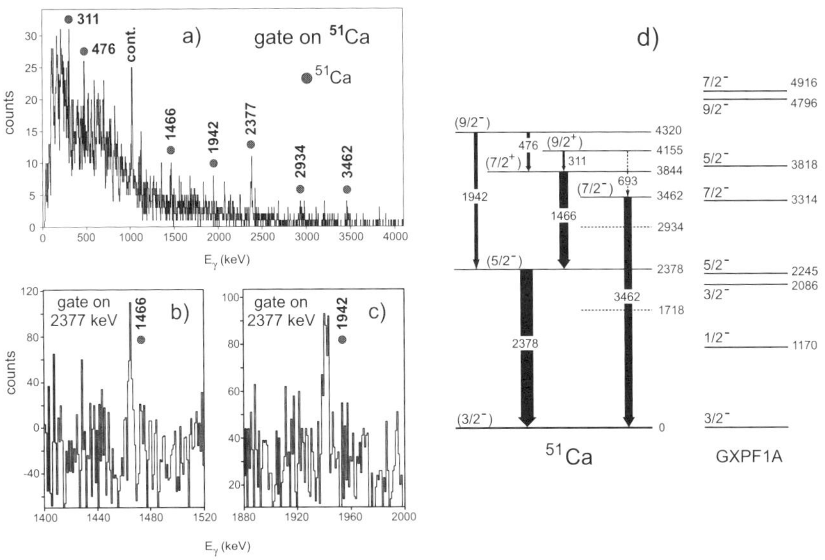

Fig. 1. (a): Gamma-ray spectrum measured in the thin-target experiment with a gate on ^{51}Ca reaction products identified at the focal plane of the PRISMA spectrometer; (b) and (c): Parts of a coincidence spectrum from the thick-target Gammasphere data gated on the 2377-keV transition; (d): Comparisons between shell-model calculations with the GXPF1A hamiltonian and data for ^{51}Ca.

coincidence relationships could not be established due to the low statistics of the gamma-gamma-ejectile coincidence data. In this situation, the set of gamma-ray coincidence data obtained with the thick target experiment at Gammasphere supplied crucial complementary information.

Using the Gammasphere data set, we started our analysis from examining the coincidence spectrum gated on the most prominent gamma line observed in coincidence with the ^{51}Ca residues at 2378 keV; a partial spectrum is displayed in Figures 1b and 1c. The inspection revealed a series of weak peaks, all potential candidates for transitions in ^{51}Ca. However, only two of those satisfied the condition of being present in the spectrum gated on ^{51}Ca reaction products: these were the 1466- and 1942-keV lines. Subsequently, a double coincidence gate placed on the 1466-2378 keV pair in the prompt $\gamma\gamma\gamma$ cube displayed the 476-keV and the 311-keV gamma rays that were also seen in coincidence with the reaction products. On the basis of those findings we located a number of new states in the ^{51}Ca. The resulting ^{51}Ca level scheme is given in Figure 1d.

2.2. $^{46,48}Ar$

Among the reaction products available for spectroscopic studies were also the neutron-rich argon isotopes, ^{46}Ar and ^{48}Ar. Previous knowledge about excited states in ^{46}Ar comes from a series of studies in which various types of reactions were used: ^{14}C-induced transfer reactions on ^{48}Ca,[19] Coulomb excitation of a ^{46}Ar beam,[20] deep-inelastic reactions occurring in ^{48}Ca+^{48}Ca collisions,[21] fragmentation reactions of a ^{48}Ca beam on a thin Be target[22] and, inverse-kinematics proton scattering with a ^{46}Ar beam.[23] A number of states at energies up to 5 MeV were located including the two lowest yrast excitations 2^+ and, tentatively, 4^+ placed by the most recent study[23] at 1558(9) and 3866(14) keV, respectively.

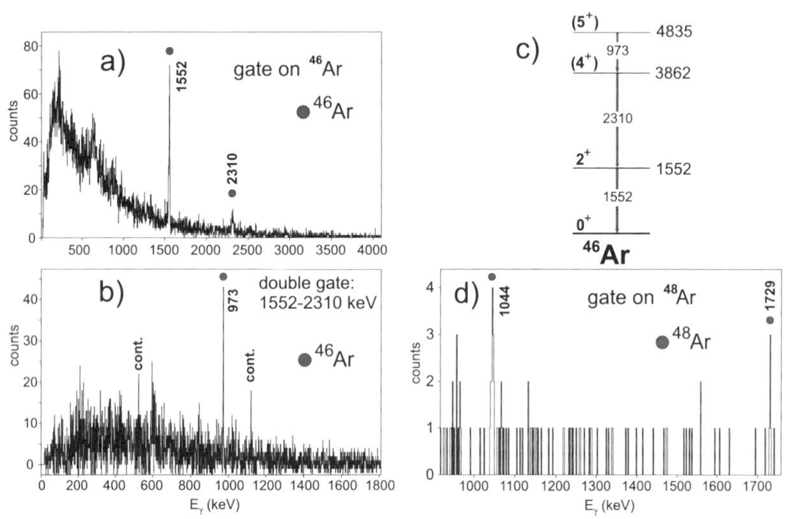

Fig. 2. (a): Gamma-ray spectrum measured in the thin-target experiment with a gate on ^{46}Ar reaction products; (b): Coincidence spectrum from the thick-target Gammasphere data doubly gated on the 1552- and 2310-keV transitions; (c): Partial level scheme of ^{46}Ar showing levels located in the present work; (d): Gamma-ray spectrum measured in the thin-target experiment with a gate on ^{48}Ar reaction products.

In the present work, we started the analysis by displaying the gamma-ray spectrum measured by the CLARA array in coincidence with ^{46}Ar products: it is shown in Figure 2a. This spectrum exhibits a strong 1552-keV peak that must correspond to the $2^+ \to 0^+$ transition, as well as a 2310-keV line. The latter was also seen in the 1552-keV gated Gammasphere data.

These observations form the basis for a 1552-2310 keV cascade in ^{46}Ar which deexcites a level located at 3862 keV. This level is probably identical with the excitation identified at 3866(16) keV in Ref. 23 and the observation of this state in the present study strengthens its previously suggested 4^+ assignment. The spectrum double-gated on the 1552- and 2310-keV lines in the prompt $\gamma\gamma\gamma$ cube is given in Figure 2b. It displays a 973-keV transition very likely feeding the state at 3862 keV from a level at 4835 keV. The identified levels are shown in the scheme proposed in Figure 2c.

The most exotic Ar isotope for which the CLARA gamma-ray spectrum gated with PRISMA showed meaningful lines, was ^{48}Ar. No excited states have been reported in the literature for this nucleus. The spectrum, which is presented in Figure 2d, clearly displays two lines at 1044 and 1729 keV with the 1044-keV transition showing higher intensity. Unfortunately, the analysis of the Gammasphere coincidence data, aimed at establishing potential coincidence relationships between these two newly observed lines, was inconclusive.

3. Discussion

The yrast structure identified in the present work for ^{51}Ca, has to be interpreted in a somewhat speculative way since no rigorous spin-parity assignments could be made for any of the states. Very often in such a case shell model calculations can provide very good guidance. Here, however, the situation is more difficult, since some of the states must arise from proton excitations across the $Z = 20$ gap, whereas the presently available shell model calculations operate only in the fp shell. Thus, the analysis has to take into account the existence of both types of excitations: the states involving promotion of neutrons may be interpreted with the help of shell model calculations; in turn, to understand the proton excitations, which have opposite parity, one should apply arguments based on systematics and decay properties.

Recently, a new effective interaction for the full fp shell, labeled GXPF1A, was developed by Honma et al.,[10] and shell-model calculations with this interaction were found to be quite successful in describing the properties of neutron-rich nuclei around the $N = 32$ subshell closure. One could then expect that the same interaction will also account well for the yrast states arising from excitations within the fp shell in the nuclei of interest here. The issue was all the more interesting that, in ^{51}Ca, the low-lying yrast states should involve the promotion of a neutron in the $\nu p_{1/2}$ and $\nu f_{5/2}$ orbitals, and the location of the latter at $Z{\sim}20$ is of great importance as it determines the presence or absence of a shell gap in ^{54}Ca.

3.1. ^{51}*Ca*

Considerations based on the results of the full *fp* shell model calculations for ^{51}Ca as well as on systematics and decay properties lead us to the interpretation of the identified structure that is presented in Figure 1d. There is practically no doubt that a strongly populated level at 2378 keV is the first yrast state with spin-parity assignment of $5/2^-$, arising from the $\nu p^2_{3/2}p_{1/2}$ coupling. The next yrast states predicted by the shell model to be located above the $5/2^-$ excitation, are: the $7/2^-$ state dominated by the $\nu p^4_{3/2}f^{-1}_{7/2}$ configuration and the $9/2^-$ level corresponding to the $\nu p^2_{3/2}f_{5/2}$ coupling. We associate those excitations with the levels located at 3462 and 4320 keV, respectively. Analyses of the location and decay properties of the states placed at 3844 and 4154 keV lead to the conclusion that they must arise from the promotion of a proton across the $Z = 20$ gap. A level at 4154 keV is probably the $9/2^+$ excitation of the $\pi f_{7/2}s^{-1}_{1/2}\nu p^{-1}_{3/2}$ type, similar to the $9/2^+$ state at 4018 keV in ^{49}Ca.[24] In turn, the feeding and decay pattern of a state at 3844 keV is in line with the $7/2^+$ spin-parity assignment.

While the overall agreement between experiment and theory for the *fp* shell states in ^{51}Ca is satisfactory, one can inspect in detail the behavior of the experimental and calculated states involving the $\nu f_{5/2}$ orbital in the ^{51}Ca and ^{53}Ti isotones. In ^{51}Ca, a state that involves in its main configuration an $f_{5/2}$ neutron is the $9/2^-$ state which is located at 4320 keV. In ^{53}Ti, the excitation that is dominated by the $\pi f^2_{7/2}\nu p^2_{3/2}f_{5/2}$ configuration is the $21/2^-$ state at 6056 keV.[25] In ^{53}Ti the agreement with the calculated value of 6107 keV is very good, ΔE=51 keV. The deviation, however, becomes larger for ^{51}Ca, ΔE=476 keV. At the same time, in these isotones, the calculations reproduce relatively well the yrast states involving a neutron on the $p_{1/2}$ single particle state: these are the $5/2^-$ and $17/2^-_1$ levels in ^{51}Ca and ^{53}Ti, respectively. Thus, the observed behavior of the states with predominant $p_{1/2}$ and $f_{5/2}$ configurations indicates that the $p_{1/2} - f_{5/2}$ energy difference in Ca nuclei might be somewhat smaller than that predicted by the GXPF1A interaction. The GXPF1A calculations correspond to a gap of 3.6 MeV between the two single particle states at $N = 34$ and $Z = 20$. Lowering this energy gap by approx. 0.5 MeV in Ca nuclei would result in a very good description of the $9/2^-$ state in ^{51}Ca. Calculations with this slightly reduced gap still predict the 2^+ energy in ^{54}Ca to lie at a rather high energy of approx. 2.5 MeV.

3.2. $^{46,48}Ar$

The analysis of the experimental data for ^{46}Ar, presented here, identified a cascade of three consecutive gamma rays with energies 1552, 2310 and, 973 keV. There is no doubt that the 1552-keV gamma ray corresponds to the $2^+ \to 0^+$ transition. As a matter of fact, the precise energy of this transition has been a matter of debate. The value obtained in the present study, 1552.3(3) keV, resolves the issue.

The second transition, with an energy of 2310.0(4) keV, deexcites a level at 3862 keV. It is very likely that this level corresponds to the excitation observed in Ref. 23 at 3866(16) keV for which the assignment of either 2^+, or 4^+ quantum numbers was suggested. Since deep-inelastic processes, which are mostly responsible for the production of ^{46}Ar in the present study, preferentially populate yrast states, our finding favors the 4^+ spin-parity assignment for the level at 3862 keV. The 973.2(3)-keV gamma ray is the third transition in the cascade. It originates from a state at 4835 keV which is probably the next yrast excitation. Indeed, the 5^+ state arising from the promotion of a neutron across the $N = 28$ shell gap is expected around 5 MeV and we associate the 4835 keV level with such an excitation.

No excited states have been reported in literature for the ^{48}Ar nucleus. Our data indicate the presence of the two transitions in this nucleus: at respective energies of 1044 and 1729 keV. The 1044-keV gamma ray, with its higher intensity, is a perfect candidate for the $2^+ \to 0^+$ ground state transition. Its energy agrees very well with the energies of the corresponding $2^+ \to 0^+$ transitions in the $N = 30$ ^{50}Ca and ^{52}Ti isotones which are 1027 and 1050 keV, respectively.

4. Conclusion

Using deep-inelastic reactions in two complementary experiments, a thick target gamma-gamma and a reaction product-gamma coincidence measurement, the yrast structure of ^{51}Ca has been located. Also, new states in ^{46}Ar and new gamma rays in ^{48}Ar were identified. Shell model calculations with the GXPF1A effective interaction reproduce the fp shell neutron excitations in ^{51}Ca reasonably well. It was noted, however, that the deviation between theory and experiment for states with predominant $\nu f_{5/2}$ increases when comparing ^{53}Ti with ^{51}Ca. It may be that the $p_{1/2} - f_{5/2}$ energy difference in Ca nuclei is somewhat smaller than that predicted by the

GXPF1A Hamiltonian. The results, however, do not contradict the presence of a subshell closure at $N = 34$ in the Ca nuclei.

Acknowledgments

This work was supported by the US Department of Energy, Office of Nuclear Physics, under Contract No. DE-AC02-06CH11357, by US National Science Foundation Grants Nos. PHY-06-06007, and PHY-0456463, by EURONS European Commission contract no. 506065, and by Polish Scientific Committee Grant No. 1PO3B 059 29.

The authors thank the ATLAS and Tandem-ALPI operating staffs for the efficient running of the accelerators and John Greene for preparing the targets used in the measurement.

References

1. J. I. Prisciandaro *et al.*, *Phys. Lett.* **B510**, 17 (2001).
2. D. E. Appelbe *et al.*, *Phys. Rev.* **C67**, 034309 (2003).
3. R.V.F. Janssens *et al.*, *Phys. Lett.* **B546**, 55 (2002).
4. D. C. Dinca *et al.*, *Phys. Rev.* **C71**, 041302(R) (2005).
5. A. Gade *et al.*, *Phys. Rev.* **C74**, 021301(R) (2006).
6. F. Perrot *et al.*, *Phys. Rev.* **C74**, 014313 (2006).
7. A. Huck *et al.*, *Phys. Rev.* **C31**, 2226 (1985).
8. R. A. Ricci and P. R. Maurenzig, *Riv. Nuovo Cim.* **1**, 291 (1969).
9. T. Otsuka *et al.*, *Phys. Rev. Lett.* **95**, 232502 (2005).
10. M. Honma *et al.*, *Eur. Phys. J.* **A25**, suppl. 1, 499 (2005).
11. S. N. Liddick *et al.*, *Phys. Rev. Lett.* **92**, 072502 (2004).
12. S. N. Liddick *et al.*, *Phys. Rev.* **C70**, 064303 (2004).
13. B. Fornal *et al.*, *Phys. Rev.* **C70**, 064304 (2004).
14. S. Zhu *et al.*, *Phys. Lett.* **B650**, 135 (2007).
15. I. Y. Lee, *Nucl. Phys.* **A520**, 641c (1990).
16. A. M. Stefanini *et al.*, *Nucl. Phys.* **A701**, 217c (2002).
17. A. Gadea *et al.*, *Eur. Phys. J.* **A20**, 193 (2004).
18. S. Szilner *et al.*, *Phys. Rev.* **C76**, 024604 (2007).
19. W. Mayer *et al.*, *Phys. Rev.* **C22**, 2449 (1980).
20. H. Scheit *et al.*, *Phys. Rev. Lett.* **77**, 3967 (1996).
21. B. Fornal *et al.*, *Eur. Phys. J.* **A7**, 147 (2000).
22. Zs. Dombradi *et al.*, *Nucl. Phys.* **A727**, 195 (2003).
23. L. A. Riley *et al.*, *Phys. Rev.* **C72**, 024311 (2005).
24. R. Broda *et al.*, *Acta Phys. Pol.* **B36**, 1343 (2005).
25. B. Fornal *et al.*, *Phys. Rev.* **C72**, 044315 (2005).

Nuclear Structure Information from ^{208}Pb(p, p') via Isobaric Analog Resonances in ^{209}Bi

A. HEUSLER[1] *, P. VON BRENTANO[2], T. FAESTERMANN[3], G. GRAW[4], R. HERTENBERGER[4], J. JOLIE[2], R. KRÜCKEN[3], K. H. MAIER[5], D. MÜCHER[2], N. PIETRALLA[6], F. RIESS[4], V. WERNER[7], H.-F. WIRTH[3]

[1] *Max-Planck-Insititut für Kernphysik, Heidelberg, Germany,*
[2] *Institut für Kernphysik, Universität Köln,*
[3] *Technische Universität München,* [4] *Sektion Physik der Universität München,*
[5] *H. Niewodniczański Institute of Nuclear Physics, Kraków, Poland,*
[6] *Institut für Kernphysik, Technische Universität Darmstadt,*
[7] *Wright Nuclear Structure Laboratory, Yale University, USA*

We studied the inelastic proton scattering on the doubly magic nucleus ^{208}Pb via isobaric analog resonances (IAR) in ^{209}Bi. In the proton decay of the intruder resonance $j_{15/2}$ several states with positive parity are identified.

The facility at München consists of an electrostatic Tandem van de Graaff accelerator and a magnetic spectrograph with a quadrupole at the entrance and three dipoles. The resolution of the proton detector, as measured by the gaussian width of a peak, is better than 3 keV for protons of typical 10 MeV. The large solid angle of 10 msr covers about 10% of the spectrum. Data taking rates up to 100 kHz are supported. The energy stability of the accelerator is better than the resolution of the detector by an order of magnitude. We have a stable Stern-Gerlach hydrogen source delivering proton beams of μA. Altogether the Q3D magnetic spectrograph allows fast measurements of highly linear proton spectra with low background and good resolution. A typical run takes 20 minutes.

In the inelastic proton scattering on ^{208}Pb we covered all seven IAR in ^{209}Bi with proton energies from 15 – 18 MeV and some off-resonance energies.[1] In total we took 200 spectra of 1 MeV length, each with up to 100 peaks. With a resolution of about 3 keV and a peak-to-valley ratio up to 500:1, 10 keV doublets differing in cross section by a factor 100 can be disentangled.

*E-mail: A.Heusler@mpi-hd.mpg.de

Fig. 1. Spectra of ^{208}Pb(p, p') at E_p = 16.355, 16.495, 16.950 MeV corresponding to the $j_{15/2}$, $d_{5/2}$, $s_{1/2}$ IAR, respectively.

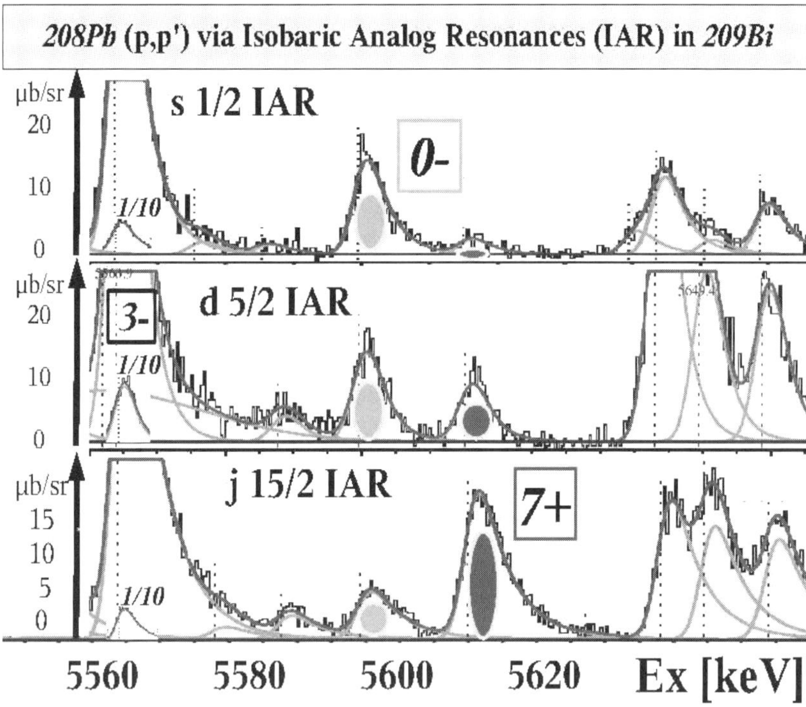

Fig. 1 shows three spectra with a length of only 100 keV but 10 states. They are selectively excited by the $j_{15/2}$, $d_{5/2}$, $s_{1/2}$ IAR in ^{209}Bi. Next to the strong 3^- state at left a 0^- state in ^{208}Pb is identified. It is the second 0^- state and consists by 90% of the configuration $d_{5/2}f_{5/2}$ with a 10% admixture of $s_{1/2}p_{1/2}$. The strong excitation of a state at Ex = 5614 keV state is observed for the first time. The selective excitation by the $j_{15/2}$ IAR indicates positive parity. On the $j_{15/2}$ resonance the maximum cross section is about 20 μb/sr and decreases to less than 1 μb/sr off-resonance. We assign spin 7^+ and a rather pure $j_{15/2}p_{3/2}$ configuration.

We encounter several experimental problems. Not all spectra are so easy to interpret as this example. The three humps at the right side of Fig. 1 contain already 5 states, one doublet has only 0.5 keV distance. A rather trivial but annoying problem arises from the contaminations by light nuclei. In the magnetic spectrograph they are recognized by the kinematic shift with scattering angle and the kinematic broadening with slit opening.

Fig. 2. Spectrum of ^{208}Pb(p, p') at $E_p = 15.720\,\text{MeV}$ corresponding to the $i_{11/2}$ IAR.

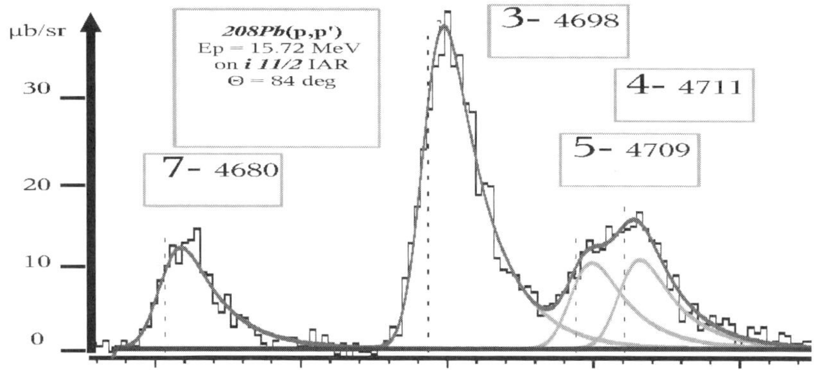

The resolution of 3 keV is well documented by Fig. 2 with the separation of a 2 keV doublet in only 10 keV distance to a stronger line. The low background allows to measure cross section as small as 1 μb/sr within 20 minutes.

A new problem arises from the scattering of the protons on the electrons in the lead target. The binding energy of the M-electrons with about 2.9 keV is comparable to the resolution (Fig. 3) and finally sets the limit to the resolution of any magnetic spectrograph. The L-electrons with 14 keV produce satellites to each line. For strong lines like the 4481 line raising by a factor 500 above the background, up to three satellites are found by the peak fitting program. They correspond to the separation of 1, 2 and 3 or more L-electrons with intensities of a few %. Fig. 4 shows a spectrum of three strong lines in ^{208}Pb each with a satellite from L-electrons. Nevertheless a weak line at $E_x = 3947\,\text{keV}$ from ^{208}Pb is clearly resolved by the peak fitting program. Namely the line is not broadened as for the satellites. Again the peak-to-valley ratio is about 500:1.

Let me remind you to the reaction mechanism. Inelastic proton scattering via IAR can be considered as equivalent to a neutron pickup reaction on a target in an excited state. Consider a target of ^{209}Pb where all nuclei reside in the third excited state with spin $j_{15/2}$ (Fig. 5). By picking a neutron from the $f_{5/2}$ shell a particle-hole state $j_{15/2}f_{5/2}$ with spin between 5^+ and 10^+ is created.

Now replace one of the 45 excess neutrons in ^{209}Bi by a proton to obtain the isobaric analog of the $j_{15/2}$ state (Fig. 6). The resonance can be created by adjusting the proton beam with 16.5 MeV taking care of

Fig. 3. Spectrum of ^{208}Pb(p, p') at E_p = 14.920 MeV corresponding to the $g_{9/2}$ IAR. Several satellites to the 4481 line in ^{208}Pb from the knock-out of L-electrons are seen.

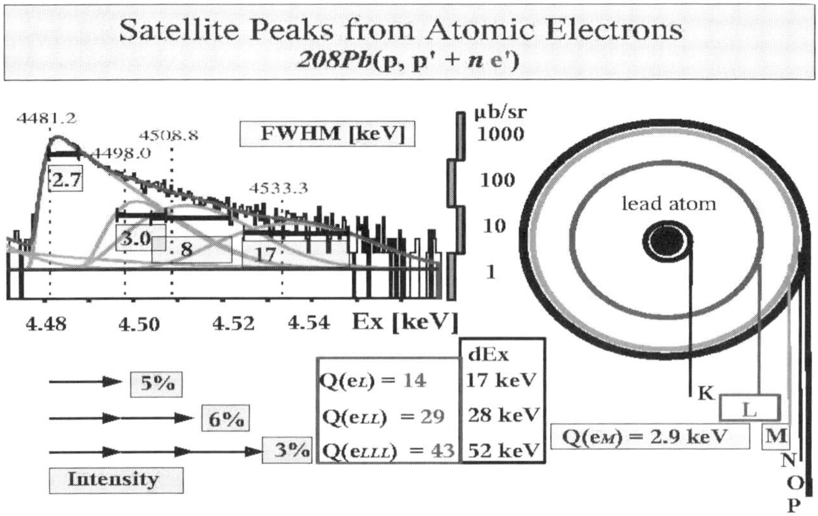

Fig. 4. Spectrum of ^{208}Pb(p, p') at E_p = 14.920 MeV corresponding to the $g_{9/2}$ IAR. The line shape distinguishes the 3947 line in ^{208}Pb from the satellite to the 3920 line.

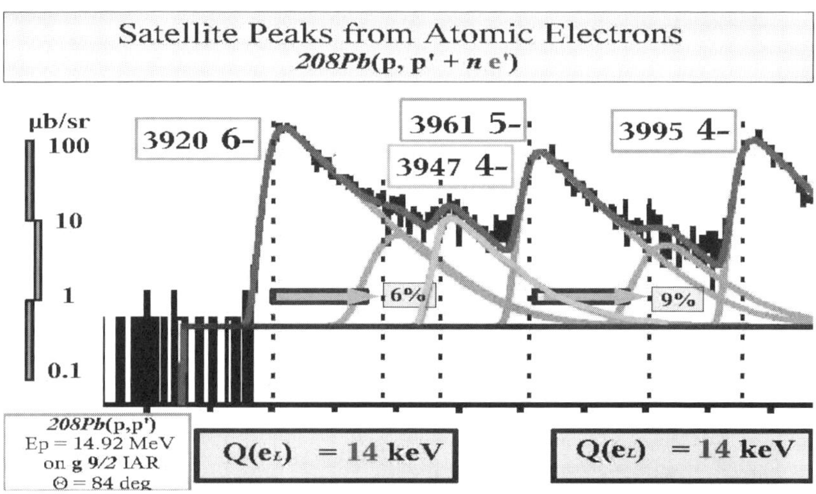

the small width of the resonance with less than 200 keV. In the proton decay of the IAR the same neutron particle-hole configurations are excited as by the neutron pickup reaction. An important difference, however, is

Fig. 5. The reaction $^{209}\mathrm{Pb}^*(d,\ p)^{208}\mathrm{Pb}^*$.

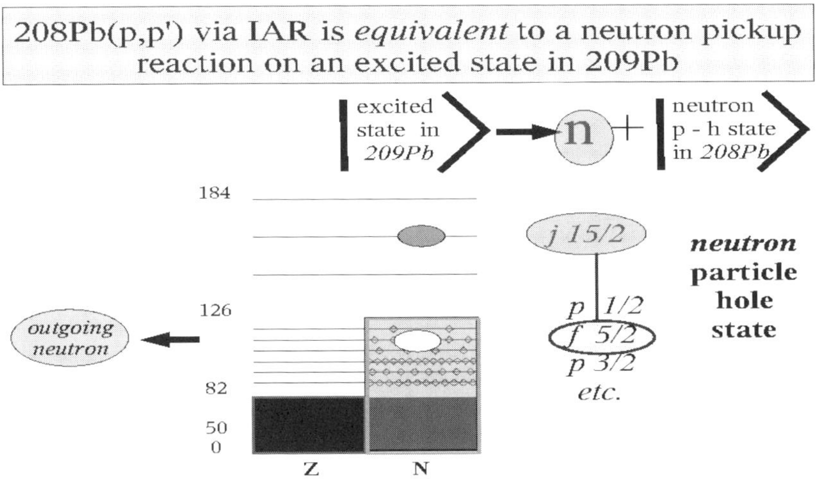

Fig. 6. The reaction $^{208}\mathrm{Pb}(p,\ p')$ via IAR in $^{209}\mathrm{Bi}$.

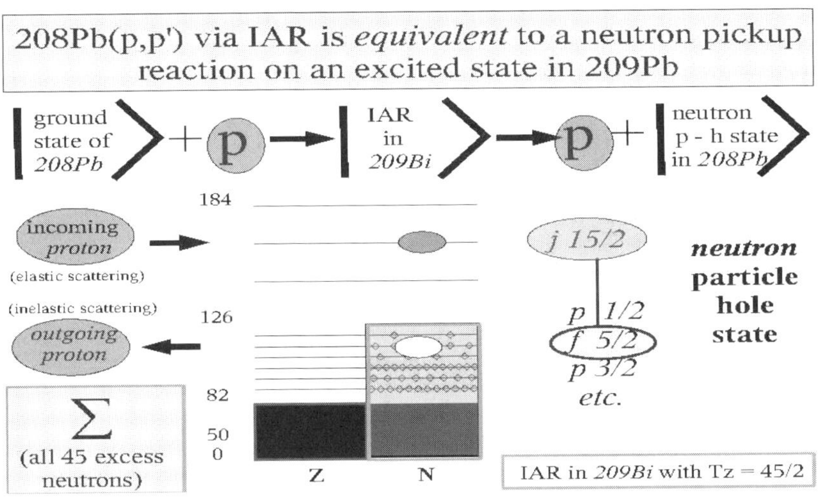

the coherent production of all configurations. Therefore the angular distributions are sensitive to the interferences between all neutron particle-hole configurations. By careful measurement of the angular distribution for some state, the amplitudes of all neutron particle-hole configurations can be determined where the particle corresponds to the decaying IAR.

Fig. 7. Neutron particle-hole configurations excited by IAR in ^{209}Bi for $E_x < 7\,\mathrm{MeV}$.

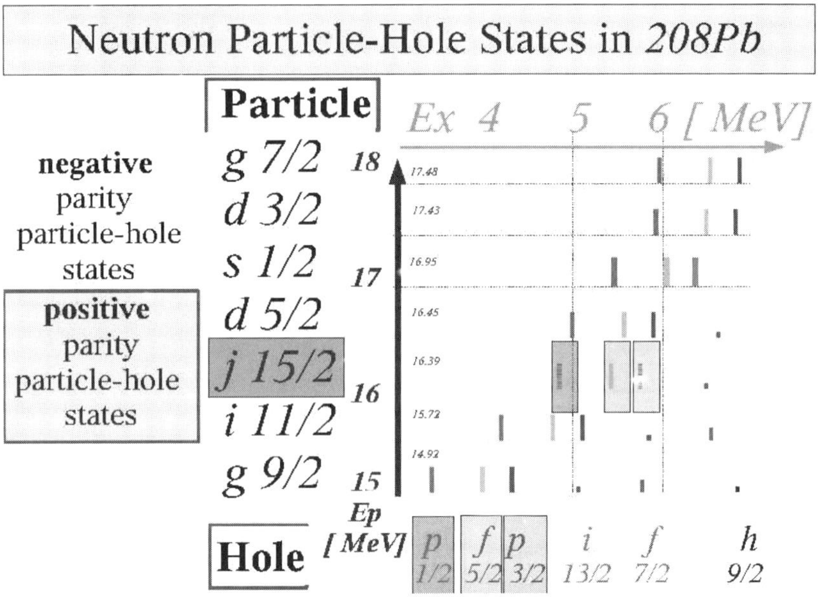

Fig. 7 shows all configurations which can be excited by the seven IAR in ^{209}Bi. They are arranged according to the excitation energy up to 7 MeV and the proton bombarding energy from 15 to 18 MeV. In each IAR the same neutron holes are created, the first particle-hole configurations with a $p_{1/2}$ hole coupled to the particle from the IAR, the second group with a $f_{5/2}$ hole, then $p_{3/2}$ etc. From the angular distribution the amplitudes of all neutron particle-hole configurations are obtained. The most extreme example I encountered is the lowest 6^- state ($E_x = 3919\,\mathrm{keV}$) with a dominant configuration $g_{9/2}f_{5/2}$ and about 1% admixture of the configurations $g_{9/2}p_{3/2}$, $g_{9/2}f_{7/2}$, and definitively $g_{9/2}h_{9/2}$.

All IAR populate negative parity states and only the $j_{15/2}$ IAR as the intruder populates positive parity states. So in principle it should be easy to identify positive parity states. Yet because of the high angular momentum barrier, the mean cross section on the $j_{15/2}$ IAR is 50 times smaller than on the $d_{5/2}$ IAR. In addition the separation of the $j_{15/2}$ and $d_{5/2}$ IAR is less than half the width of each resonance. With the Q3D magnetic spectrograph weak lines close to 100 times stronger lines can be disentangled in only 10 keV distance. By this ability of the Q3D spectrograph we identified about 30 states with positive parity up to excitation energies of 6 MeV.

The three lowest states contain the main strength of the configuration $j_{15/2}p_{1/2}$ and are rather well known (Fig. 8). The states with the dominant proton configuration $h_{9/2}h_{11/2}$ are recognized by admixtures of configurations with the $j_{15/2}$ particle. Most states with dominant configurations $j_{15/2}f_{5/2}$ and $j_{15/2}p_{3/2}$ are not known before. The centroid energies and the total strengths of the three configurations agree rather well with the prediction by the shell model without residual interaction.

By comparison to states with a $p_{3/2}$ hole and a $g_{9/2}$ or $i_{11/2}$ particle,[1] the structure of the 5614 7^+ state (marked in Fig. 8) is determined as a rather pure configuration where a $p_{3/2}$ neutron hole is coupled to the $|15/2^- >$ state in ^{209}Pb (Eq. 1). Bohr& Mottelson[2] determined the structure of the $15/2^-$ state in ^{209}Pb as the $j_{15/2}$ single particle configuration with a 25% admixture of the configuration where the $g_{9/2}$ particle couples to the lowest 3^- state in ^{208}Pb which is described as a complicated mixture of 1p-1h configurations.

$$\begin{aligned} |^{208}\text{Pb } 5614\, 7^+ > &\approx |^{209}\text{Pb}(15/2^-) \otimes p_{3/2} > = \qquad (1) \\ 0.85\, &|^{208}\text{Pb}(g.s.) \otimes j_{15/2} \;\otimes p_{3/2} > + \\ 0.52\, &|^{208}\text{Pb}(2614\, 3^-) \otimes g_{9/2} \;\otimes p_{3/2} > = \\ 0.85\, |^{208}\text{Pb}(g.s.) \otimes \mathbf{1p} \otimes \mathbf{1h} > \; &+ \sum |^{208}\text{Pb}(g.s.) \otimes \mathbf{2p} \otimes \mathbf{2h} > \end{aligned}$$

We conclude that the 5614 7^+ state in ^{208}Pb consists by only 75% of the 1p-1h configuration $j_{15/2}p_{3/2}$ and has 25% admixture of 2p-2h configurations involving the $g_{9/2}$ particle.

Summary. We have measured spectra of the inelastic proton scattering on ^{208}Pb via IAR in ^{209}Bi with a resolution of 3 keV using the Q3D magnetic spectrograph at München. The low background with a peak-to-valley ratio up to 500:1 allows to identify 150 states below Ex = 6.1 MeV with cross sections as small as 0.1 μb/sr. In the region 5.5 – 5.9 MeV the mean level spacing is about 10 keV, but most levels are resolved by the high resolution. The energies are determined with absolute uncertainties down to 100 eV for strong lines.

Below Ex = 6.1 MeV, the shell model predicts 70 negative and 50 positive parity states. We identify slightly more states. For spin 5^+ to 10^+ the shell predicts 36 states. We identify about 30 states with major configurations $j_{15/2}p_{1/2}$, $j_{15/2}f_{5/2}$, $j_{15/2}p_{3/2}$. In addition the states with the dominant proton particle-hole configuration $h_{9/2}h_{11/2}$ are better identified. Twelve positive parity states were not known before.

For the first time, experimental evidence is given for the mixing between

Fig. 8. Positive parity states in ^{208}Pb excited by the proton decay of the intruder IAR in ^{209}Bi with spin $j_{15/2}$.

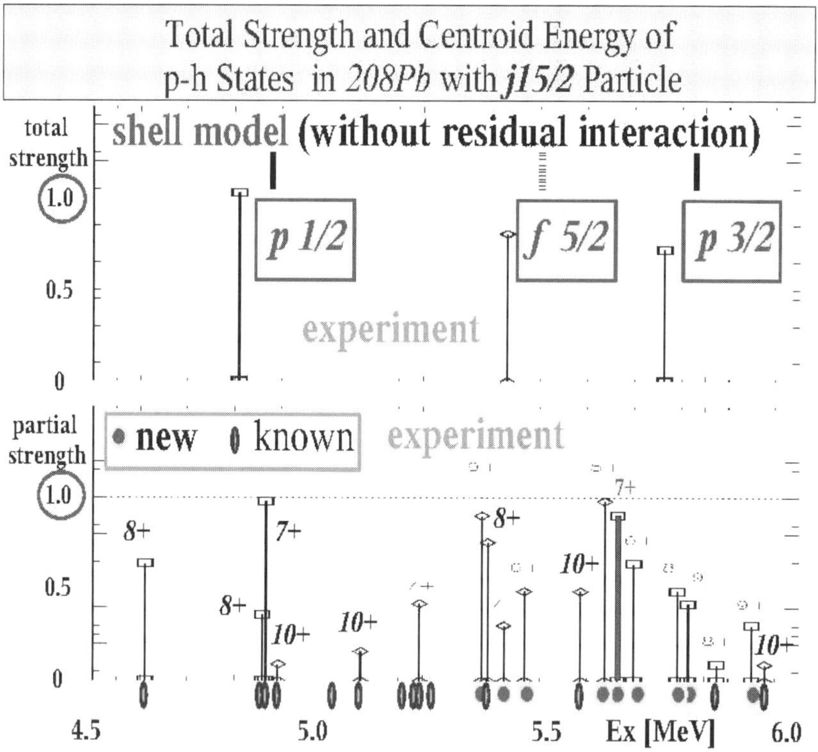

1p-1h and 2p-2h configurations. We have plans for a beam time with the Q3D magnetic spectrograph at München to improve the statistics and enhance the parity identification. What we urgently need, are more precise theoretical calculations of positive parity states in ^{208}Pb including 1p-1h and 2p-2h configurations. (*In the discussion B. A. Brown pointed out that there exist such calculations.*[3])

References

1. A. Heusler, G. Graw, R. Hertenberger, F. Riess, H.-F. Wirth, T. Faestermann, R. Krücken, J. Jolie, D. Mücher, N. Pietralla, P. von Brentano, *Phys. Rev. C* **74**, p. 03403 ((2006)).
2. A. Bohr and Ben R. Mottelson, *Nuclear Structure* (W. A. Benjamin, New York, (1969)).
3. B. A. Brown, *Phys. Rev. Lett.* **85**, p. 5300 ((2000)).

FROM THE QUARK SHELL MODEL TO THE NUCLEAR SHELL MODEL

J. N. GINOCCHIO *

Los Alamos National Laboratory,
Los Alamos, New Mexico 87545, USA
**E-mail: gino@lanl.gov*

The quark shell model has been successful in describing properties of hadrons. Because of color the quark shell model with 3n valence quarks has many more states which are singlet in color than the nuclear shell model with n valence nucleons. However, the quark interaction has been shown to favor two quarks coupled to spin zero and isospin zero and color $\bar{3}$, called diquarks. We show that the color singlet states in the quark shell model which have the maximal number of diquarks consistent with the Pauli symmetry are in one to one correspondence with the states of the nuclear shell model. We also investigate the implications of the quark interactions on the nuclear shell model interaction.

Keywords: quarks, diquarks, nuclear shell model, pairing.

1. Introduction

The spectra of hadrons exhibits correlations of pairs of quarks with spin and isospin quantum numbers S=T=0 and color $\bar{3}$ which are called diquarks.[1,2] The third quark, which has color 3, is then combined with the diquark so that the composite hadron is antisymmetric and a color singlet with T=S= $\frac{1}{2}$. Pairs of quarks with correlations S=T=1 and color $\bar{3}$ are higher in energy as evidenced by the energy difference between the delta resonance and the nucleon and as confirmed by lattice calculations.[3]

2. Quark Shell Model with Diquark Pairing

The hadron spectra exhibits approximate spin symmetry.[4] Also the quarks are confined in a deep potential well. Therefore we expect that the quark shell model potential will be well approximated by a harmonic oscillator potential and the single-particle energies in a major shell will be quasi-degenerate.

The valence quark shell model space limited to harmonic oscillator quantum number N can be classified by the irreducible representations (IR) of the unitary group in 24Ω dimensions, $U(24\Omega)$, where 24Ω is the total number of single valence quark states for a given N. That is, $2 \times 2 \times 3 \times 2\Omega$, since there are 2 spin states, 2 isospin states, and 3 color states for each quark, and 2Ω orbital states where $2\Omega = \sum_i (2\ell_i + 1)$ and the sum is over all the orbital angular momentum allowed for N. The states of the quark shell model are in the totally antisymmetric IR of $U(24\Omega)$ labeled by the Young diagram, $[1^{3n}]$, for a system with $3n$ quarks.

The physical states are those which are color singlets. The states with definite color are classified with respect to the group chain $U(24\Omega) \supset U(8\Omega) \times SU(3)$ where $U(8\Omega)$ classifies the states with a definite color and $SU(3)$ is the color group. The states which are singlets with respect to color will be in the IR $[3^n]$ of $U(8\Omega)$.

The states in the nuclear shell model are also classified by $U(8\Omega)$ but they belong to the IR $[1^n]$ and hence there will be many more physical quark states than nucleon states. However, if we use diquark pairing we can restrict the quark shell model space further.

The quark creation operators, $q^\dagger_{N,\ell,m,m_s,m_t,\alpha}$, create a quark in a spherical shell model space with harmonic oscillator quantum number N, orbital angular momentum ℓ, orbital angular momentum projection m, spin projection m_s, isospin projection m_t, and color α. We consider a diquark pair defined as

$$A^\dagger_{\bar{\alpha}} = \sum_{i,\alpha,\beta} \epsilon_{\alpha,\beta,\bar{\alpha}} \sqrt{2\ell_i + 1}\, [q^\dagger_{N,\ell_i,\alpha} q^\dagger_{N,\ell_i,\beta}]^{(0,0,0)}, \tag{1}$$

where $\epsilon_{\alpha,\beta,\bar{\alpha}}$ is the three dimensional color antisymmetric matrix, $[\dots]^{(0,0,0)}$ means coupled to angular momentum, spin, and isospin zero, and the sum is over all the orbital angular momentum in the harmonic oscillator major shell with quantum number N.

The special orthogonal subgroup, $SO(8\Omega)$, is a subgroup which leaves this pair invariant, $U(8\Omega) \supset SO(8\Omega) \times U(1)$, where the U(1) generator is the number operator.[5] The IR of $SO(8\Omega)$ with the most diquark pairs is the (0,...,1,0,...) IR, where the 1 is in the n th position. Furthermore this IR has the same dimension as the nuclear shell model with n nucleons with single-nucleon quantum numbers N, ℓ, m, m_s, m_t, and also has the same orbital angular momentum, spin and isospin content. The (unnormalized) states of this IR for 3n quarks are[5–8]

$$|n, v_0 = n, \lambda_1 \ldots \lambda_n >= \sum_{\alpha_i} A^{\dagger}_{\bar{\alpha}_1} \ldots A^{\dagger}_{\bar{\alpha}_n} q^{\dagger}_{\lambda_1 \alpha_1} \cdots q^{\dagger}_{\lambda_n, \alpha_n} |0>, \tag{2}$$

where λ_i refers to the quantum numbers $N, \ell_i, m_i, m_{s_i}, m_{t_i}$ and $|0>$ is the closed shell of occupied non-valence quarks. Thus states which correspond to n nucleons are the states with n diquark pairs and n quarks with each pair joined with a quark to form a color singlet with the remaining quantum numbers being free. Since there are n unpaired quarks, this state has seniority, v_0, equal to n.

3. Additional Pairing

We note that pairs in Eq. (2) commute with each other and therefore the product of pairs is symmetric in the color quantum numbers. Thus the unpaired quarks are also symmetric in the color quantum number. Hence these quarks can not form diquarks because the diquarks are antisymmetric in color. However there could be pairing between quarks with $S = 0, T = 1$ and belonging to the color symmetric 6 IR.

We know that pairing in nuclei occurs for pairs of nucleons with $S = 0, T = 1$. Thus we consider the state with this type of pairing,

$$B^{\dagger}_{M_t, \alpha_1, \alpha_2} = \sum_i \sqrt{2\ell_i + 1}\, [q^{\dagger}_{N, \ell_i, \alpha_1} q^{\dagger}_{N, \ell_i, \alpha_2}]^{(0,0,1)}_{M_t}, \tag{3}$$

where M_t is the total isospin projection This pair is left invariant by the symplectic subgroup, $Sp(4\Omega)$, $SO(8\Omega) \supset Sp(4\Omega) \times SU_T(2)$, where $SU_T(2)$ is the isospin subgroup.

Finally, the pairs with $S = 1, T = 0$ symmetric color, and the pairs with $S = 1, T = 1$ and antisymmetric color are left invariant by the subgroup $SO(2\Omega)$, $Sp(4\Omega) \supset SO(2\Omega) \times SU_S(2)$, where $SU_S(2)$ is the spin subgroup.

The state with the maximal $S = 0, T = 0$ and $S = 1, T = 0$ pairing is

$$|n, v_0 = n, v_1 = 0 > = \sum_{\alpha_i} A^{\dagger}_{\bar{\alpha}_1} A^{\dagger}_{\bar{\alpha}_2} \ldots A^{\dagger}_{\bar{\alpha}_{n-1}} A^{\dagger}_{\bar{\alpha}_n} B^{\dagger}_{1, \alpha_1, \alpha_2} \cdots B^{\dagger}_{1, \alpha_{n-1}, \alpha_n} |0>, \tag{4}$$

for even nuclei and

$$|n, v_0 = n, v_1 = 1, \lambda_n > = \sum_{\alpha_i} A^{\dagger}_{\bar{\alpha}_1} A^{\dagger}_{\bar{\alpha}_2} \ldots A^{\dagger}_{\bar{\alpha}_{n-2}} A^{\dagger}_{\bar{\alpha}_{n-1}} A^{\dagger}_{\bar{\alpha}_n} B^{\dagger}_{1, \alpha_1, \alpha_2} \cdots B^{\dagger}_{1, \alpha_{n-2}, \alpha_{n-1}} q^{\dagger}_{\lambda_n, \alpha_n} |0>, \tag{5}$$

for odd nuclei. We have assumed that the valence nucleons are neutrons and thus set $M_1, M_2 \ldots = 1$.

The $S = 1, T = 0$ and $S = 0, T = 1$ pairing interactions can not occur in non-exotic hadrons because they are symmetric in color. Therefore their strength is not known empirically. However, lattice calculations suggest that the interactions inducing these pairing are strongly repulsive.[3]

4. Quark Pairing and Nuclear Pairing

Nuclear pairing is observed in the binding energies of pairs of neutrons in the Calcium, Nickel, and Tin isotopes.[9] These binding energies are fit very well with $S = 0, T = 1$ pairing between nucleons. Our goal for the future is to try to fit these binding energies with the quark pairing interactions discussed in the previous section and, from these fits, determine the strength of these quark pairing interactions and to see if the couplings are consistent with what is known about these quark coupling from hadronic empirical evidence and from lattice calculations.

The quark couplings are much stronger than nuclear couplings. However, the $S = T = 0$ coupling is attractive while the $S = 0, T = 1$ coupling is repulsive and the partial cancellation of these large numbers may produce the observed nuclear couplings.

5. The Quark Shell Model and Nucleon Clustering

When the original idea of a quark shell model with diquark pairing[6,7] was introduced there was criticism that the 3n quarks would not be localized into clusters of 3 quarks.[5,8,10] This would not be consistent with pickup reactions which measure spectroscopic factors close to unity.[5,8] Estimates were made of the overlap of a nucleon in a nuclear shell model with the composite nucleons in the quark shell model with diquark pairing. The estimates turned out to be very small, thereby delivering a fatal blow to the quark shell model with diquark pairing.

The resurrected version of the diquark shell model in this paper differs from the original in that we assume that the quark shell model potential has spin symmetry and hence the spin overlap is unity, as opposed to the earlier version in which the quarks are confined to a single-j shell. Furthermore the quark shell model potential is very deep and confining and so has the symmetries of the spherical harmonic oscillator. Thus the diquark is correlated over a full harmonic oscillator shell involving many shell model orbitals. The spatial overlap within this correlated model needs to be calculated.

6. Summary and the Future

We have revisited the quark shell model with diquark pairing. In contrast to the original study, we propose a quark shell model which has spin symmetry and the degeneracies of the spherical harmonic oscillator. The diquark will then have multi-shell correlations. The effective nucleon will be composed of a correlated diquark and a quark coupled to a color singlet. The spatial overlap of these effective nucleons with a nucleon in the nuclear shell model needs to be calculated in order to determine if the diquark shell model can reproduce the spectroscopic factors observed in nuclei.

In addition to the diquark pairing we have considered additional pairing interactions which can not exist in non-exotic baryons. We speculate that these pairing interactions may account for the nuclear pairing observed in nuclei. The detailed calculations needed to make this connection are being carried out.

7. Acknowledgments

The author would like to thank Professor Talmi and Professor Leviatan for discussions. This research is supported by the U. S. Department of Energy under contract W-7405-ENG-36.

References

1. R. L. Jaffe, Phys. Rep. **409**, 1 (2005).
2. F. Wilczek, *hep-ph/0409168* (2006).
3. C. Alexander, Ph. de Forcand, B. Lucini, *hep-lat/0609004* (2006).
4. P.R. Page, T. Goldman, J. N. Ginocchio, Phys. Rev. Lett. **86**, 204 (2001).
5. I. Talmi, in *Trends in Nuclear Physics*, ed. by P. Kienle, R. A. Ricci, A. Rubbino (North Holland, New York, 1989).
6. H.R. Petry, *et. al.*, Phys. Lett. **B159**, 363 (1985).
7. A. Arima, K. Yazaki, H. Bohr , Phys. Lett. **B183**, 131(1987).
8. I. Talmi, Phys. Lett. **B 205**, 140 (1988).
9. I. Talmi, *Simple Models of Complex Nuclei* (Harwood, Switzerland, 1993).
10. K. Suzuki and K. T. Hecht, Phys. Lett. **B 232**, 159 (1989).

LIFETIME MEASUREMENTS IN ^{67}SE AND ^{67}AS MIRROR PAIR: A TEST OF ISOSPIN SYMMETRY BREAKING

R. ORLANDI*, G. DE ANGELIS, F. DELLA VEDOVA, A. GADEA, N. MĂRGINEAN, D. R. NAPOLI, F. RECCHIA, J. J. VALIENTE-DOBÓN

Laboratori Nazionali di Legnaro dell'INFN
v.le Università, 2 Legnaro (PD) 35020 Italy
**E-mail: Riccardo.Orlandi@lnl.infn.it*
www.lnl.infn.it

F. BRANDOLINI, E. FARNEA, S. LENZI, S. LUNARDI, D. MENGONI, C. A. UR

INFN Sezione di Padova and Dipartimento di Fisica, Università di Padova

K. T. WIEDEMANN

University of São Paulo, São Paulo, Brazil

E. SAHIN

Laboratori Nazionali di Legnaro dell'INFN, Legnaro (PD), Italy
Istanbul University, Istanbul, Turkey

A. BRACCO, S. LEONI

INFN Sezione di Milano e Dipartimento di Fisica dell'Universitá di Milano, Italy

D. TONEV

Institute for Nuclear Research and Nuclear Energy, BAS, Sofia, Bulgaria

R. WADSWORTH, B. S. NARA SINGH

University of York, York, U.K.

D. G. SARANTITES, W. REVIOL, C. J. CHIARA, O. L. PECHENAYA

Washington University, St. Louis, MO, 63130, U.S.A.

C. J. LISTER, M. CARPENTER, J. GREENE, D. SEWERYNIAK and S. ZHU

Argonne National Laboratory, Argonne, IL, 60439, U.S.A.

The comparison of analogue E1 transitions of mirror nuclei offers a valuable tool in the investigation of isospin symmetry breaking. Differences in the measured B($E1$) reduced strengths in fact can be interpreted as the signature of isospin mixing. In this work, the isospin $T_z = 1/2$ nuclei ^{67}As and ^{67}Se were investigated. The nuclei were produced via the fusion evaporation reaction of ^{32}S on ^{40}Ca, at the ATLAS accelerator at Argonne National Laboratory. Gammasphere, Microball and the neutron wall arrays were employed for output channel selection.

The lifetime of the mirror $9/2^+$ states has been measured using the centroid shift method. A confirmation of the validity of this technique is given by the excellent reproduction of the known lifetime of the $9/2^+$ state in ^{69}As, previously determined via a recoil distance experiment. From a preliminary comparison of the extracted B($E1$), our results indicate that isospin symmetry in these nuclei is not significantly broken. Further comparison with large scale shell model calculations is however still required.

Keywords: Gamma spectroscopy; isospin symmetry; lifetime measurements; centroid shift method.

1. Introduction

Mirror nuclei provide a privileged viewpoint to test our understanding of isospin symmetry. In particular, ^{67}As and ^{67}Se lie at the frontier of current spectroscopic knowledge on $N = Z \pm 1$ nuclei near ^{100}Sn, and are good candidates for the study of isospin symmetry breaking. Several models[1–3] in fact predict the size of isospin mixing, which results from isospin symmetry breaking, to increase with nuclear mass, and to be maximum at N=Z.

Isospin symmetry breaking can be searched experimentally by comparing the B($E1$) reduced transition strengths of analogue states in nuclei with $T_z = 1/2$, where T_z is the projection of the isospin. If the symmetry is not broken, and in the long wavelength limit,[4] $E1$ transitions should exhibit equal reduced strengths, since they are purely isovector in nature and therefore proportional to T_z. Mixing of the isospin with $T_z = 3/2$ induces non-diagonal matrix elements which combine with opposite sign in mirror nuclei. This extra contribution, often referred to as the "induced isoscalar term", in one of the nuclei intensifies, while in the other quenches, the $E1$ transition.

Prior to this experiment, the available information on ^{67}As was provided mainly by the work of Jenkins *et al*,[5] and included a value for the lifetime of the first excited $9/2^+$ state, 12(2) ns. For ^{67}Se, only four transitions were previously known, the 647, 717, 915 and 1226 keV, observed with Euroball at LNL.[6] The observed states comprised the mirror $9/2^+$, with an energy of 1365 keV; the measurement of its lifetime was the main objective of this experiment.

2. Experiment

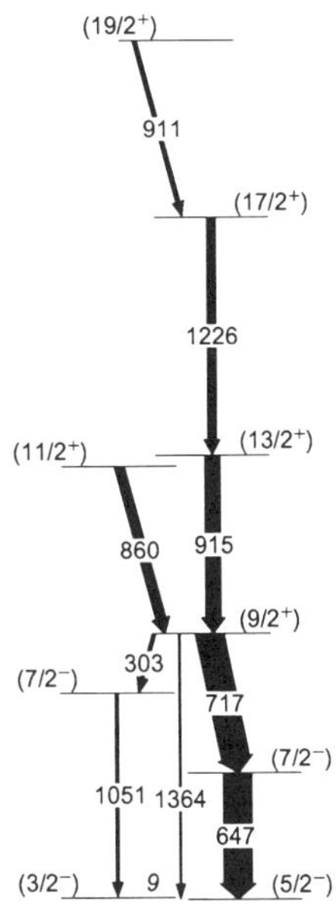

Fig. 1. Partial level scheme of ^{67}Se

The nuclei of interest were produced in the fusion evaporation reactions ^{40}Ca(^{32}S, αn)^{67}Se and ^{40}Ca(^{32}S, αp)^{67}As. The 90-MeV, pulsed ^{32}S beam was provided by the ATLAS accelerator at Argonne National Laboratory. The target consisted of 550μg/cm^2 of ^{40}Ca evaporated onto a thick Au backing of 10mg/cm^2, with a 30 μgcm^{-2} front layer to prevent oxidation. The high-selectivity required to discern, amidst a welter of radiation, the information relative to ^{67}Se and ^{67}As was obtained via the concomitant employment of the Gammasphere Ge array, the Neutron Shell and the Microball array. These detector arrays were employed for the detection of γ rays, neutrons and light charged particles, respectively, necessary for the identification of the output channels.

Five new γ-ray transitions were added to the level scheme of ^{67}Se[7] shown in Fig. 1. The observation of the 304 and 1365 keV lines was particularly important because indispensable for the determination of the branching ratios of the γ rays de-exciting the $9/2^+$ state. The limited statistics for the ^{67}Se channel did not allow, however, a determination of the transitions multipolarities. The spin and parity assignments of the excited states rely therefore only on the comparison with analogue states in ^{67}As.

The time spectra of the Gammasphere Ge detectors were aligned and calibrated using the known period of successive beam pulses. Following from the symmetry of the mirror pair, a lifetime of approximately 10-15 ns was expected for $9/2^+$ state in ^{67}Se. It came however as a surprise that not only in this nucleus, but also in ^{67}As, no sign of such long lifetime was visible in the current data. The Ge time spectra obtained by gating above and below the isomeric state in ^{67}As, and with the same gates but in reversed order, are shown in Fig. 2(a) and 2(b) respectively. The slopes of these curves do not differ significantly from those belonging to the prompt curves displayed in Fig. 2(c) and 2(d). No indication of a 12 ns lifetime can be seen. The reason for this discrepancy with the value given by Jenkins[5] is not clear. Unfortunately, no excited state of similar lifetime was sufficiently populated among the other reaction products, to attempt a comparison with another long-lived state.

The centroids of the distribution 2(a) and 2(b) however show a detectable shift in the position of their centroids, a shift six times larger than the shift between the prompt distributions 2(c) and 2(d). This same effect is consistently observed for several other γ ray pairs: pairs of gates chosen across the isomer (delayed) show overall larger shifts than gates chosen either both above or both below the $9/2^+$ state (prompt).

These shifts of the time distributions were used to determine the lifetime of the $9/2^+$ state, both in ^{67}As and in ^{67}Se, according to the theory of the Centroid Shift Method.[8] By definition, a sufficiently long lifetime causes a shift in the centroid of the time distribution, from the position of the prompt; the shift is equivalent to the lifetime of the state. The inversion of the ordering of the γ rays produces an equally large shift, but in the opposite direction. In the absence of any other effects, the difference between the centroids of curves 2(a) and 2(b) corresponds to twice the lifetime of the state between them.

To assess the applicability of this technique to the data collected, the Centroid Shift Method was tested against the reproduction of the previously known lifetime of the $9/2^+$ state in ^{69}As. Its value was determined via a recoil distance measurement by Hellmeister *et al.*[9] to be 1.94(5)ns. The results, both for ^{69}As and for ^{67}As, are presented in the next section. Due to the limited statistics, only a preliminary estimate will be given for ^{67}Se.

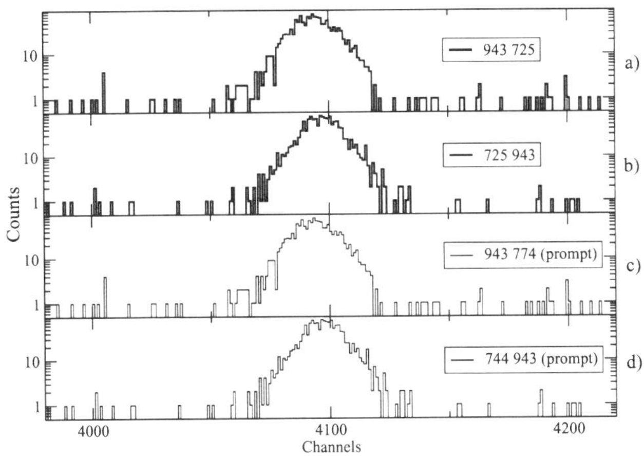

Fig. 2. Time distribution for two different pairs of gates in ^{67}As. The curves (a) and (b) were produced by lines above and below the $9/2^+$ state, 943 and 725 keV, while the gates for curves (c) and (d) both lie above this state. The centroids of the shown time distributions correspond, respectively, to channels 4094.97(25), 4098.05(24), 4096.07(28) and 4096.39(27). The prompt is expected at channel 4096, and the calibration is of 0.56 ns/channel.

3. Results and Discussion

Fig. 3 displays the centroid positions obtained from several pairs of gates in ^{69}As. For clarity, its partial level scheme, taken from Bruce *et al.*[10] is also included in the figure. The first ten pairs of gates were taken across the $9/2^+$ isomeric state, while the last three pairs are prompt. The shifts for the delayed pairs, coherently larger than the prompt, manifest a real physical origin. In general it can be seen that, within statistical fluctuations, the shifts are larger for the pairs which, below the isomer, include the 862 keV instead of the 1304 keV line. The lifetime extracted from the first five and the second five pairs are, respectively, 1.94(10) and 2.40(20) ns, where the quoted error is purely statistical.

Such an effect is expected when low energy lines are involved. It is a well known fact that at low energies the longer rise time of the Ge signal is not fully corrected for by the constant fraction discriminators. This effect

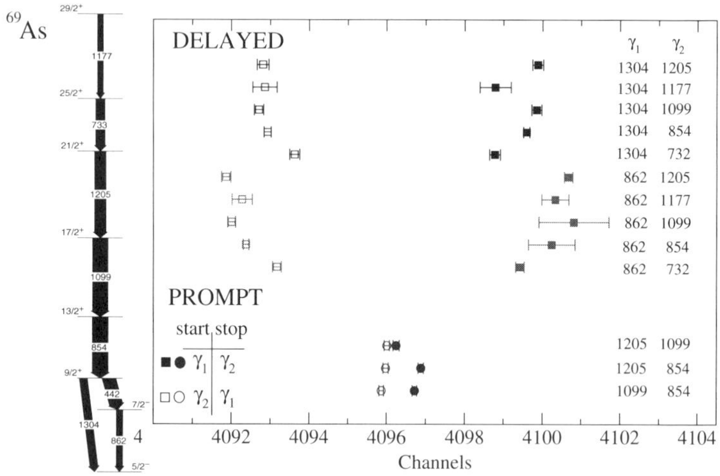

Fig. 3. Centroids of γ-γ time distributions of selected transition pairs in ^{69}As (see text for details). The partial level scheme of ^{69}As, shown on the left hand side, was taken from Bruce *et al.*[10]

induces an extra shift in the centroid of the time distribution, translating into a larger apparent lifetime. The last two pair of prompt gates, 1205-854 and 1099-854, as it can be expected exhibit a shift larger than the high energy pair 1205-1099. These shifts give an estimate of the effect due to the lower energy, 0.28(3) ns. Subtracting it from the value obtained from the set of pairs including the 862 keV transition, yields a lifetime of 2.1(2) ns. This lifetime, together with 1.94(10) ns, is in excellent agreement with the published value, and was taken as a proof of the validity of this method.

Fig. 4 presents the results obtained for ^{67}As. Although the statistical fluctuations are larger than in the case of ^{69}As, the delayed coincidences show clearly a larger shift. The effect was not attributed to the energy of the transitions since it is also seen at high energies (e.g. 1229 and 1035 keV), and since the prompt pairs of very similar energies do not exhibit such a large shift (consider for example the 943-774 and the 725-697). Furthermore, with the exception of the last pair of gates, which were taken from ^{66}Ge, all the transitions belong to ^{67}As.

The average prompt contribution was estimated to be 0.20(5) ns. Subtracting this value from the weighted average of all the delayed pairs, the resulting lifetime for the $9/2^+$ state in ^{67}As is 0.7(1) ns. Again, the quoted error is only statistical.

Although the analysis is still in progress, the data from ^{67}Se reveal shifts larger than for ^{67}As. Taking into account the energy correction estimated from the prompt gates in ^{67}As, the $9/2^+$ state in ^{67}Se has a lifetime of approximately 1.3(4) ns.

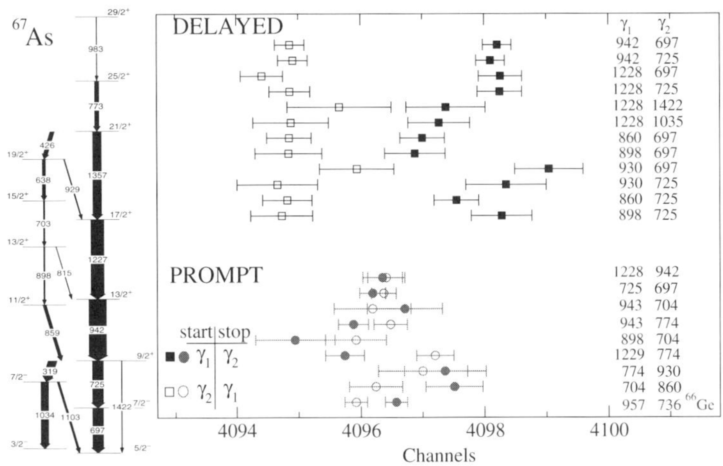

Fig. 4. Centroids of γ-γ time distributions of selected transition pairs in ^{69}As (see text for details). The partial level scheme of ^{67}As, shown on the left hand side, was taken from Jenkins *et al.*[5]

The lifetimes measured from the centroid shift were combined with the measured branching ratios of the transitions de-exciting the $9/2^+$ state to determined the B($E1$) reduced strengths. The data for ^{67}As confirms the branching ratios measured by Jenkins *et al.*[5] In ^{67}Se, the measured branching ratios for the 304, 717 and 1365 keV transitions are, respectively, 0.10(4), 0.84(9) and 0.06(4). Due to the limited statistics, for the 1365 the measured value is mostly an upper limit.

Assuming that the lifetime of ^{67}Se will be confirmed by further analysis, a preliminary comparison of the B($E1$) reduced strengths can be attempted. The $9/2^+$ state decays in both nuclei via one M2 transition to the ground state, and via two E1 transition to the first and second excited $7/2^-$ states. The B($E1$) to the first excited $7/2^-$, corresponding to the 725 and 717 transitions in, respectively, ^{67}As and ^{67}Se, are in excellent agreement. Their values, 1.3(2) 10^{-6} and 1.0(3) 10^{-6} w. u. are consistent with isospin symmetry conservation. A difference can be seen, instead, in the B($E1$) strengths for the decay of the $9/2^+$ to the second excited $7/2^-$ state, which were extracted to be 8.1(6) 10^{-6} and 1.7(8) 10^{-6} w. u.

It is yet not clear whether the size of the discrepancy in the second case is a sign of isospin symmetry breaking, originating possibly from the different structure of the $7/2^-$ states,[1] or whether it is compatible with no violation of isospin symmetry. To clarify this issue, these results need to be compared with large-scale shell-model calculations. Some initial results have been recently published by Hasegawa *et al.*[11]

4. Conclusion

The successful reproduction of the known lifetime of the $9/2^+_{\rm I}$ state in ^{69}As indicates that the sensitivity of Ge timing in the present configuration is compatible with the lifetime suggested by the centroid shifts for the ^{67}Se and ^{67}As mirror pair. The preliminary values for these states are 0.7(1) ns and 1.3(4) ns respectively. The initial comparison of the B($E1$) reduced strengths suggest that no major violation of isospin symmetry is present, but further theoretical information needs to be provided by shell-model calculations.

This work was supported by the U.S. Department of Energy, Office of Nuclear Physics, under Contract No. DE-AC02-06CH11357 and Grant No. DE-FG02-88ER-40406.

References

1. J. Dobaczewski *et al.*, Phys. Lett. B345 (1995) 181.
2. G. Colò *et al.*, Phys. Rev. C **52** (1995) R1175.
3. A. F. Lisetskiy *et al.*, Phys. Rev. Lett. **89** (2002) 12502.
4. D.H. Wilkinson, *Isospin in Nuclear Physics* (North-Holland Publishing Company, Amsterdam, 1969
5. D. G. Jenkins *et al.*, Phys. Rev. C **64** (2001) 64311.
6. G. de Angelis *et al.*, Proceedings of ENAM 2001, (2001).
7. K. T. Wiedemann *et al.*, LNL Annual Report 2006 (2007) 19.
8. H. Mach *et al.*, J. Phys. G **31** (2005) S1421.
9. H. P. Hellmeister *et al.*, Phys. Rev. C **17** (1978) 2113.
10. A. M. Bruce *et al.*, Phys. Rev. C **62** (2000) 027303.
11. M. Hasegawa *et al.*, Phys. Lett. B 617 (2005) 150.

THE ^{120}Sn(p,t) ^{118}Sn REACTION: LEVEL STRUCTURE OF ^{118}Sn AND MICROSCOPIC DWBA CALCULATIONS

P. GUAZZONI* and L. ZETTA

Dipartimento di Fisica dell'Università, and I.N.F.N., Via Celoria 16, I-20133 Milano, Italy
E-mail: paolo.guazzoni@mi.infn.it, luisa.zetta@mi.infn.it

A. COVELLO and A. GARGANO

Dipartimento di Scienze Fisiche, Università di Napoli Federico II, and I.N.F.N., Complesso Universitario di Monte S. Angelo, Via Cintia, I-80126 Napoli, Italy
E-mail: aldo.covello@na.infn.it, angela.gargano@na.infn.it

B. F. BAYMAN

School of Physics and Astronomy, University of Minnesota, Minneapolis, MN55455, USA
E-mail: bayman@phys.umn.edu

G. GRAW and R. HERTENBERGER

Sektion Physik der Universität München, D-85748, Garching, Germany
E-mail: gerhard.graw@physik.uni-muenchen.de, ralf.hertenberger@physik.uni-muenchen.de

T. FAESTERMANN and H-F. WIRTH

Physik Department, Technische Universität München, D-85748 Garching, Germany
E-mail: thomas.faestermann@ph.tum.de, hans-friedrich.wirth@ph.tum.de

M. JASKÓLA

Soltan Institute for Nuclear Studies, Hoza Street 69, Warsaw, Poland
E-mail: marian.jaskola@fuw.edu.pl

The ^{120}Sn(p,t)^{118}Sn reaction has been studied in a high-resolution experiment at an incident energy of 21 MeV. Differential cross-sections for 37 transitions to levels of ^{118}Sn up to an excitation energy of about 3.6 MeV have been measured. Cluster and microscopic DWBA analysis and shell model calculations have been carried out.

1. Introduction

The tin isotopes, ranging from ^{112}Sn to ^{124}Sn, offer an excellent opportunity for detailed theoretical and experimental investigations. The Z=50 proton shell is closed and we can assume that the low-lying states are mostly formed by excitation of neutrons in the five active orbitals $0g_{7/2}$, $1d_{5/2}$, $2s_{1/2}$, $1d_{3/2}$, $0h_{11/2}$. Two-neutron transfer reactions, such as (p,t), are very sensitive to the predominant neutron structure of the isotopic sequence $N = 60$ to $N = 74$ and may be profitably used to provide spectroscopic information for these nuclei.

In recent years we have undertaken a systematic study of tin isotopes via the (p,t) reaction in high-resolution experiments, at the Munich HVEC MP Tandem. In previous papers[1–3] we reported the results concerning the ^{122,116,112}Sn(p,t) ^{120,114,110}Sn reactions. This contribution extends our study to the ^{120}Sn$(p,t)^{118}$Sn reaction. The level structure of ^{118}Sn has been previously investigated by different kinds of experimental measurements and the relevant results are reported in the NDS compilation.[4]

We have accurately measured the differential cross sections of 37 transitions to levels of the ^{118}Sn residual nucleus, determined the angular momentum transfers, and assigned spins and parities to 41 levels.

In connection with the experimental work, we have carried out a shell-model study of ^{118}Sn making use of a realistic effective interaction derived from the CD-Bonn nucleon-nucleon (NN) potential.[5]

Preliminary DWBA microscopic calculations of cross section angular distributions for the ground and the lowest 2^+ state of ^{118}Sn have been performed, using two-neutron spectroscopic amplitudes for ^{118}Sn and the ^{120}Sn target nucleus obtained from our shell-model calculations.

2. Experimental Procedure and Results

The ^{120}Sn$(p,t)^{118}$Sn reaction has been measured at high resolution using the 21 MeV proton beam from the Munich HVEC MP Tandem accelerator. The 114 μg/cm^2 thick ^{120}Sn isotopically-enriched (99.6%) target was evaporated on a 6 μg/cm^2 carbon backing. The reaction products were momentum separated by the Q3D magnetic spectrograph and detected by the 1.8 m long focal plane detector for light ions.[6] The energy resolution was 8 keV full width at half maximum. The uncertainty in our quoted energies is estimated to be 3 keV. Absolute cross sections were measured with an uncertainty of $\sim$ 15%. Table 1 shows the results obtained in the present experiment. The integrated cross sections reported are calculated in the interval $10^o \leq \theta_{lab} \leq 65^o$.

Table 1. Present Experiment

E_{exc} (MeV)	J^π	σ_{int} (μb)	E_{exc} (MeV)	J^π	σ_{int} (μb)
0.0	0^+	2250 ± 14	3.137	0^+	3 ± 1
1.230	2^+	613 ± 12	3.218	0^+	7 ± 1
1.758	0^+	32 ± 2	3.228	2^+	6 ± 1
2.043	2^+	4 ± 1	3.237	0^+	1 ± 0.5
2.057	0^+	42 ± 2	3.252	6^+	4 ± 1
2.280	4^+	28 ± 2	3.274	$1^- + 3^-$	15 ± 1
2.323	$5^-+3^-+2^+$	415±6	3.309	$1^- + 2^+$	6 ± 1
2.403	2^+	16 ± 1	3.344	3^-	8 ± 1
2.489	4^+	71 ± 2	3.355	0^+	12 ± 1
2.497	0^+	19 ± 1	3.375	5^-	38 ± 2
2.575	7^-	25 ± 1	3.395	5^-	7 ± 1
2.677	2^+	44 ± 2	3.427	3^-	9 ± 1
2.734	4^+	119 ± 3	3.463	2^+	19 ± 1
2.879	5^-	8 ± 1	3.524	2^+	9 ± 1
2.904	2^+	25 ± 1	3.541	3^-	4 ± 1
2.930	0^+	11 ± 1	3.559	6^+	5 ± 1
2.963	5^-	114 ± 3	3.585	2^+	43 ± 2
3.057	2^+	6 ± 1	3.597	2^+	10 ± 1
3.108	7^-	1 ± 0.3			

3. Cluster DWBA Calculation

A direct one-step (p,t) transfer reaction on an even-even 0^+ target nucleus populates only natural parity states, with a unique L-transfer value, in the hypothesis that the two neutrons are transferred with the relative angular momentum zero. The spin and parity of the observed levels are J_f=L and π_f=$(-1)^L$.

A DWBA analysis of the experimental differential cross sections has been carried out assuming a semimicroscopic dineutron cluster pickup mechanism. The calculations have been performed in finite range approximation using the computer code TWOFNR[7] and a proton dineutron interaction potential of Gaussian form $V(r_{p2n}) = V_0 \exp -(r_{p2n}/\xi)^2$ with $\xi = 2$ fm. The optical model parameters used were the same as those used for the analysis of the ^{122,116,112}Sn(p,t) reactions.[1–3]

The cluster DWBA analysis is used to provide (p,t) angular distributions to guide the attribution of L values and consequently J^π values to individual final states. Examples of typical analyses for L=0 and L=2 transfers are reported in Fig. 1.

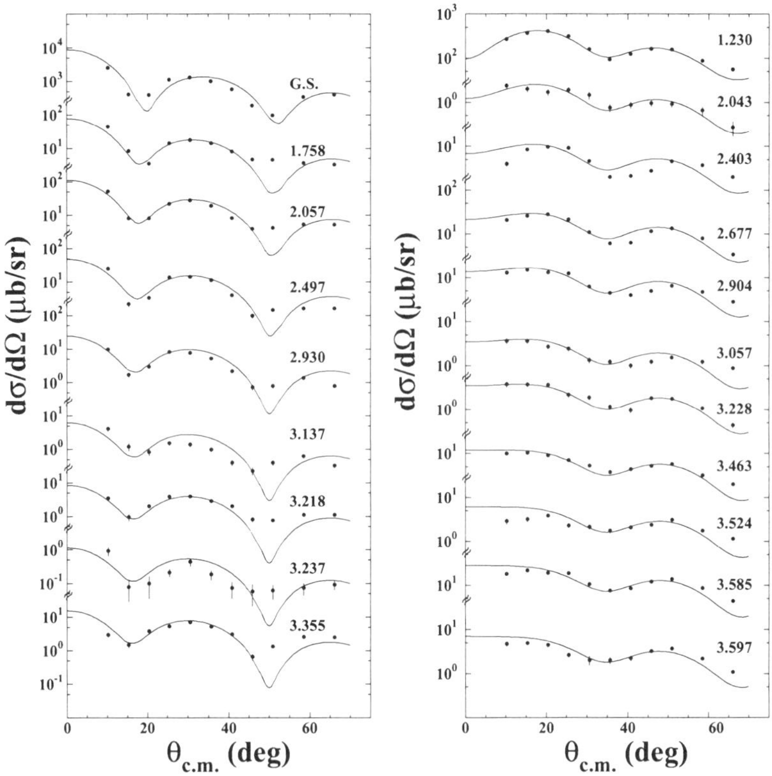

Fig. 1. Cluster DWBA calculations for the 0^+ (left) and 2^+ (right) states.

Spin-parity assignments have been done for all the observed levels. In particular, with respect to the adopted levels,[4] 17 confirmations, 13 new assignments have been made and 4 ambiguities removed. Two unresolved doublets and one triplet have been observed, giving 6 confirmations, and 1 new assignment.

4. Theoretical analysis

4.1. *Shell model calculation*

Our shell model calculation for ^{118}Sn has been performed assuming ^{132}Sn as a closed core, with the 14 valence neutron holes occupying the five levels $0g_{7/2}$, $1d_{5/2}$, $1d_{3/2}$, $2s_{1/2}$, and $0h_{11/2}$ of the 50-82 shell.

The two body effective interaction has been derived from the CD-Bonn

NN potential,[5] the short-range repulsion of the latter being renormalized by integrating out the high-momentum modes down to a cutoff momentum Λ. This procedure leads to a low-momentum potential $V_{\rm low-k}$[8] which can be used directly as input for the calculation of the effective interaction $V_{\rm eff}$ within a folded-diagram method.[9] A brief outline of our derivation of $V_{\rm eff}$ is as follows. We first construct the $V_{\rm low-k}$ using a cutoff momentum $\Lambda = 2.1\,{\rm fm}^{-1}$ and then calculate the so-called $\hat{Q}$-box including diagrams up to second order in $V_{\rm low-k}$. The computation of these diagrams is performed by using an harmonic-oscillator basis with $\hbar\omega = 7.88$ MeV and inserting intermediate states composed of particle and hole states restricted to the two major shells above and below the $Z = 50$, $N = 82$ Fermi surface. Finally, the effective interaction is obtained by summing up the $\hat{Q}$-box folded diagram series by means of the Lee-Suzuki iteration method.[10] It is worth noting that in the present calculation, where ^{118}Sn is described as 14 neutron holes with respect to ^{132}Sn, the matrix elements of the hole-hole effective interaction are needed. In this case, the calculation of the $\hat{Q}$-box diagrams is somewhat different from that for particles. A description of the derivation of the hole-hole effective interaction is given in Ref. [11].

Our adopted values for the single-hole (SH) energies are (in MeV): $\epsilon^{-1}_{g_{7/2}} = 2.8$, $\epsilon^{-1}_{d_{5/2}} = 2.155$, $\epsilon^{-1}_{s_{1/2}} = 0.85$, $\epsilon^{-1}_{d_{3/2}} = 1.2$, and $\epsilon^{-1}_{h_{11/2}} = 0.0$. They have been determined by reproducing the experimental yrast states in ^{119}Sn which have angular momentum and parity corresponding to the SH levels.

The energy matrices are set up and diagonalized by means of an approach which makes use of the seniority scheme and is based on a chain calculation across nuclei differing by two in nucleon number. A description of this method, which we call chain-calculation method (CCM), can be found in Refs. [1,12]. Consistently with our previous study of ^{114}Sn and ^{120}Sn, we have included here states with seniority $v \leq 4$.

In Table 2, the excitation energies measured in the present experiment are compared with the calculated values. We have excluded the 3^- and 1^- states, for which the discrepancies between experiment and theory are well above 1 MeV. The description of these states seems to be beyond the scope of the present calculation. This is likely to be due to the significant role played by configurations outside the chosen model space. However, also for several other states, as we see from Table 2, the agreement between theory and experiment cannot be considered completely satisfactory. The discrepancies range from few tens of keV to about 1 MeV, the measured values being in almost all cases overestimated by the theory. The only

Table 2. Comparison of calculated energies with those obtained from the present experiment. See text for details.

J^π	E_{calc} (MeV)	E_{expt} (MeV)	J^π	E_{calc} (MeV)	E_{expt} (MeV)
0^+	0.000	0.000	4^+	2.564	2.280
	1.805	1.758		2.855	2.489
	2.090	2.057		3.198	2.734
	3.077	2.497			
	3.385	2.930	6^+	2.900	3.559
	3.700	3.137			
	3.946	3.218	5^-	2.427	2.323
	4.215	3.237		2.728	2.879
	4.262	3.355		3.218	2.963
				3.388	3.375
2^+	1.641	1.230		3.720	3.395
	2.483	2.043			
	2.502	2.323	7^-	2.546	2.575
	2.895	2.403		3.245	3.108
	3.181	2.677			
	3.275	2.904			
	3.449	3.057			
	3.633	3.228			
	3.799	3.309			
	3.938	3.463			
	4.039	3.524			
	4.228	3.585			
	4.201	3.597			

exceptions are the first 7^- state, which is predicted however to lie at ~ 30 keV below its experimental counterpart, and the first 6^+ state at 2.900 MeV to be compared with the experimental energy 3.559 MeV. In this connection, it should be mentioned that a 6^+ state at 2.999 MeV is reported in [4]. As a general remark, we may recall that a down-shift of the calculated levels is generally produced by the inclusion of $v > 4$ components.

4.2. *Microscopic DWBA calculations*

The microscopic calculation of the (p,t) transfer has been also done with the reaction code TWOFNR.[7] We use the wave functions generated in the shell model calculations of Sec. 4.1 for the ^{120}Sn ground state and for each ^{118}Sn final state and the simplest one-step transfer theory.[13,14] The wave function of the pair of transferred neutrons is:

$$\sum_{n_1,\ell_1,j_1;n_2,\ell_2,j_2} S^J_{n_1,\ell_1,j_1;n_2,\ell_2,j_2} \left\{ \begin{array}{c} \frac{\left[\psi^{n_1,\ell_1,j_1}(\mathbf{r}_1,\sigma_1)\psi^{n_2,\ell_2,j_2}(\mathbf{r}_2,\sigma_2)\right]^J_M}{2(1+\delta_{n_1,n_2}\delta_{\ell_1,\ell_2}\delta_{j_1,j_2})} \\ - \frac{\left[\psi^{n_1,\ell_1,j_1}(\mathbf{r}_2,\sigma_2)\psi^{n_2,\ell_2,j_2}(\mathbf{r}_1,\sigma_1)\right]^J_M}{2(1+\delta_{n_1,n_2}\delta_{\ell_1,\ell_2}\delta_{j_1,j_2})} \end{array} \right\} \quad (1)$$

The spectroscopic amplitudes, $S^J_{n_1,\ell_1,j_1;n_2,\ell_2,j_2}$, are calculated from the target and residual wave-functions. To calculate the form-factor for the reaction, it is necessary to extract that part of the above two-neutron wave-function in which the neutrons have the same relative and spin wave function that they have in the outgoing triton. The zero-range one-step DWBA approach used here should give an accurate description of the shapes of the angular distributions, and the relative cross sections for different residual states of different angular momenta, but it is unable to calculate their absolute cross sections. Therefore a single multiplicative factor must be chosen before the DWBA output can be compared to experimental data. We have chosen this factor to produce the best visual fit with the ground state cross section, as shown in Fig. 2. The (p,t) cross section for the excitation of the lowest 2^+ state is, however, overpredicted by the theory and the 2^+_1 angular distribution has to be scaled down by a factor 2. This discrepancy shows that there are properties of the ^{120}Sn and ^{118}Sn wave functions that are not adequately described by the shell model and illustrates the power of the (p,t) reaction in revealing details of nuclear structure.

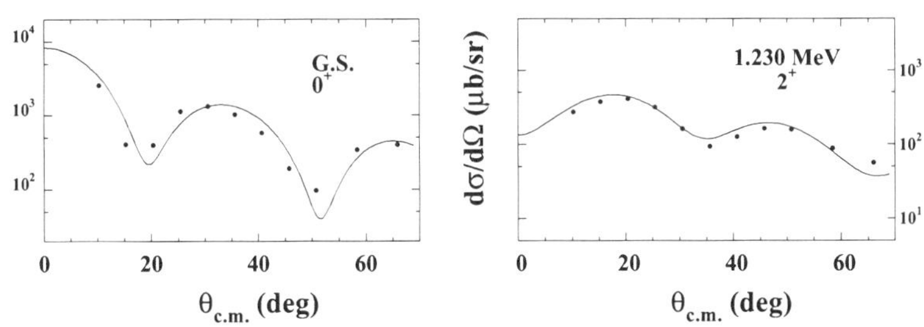

Fig. 2. Microscopic DWBA calculations for the G.S. and lowest 2^+ state.

5. Summary

The ^{120}Sn$(p,t)^{118}$Sn reaction has been studied in a high resolution experiment at 21 MeV. Angular distributions for 37 transitions to ^{118}Sn levels up to $E_x = 3.597$ MeV have been measured. Spin and parity assignments have been done for all the observed levels, thanks to a finite range DWBA analysis performed in the framework of a semimicroscopic dineutron cluster pickup mechanism.

Zero-range one-step DWBA microscopic calculations of differential cross sections for the ground state and first excited 2^+ state of ^{118}Sn have been performed using two-neutron spectroscopic amplitudes obtained from a shell-model study of ^{120}Sn and ^{118}Sn. The shell-model calculations have been carried out using a realistic effective interaction derived from the CD-Bonn nucleon-nucleon potential.

We have reported here some preliminary results of this study. Our final results and a detailed comparison with the experimental data will be presented in a forthcoming publication.

References

1. P. Guazzoni, M. Jaskóla, L. Zetta, A. Covello, A. Gargano, Y. Eisermann, G. Graw, R. Hertenberger A. Metz, F. Nuoffer, and G. Staudt, *Phys. Rev. C* **60**, 054603 (1999).
2. P. Guazzoni, L. Zetta, A. Covello, A. Gargano, G. Graw, R. Hertenberger, H.-F. Wirth, and M. Jaskóla, *Phys. Rev. C* **69**, 024619 (2004).
3. P. Guazzoni, L. Zetta, A. Covello, A. Gargano,B. F. Bayman, G. Graw, R. Hertenberger, H.-F. Wirth, and M. Jaskóla, *Phys. Rev. C* **74**, 054605 (2006).
4. K. Kitao, *Nuclear Data Sheets* **75**, 99 (1995).
5. R. Machleidt, *Phys. Rev. C* **63**, 024001 (2001).
6. E. Zanotti, M. Bisenberger, R. Hertenberger, H. Kader, G. Graw, *Nucl. Instrum. Methods Phys. Res. A* **310**, 706 (1991).
7. M. Igarashi, computer code TWOFNR (1977) unpublished.
8. Scott Bogner, T. T. S. Kuo, L. Coraggio, A. Covello, and N. Itaco, *Phys. Rev. C* **65**, 051301(R) (2002).
9. T. T. S. Kuo, S. Y. Lee, and K. F. Ratcliff, *Nucl. Phys. A* **176**, 62 (1971).
10. K. Suzuki and S. Y. Lee, *Prog. Theor. Phys.* **64**, 2091 (1980).
11. L. Coraggio.A. Covello, A. Gargano, N. Itaco, and T. T. S. Kuo, *J. Phys. G* **26**, 1697 (2000).
12. A.Covello, F. Andreozzi, L. Coraggio, A. Gargano, A. Porrino, in *Contemporary Nuclear Shell Models*, Lecture Notes in Physics Vol. 482 (Springer-Verlag, Berlin,1997).
13. B. F. Bayman and A. Kallio, *Phys. Rev.* **156**, 1126 (1967).
14. D. N. Mihailidis, N. M. Hintz, A. Sethi, E. J. Stephenson, *Phys. Rev. C* **64**, 054608 (2001).

SECTION IV

COLLECTIVE MODES OF NUCLEAR EXCITATION

ISOVECTOR VALENCE SHELL EXCITATIONS IN VIBRATIONAL NUCLEI

N. PIETRALLA*, T. AHN, O. BURDA

Institut für Kernphysik, Technische Universität Darmstadt,
64289 Darmstadt, Germany
**E-mail: pietralla@ikp.tu-darmstadt.de*

G. RAINOVSKI

St. Kliment Ohridski University of Sofia,
1164 Sofia, Bulgaria

Mixed-symmetry states have been studied in various vibrational nuclei by electromagnetic probes using the techniques of electron scattering and Coulomb excitation. An inelastic electron scattering experiment on ^{94}Mo has been performed at the Darmstadt S-DALINAC. The technique is particularly sensitive to nuclear one-phonon excitations and shown to be a valuable tool for investigating one- and two-phonon states with mixed proton-neutron symmetry. A series of Coulomb excitation experiments has been performed at Argonne National Laboratory. Absolute transition strengths in ^{134}Xe are derived. The main fragment of the one-phonon $2^+_{1,\mathrm{ms}}$ mixed symmetry state has been identified from the $2^+ \rightarrow 2^+_1$ $M1$ strength distribution.

Keywords: mixed-symmetry states, electron scattering, Coulomb excitation

1. Introduction

The proton-neutron interaction in the nuclear valence shell has been known for a long time as the driving force for the evolution of nuclear structure. This has been discussed in many ways, e.g. by Casten[1] in terms of the evolution of collectivity as a function of the product of valence proton and neutron numbers N_pN_n. More recently Otsuka et al.[2] have identified the proton-neutron interaction as being responsible for the evolution of shell structure. Therefore, it is interesting to study those nuclear excitations that are most sensitive to the proton-neutron interaction in the valence shell. One class of examples are collective isovector valence shell excitations that are frequently called mixed-symmetry states (MSSs) in the terminology of the interacting boson model (IBM).

In the sd-IBM-2[3] pairs of valence protons and neutrons are approximated as bosons that can have angular momentum $L = 0$ or $L = 2$. The bosons are treated as elementary particles that can be either proton bosons or neutron bosons. The concept of F spin[4,5] was introduced to quantify the symmetry of multiboson wave functions formed by N_π (N_ν) proton (neutron) bosons with respect to the pairwise exchange of boson labels. Proton bosons are assigned an F spin of $F = 1/2$ with an F spin projection of $F_z = +1/2$ while neutron bosons are assigned a F spin value of $F = 1/2$ with an F spin projection of $F_z = -1/2$. The concept of F spin for bosons in the IBM-2 is analogous to the concept of isospin for nucleons.

Boson states with F spin $F = F_{\text{max}} = (N_\pi + N_\nu)/2$ are called full-symmetry states (FSSs). These states are symmetric with respect to the pairwise interchange of proton and neutron labels. States with $F < F_{\text{max}}$ are called mixed-symmetry states and contain antisymmetric parts with respect to the pairwise interchange of proton and neutron labels.[6] It is desirable to study the properties of mixed-symmetry states since their isovector character can yield information on the proton-neutron interaction in the valence shell of nuclei.

One of the properties of FSSs is that $M1$ transitions are forbidden between them.[7] The antisymmetric nature of the wavefunction of MSSs allows for $M1$ transitions connecting MSSs to FSSs to exist, the transition matrix element being on the order of 1 μ_N. The presence of an M1 transition between these states can be used for identifying MSSs experimentally.

The first observation of a nuclear MSS was made by Richter and his group in electron scattering experiments on the deformed nucleus ^{156}Gd at Darmstadt in 1983.[8] A strong $M1$ excitation to a 1^+ state close to 3 MeV excitation energy, the scissors mode, was observed. The scissors mode has subsequently been studied mainly in electron scattering and photon scattering experiments on deformed nuclei.[9,10] For these two techniques the population mechanism from the 0^+ ground state to the 1^+ scissors mode coincides with the signature for mixed-symmetry character, i.e. the strong $M1$ transition. Systematic data are available for the deformed nuclei in the mass region of the rare-earth elements.[11,12]

In order to obtain a microscopic understanding of the formation of MS structures, the deformed nuclei are less favorable due to their large valence spaces than vibrational nuclei near shell closures. Therefore, there was recently a surge of interest in studying mixed-symmetry structures in vibrational nuclei.[13]

Experimentally MSSs have been identified in a number of vibrational nuclei predominantly in the mass regions $A = 90$ and $A = 130$, e.g. Refs.[14–23] The best studied example is ^{94}Mo. One and two phonon MSSs in ^{94}Mo have been identified from large absolute $M1$ transition strengths obtained in a variety of experiments.[15–17,20,23] Experimental procedures for identifying MSSs of vibrational nuclei differ from those for deformed nuclei because in vibrational nuclei the population mechanism to MSSs does not coincide with the identifying large $M1$ transition. It is interesting to identify experimental methods that are most sensitive to MSSs of vibrational nuclei. A method, applicable to radioactive nuclei that can be made available as ion beams, would be most useful.

In this contribution we will report on our recent experiments on MSSs of vibrational nuclei using electron scattering and Coulomb excitation. An electron scattering experiment was done at the S-DALINAC facility at the Technische Universität Darmstadt to measure form factors for states in ^{94}Mo. The knowledge of the form factors as a function of momentum transfer represents complementary information to γ-ray spectroscopy in investigating the wavefunctions of the various states. Projectile Coulomb excitation is another technique for identifying and investigating the main fragments of the one-phonon MSS. The potential of this method has recently been demonstrated in an experiment on ^{138}Ce.[24] In a following series of projectile Coulomb excitation experiments, the same technique was used to study all stable open-shell nuclei of the Ce ($Z = 58$) and Xe ($Z = 54$) isotopic chains. Below we present the data that allow us to identify the main fragment of the one-phonon MSS in ^{134}Xe. The preliminary results for the identification of the $2^+_{1,\mathrm{ms}}$ state coming from the measurement of relative Coulomb excitation cross sections and angular distributions of γ rays will be presented.

2. Experiment

In this section we will present the observables from electron scattering and projectile Coulomb excitation and comment on their sensitivity to the one-quadrupole phonon MSS, the $2^+_{1,\mathrm{ms}}$ state.

2.1. *Electron scattering*

An electron scattering experiment was carried out with the Lintott spectrometer[25] at the S-DALINAC facility at TU Darmstadt. An electron beam with an energy of 70 MeV and beam intensity of 2 μA was shot

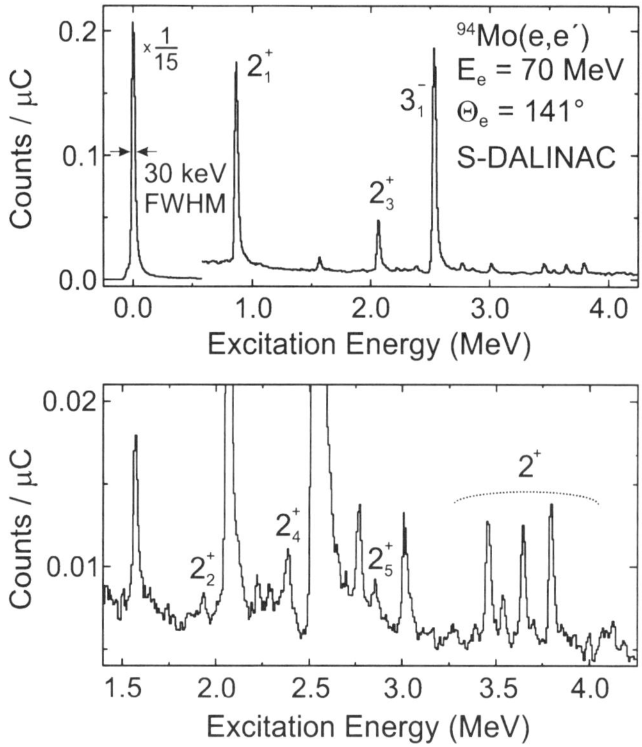

Fig. 1. An spectrum is shown for scattered electrons from ^{94}Mo at a scattering angle $\theta = 141°$ and an incident electron beam energy of 70 MeV. The upper spectrum shows the full measured energy range from 0 to 4 MeV. The lower spectrum shows the energy range from 1.5 MeV to 4 MeV showing the lower intensity peaks. Data from Ref.[26]

on a 9.7 mg/cm^2 thick ^{94}Mo target. Data were taken for scattering angles $\theta = 93° - 141°$. Using the dispersion matching mode, an energy resolution of 30 keV was achieved.

The energy spectrum of the scattered electrons for a scattering angle of $\theta = 141°$ is shown in Fig. 1. At the top of Fig. 1, the whole range of measured energies from 0 to 4 MeV can be seen. The lowest energy peak corresponds to the elastically scattered electrons. It is suppressed by a factor of 15 for better visibility of the other peaks. The peaks for inelastically scattered electrons corresponding to the excitations of the 2^+_1, 2^+_3, and 3^-_1 states can also be seen as very pronounced peaks. The 2^+_3 state of ^{94}Mo has previously been identified as the $2^+_{1,\mathrm{ms}}$ isovector one-quadrupole phonon excitation of

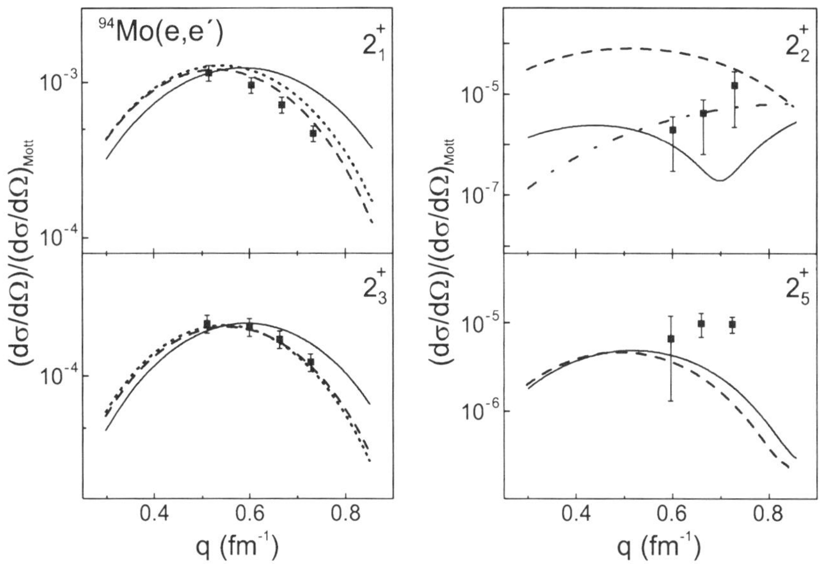

Fig. 2. Form factors for 2^+ states of ^{94}Mo from measured differential cross sections. The left column shows the mainly one-phonon 2_1^+ and 2_3^+ states and the right column the mainly two-phonon 2_2^+ and 2_5^+ states. Fits using the IBM-2, shell model with $V_{\mathrm{low-k}}$, and QPM are shown in the dotted, dashed, and solid lines, respectively. From Ref.[26]

the valence shell.[15] It is obvious that electron scattering is well suited for making precision studies for this class of states. In the lower figure, the smaller, but still clearly visible peaks, corresponding to the excitation of higher lying 2^+ states, are shown with the peaks corresponding to known 2^+ states labeled.

By measuring the electron scattering intensities one can measure the differential cross section as a function of momentum transfer and by dividing out the Mott cross section, one obtains the form factor. The form factors for one and two phonon 2^+ states in ^{94}Mo calculated from measured differential cross sections are shown in Fig. 2. Fits for the form factors have been done with the IBM-2, shell model using the low-momentum renormalized interaction $V_{\mathrm{low-k}}$,[27] and the quasiparticle phonon model (QPM), e.g. Ref. [28] They are also shown in Fig. 2. It can be seen that the model fits are able to describe the measured form factors for the one-phonon states quite well. The similarity of the form factors for the 2_1^+ and 2_3^+ state of ^{94}Mo presents further evidence that the latter is a one-phonon state, here of isovector character instead of isoscalar character like the 2_1^+ state.

2.2. *Coulomb excitation*

Coulomb excitation is a well suited technique for investigating low-lying collective states in nuclei. We have used the technique of Coulomb excitation in inverse kinematics for identifying the main fragment of the one-quadrupole phonon MSS, the $2^+_{1,\mathrm{ms}}$ state, of ^{134}Xe. The experiment was performed at Argonne National Laboratory using a beam of ^{134}Xe ions accelerated by the ATLAS heavy-ion accelerator to an energy of 435 MeV, which corresponds to 82% of the Coulomb barrier. The ions impinged on a 1.0 mg/cm^2 thick ^{12}C target. Gamma rays emitted from the deexcitation of nuclei were detected with the Gammasphere germanium detector array, which consisted of 101 high-purity germanium detectors. Gamma rays with 1-fold multiplicity and higher were recorded. A spectrum obtained by requiring a coincidence with the $2^+_1 \to 0^+_1$ transition is shown in Fig. 3. The coincidence intensities vary by about four orders of magnitude. The peaks in the spectrum corresponding to γ rays for transitions from the second, third, and fourth 2^+ states to the 2^+_1 state can be seen. We draw attention to the pronounced peak from the $2^+_3 \to 2^+_1$ transition. This is the identifying $M1$ transition for the $2^+_{1,\mathrm{ms}}$ state of ^{134}Xe as is discussed below.

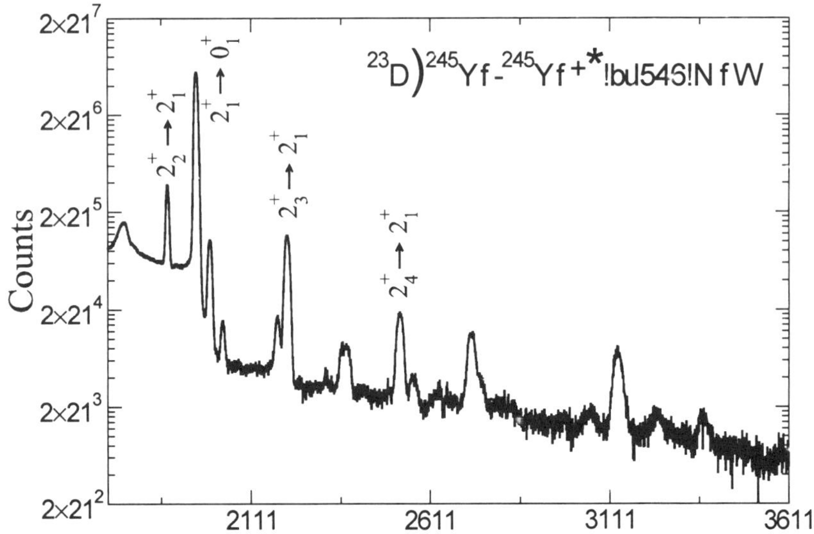

Fig. 3. Gamma-ray coincidence spectrum of a gate on the $2^+_1 \to 0^+_1$ transition in ^{134}Xe. Data taken at Argonne National Laboratory using the Gammasphere detector array.

For the data analysis, detectors located at the same polar angle θ with respect to the beam axis were grouped into rings. That way Gammasphere was subdivided into 16 different rings. The peak areas for the spectra belonging to each ring were measured, which allowed one to measure the angular distribution of γ rays emitted in the reaction. The measured intensities were corrected for efficiency as well as relativistic effects. Relative Coulomb excitation cross sections can be measured from the total γ-ray yield coming from the transitions of a given state. Using the known value $B(E2; 0_1^+ \rightarrow 2_1^+) = 15.3(11)$ W.u., one can derive the absolute $B(E2)$ ground state excitation strengths to the other observed 2^+ states. This procedure yields absolute lifetimes for the observed $2_{2,3,4}^+$ states. In order to deduce the $2^+ \rightarrow 2_1^+$ $M1$ strength distribution we have to determine the multipolarities of these transitions. The alignment of the initial states were deduced from the measured angular distributions for the $E2$ ground state decays. From an initial state's alignment one can deduce the multipole mixing ratio from the angular distribution of a possibly mixed transition,

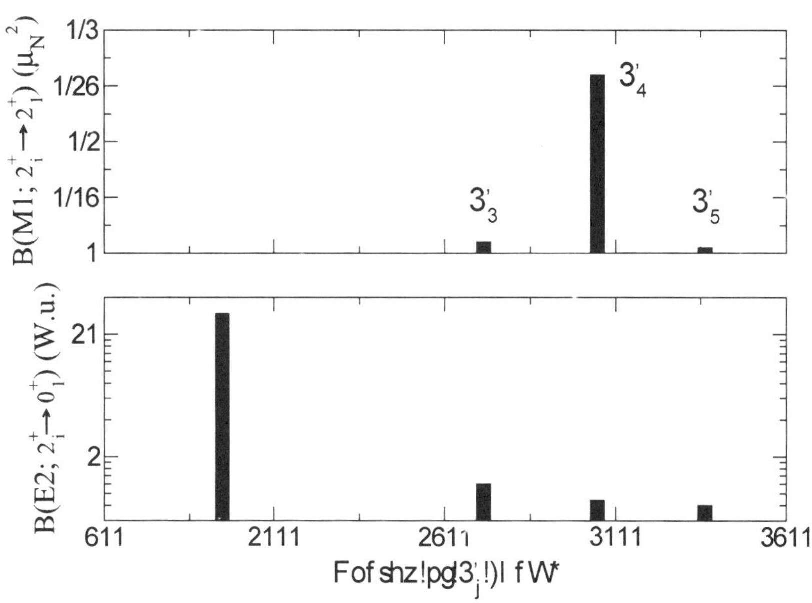

Fig. 4. Preliminary transition strength distributions for ^{134}Xe showing measured $M1$ transition strengths in the upper plot and $E2$ transition strengths in the lower. The relative uncertainties on the data points are estimated to be on the order of 20% at the present stage of the data analysis.

e.g. deduce the $E2/M1$ multipole mixing ratio for a $2^+ \rightarrow 2_1^+$ transition.

By knowing the $B(E2; 0_1^+ \rightarrow 2_i^+)$ values from the Coulomb excitation cross sections, measured multipole mixing ratios, and branching ratios, the remaining $B(E2; 2_i^+ \rightarrow 2_1^+)$ and $B(M1; 2_i^+ \rightarrow 2_1^+)$ transition strengths can be calculated. The preliminary results for the transition strengths in ^{134}Xe are shown in Fig. 4. We can identify the 2_3^+ state as the main fragment of the $2_{1,ms}^+$ state from the large value for the $M1$ transition strength connecting it to the 2_1^+ state. Also the $E2$ decay of the 2_3^+ state to the 0_1^+ ground state is weakly collective with a decay strength of approximately 0.4 W.u.

3. Discussion

We would like to confront the new data on ^{134}Xe with data obtained previously from experiments on neighboring nuclei. The $2_{1,ms}^+$ state has been measured in two other $N = 80$ isotones, namely ^{136}Ba[14] and ^{138}Ce.[24] A plot of energies of the 2_1^+ states and known one-phonon $2_{1,ms}^+$ states for a series of $N = 80$ isotones is shown in Fig. 5. It is interesting to observe that the $2_{1,ms}^+$ state increases in energy as one moves to higher proton number. In contrast, the energy of the 2_1^+ state decreases as a function of proton number moving away from the $Z = 50$ closed shell. Further distance from the $Z = 50$ shell closure results in larger valence spaces and therefore in an increase in collectivity. The separation in energy of the $2_{1,ms}^+$ state and the 2_1^+ state can give us a measure of the proton-neutron quadrupole-quadrupole interaction in the valence shell.[29]

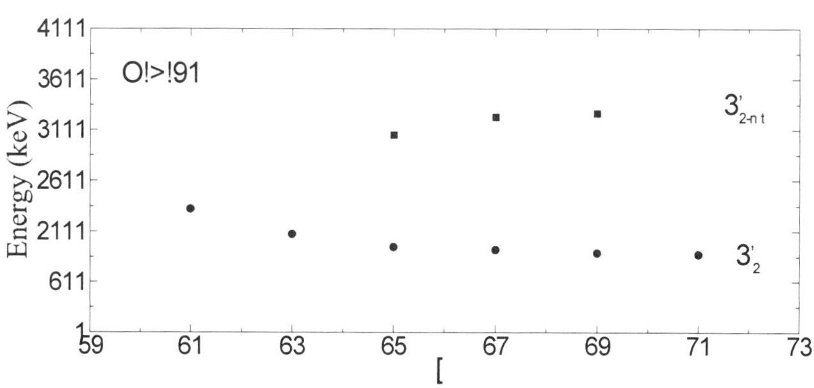

Fig. 5. The energies of the 2_1^+ states as well as the currently known $2_{1,ms}^+$ states for the $N = 80$ isotones are plotted as a function of proton number Z.

4. Outlook

The data presented above show the capability of electron scattering and projectile Coulomb excitation for contributing to the investigation of the one-quadrupole phonon $2^+_{1,\mathrm{ms}}$ state of vibrational nuclei. Form factors from electron scattering on the $2^+_{1,\mathrm{ms}}$ state are sensitive not only to the $E2$ excitation strength, but also to the one-phonon character of the state. It will be interesting to apply this method to the $N = 52$ isotone ^{92}Zr, where the $2^+_{1,\mathrm{ms}}$ state is known to be isospin-polarized in favor of proton contributions[27] and to mix with the two-phonon structure.[22]

Though MSSs have been studied in a number of stable nuclei, there has been no firm identification of a MSS in an unstable nucleus. The technique of Coulomb excitations in inverse kinematics holds great promise for measuring transition strengths for the identification of MSSs in unstable nuclei as well as in nuclei where natural abundances are very small.

Acknowledgments

We would like to thank R.V.F. Janssens, C.J. Lister, M.P. Carpenter, and S. Zhu at Argonne National Laboratory for their collaboration on the Coulomb excitation experiment. We also thank P. von Neumann-Cosel and A. Richter for their contributions to the electron scattering experiment on the nucleus ^{94}Mo at Darmstadt. This work was supported by the DFG under grants SFB 634 and Pi 393/2-1.

References

1. R.F. Casten, Nucl. Phys. A **443**, 1 (1985).
2. T. Otsuka, T. Matsuo, and D. Abe, Phys. Rev. Lett. **97**, 162501 (2006).
3. F. Iachello and A. Arima, *The interacting boson model*, (Cambridge Univ. Press, Cambridge, 1987).
4. A. Arima, T. Otsuka, F. Iachello, and I. Talmi, Phys. Lett. B **66**, 205 (1977).
5. T. Otsuka, A. Arima, and F. Iachello, Nucl. Phys. A **309**, 1 (1978).
6. F. Iachello, Nucl. Phys. A **358**, 89c (1981); Phys. Rev. Lett. **53**, 1427 (1984).
7. P. Van Isacker, K. Heyde, J. Jolie, and A. Sevrin, Ann. Phys. (NY) **171**, 253 (1986).
8. D. Bohle, A. Richter, W. Steffen, A.E.L. Dieperink, N. LoIudice, F. Palumbo, and O. Scholten, Phys. Lett. **B137**, 27 (1984).
9. A. Richter, Prog. Part. Nucl. Phys. **34**, 261 (1995).
10. U. Kneissl, H.H. Pitz, and A. Zilges, Prog. Part. Nucl. Phys. **37**, 349 (1996).
11. N. Pietralla, P. von Brentano, R.-D. Herzberg, U. Kneissl, N. LoIudice, H. Maser, H.H. Pitz, and A. Zilges, Phys. Rev. C **58**, 184 (1998).
12. J. Enders, H. Kaiser, P. von Neumann-Cosel, C. Rangacharyulu, A. Richter, Phys. Rev. C **59**, R1851 (1999).

13. N. Pietralla, P. von Brentano, A. Lisetskiy, Prog. Part. Nucl. Phys., *submitted for publication.*
14. N. Pietralla, D. Belic, P. von Brentano, C. Fransen, R.-D. Herzberg, U. Kneissl, H. Maser, P. Matschinsky, A. Nord, T. Otsuka, H.H. Pitz, V. Werner, and I. Wiedenhöver, Phys. Rev. C **58**, 796 (1998).
15. N. Pietralla, C. Fransen, D. Belic, P. von Brentano, C. Frießner, U. Kneissl, A. Linnemann, A. Nord, H.H. Pitz, T. Otsuka, I. Schneider, V. Werner, and I. Wiedenhöver, Phys. Rev. Lett. **83**, 1303 (1999).
16. N. Pietralla, C. Fransen, P. von Brentano, A. Dewald, A. Fitzler, C. Frießner, and J. Gableske, Phys. Rev. Lett. **84**, 3775 (2000).
17. C. Fransen, N. Pietralla, P. von Brentano, A. Dewald, J. Gableske, A. Gade, A. Lisetskiy, and V. Werner, Phys. Lett. B **508**, 219 (2001).
18. N. Pietralla, C.J. Barton III., R. Krücken, C.W. Beausang, M.A. Caprio, R.F. Casten, J.R. Cooper, A.A. Hecht, H. Newman, J.R. Novak, and N.V. Zamfir, Phys. Rev. C **64**, 031301 (2001).
19. V. Werner, D. Belic, P. von Brentano, C. Fransen, A. Gade, H. von Garrel, J. Jolie, U. Kneissl, C. Kohstall, A. Linnemann, A.F. Lisetskiy, N. Pietralla, H.-H. Pitz, M. Scheck, K.-H. Speidel, F. Stedile, and S.W. Yates, Phys. Lett. B **550**, 140 (2002).
20. C. Fransen, N. Pietralla, Z. Ammar, D. Bandyopadhyay, N. Boukharouba, P. von Brentano, A. Dewald, J. Gableske, A. Gade, J. Jolie, U. Kneissl, S.R. Lesher, A.F. Lisetskiy, M.T. McEllistrem, M. Merrick, H.H. Pitz, N. Warr, V. Werner, and S.W. Yates, Phys. Rev. C **67**, 024307 (2003).
21. C. Fransen, N. Pietralla, A.P. Tonchev, M.W. Ahmed, J. Chen, G. Feldman, U. Kneissl, J. Li, V.N. Litvinenko, B. Perdue, I.V. Pinayev, H.-H. Pitz, R. Prior, K. Sabourov, M. Spraker, W. Tornow, H.R. Weller, V. Werner, Y.K. Wu, and S.W. Yates, Phys. Rev. C **70**, 044317 (2004).
22. C. Fransen, V. Werner, D. Bandyopadhyay, N. Boukharouba, S.R. Lesher, M.T. McEllistrem, J. Jolie, N. Pietralla, P. von Brentano, and S.W. Yates, Phys. Rev. C **71**, 054304 (2005).
23. P. von Neumann-Cosel, N. T. Botha, O. Burda, J. Carter, R. W. Fearick, S. V. Förtsch, C. Fransen, H. Fujita, M. Kuhar, A. Lenhardt, R. Neveling, N. Pietralla, V. Yu. Ponomarev, A. Richter, E. Sideras-Haddad, F. D. Smit, and J. Wambach, AIP Conf. Proc. **819**, 611 (2006).
24. G. Rainovski, N. Pietralla, T. Ahn, C.J. Lister, R. V. F. Janssens, M. P. Carpenter, S. Zhu, and C. J. Barton III, Phys. Rev. Lett. **96**, 122501 (2006).
25. A.W. Lenhardt, U. Bonnes, O. Burda, P. von Neumann-Cosel, M. Platz, A. Richter, and S. Watzlawik, Nucl. Instr. Meth. A **562**, 320 (2006).
26. O. Burda et al., *submitted for publication.*
27. J.D. Holt, N. Pietralla, J.W. Holt, T.T.S. Kuo, and G. Rainovski, *submitted for publication.*
28. N. Lo Iudice and Ch. Stoyanov, Phys. Rev. C **62**, 047302 (2000).
29. K. Heyde and J. Sau, Phys. Rev. C **33**, 1050 (1986).

SOFT ELECTRIC DIPOLE MODES IN HEAVY NUCLEI: SOME SELECTED EXAMPLES

P. VON NEUMANN-COSEL

Institut für Kernphysik, Technische Universität Darmstadt,
Schlossgartenstr. 9, 64289 Darmstadt, Germany
**E-mail: vnc@ikp.tu-darmstadt.de*

Two examples of our recent work on the structure of low-energy electric dipole modes are presented. The first part discusses attempts to unravel the structure of the pygmy dipole resonance (PDR) and possible evidence for a vortex-type electric dipole mode below the giant dipole resonance (GDR) in ^{208}Pb with a high-resolution measurement of the $(\vec{p},\vec{p}')$ reaction under 0°. The second part presents the systematics of the PDR in stable tin isotopes deduced from high-resolution (γ,γ') experiments and its comparison to studies of the exotic neutron-rich isotopes ^{130,132}Sn at GSI using Coulomb breakup.

Keywords: Soft dipole resonances; ^{208}Pb$(\vec{p},\vec{p}')$ at 0°; ^{112,120}Sn(γ,γ'), PDR systematics in Sn isotopes

1. Introduction

The nature of low-energy electric dipole modes is presently a focus of nuclear structure research. The first part of this contribution discusses attempts to unravel the structure of the PDR and possible evidence for a vortex-type electric dipole mode below the GDR in ^{208}Pb. Based on the knowledge of the low-energy strength distribution[1] one can investigate E1 transitions in high-resolution (p,p′) experiments under 0° by Coulomb excitation.[2] Angular distributions of the low-energy modes exhibit features distinct from the GDR. Signatures may also be found in polarization observables.

Considerable effort has been put into establishing the properties of the PDR below particle threshold using the (γ,γ') reaction, where the $Z = 20$,[3] $N = 82$[4] and $Z = 82$, $N = 128$[1] shell closures have been investigated. The second part of this contribution presents for the first time systematics for the stable $Z = 50$ tin isotopes combining recent measurements of ^{112,120}Sn at the S-DALINAC[5] with older data taken at the Gent linac.[6] These data provide a link to recent results on the exotic neutron-rich isotopes ^{130,132}Sn at GSI from Coulomb breakup.[7] The tin isotope chain is of particular interest since it serves as a reference case for theory and its systematics have been investigated in a variety of microscopic models.[8–14]

2. Nature of soft dipole modes in ^{208}Pb

2.1. *Proton scattering as an experimental tool*

The $E1$ strengths in ^{208}Pb has been extensively studied by the $^{208}\text{Pb}(\gamma,\gamma)$ reaction as well as the $^{208}\text{Pb}(\gamma,n)$ reaction for excitations above neutron threshold. It is known that a considerable amount of $E1$ strength is found well below the GDR located around the neutron threshold. The low-lying $E1$ strength distribution has been precisely studied by the $^{208}\text{Pb}(\gamma,\gamma)$ reaction at Darmstadt.[1] The distribution is well described by a microscopic calculation within the quasi-particle phonon model (QPM) including coupling up to three phonons. The close agreement of the QPM predictions with the experimental data permits an interpretation of the PDR structure to arise from an oscillation of excess neutrons relative to an approximately isospin-saturated core. Thus, the origin of the phenomenon in stable nuclei seems to be similar - although naturally less pronounced on a quantitative level - to recent observations of the PDR in exotic neutron-rich nuclei.[7,15]

Another unique feature of the $E1$ response was deduced from the velocity distributions of the transition currents.[1] While the properties of the GDR are rather consistent with the conventional description that protons and neutrons oscillate out of phase, a snapshot of the velocity distribution of low-lying states between 6.5 and 10.5 MeV shows the signature of vortex-type collective motion. Such a toroidal dipole resonance corresponds to a transverse zero-sound wave.[16–18] Besides a recently reported orbital magnetic quadrupole twist-mode[19] it may constitute another case of an excitation mode whose existence directly invalidates hydrodynamical interpretations of collective motion.

In intermediate-energy (p,p′) scattering at very forward angles, E1 transitions are prominently excited by the Coulomb interaction. Calculations of ^{208}Pb(p,p′) cross sections have been performed with the code DWBA05 for the excitation of 1^- states at $E = 295$ MeV using wave functions of the QPM calculations described in Ref.[1] Representative examples for prominent excitations in the GDR and PDR region and of a transition with toroidal character (measured through the so-called vorticity[20]) are presented in Fig. 1. In the forward angle region striking differences are observed due to their different coupling to the dominantly longitudinal virtual pho-

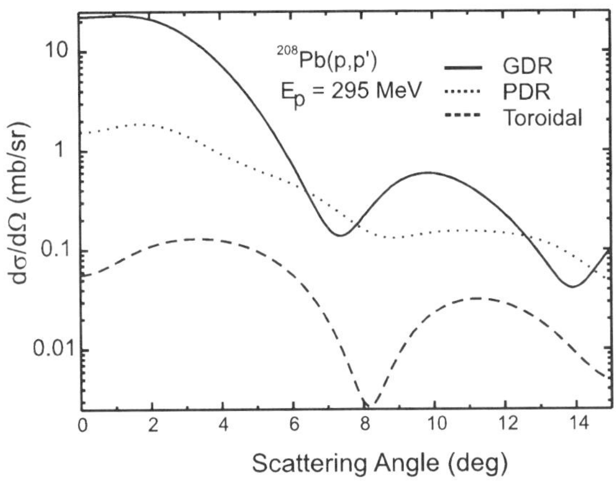

Fig. 1. DWBA predictions of angular distributions of E1 transitions with GDR, pygmy or toroidal nature excited in the ^{208}Pb(p,p$'$) reaction at $E_p = 300$ MeV.

ton field in Coulomb excitation. The effects are particularly pronounced for the toroidal mode because of its transverse nature. (However, the predicted cross sections are very small). Thus, the Coulomb-nuclear interference region provides a unique tool to test the structure of E1 transitions.

2.2. *A high-resolution ^{208}Pb($\vec{p},\vec{p}'$) experiment under 0°*

Recently, a group at RCNP Osaka has succeeded to perform inelastic proton scattering experiments under 0° combined with dispersion matching to achieve an energy resolution $\Delta E/E \leq 10^{-5}$. Examples of such data are discussed in Ref.[2] Besides Coulomb excitation, the spectra are dominated at low energies by the $\Delta L = 0$ spin-flip M1 resonance. In November 2006 a study of the ^{208}Pb($\vec{\text{p}},\vec{\text{p}}'$) reaction at $E = 295$ MeV was performed. A background-subtracted spectrum is displayed in Fig. 2. In the giant resonance region, prominent excitation of the GDR is observed. Because of the excellent energy resolution $\Delta E = 25 - 30$ keV (FWHM) pronounced fine structure is visible, a phenomenon now established as a global feature of giant resonances.[21,22]

Data have also been taken at finite scattering angles up to 10°. The angular distributions should allow a distinction between spin-flip M1 transitions excited through the spin-isospin part of the effective NN interaction and dominantly Coulomb-excited E1 transitions. Additional information from polarization variables is useful in two ways: With a complete set (e.g.

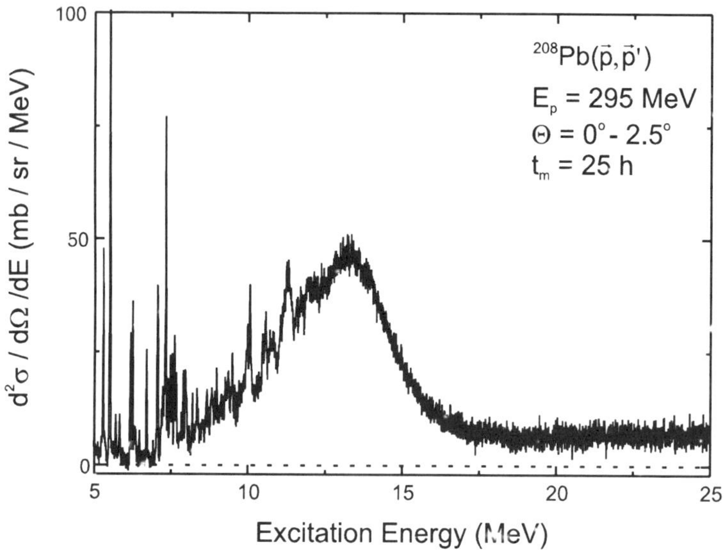

Fig. 2. Spectrum of the ^{208}Pb(p,p′) reaction measured at RCNP at $E_p = 295$ MeV and $\Theta_p = 0°$ after background subtraction.

polarization transfer coefficients D_{SS}, D_{NN}, D_{LL}) it is possible to distinguish spin-flip from non-spin-flip transitions in a model-independent way.[24] Furthermore, there may be a signature of the toroidal E1 mode in asymmetries at very forward angles.

A zoom on the low-energy region is shown in Fig. 3. The arrows indicate transitions also identified in the (γ, γ') data.[1] Essentially all prominent dipole transitions observed in the latter experiment are excited in the present work. An estimate of the cross sections for these transitions based on the semiclassical theory of Coulomb excitation[23] demonstrates that the observed cross sections at 0° are indeed due to the virtual photon interaction.

3. Systematics of the PDR in Sn isotopes

3.1. *Experiments*

Experiments on ^{112}Sn and ^{120}Sn were performed at the S-DALINAC with the nuclear resonance fluorescence (NRF) technique using electron energies of 5.5, 7.0, 9.5 MeV for ^{112}Sn and 7.5, 9.1 MeV for ^{120}Sn to generate bremsstrahlung. The maximum photon energies are below the neutron sep-

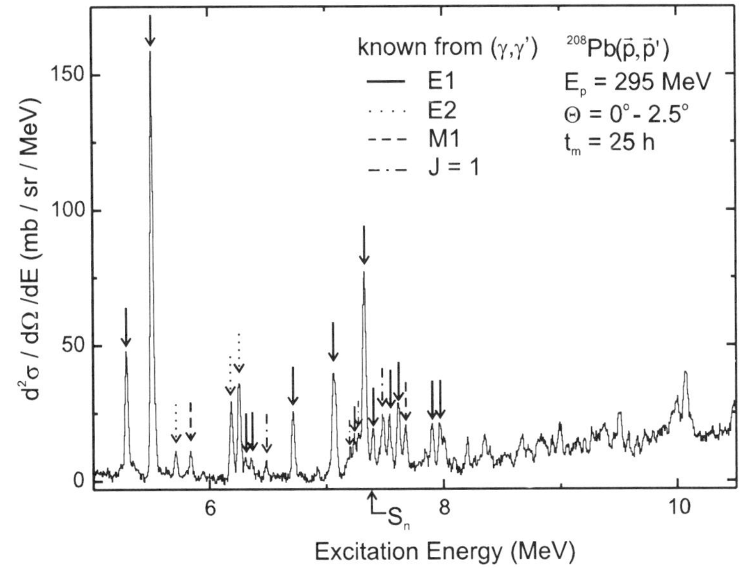

Fig. 3. Extended view of the low-energy region of Fig. 2.

aration energies of both nuclides to avoid the production of neutrons from (γ, n) reactions, which would lead to a significant increase of the background in the spectra. A detailed description of the experimental setup can be found in Ref.[25] Targets consisted of about 2 g highly enriched ($>$ 90%)) ^{112}Sn and ^{120}Sn sandwiched between two layers of boron. Well known transitions in ^{11}B were used to determine the photon flux and for the energy calibration. In Fig. 4 (γ, γ') spectra at 130° and $E_0 = 9.5$ MeV for ^{112}Sn, respectively $E_0 = 9.1$ MeV for ^{120}Sn, are shown. Significant differences are suggested by the data as the strength in ^{120}Sn seems to be much more fragmented.

3.2. *B(E1) strength*

Reduced B(E1) transition strengths were extracted for ^{112,120}Sn as explained e.g. in Ref.[26] All dipole transitions were assumed to have E1 character. Feeding effects were corrected utilizing the comparison of results obtained at different endpoint energies. In Fig. 5 the extracted B(E1) distributions are shown between 4 and 9 MeV compared to those in ^{116}Sn and ^{124}Sn measured previously by Govaert et al.[6] Note that the prominent

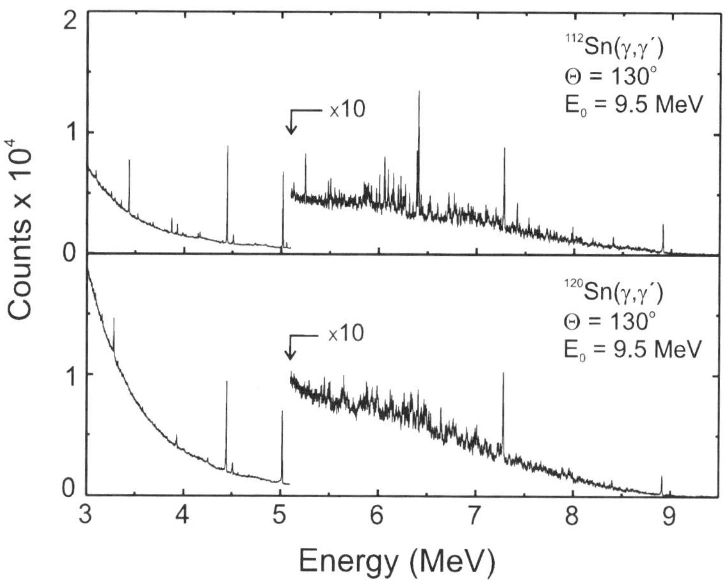

Fig. 4. Sample spectra of the ^{112,120}Sn(γ, γ') reaction measured at the S-DALINAC.

transitions resulting from the population of the two-phonon states[27,28] lie below 3.5 MeV and are therefore not shown.

All distributions show a concentration of strength between 6 and 7 MeV believed to represent the main part of the PDR. However, there is also non-negligible strength at higher energies which varies between the isotopes. The different fragmentation pattern with smaller individual strengths in ^{120}Sn already indicated by the spectrum (Fig. 4) is clearly visible.

3.3. *Systematics*

In Fig. 6 we compare the experimentally available data on the PDR in Sn isoptes with theoretical calculations for the total B(E1) strength (r.h.s.) and centroid energies (l.h.s.). The stars and open triangles show results of quasiparticle phonon model (QPM) calculations based on a Woods-Saxon ground state and a separable multipole force for the residual interaction[1] and on a self-consistent HFB approach,[9] respectively. The open circles display the predictions of a fully self-consistent relativistic QRPA approach based on the relativistic Hartree-Bogoliubov model, using an interaction

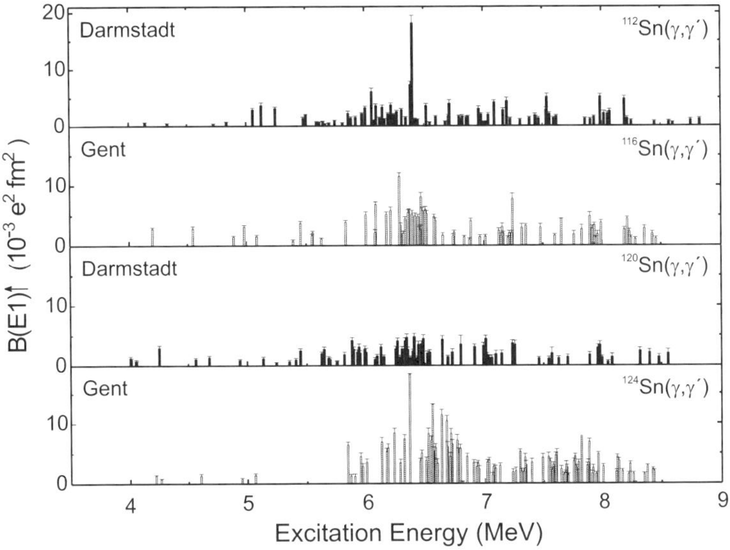

Fig. 5. B(E1) strength distributions in ^{112,116,120,124}Sn up to $E_x = 9$ MeV. The data for ^{112,120}Sn are from the present work, those for ^{116,124}Sn from.[6]

with density-dependent meson-nucleon couplings (DD-ME2).[11] The measured low-lying B(E1) strength in ^{112}Sn from the present study appears to be in rather good agreement with both the non-relativistic QPM and the relativistic QRPA. However, in heavier Sn isotopes, the results of two theoretical approaches diverge. The relativistic QRPA overpredicts the strength and centroid energies in stable isotopes but roughly agrees with the experimental results in exotic Sn isotopes studied with Coulomb dissociation,[7] while the QPM results reasonably agree for the stable isotopes but is far below the data for ^{130,132}Sn.

3.4. *Fluctuation analysis*

A statistical analysis of the PDR in $N = 82$ nuclei[29] suggests that the level density in the region of the PDR is very high. This might lead to unresolved strength from levels overlapping due to the finite energy resolution of the Ge detectors. These contributions to the E1 response can be extracted with a fluctuation analysis.[30] Application of this technique to (γ, γ') spectra is discussed e.g. in Refs.[31,32] The analysis depends on a knowledge of the 1^- level density which is experimentally unknown. Therefore, different level

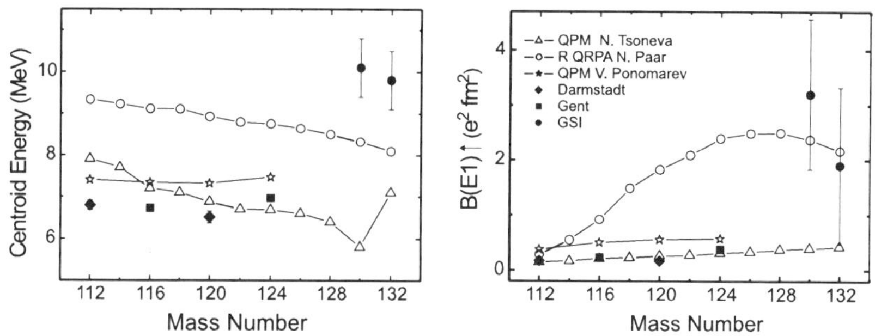

Fig. 6. Systematics of the energy centroid and summed B(E1) strength of the PDR in the Sn isotope chain including the data from Fig. 5 and[7] for ^{130,132}Sn.

density models based on the backshifted Fermi-gas approach[33,34] and microscopic HF-BCS calculations[35] have been used. Figure 7 shows the shape of the non-resonant background based on the different level density models for the case of ^{120}Sn. Results for the ^{112}Sn spectrum are quite similar.

A comparison of the analysis of discrete transitions to the present one shows an increase of the total B(E1) strength of about 35% due to unresolved levels when integrating up to E_x = 8 MeV. At higher energies uncertainties in the present method get too large because the fluctuation signal is too small. On the other hand, as demonstrated in Fig. 7, the dependence on the choice of the level density model is weak.

4. Concluding remarks

High-resolution intermediate-energy $\vec{p}, \vec{p}'$) scattering under 0° is a promising new approach to elucidate the structure of low-energy E1 transitions in stable nuclei. In particular, it may provide a way to overcome a limitation of the present experiments on the systematics of the PDR in the Sn isotope chain. Here, the comparison of the (γ, γ') results for stable nuclei and the Coulomb breakup for unstable nuclei is complicated by the fact that the former measures strength below the particle threshold only, while the latter method can be applied above threshold only. In view of the drastic differences predicted for the PDR centroid in relativistic and non-relativistic calculations it is presently unclear whether complete strength distributions of the PDR have been measured in either experiment. As demonstrated in Fig. 2 the (p,p′) reaction provides the complete response from the low-energy region to the giant resonance. However, further analysis has to show

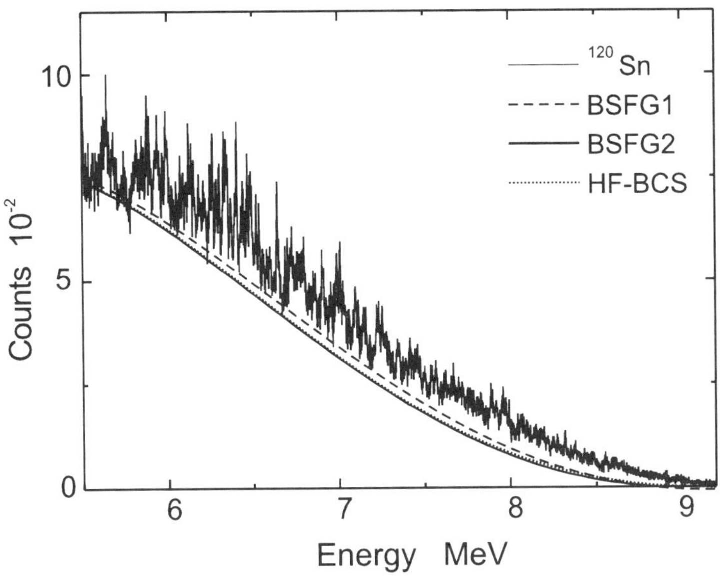

Fig. 7. Background in the ^{120}Sn(γ, γ') spectrum determined with a fluctuation analysis[30] based on the level density models of Ref.[33] (dashed line), Ref.[34] (solid line) and Ref.[35] (dotted line).

how well E1 and M1 transitions can be distinguished. An alternative will be measurement with the low-energy tagger NEPTUN at the S-DALINAC presently under construction, which is designed to cover the energies from threshold to the giant resonance region with good resolution.[36]

Acknowledgments

I am grateful for the contributions of J. Enders, Y. Fujita, H. Lenske, B. Özel, N. Paar, I. Poltoratska, V.Yu. Ponomarev, A. Richter, D. Savran, A. Tamii and N. Tsoneva to the results presented. The experiments described in Sect. 2 have been performed in a TU Darmstadt / iThemba LABS / University Kyoto / University Osaka / RCNP Osaka / Wits University collaboration. This work was supported by the DFG under contracts SFB 634 and 446 JAP 113/267/0-2.

References

1. N. Ryezayeva *et al.*, *Phys. Rev. Lett.* **89**, 272502 (2002).

2. A. Tamii *et al.*, *Nucl. Phys. A* **788**, 53c (2007).
3. T. Hartmann *et al.*, *Phys. Rev. Lett.* **85**, 274 (2000) (2000); erratum **86** 4981 (2001).
4. S. Volz *et al.*, *Nucl. Phys. A* **779**, 1 (2006).
5. B. Özel *et al.*, *Nucl. Phys. A* **788**, 385c (2007).
6. K. Govaert *et al.*, *Phys. Rev. C* **57**, 2229 (1998).
7. P. Adrich *et al.*, *Phys. Rev. Lett.* **95**, 132501 (2005).
8. D. Sarchi, P. F. Bortignon and G. Colò, *Phys.Lett. B* **601**, 27 (2004).
9. N. Tsoneva, H. Lenske and Ch. Stoyanov, *Nucl. Phys. A* **731**, 273 (2004).
10. D. Vretenar, T. Nikšić, N. Paar and P. Ring, *Nucl. Phys. A* **731**, 281 (2004).
11. N. Paar, T. Nikšić, D. Vretenar and P. Ring, *Phys. Lett. B* **606**, 288 (2005).
12. S. P. Kamerdzhiev, *Phys. At. Nucl.* **69**, 1110 (2006).
13. J. Piekarewicz, *Phys. Rev. C* **73**, 044325 (2006).
14. J. Terasaki and J. Engel, *Phys. Rev. C* **74**, 044301 (2006).
15. A. Leistenschneider *et al.*, *Phys. Rev. Lett.* **86**, 5442 (2001).
16. S. I. Bastrukov, S. Misiçu and A. V. Sushkov, *Nucl. Phys. A* **562**, 191 (1993).
17. E. B. Balbutsev, I. V. Molodtsova and A. V. Unzhakova, *Europhys. Lett.* **26**, 499 (1994).
18. S. Misiçu, *Phys. Rev. C* **73**, 024301 (2006).
19. P. von Neumann-Cosel *et al.*, *Phys. Rev. Lett.* **82**, 1105 (1999).
20. D. G. Ravenhall and J. Wambach, *Nucl. Phys. A* **475**, 468 (1987).
21. A. Shevchenko *et al.*, *Phys. Rev. Lett.* **93**, 122501 (2004).
22. Y. Kalmykov *et al.*, *Phys. Rev. Lett.* **96**, 012502 (2006).
23. C. A. Bertulani and G. Baur, *Phys. Rep.* **163**, 299 (1988).
24. A. Tamii *et al.*, *Phys. Lett. B* **459**, 61 (1999).
25. P. Mohr *et al.*, *Nucl. Instrum. Meth. A* **423**, 480 (1999).
26. U. Kneissl, N. Pietralla and A. Zilges, *J. Phys. G* **32**, R217 (2006).
27. J. Bryssinck *et al.*, *Phys. Rev. C* **59**, 1930 (1999).
28. I. Pysmenetska *et al.*, *Phys. Rev. C* **73**, 017302 (2006).
29. J. Enders *et al.*, *Nucl. Phys. A* **741**, 3 (2004).
30. P. G. Hansen, B. Jonson, and A. Richter, *Nucl. Phys. A* **518**, 13 (1990).
31. J. Enders, N. Huxel, P. von Neumann-Cosel, and A. Richter, *Phys. Rev. Lett.* **79**, 2010 (1997).
32. N. Huxel *et al.*, *Nucl. Phys. A* **645**, 239 (1999).
33. T. Rauscher, F.-K. Thielemann, and K.-L. Kratz, *Phys. Rev. C* **56**, 1613 (1997).
34. T. von Egidy and D. Bucurescu, *Phys. Rev. C* **72**, 044311 (2005); erratum **73**, 049901 (2006).
35. P. Demetriou and S. Goriely, *Nucl. Phys. A* **695**, 95 (2001).
36. K. Lindenberg, Doctoral thesis D17, TU Darmstadt (2007); and to be published.

THE STRUCTURE OF THE PYGMY DIPOLE RESONANCE

D. SAVRAN and A. ZILGES*

TU Darmstadt, Schlossgartenstr. 9
64289 Darmstadt, Germany
**New address: Universität zu Köln, Zülpicher Str. 77*
50937 Köln, Germany
E-mail: zilges@ikp.uni-koeln.de
www.zilges.de

Different experimental probes have been used to investigate particle bound electric dipole excitations below 10 MeV in semi-magic atomic nuclei where a collective excitation mode often denoted as Pygmy Dipole Resonance has been predicted. An interesting discrepancy in the E1 response measured with photons and α-particles is observed which may point to a splitting of the E1 dipole strength into two parts. A new experimental setup which measures the photoresponse in a wide energy range using a high resolution photon tagger will become operational soon.

Keywords: Pygmy Dipole Resonance, Photon Scattering, Alpha Scattering

1. Introduction: E1 Strength in Atomic Nuclei

The observation of electric dipole strength can be a sign of a dynamic breaking of the symmetry between protons and neutrons in atomic nuclei. Because the Isovector Electric Giant Dipole Resonance (GDR) exhausts the E1 sum rule nearly completely it was surprising when first signs of additional E1 strength below the GDR in the energy range between about 5 and 12 MeV were found.[1–4] This strength is usually denoted as Pygmy Dipole Resonance (PDR). Real photons are an ideal tool to measure dipole strength distributions below as well as above the particle threshold because the excitation mechanism is well understood and photons selectively populate the strongest dipole excitations (i.e. those with the largest γ-groundstate decay widths) from the groundstate of a nucleus. Below the particle threshold the strength can be derived from Nuclear Resonance Fluorescence (NRF) experiments whereas above the threshold the particle decay channel has to be detected (see fig. 1). In principle virtual photons exchanged in Coulomb interaction between two charged particles can be used in both energy regimes too. This allows the investigation of the photoresponse of radioactive nuclei as well.[4]

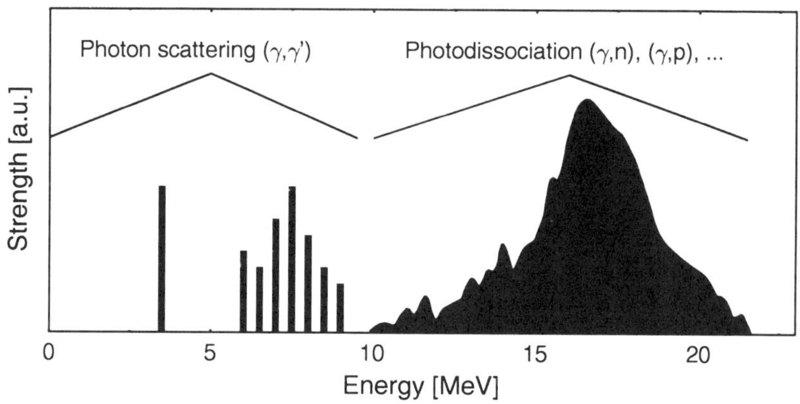

Fig. 1. Schematic distribution of E1 strength in a spherical nucleus. The fine structure in the tail of the GDR is usually not well known.

In the following section we will summarize the experimental results obtained on the PDR in the last decade using real photons. We will show that experiments with complementary probes are needed to get a better understanding of the structure of the excitation mode. Section 3 presents first result on the PDR obtained with hadronic probes, namely in α scattering experiments. Finally we will present an experimental approach using tagged photons which may allow to understand the relation of E1 strength observed below and above the particle threshold in the near future.

2. Results from photon scattering experiments

The advent of powerful experimental setups for (γ, γ') experiments has lead to a wealth of measured dipole strength distributions up to the particle threshold for nuclei in various regions of the chart of nuclides. The present experimental status for the summed E1 strength below the neutron threshold measured in high resolution photon scattering experiments is summarized in fig. 2 and includes results from studies described in refs.[2,3,5–13] It shows that up to one percent of the E1 Isovector Energy Weighted Sum Rule (IVEWSR) is exhausted by discrete lines stemming from 1^- states observed in this energy region. Due to the high level density and unobserved weak γ branchings to excited states even some additional strength may be hidden in the background (see contribution by R. Schwengner in these proceedings).

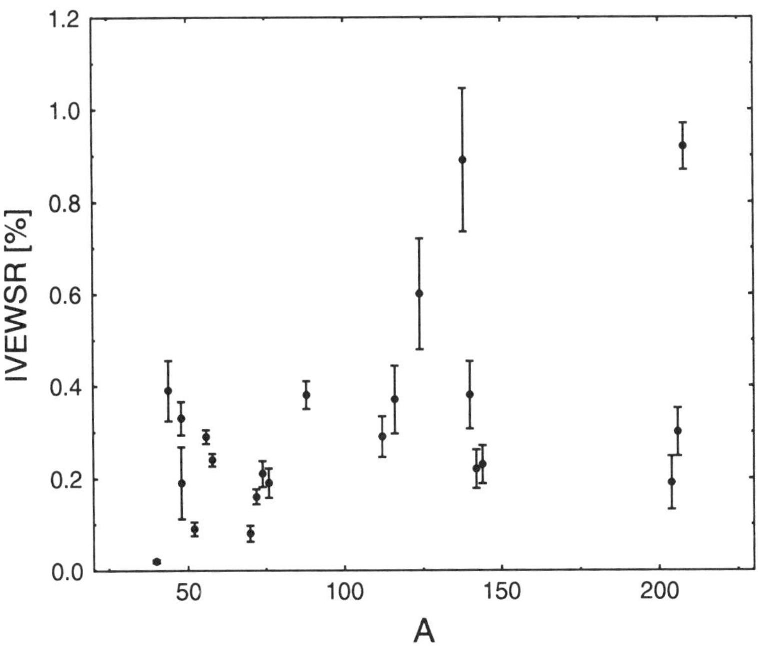

Fig. 2. Summed electric dipole strength detected in discrete lines below the neutron threshold given as the relative exhaustion of the isovector energy weighted sum rule. The data has been taken from refs.[2,3,5–13]

As one example the detailed E1 strength distribution in five stable semi-magic N=82 isotones derived from (γ, γ') experiments at the Superconducting Linear Accelerator S-DALINAC at TU Darmstadt is shown in fig. 3. The left panel shows the observed 1^- states and their reduced $B(E1)\uparrow$ strength. The use of high resolution HPGe detectors ($\Delta E/E$ approx. 0.1% in the relevant energy range) to detect the photons in the exit channel enables a detailed state-by-state analysis of spins, parities, and absolute transition strengths. For the right panel the strength distribution has been folded by a Lorentzian with a width of 500 keV, an idea originally proposed by F. Iachello to emphasize gross scales of the distribution.[11] These smoothed out distributions show two or even thress "bumps" in the 5-8 MeV range for some nuclei. This may be a first hint that the states in this energy region split up into two energetically separated groups. Please note that the lowest lying bump around 3-4 MeV stems from the well understood two phonon $(2^+ \otimes 3^-)_{1^-}$ excitation.

E1 strength on top of the low energy tail of the GDR has been observed in the neutron rich radioactive nuclei ^{130}Sn and the doubly magic N=82 nucleus ^{132}Sn in experiments at the FRS/LAND setup at GSI also.[14] However, here the strength has been detected above the neutron threshold (at about 10 MeV excitation energy) and the connection to the strength observed in (γ,γ') experiments below the threshold in the stable N=82 isotones is still unclear.

The experimental findings from photon scattering experiments triggered numerous theoretical studies in various model approaches ranging from rather simple collective models to sophisticated QRPA calculations including phonon coupling and the coupling to the continuum.[15-23] Many of the calculations generate E1 strength in the right energy range. But one surprising result of some microscopic models is the charge transition densities of some selected states building the PDR. These transition densities point to a dominant isoscalar character.[6,22,24] To investigate the isospin structure of the mode and to further look into the possible splitting of the Pygmy Dipole Resonance discussed above we started a series of experiments using α particles as a complementary hadronic probe to excite the 1^- states.

3. Results from α scattering experiments

The method of α scattering at forward angles and energies in the 100 MeV range is a well established tool to investigate especially isoscalar excitations in atomic nuclei.[25] The applicability of α scattering to study bound states in nuclei is usually restricted to some favourable cases by the limited energy resolution. In addition one looses the strong spin selectivity of photons where mainly dipole and quadrupole transitions are induced in the entrance channel and observed in the exit channel. An additional coincident measurement of the γ decay can improve this situation as has been shown in pioneering experiments by Poelhekken et al.[26] The simultaneous determination of excitation and decay energies allows to isolate different decay channels. By selecting the decays to the ground state one gains again the selectivity to states with large ground state decay width.

Recently the (α,α') setup at the Big-Bite-Spectrometer BBS at the AGOR cyclotron of the KVI Groningen has been extended by an array of

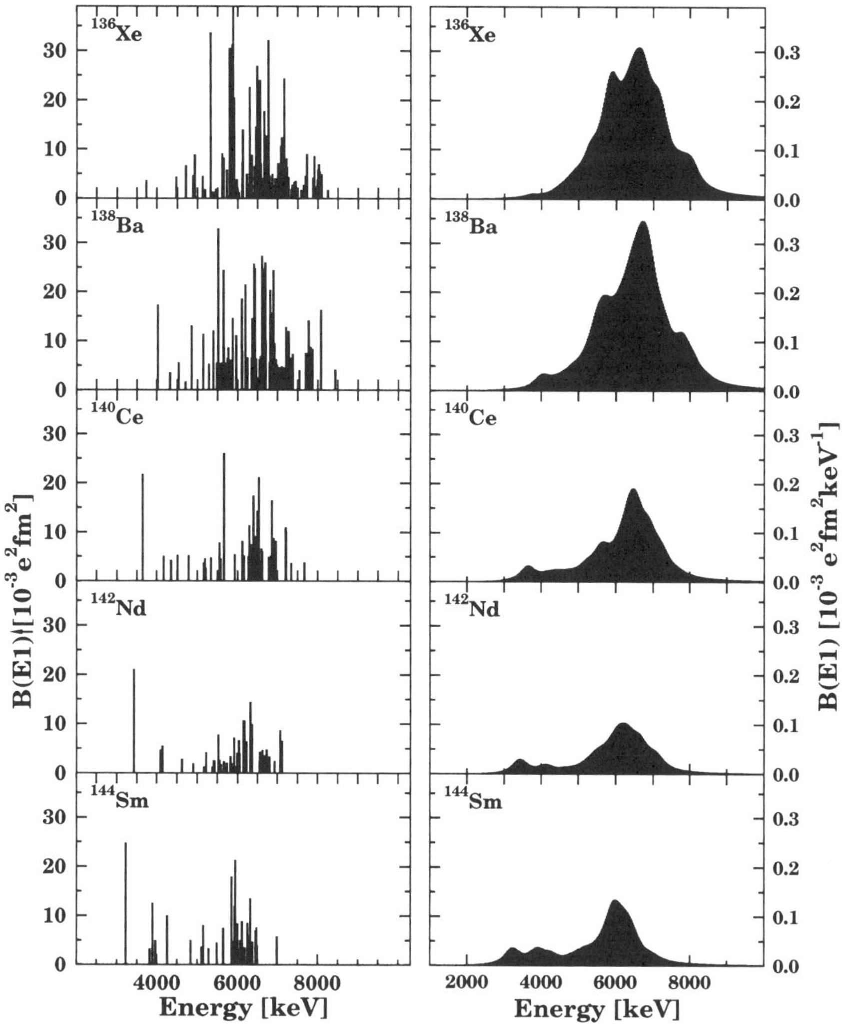

Fig. 3. E1 strength distribution of the five investigated stable N=82 nuclei (left panel) and the same distribution folded with a Lorentzian with a width of 500 keV (right panel). Please note that the data for ^{136}Xe stem from a new improved experiment compared to the preliminary data shown which has been shown in ref.[13]

high resolution HPGe detectors.[28] This new setup enables to analyze the excitation in the BBS and observe the coincident γ decays with highest resolution. Thus one combines the properties of excitation with α particles,

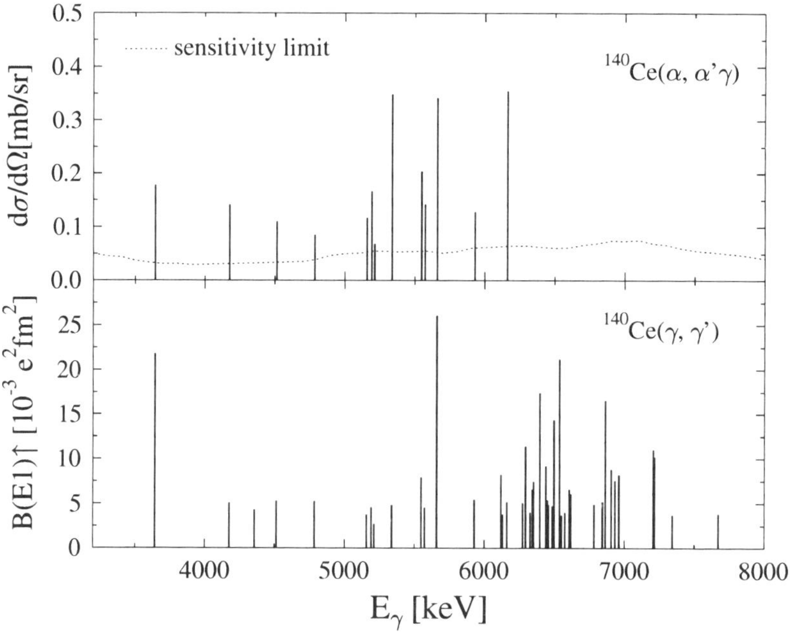

Fig. 4. Different excitation patterns of the PDR with photons and α particles in the nucleus ^{140}Ce.[27] Please note that the upper part shows cross sections and the lower part B(E1)↑ values.

e.g. the isospin selectivity, with the spin filter property and the high energy resolution of γ spectroscopy. First results on ^{140}Ce and ^{138}Ba were very surprising[27] as can be seen in fig. 4 for ^{140}Ce: Whereas the 1^- states in the energy region between about 5.5 and 6.5 MeV show a similar excitation pattern in (γ, γ') and $(\alpha, \alpha'\gamma)$ experiments, the states at higher energies are observed in (γ, γ') experiments only. A possible explanantion would be a different isosopin character of the states, however the observed rather abrupt transition is not expected. Some theoretical calculations seem to predict a certain splitting of the PDR.[23,24] It is obvious that additional experimental studies as well as improved theoretical calculations are highly mandatory to get a deeper insight into this phenomenon.

4. The high resolution photon tagger facility NEPTUN

An ideal experiment using electromagnetic probes should enable to measure the photoresponse from the lowest energies up to the GDR region with the highest possible energy resolution. We have therefore designed and setup the photon tagger system NEPTUN at S-DALINAC. With NEPTUN each bremsstrahlung photon produced in the energy range from about 8-20 MeV is marked ("tagged") with its energy. This is realized by analyzing the remaining energy of the electron after the bremsstrahlung process using a magnet spectrometer. The energy resolution will finally be better than 0.25%, i.e. one can determine the excitation energy of the nucleus with this precision. Beside this excellent energy resolution NEPTUN will provide a high flux of tagged photons of about $10^4 \gamma/s \cdot keV$ by using fast scintillating fibers as focal plane detectors for the electrons. Depending on the energy the decay channel will be either photons (below the neutron separation energy S_n) or neutrons (above S_n) which are measured with an adequate detection system, respectively.

Fig. 5. The photon tagger system NEPTUN at S-DALINAC. The electron beam comes from the right, hits the radiator target and is deflected vertically by the magnet on the left side. The electrons are analyzed in the focal plane chamber.

The rather low energy of the inital electrons lead to a rather large opening angle of the electrons after hitting the radiator target which makes the necessary spectroscopy of the electrons rather difficult. This problem can be reduced by a so called clam shell design of the dipole magnet. The present setup at S-DALINAC is shown in fig. 5.[29] The deflection of the electrons in the magnet system is in vertical direction. One can see the photomultipliers sticking out of the focal plane chamber which are connected to the scintillating fibres detecting the electrons. First test experiments showed very promising results. After an extension of the focal plane the tagger will allow to cover a γ energy region of 1-2 MeV in a experiment. The full NEPTUN setup will become operational in 2008.

Acknowledgments

This work has been stimulated by discussions with many colleagues. We thank A.M. van den Berg, M.N. Harakeh and H.J. Wörtche for their invaluable support during the experiments at KVI Groningen. We thank M. Büssing, M. Elvers, J. Endres, M. Fritzsche, J. Hasper, L. Kern, K. Lindenberg, S. Müller, K. Sonnabend, and S. Volz for their outstanding engagement which enabled the realization of the experiments at the S-DALINAC. This work was supported by the Deutsche Forschungsgemeinschaft (contract SFB 634), by the BMBF (contract 06DA129I) and and was performed as part of the research program of the Stichting FOM with financial support from the Nederlandse Organisatie voor Wetenschappelijk Onderzoek. The research has further been supported by the EU under EURONS contract no. RII3-CT-2004-506065 in the 6th framework program.

References

1. G. A. Bartholomew, E. D. Earl, A. J. Ferguson, J. W. Knowles, M. A. Lone, Adv. Nucl. Phys. **7** (1973) 229.
2. K. Govaert, F. Bauwens, J. Bryssinck, D. De Frenne, E. Jacobs, W. Mondelaers, L. Govor, V. Y. Ponomarev, Phys. Rev. C **57** (1998) 2229.
3. A. Zilges, S. Volz, M. Babilon, T. Hartmann, P. Mohr, K. Vogt, Phys. Lett. B **542** (2002) 43.
4. T. Aumann, Eur. Phys. Journal A **26** (2005) 441.
5. A. Jung, S. Lindenstruth, H. Schacht, B. Starck, R. Stock, C. Wesselborg, R.-D. Heil, U. Kneissl, J. Margraf, H. H. Pitz, F. Steiper, Nucl. Phys. **A584** (1995) 103.
6. N. Ryezayeva, T. Hartmann, Y. Kalmykov, H. Lenske, P. von Neumann-Cosel, V. Y. Ponomarev, A. Richter, A. Shevchenko, S. Volz, J. Wambach, Phys. Rev. Lett. **89** (2002) 272502.
7. N. Pietralla, Z. B. V. N. Litvinenko, S. Hartman, F. F. Mikhailov, I. V. Pinayev, G. S. M. W. Ahmed, J. H. Kelley, S. O. Nelson, R. Prior, K. Sabourov, A. P. Tonchev, H. R. Weller, Phys. Rev. Lett. **88** (2002) 012502.
8. J. Enders, P. von Brentano, J. Eberth, A. Fitzler, C. Fransen, R.-D. Herzberg, H. Kaiser, L. Käubler, P. von Neumann-Cosel, N. Pietralla, V. Y. Ponomarev, A. Richter, R. S. I. Wiedenhöver, Nucl. Phys. **A724** (2003) 243.
9. T. Hartmann, M. Babilon, S. Kamerdzhiev, E. Litvinova, D. Savran, S. Volz, A. Zilges, Phys. Rev. Lett. **93** (2004) 192501.

10. L. Käubler, H. Schnare, R. Schwengner, H. Prade, F. Dönau, P. von Brentano, J. Eberth, J. Enders, A. Fitzler, C. Fransen, M. Grinberg, R.-D. Herzberg, H. Kaiser, P. von Neumann-Cosel, N. Pietralla, A. Richter, G. Rusev, C. Stoyanov, I. Wiedenhöver, Phys. Rev. C **70** (2004) 064307.
11. S. Volz, N. Tsoneva, M. Babilon, M. Elvers, J. Hasper, R.-D. Herzberg, H. Lenske, K. Lindenberg, D. Savran, A. Zilges, Nucl. Phys. **A779** (2006) 1.
12. G. Rusev, R. Schwengner, F. Dönau, M. Erhard, S. Frauendorf, E. Grosse, A. R. Junghans, L. Kaubler, K. Kosev, L. K. Kostov, S. Mallion, K. D. Schilling, A. Wagner, H. von Garrel, U. Kneissl, C. Kohstall, M. Kreutz, H. H. Pitz, M. Scheck, F. Stedile, P. von Brentano, C. Fransen, J. Jolie, A. Linnemann, N. Pietralla, V. Werner, Phys. Rev. C **73** (2006) 044308.
13. U. Kneissl, N. Pietralla, A. Zilges, J. Phys. G **32** (2006) R217.
14. P. Adrich, A. Klimkiewicz, M. Fallot, K. Boretzky, T. Aumann, D. Cortina-Gil, U. Datta Pramanik, Th. W. Elze, H. Emling, H. Geissel, M. Hellström, K. L. Jones, J. V. Kratz, R. Kulessa, Y. Leifels, C. Nociforo, R. Palit, H. Simon, G. Surowka, K. Sümmerer, W. Walus, Phys. Rev. Lett. **95** (2005) 132501.
15. J. Chambers, E. Zaremba, J. P. Adams, B. Castel, Phys. Rev. C **50** (1994) R2671.
16. A. M. Oros, K. Heyde, C. De Coster, B. Decroix, Phys. Rev. C **57** (1998) 990.
17. D. Sarchi, P. F. Bortignon, G. Colo, Phys. Lett. B **601** (2004) 27.
18. S. Goriely, E. Khan, M. Samyn, Nucl. Phys. **A739** (2004) 331.
19. N. Tsoneva, H. Lenske, Ch. Stoyanov, Phys. Lett. B **586** (2004) 213.
20. N. Paar, T. Nikšić, D. Vretenar, P. Ring, Phys. Rev. B **606** (2005) 288.
21. J. Terasaki, J. Engel, Phys. Rev. C **74** (2006) 044301.
22. V. Tselyaev, J. Speth, F. Grummer, S. Krewald, A. Avdeenkov, E. Litvinova, G. Tertychny, Phys. Rev. C **75** (2007) 014315.
23. G. Tertychny, V. Tselyaev, S. Kamerdzhiev, F. Grummer, S. Krewald, J. Speth, A. Avdeenkov, E. Litvinova, Phys. Lett. B **647** (2007) 104.
24. D. Tarpanov, C. Stoyanov, N. Van Giai, V. Voronov, Phys. Atom. Nucl. **70** (2006) 1402.
25. M. N. Harakeh, A. van der Woude, *Giant Resonances*, Oxford University Press, 2001.
26. T. D. Poelhekken, S. K. B. Hesmondhalgh, H. J. Hofmann, A. van der Woude, M. N. Harakeh, Phys. Lett. B **278** (1992) 423.
27. D. Savran, A. M. van den Berg, M. N. Harakeh, J. Hasper, A. Matic, H. J. Wörtche, A. Zilges, Phys. Rev. Lett. **97** (2006) 172502.
28. D. Savran, A. M. van den Berg, M. N. Harakeh, K. Ramspeck, H. J. Wörtche, A. Zilges, Nucl. Instr. and Meth. A **564** (2006) 267.
29. K. Lindenberg, *Design and Construction of the Low-Energy Photon Tagger NEPTUN*, PhD thesis, TU Darmstadt, 2007.

DIPOLE-STRENGTH DISTRIBUTIONS UP TO THE GIANT DIPOLE RESONANCE DEDUCED FROM PHOTON SCATTERING

R. SCHWENGNER*, G. RUSEV, N. BENOUARET[1], R. BEYER, F. DÖNAU, M. ERHARD, E. GROSSE, A.R. JUNGHANS, K. KOSEV, J. KLUG, C. NAIR, N. NANKOV[2], K.D. SCHILLING, A. WAGNER

Institut für Strahlenphysik, Forschungszentrum Dresden-Rossendorf, PF 510119, 01314 Dresden, Germany
** E-mail: R.Schwengner@fzd.de*

[1] *On leave from Faculté de Physique, Université des Sciences et de la Technologie d'Alger, 16111 Bab-Ezzouar-Alger, Algerie*

[2] *On leave from Institute for Nuclear Research and Nuclear Energy, BAS, 1784 Sofia, Bulgaria*

Dipole-strength distributions up to the neutron-separation energies of the even-mass Mo isotopes from ^{92}Mo to ^{100}Mo and of the N=50 isotones ^{88}Sr, ^{89}Y, ^{90}Zr have been investigated in photon-scattering experiments using the bremsstrahlung facility at the superconducting electron accelerator ELBE of the Forschungszentrum Dresden-Rossendorf. A measurement using polarised bremsstrahlung impinging on ^{88}Sr revealed that all resolved transitions with energies greater than 6 MeV in this nuclide except for one are $E1$ transitions. The intensity distributions obtained from the measured spectra after a correction for detector response and a subtraction of atomic background in the target contain a continuum part in addition to the resolved peaks. It turns out that the dipole strength in the resolved peaks amounts to about 30% of the total dipole strength while the continuum contains about 70%. In order to estimate the distribution of inelastic transitions and to correct the ground-state transitions for their branching ratios simulations of γ-ray cascades were performed. The photoabsorption cross sections obtained in this way connect smoothly to (γ, n) cross sections and give novel information about the strength on the low-energy tails of the Giant Dipole Resonances below the neutron-separation energies. The experimental cross sections are compared with predictions of a Quasiparticle-Random-Phase Approximation in a deformed basis. The calculations describe the experimentally observed increase of the dipole strengths with increasing neutron number of the Mo isotopes as a consequence of increasing nuclear deformation.

Keywords: Photon scattering; Photoabsorption cross section; Quasiparticle-Random-Phase Approximation

1. Introduction

The detailed understanding of the response of atomic nuclei to photons has received increasing attention in recent years. Information on the dipole strength at the low-energy tail of the Giant Dipole Resonance (GDR) is important for an estimate of the effect of high temperatures during the formation of heavy elements in the cosmos. In particular, reaction rates in the so-called p-process are influenced by the behaviour of dipole-strength distributions close to the neutron-separation energy.[1] This behaviour may be affected by excitations like the Pygmy Dipole Resonance (see, e.g., Refs.[2–4]).

So far, estimates of the dipole strength obtained from calculations within a Quasiparticle-Random-Phase Approximation for spherical nuclei with a phenomenological implementation of nuclear deformation have been used for astrophysical applications.[5,6] A systematic investigation of the dipole strength with varying nucleon numbers and, thus, varying properties like deformation is mandatory for the improvement of models that are used for modelling processes for the production of heavy elements in the cosmos. Dipole-strength distributions up to the neutron-separation energies have been studied for only few nuclides in experiments with bremsstrahlung (see, e.g., Ref.[7] and Refs. therein). The new bremsstrahlung facility[8] at the superconducting electron accelerator ELBE of the research centre Dresden-Rossendorf opens up the possibility to study the dipole response of the stable nuclei with even the highest neutron-separation energies in photon-scattering experiments.

In order to obtain information on the influence of nuclear properties on the dipole response we have studied the chain of the even-mass Mo isotopes from ^{92}Mo to ^{100}Mo and of the N=50 isotones ^{88}Sr, ^{89}Y, ^{90}Zr. In the following, we describe especially the results of our experiments on the Mo isotopes.

2. Photon-scattering experiments

Photon-scattering experiments on ^{92}Mo, ^{94}Mo ^{96}Mo ^{98}Mo and ^{100}Mo were carried out using the bremsstrahlung facility at the electron accelerator ELBE of the Forschungszentrum Rossendorf.[8] Bremsstrahlung was produced by irradiating a 6 mg/cm^2 thick niobium radiator with a continuous-wave electron beam of a kinetic energy of 13.2 MeV and an average current of 600 μA. Gamma rays scattered from the target were measured with four high-purity germanium detectors of 100% efficiency relative to a 3 in × 3 in NaI detector, two of them placed at 90° and the other two at 127° relative to the incident photon beam in order to deduce angular distributions of the γ rays. All detectors are equipped with escape-suppression shields made of bismuth-germanate scintillation detectors. The five nuclides were studied under identical experimental conditions. Spectra measured during the irradiation of the Mo isotopes for about 60 hours are shown in Fig. 1 together with the photon flux available in these experiments. These spectra

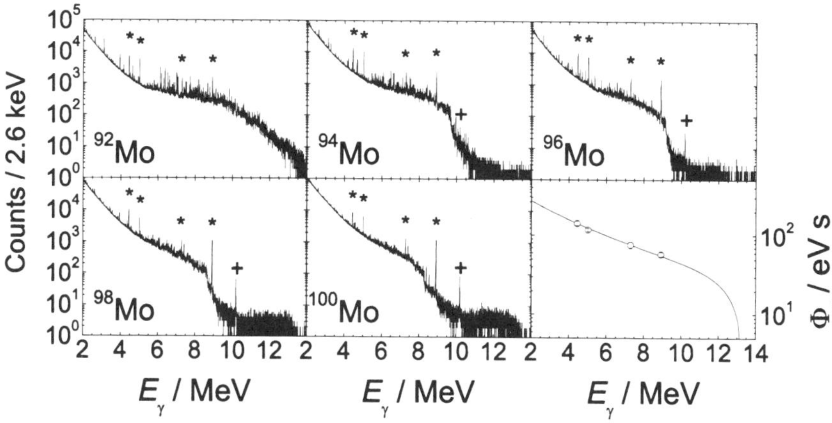

Fig. 1. Spectra of γ-rays scattered from ^{92}Mo, ^{94}Mo, ^{96}Mo, ^{98}Mo and ^{100}Mo, respectively, at an electron energy of $E_{\mathrm{kin}} = 13.2$ MeV and an angle of 127° relative to the incident photon beam. Stars mark transitions in ^{11}B which was combined with the target. The cross marks a transition following the ^{73}Ge(γ, n) reaction in the detector material. The right bottom panel shows the photon flux as obtained from the approach given in Ref.[9] adjusted to an absolute scale by means of levels with known integrated cross sections in ^{11}B.

show that (i) in average the intensities of the transitions decrease with increasing neutron number N while the number of transitions increases and (ii) the γ-ray intensities drop at the neutron-separation energies S_n because the deexcitation via emission of neutrons dominates above the threshold for the (γ, n) reaction. The high level density and Porter-Thomas fluctuations of the decay widths have the consequence that many transitions will not be observed as resolved peaks but instead form a continuum.

In order to estimate the strength in the continuum we performed GEANT3[10] simulations of the non-resonant background caused by atomic processes in the target. The comparison of the experimental spectra with the respective atomic backgrounds reveals that the resolved peaks comprise only about 30% of the total dipole strength while the continuum parts contain about 70%. The relevant intensity distributions of the scattered photons used for the further analysis were obtained from a subtraction of the atomic background from the experimental spectra.

The subtracted spectra contain the ground-state (elastic) transitions and in addition, transitions to lower-lying excited states (inelastic transitions) as well as transitions from these states to the ground state (cascade transitions). The different types of transitions cannot be clearly distinguished. However, for the determination of the absorption cross section only the intensities of the ground-state transitions are needed. This means that contributions of inelastic and cascade transitions have to be removed from the spectra. We corrected the intensity distributions by simulating γ-ray cascades[11] from the levels in the whole energy range analogous to the strategy of the statistical DICEBOX code.[12] Spectra of γ-ray cascades were generated for groups of levels in 100 keV bins. Examples are given in the left panel of Fig. 2. These spectra resemble strongly the ones measured in experiments with tagged photons[13] indicating that the simulated quasi-monoenergetic spectra can describe experimental ones. Starting from the high-energy end of the experimental spectrum, which contains ground-state transitions only, the simulated intensities of the ground-state transitions were normalised to the experimental ones in the considered bin and the intensity distribution of the branching transitions was subtracted from the experimental spectrum. Applying this procedure step-by-step for each energy bin moving towards the low-energy end of the spectrum one obtains the intensity distribution of the ground-state transitions. This intensity distribution is compared with the uncorrected one in the middle panel of Fig. 2. The right panel of Fig. 2 shows a mean distribution of the branching ratios B_0 deduced simultaneously from the simulations of the γ-ray cascades as the ratios of the intensities of the ground-state transitions to the total intensities of all transitions depopulating particular levels in a considered energy bin. Dividing the intensities of the ground-state transitions, which are proportional to the elastic scattering cross sections $\sigma_{\gamma\gamma}$, by the corresponding branching ratios $B_0 = \Gamma_0/\Gamma$ we calculated the absorption cross sections $\sigma_\gamma = \sigma_{\gamma\gamma}/B_0$.

The absorption cross sections obtained for the five Mo isotopes before correction and after correction with the described procedure are shown in Fig. 3. They are compared with the results of (γ, n) experiments.[14] In the case of ^{92}Mo the calculated cross sections for the (γ, p) reaction are added.[15] It can be seen that the absorption cross sections obtained from our photon-scattering experiments connect smoothly to the (γ, n) data. Thus, we have identified in our experiments for the first time how the tail of the GDR

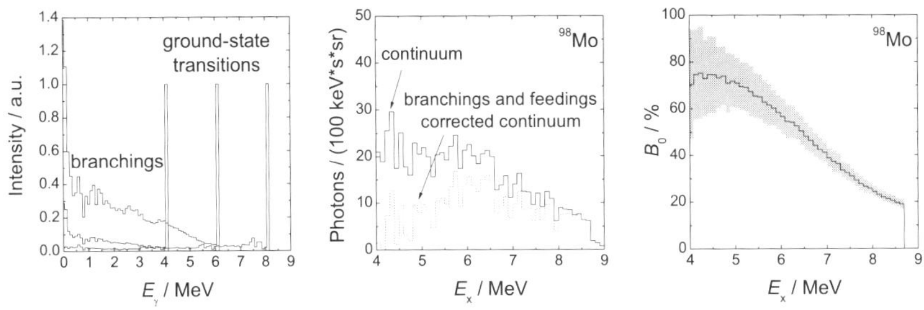

Fig. 2. Left panel: Simulated intensity distributions of transitions depopulating levels at 4, 6 and 8 MeV, respectively. Middle panel: Continuum spectrum of ^{98}Mo before (upper spectrum) and after (lower spectrum) correction for branching transitions. Right panel: Distribution of the branching ratios $B_0 = \Gamma_0/\Gamma$ for ground-state transitions.

extends across the threshold region towards low energy.

The cumulative integrated cross sections $\Sigma(E_x) = \sum_i^{E_x} \sigma_\gamma(E_i)\,\Delta E$ normalised to the Thomas-Reiche-Kuhn sum rule $\Sigma_{\text{TRK}} = 60\,NZ/A$ MeV mb[16] are shown for all five isotopes in Fig. 4. One observes an increase of $\Sigma(E_x)/\Sigma_{\text{TRK}}$ with increasing neutron number. In order to gain information about the origin of this increase we compare the experimental results with predictions of a Quasiparticle-Random-Phase-Approximation (QRPA) in a deformed basis.

3. QRPA calculations in a deformed basis

The QRPA calculations used a deformed modified oscillator basis, a separable dipole-plus-octupole interaction, and the nuclear selfconsistency approach.[17] The possible contamination of the calculated $E1$ strength with spurious motion of the centre-of-mass is completely removed by means of the suppression method described in Ref.[18] The relative strength of the isovector part of the dipole-plus-octupole interaction was adjusted such that it reproduces the position of the maximum of the GDR. The model is described in detail in Ref.[19] Similar QRPA calculations were also performed in our recent study of the magnetic-dipole ($M1$) strength in Mo isotopes.[20] The $\Sigma(E_x)/\Sigma_{\text{TRK}}$ calculated for the five isotopes using the deformation parameters given in Ref.[20] are compared with the results of calculations without deformation in Fig. 5. One sees that the calculations with deformation reproduce qualitatively the increase of the experimental $\Sigma(E_x)/\Sigma_{\text{TRK}}$ with increasing neutron number. In contrast, the results of the calculations applying spherical shapes to all isotopes do not show any remarkable change of $\Sigma(E_x)/\Sigma_{\text{TRK}}$ with changing neutron number. This shows that the increase of the dipole strength when going from the spherical ^{92}Mo to the heavy isotopes ^{98,100}Mo is related to the increasing deformation.

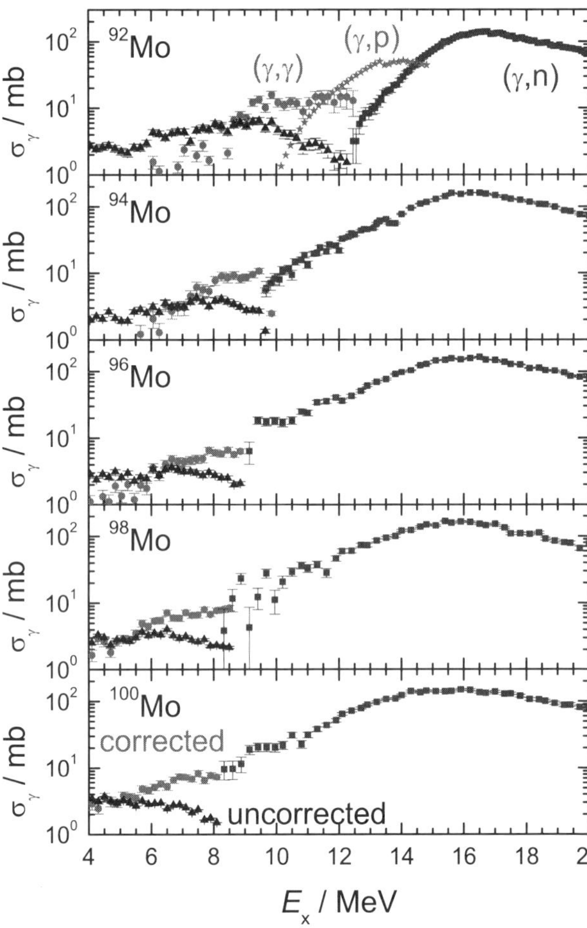

Fig. 3. Comparison of the photoabsorption cross sections determined from our photon-scattering experiments and from the measurements of the (γ, n) reaction. The triangles mark the uncorrected data while the circles show the result of the correction described in the text. The (γ, n) data are shown as boxes.

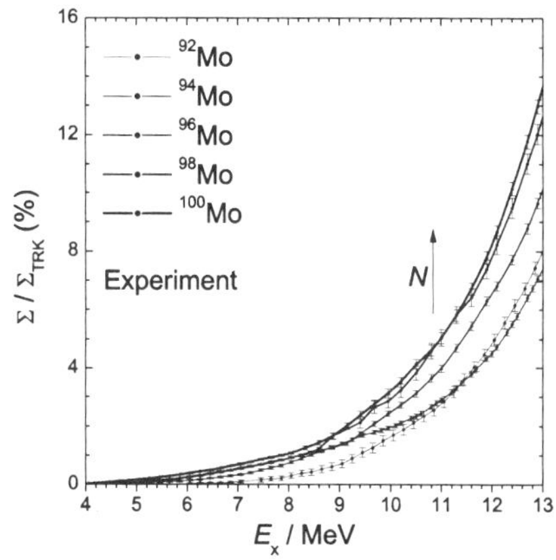

Fig. 4. Cumulative normalised integrated cross sections $\Sigma(E_x)/\Sigma_{\mathrm{TRK}}$ deduced from the present experiments.

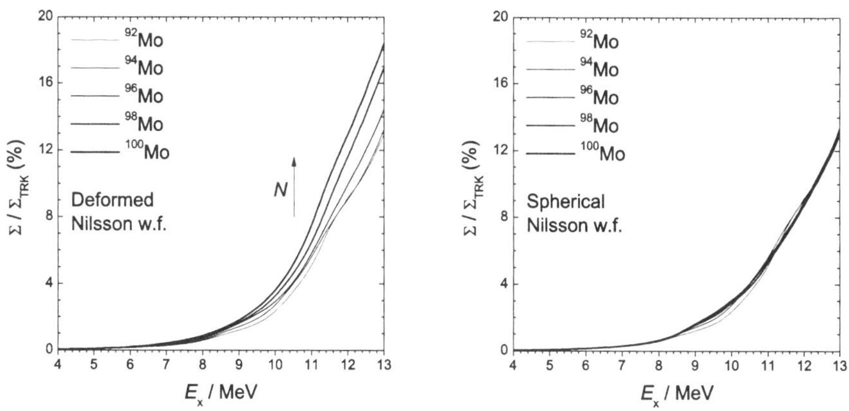

Fig. 5. Cumulative normalised integrated cross sections $\Sigma(E_x)/\Sigma_{\mathrm{TRK}}$ predicted by QRPA calculations using deformed wave functions (left panel) and spherical wave functions (right panel).

4. Summary

Dipole-strength distributions of even-mass Mo isotopes have been studied at the photon-scattering facility of the ELBE accelerator. Photoabsorption cross sections were deduced by means of simulations of γ-ray cascades which allowed us to estimate the intensities of branching and feeding transitions in the spectra. The obtained absorption cross sections connect smoothly with the low-energy tails of the Giant Dipole Resonances. QRPA calculations in a deformed basis ascribe the observed increase of the dipole-strength with the neutron number to the increasing deformation.

This work was supported by Deutsche Forschungsgemeinschaft under contract DO 466/1-2.

References

1. M. Arnould and S. Goriely, *Phys. Rep.* **384**, 1 (2003).
2. G. A. Bartholomew, E. D. Earle, A. J. Ferguson, J. W. Knowles and M. A. Lone, *Adv. Nucl. Phys.* **7**, 229 (1972).
3. S. Goriely, *Phys. Lett. B* **436**, 10 (1998).
4. A. Zilges, M. Babilon, T. Hartmann, D. Savran, and S. Volz, *Progr. Part. Nucl. Phys.* **55**, 408 (2005).
5. S. Goriely and E. Khan, *Nucl. Phys. A* **706**, 217 (2002).
6. S. Goriely, E. Khan, and M. Samyn, *Nucl. Phys. A* **739**, 331 (2004).
7. U. Kneissl, N. Pietralla, and A. Zilges, *J. Phys. G: Nucl. Part. Phys* **32**, R217 (2006).
8. R. Schwengner, R. Beyer, F. Dönau, E. Grosse, A. Hartmann, A. R. Junghans, S. Mallion, G. Rusev, K. D. Schilling, W. Schulze, and A. Wagner, *Nucl. Instr. Meth. A* **555**, 211 (2005).
9. G. Roche, C. Ducos, and J. Proriol, *Phys. Rev. A* **5**, 2403 (1972).
10. CERN Program Library Long Writeup W5013, Geneva 1993, unpublished.
11. G. Rusev, Dissertation, Technische Universität Dresden, 2007, unpublished.
12. F. Bečvář, *Nucl. Instr. Meth. A* **417**, 434 (1998).
13. R. Alarcon, R. M. Laszewski, A. M. Nathan, and S. D. Hoblit, *Phys. Rev. C* **36**, 954 (1987).
14. H. Beil, R. Bergère, P. Carlos, A. Leprêtre, A. De Miniac, and A. Veyssière, *Nucl. Phys. A* **227**, 427 (1974).
15. T. Rauscher and F.-K. Thielemann, *At. Data Nucl. Data Tables* **88**, 1 (2004).
16. P. Ring and P. Schuck, in *The Nuclear Many Body Problem*, (Springer, New York, 1980).
17. H. Sakamoto and T. Kishimoto, *Nucl. Phys. A* **501**, 205 (1989).
18. F. Dönau, *Phys. Rev. Lett.* **94**, 092503 (2005).
19. F. Dönau, G. Rusev, R. Schwengner, A. R. Junghans, K. D. Schilling, and A. Wagner, *Phys. Rev C* (2007), in print.
20. G. Rusev, R. Schwengner, F. Dönau, M. Erhard, S. Frauendorf, E. Grosse, A.R. Junghans, L. Käubler, K. Kosev, L.K. Kostov, S. Mallion, K.D. Schilling, A. Wagner, H. von Garrel, U. Kneissl, C. Kohstall, M. Kreutz, H.H. Pitz, M. Scheck, F. Stedile, P. von Brentano, C. Fransen, J. Jolie, A. Linnemann, N. Pietralla, V. Werner, *Phys. Rev. C* **73**, 044308 (2006).

CRITICAL POINT BEHAVIOUR OF ^{224}Ra AND ^{224}Th

P. G. BIZZETI* and A. M. BIZZETI–SONA

Dipartimento di Fisica, Università di Firenze,
and Istituto Nazionale di Fisica Nucleare, Sezione di Firenze
Via G. Sansone 1, I 50019 Sesto Fiorentino (Firenze), Italy
** E-mail: bizzeti@fi.infn.it*
www.fi.infn.it

The level scheme and transition amplitudes in the transitional nuclei ^{224}Ra and ^{224}Th are discussed in the frame of a collective model for quadrupole–octupole excitation. These two nuclei appear to be close to the critical point of a phase transition from spherical to stable reflection asymmetric shape.

Keywords: Octupole–Quadrupole Excitations; Phase transitions

1. Introduction

For many of us,it was a real surprise when Iachello showed[1,2] that a simple description is possible for some of the nuclei in the middle of the transitional regions between IBA symmetries U(5) and O(6) or U(5) and SU(3). As you know, two new symmetries – the exact symmetry E(5) and the quasi–symmetry X(5) – correspond to the critical point of the phase transition, where the potential energy has not a sharp minimum (neither at $\beta = 0$ nor at a definite deformation), but can be approximated by a square well potential extending from $\beta = 0$ to some limit β^w.

Evidence of the phase transition is provided by a plot of the order parameter $E(4^+)/E(2^+)$ as a function of a proper driving parameter, which can be the proton number or the neutron number (or perhaps a linear combination of the two). For a phase transition along the line of axial shapes, from U(5) to SU(3), the X(5) model predicts the value 2.92 for the order parameter at the critical point. There are several nuclei that correspond to this condition, and most of them have a level scheme in complete agreement with X(5) predictions.[3] The X(5) model also predicts the relative values of E2 transition strengths, but only in a minority of cases does the experimental data agree with the model.[3]

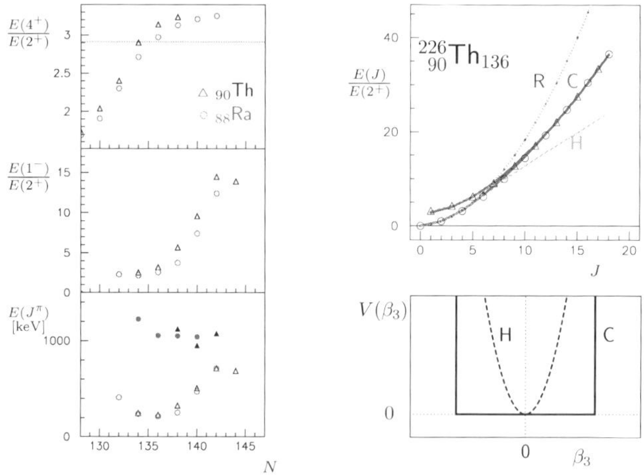

Fig. 1. Left: Indicators of quadrupole collectivity (upper panel) and of octupole collectivity (middle and lower panel) as a function of the neutron number for Ra (squares) and Th (triangles) isotopes. In the lower panel, open symbols refer to $J^\pi = 1^-$, full symbols to the band head of negative parity bands with $K \neq 0$. Right: The ground-state band of ^{226}Th, compared with the trend expected for rigid rotation (R), for harmonic octupole vibration of a quadrupole deformed system (H), and for the critical point of the phase transition (C), corresponding to the square-well potential shown in the lower panel.

It is natural to wonder, at this point, whether in nuclei with quadrupole–octupole deformation, a similar phase transition takes place also for the octupole mode and whether a simple description exists for its critical point. For us, this was a natural extension of a line of research on the octupole excitation we had been following during the last few years. A simple case corresponds to the phase transition between octupole vibrations around zero, in a nucleus that already possesses a stable (axial) quadrupole deformation, and a rigid rotation of a reflection asymmetric (but axially symmetric) nucleus. The former limit corresponds, *e.g.*, to the neutron rich Thorium and Radium isotopes.[4] The opposite limit (stable octupole deformation) is approached (but not reached) when the neutron number decreases. The evolution of the order parameter $E(1^-)/E(2^+)$ as a function of N (Fig. 1) gives evidence of such a phase transition, with a rapid variation between neutron numbers 134 and 142. It is important to note that only the $K^\pi = 0^-$ band (with band head 1^-) actually decreases in energy with decreasing neutron number, and eventually gets close to the first 2^+. Levels of the other octupole bands (as the first 2^- or the second 1^- or 3^-) remain above 1 MeV in the entire region of interest. This means that the phase transition only

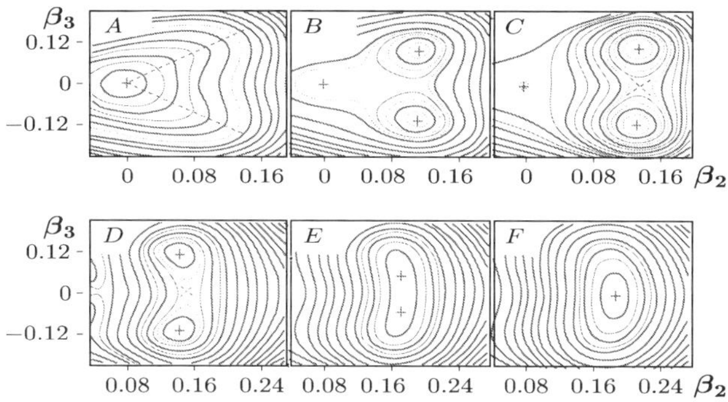

Fig. 2. Energy surfaces for axial quadrupole deformed shapes according to Nazarewicz *et al.* (adapted from Ref. 6.)

concerns the axially symmetric shapes, while non-axial degrees of freedom remain vibrational, just as in the X(5) model. We have found[5] that the level scheme – and also the transition strengths – in the isotope ^{226}Th are very well reproduced (see right part of Fig. 1) by a model assuming a constant quadrupole deformation $\beta_2 = \beta_2^0$ and the potential energy for the octupole deformation β_3 approximated by a square–well potential with impenetrable borders, extending symmetrically around zero, from $-\beta_3^w$ to $+\beta_3^w$. With respect to X(5), this model contains one more free parameter (the ratio β_3^w/β_2^0) but can also reproduce one additional band (the one with negative parity).

We must remark, however, that ^{226}Th is just at the lower border of the region of stable quadrupole deformation, characterized by $E(4^+)/E(2^+) \approx 3.3$. In the next lighter isotope ^{224}Th (and also in ^{224}Ra) we find a value of this ratio close to the value predicted by the X(5) model for the critical point. Actually, the entire positive-parity part of the ground-state band is, in ^{224}Ra as well as in ^{224}Th, in very good agreement with X(5) predictions, yet we have to reconcile this fact with the presence of a very low–lying $K^\pi = 0^-$ band, which soon merges with the positive–parity one above $J = 5$ or 7. This fact has long remained, for us, a veritable puzzle.

We think we have now found a possible explanation – and this will be the main subject of this talk. We will assume again a square well potential, but in this case we need a square well extending in two dimensions: the simplest case being for $0 < \beta_2 < \beta_2^w$ and $-\beta_3^w < \beta_3 < +\beta_3^w$. A suggestion in this sense is given also by some very old calculations by Nazarewicz and coworkers[6] (Fig. 2). The lower row of the figure shows an octupole phase transition as we have considered for ^{226}Th. At one side (D), we observe two symmetric minima of the potential energy, for values of β_2 and β_3 far from zero, while at the opposite side (F) we find β_3 oscillating around zero at a fixed value of β_2. The central panel (E) shows, for a fixed value of β_2, a potential energy almost independent of β_3 over a finite interval symmetric around zero: just what one should expect for the critical point.

The upper row shows a more complicated transition, from a stable deformation in β_2 and β_3 (C) to quadrupole and octupole vibrations around the spherical shape (A). Somewhere between this situation and the one shown in the central panel (B), one should expect to find a potential–energy well with a finite extension both in the β_2 and in the β_3 direction.

2. The model for quadrupole–octupole deformation

Here, I will not discuss in details the formalism we have introduced to analyze the quadrupole–octupole deformations (in conditions not far from the axial symmetry), which can be found in the Ref. 5. I will only mention the fact that, as in the Bohr model, we chose as the reference frame the one in which the overall tensor of inertia is diagonal. The requirement that the three products of inertia vanish in the intrinsic frame corresponds to a set of non linear equations in the intrinsic variables $a_\mu^{(\lambda)}$ describing the quadrupole and the octupole deformations. However, if we are only interested in situations close to axial symmetry, we can linearize them by neglecting second- and higher-order terms in the small non-axial amplitudes $a_\mu^{(\lambda)}$ with $\mu \neq 0$. At this point, all the amplitudes (and also the principal moments of inertia) can be expressed in terms of a new set of variables, each of which can be associated to a definite angular momentum and parity.

Up to this point, we have only summarized our previous work.[5] To proceed further, we require that the results obtained with our choice of the potential ($V(\beta_2, \beta_3) = 0$ for $0 < \beta_2 < \beta_2^w$, $|\beta_3| < \beta_3^w$ and $V = +\infty$ elsewhere) converge to those of the X(5) model when $\beta_3^w \to 0$. This requirement can be fulfilled with a proper choice of the free function $c(\beta_2, \beta_3)$ appearing in the eq.14 of Ref. 5, *i.e.*, $c(\beta_2, \beta_3) = \left[(\beta_2^2 + \beta_3^2)(\beta_2^2 + 2\beta_3^2)\right]^{-1/2}$, giving for the determinant of the matrix of inertia[5] $G \equiv g^2 = \mathrm{Det}\mathcal{G} = 288(\beta_2^2 + \beta_3^2)(\beta_2^2 + 2\beta_3^2)^4(\beta_2^2 + 5\beta_3^2)^{-2}\, \gamma_0^2 v^2 u^2 w^2$. With the Pauli procedure, we obtain the differential equation in β_2 and β_3

$$\left\{\frac{\partial^2}{\partial\beta_2^2} + \frac{\partial^2}{\partial\beta_3^2} + \epsilon - V(\beta_2, \beta_3) - \frac{J(J+1)}{3(\beta_2^2 + 2\beta_3^2)} + V_g(\beta_2, \beta_3)\right\}\Phi(\beta_2, \beta_3) = 0 ,$$

$$V_g(\beta_2, \beta_3) = \frac{1}{4g^2}\left[\left(\frac{\partial g}{\partial\beta_2}\right)^2 + \left(\frac{\partial g}{\partial\beta_3}\right)^2\right] - \frac{1}{2g}\left[\frac{\partial^2 g}{\partial\beta_2^2} + \frac{\partial^2 g}{\partial\beta_3^2}\right]$$

which is not separable and must be solved numerically. To this purpose, we have used the finite–difference method in a rectangular lattice, and found the lowest eigenvalues of the Hamiltonian matrix with the help of the ARPACK package.[7] Obviously, the choice of a rectangular shape for the square–well potential is not the only possibility. We have also explored different (and perhaps more realistic) shapes,[9] but a common drawback was the number of parameters necessary to define them. Only with a rectangular shape there is one single free parameter, the ratio β_3^w/β_2^w.

3. Results and comparison with experimental data

Our results for the level energies, in units of the energy of the first 2^+, are shown in the Fig. 3, as a function of β_3^w/β_2^w. When $\beta_3^w/\beta_2^w \to 0$, the energies of even parity levels converge (as expected) to the X(5) values, while those of the negative-parity ones diverge to $+\infty$. When β_3^w/β_2^w increases, the energies of the positive-parity states initially deviate from the X(5) values, but come back close to them when the parameter approaches 1. It is just in this region that we have found the values giving the best agreement with experimental data, $\beta_3^w/\beta_2^w \approx 0.81$ for ^{224}Ra and $= 0.85$ for ^{224}Th. A rather good agreement is obtained, for the positive parity states as well as for the negative parity ones (Fig. 4). In Fig. 5, the calculated level scheme of ^{224}Ra – including a few levels of the excited $K = 0$ band (the $s = 2$ band, in the notations of Iachello) – is compared to the experimental one. If the 0_2^+ level at 916 keV $[E(0_2^+)/E(2^+) = 10.86]$ has to be identified with the band-head of the $s = 2$ band, it is somewhat too high compared to the

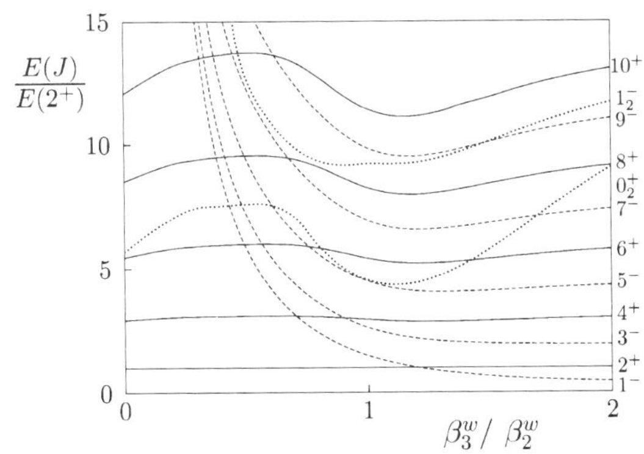

Fig. 3. Calculated excitation energies, in units of $E(2^+)$, for the positive- and negative-parity part of the ground-state band and for the first two level of the first-excited $K = 0$ band, as a function of β_3^w/β_2^w.

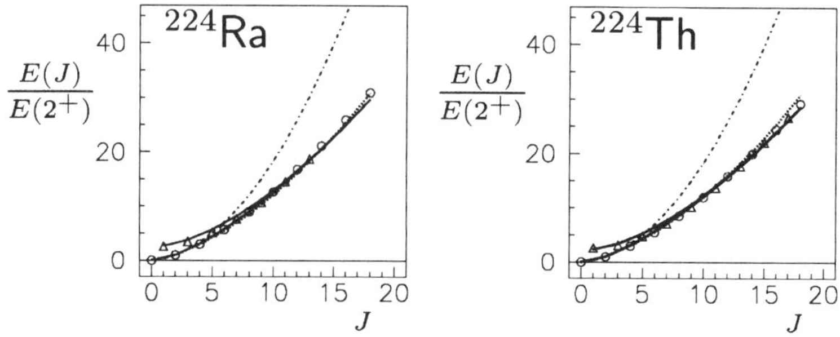

Fig. 4. Experimental excitation energies of the positive-parity levels (circles) and of the negative parity ones (triangles), as a function of J, for ^{224}Ra and ^{224}Th. The full line shows the results of the present model with $\beta_3^w/\beta_2^w = 0.81$ or $= 0.85$, respectively. Predictions of the X(5) model (dotted lines) and of a rigid-rotor model (dashed-dotted) are shown for comparison.

calculated position (but a similar discrepancy is observed also in some of the best X(5) nuclei).

Now, the last point that remains to be checked concerns the relative values of the electromagnetic matrix elements. These values can be calculated easily for the E2 transitions, as $\mathcal{M}(\text{E2}) = C_2\beta_2 Y^{(2)}$, with C_2 a common normalization factor. As for the E1 matrix elements, they all vanish in case of nuclear matter with constant charge density. By assuming electrical po-

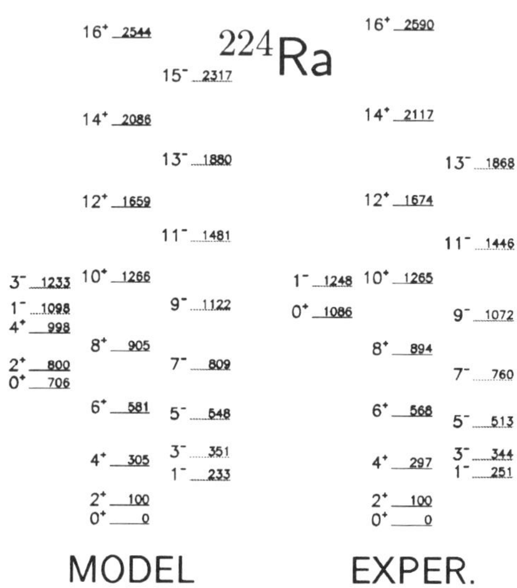

Fig. 5. Calculated and experimental level scheme of ^{224}Ra, normalized to $E(2^+) = 100$.

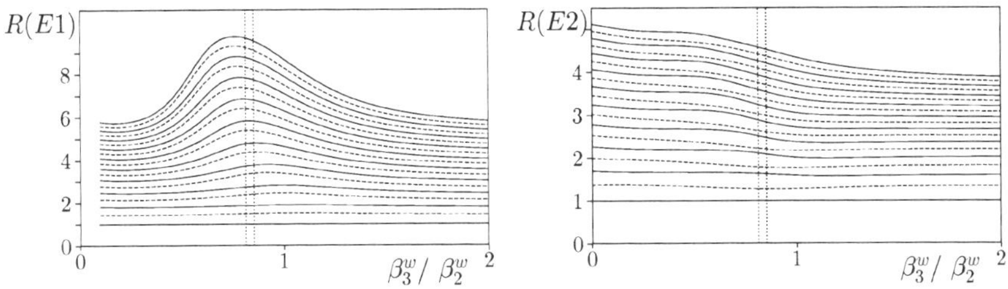

Fig. 6. Calculated ratios $R(EL) = | < J_i||\mathcal{M}(EL)||J_i - L > / < L||\mathcal{M}(EL)||0 > |$, for $L = 1$ and $L = 2$. From the bottom: for E1 transitions, $J_i = 1, 2, 3, ...18$; for E2 transitions, $J_i = 2, 3, 4, ...18$. Vertical lines correspond to $\beta_3^w/\beta_2^w = 0.81$ and $= 0.85$.

larization as a consequence of deformation, and with a constant electrical polarizability[a], one obtains $\mathcal{M}$(E1)= $C_1\beta_2\beta_3 Y^{(1)}$. Calculated values of the reduced matrix element of the E2 transitions (in units of the one for the $2^+ \to 0^+$) and of the E1 transitions (in units of the $1^- \to 0^+$) are depicted as a function of β_3^w/β_2^w in Fig. 6, where the vertical lines correspond to the adopted values of the parameter for ^{224}Ra and ^{224}Th. Experimental and calculated values of the transition amplitudes are compared in Table 1. The most significant comparison concerns the ratio of two E2 matrix elements: in the only available case, the ^{224}Ra $4^+ \to 2^+$ and $2^+ \to 0^+$ transitions, the agreement is excellent. In both nuclei we know the branching ratio of two E1 transitions from the same level ($1^- \to 0^+$ and $1^- \to 2^+$): also in these cases we find a very good agreement. We remark that, once the parameter β_3^w/β_2^w is fixed so as to reproduce the level scheme, the comparison of transition amplitudes becomes parameter free.

Finally, we know in ^{224}Th several cases of one E1 and one E2 transition from the same level. The comparison of these branching ratios is less significant, as the theoretical values include one more free parameter (the ratio of the coefficients $C1$ and $C2$ in the E1 and E2 matrix elements). This parameter has been deduced with a minimum–χ^2 fit. The experimental values of the ratio $R(E1/E2)$ of the reduced matrix elements (normalized to the W.u.) for the E1 and the E2 transition from a given level J_i are depicted in the Fig 7. The dashed and dotted lines connect calculated values with even or odd values of J_i, respectively. The agreement is not as good as for

[a]This assumption should be checked by means of microscopic calculations. Recently, Tsvenkov *et al.*[10] have reported the results obtained, in the frame of the Hartree–Foch model with Skyrme forces, for well–deformed nuclei in the Th region. They find that the electrical polarizability remains almost constant for the different states of a given nucleus but changes drastically from one isotope to the next.

Table 1. Experimental and calculated values of the ratios of reduced amplitudes for transitions with equal multipolarity (E1 or E2).

	$E1(1^- \to 2^+)/(1^- \to 0^+)$		$E2(4^+ \to 2^+)/(2^+ \to 0^+)$	
	Exper.	Calc.	Exper.	Calc.
^{224}Ra	1.52 ± 0.14	1.50	1.60 ± 0.05	1.63
^{224}Th	1.49 ± 0.26	1.50		

the ratios shown in Table 1, but at least the general trend is reproduced. For ^{224}Th, we have 9 experimental results, enough to perform a χ^2 test of agreement (with $n = 8$ degrees of freedom). The result is $\chi^2/n = 1.17$, corresponding to a confidence level of 31%. New measurements of the electromagnetic transition strengths in ^{224}Th and/or ^{224}Ra would be desirable, for a more significant test of the model.

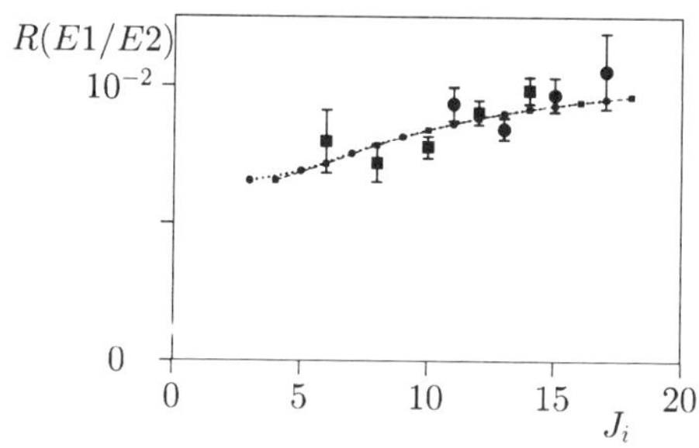

Fig. 7. Ratios of the absolute values of the transition matrix elements (normalized to the Weisskopf Unit) of stretched E1 and E2 transitions from the same level J_i of ^{224}Th, $R(E1/E2) = | < J_i||\mathcal{M}(E1)/\mathcal{M}_W(E1)||J_i - 1 > / < J_i||\mathcal{M}(E2)/\mathcal{M}_W(E2)||J_i - 2 > |$. Circles: J_i odd, squares: J_i even. The dashed and dotted lines join the calculated values.

References

1. F. Iachello, Phys. Rev. Lett. **85**, 3580 (2000).
2. F. Iachello, Phys. Rev. Lett. **87**, 052502 (2001).
3. R.M. Clark *et al.*, Phys. Rev. C **68**, 037301 (2003).
4. P.G. Bizzeti and A.M. Bizzeti–Sona, Europ. Phys. J. A **20**, 179 (2004).
5. P.G. Bizzeti and A.M. Bizzeti–Sona, Phys. Rev. C **70**, 064319 (2004).
6. W. Nazarewicz and P. Olanders, Nucl. Phys. A 441, 420 (1985)
7. http://www.caam.rice.edu/software/ARPACK/.
8. P.G. Bizzeti and A.M. Bizzeti–Sona in *Nuclear Theory 2005*, (Ed. S.Dimitrova, Sofia 2005) p. 265–280.
9. P.G.Bizzeti, A.M.Bizzeti–Sona: in "Collective Motion and Phase transitions in Nuclear Systems" (ed. A.A. Raduta, *et al.*, Singapore 2007), p.3-20.
10. A.Tsvenkov *et al.*, J. Phys. G **28**, 2187 (2002).

STUDY OF THE ^{119}Sb VIA THE ^{121}Sb$(p,t)^{119}$Sb REACTION

P. GUAZZONI and L. ZETTA*

Dipartimento di Fisica dell'Università, and I.N.F.N., Via Celoria 16, I-20133 Milano, Italy
E-mail: paolo.guazzoni@mi.infn.it, luisa.zetta@mi.infn.it

V. Yu. PONOMAREV

Institut für Kernphysik, Technische Universität Darmstadt, Schlossgartenstrasse 9, D-64289 Darmstadt, Germany
E-mail: ponomare@crunch.ikp.physik.tu-darmstadt.de

G. GRAW and R. HERTENBERGER

Sektion Physik der Universität München, D-85748, Garching, Germany
E-mail: gerhard.graw@physik.uni-muenchen.de, ralf.hertenberger@physik.uni-muenchen.de

T. FAESTERMANN and H-F. WIRTH

Physik Department, Technische Universität München, D-85748 Garching, Germany
E-mail: thomas.faestermann@ph.tum.de, hans-friedrich.wirth@ph.tum.de

M. JASKÓLA

Soltan Institute for Nuclear Studies, Hoza Street 69, Warsaw, Poland
E-mail: marian.jaskola@fuw.edu.pl

Accurate measurement of the (p,t) reaction angular distributions for the transitions to the levels of ^{119}Sb allows us to confirm or determine energies of 59 levels, 23 of which have been identified for the first time and to assign the angular momentum transfer values and a well-defined range for the J values. By using conventional Woods-Saxon potentials for the entrance proton and exit triton channel, the DWBA analysis has been performed in a finite range approximation, assuming a dineutron cluster pickup mechanism. The present (p,t) data have been supplemented by microscopic calculations in the framework of QPM, giving a reasonably good description of the experimental fragmentation of the cross sections and the absence of (p,t) strength above 2.9 MeV.

Keywords: High resolution (p,t), Cluster DWBA analysis, QPM calculation

1. Introduction

The particle-core weak coupling model is a useful spectroscopic tool for supplementig level structure information obtained by few-nucleon transfer reactions[1–4] on near magic nuclei having, outside a completely filled major shell, one unpaired nucleon, slightly bound acting as a spectator. To further investigate the systematic of this coupling we have measured the (p, t) reaction on ^{121}Sb target, an odd proton nucleus with a ground state J^{π} $= 5/2^{+}$, where the $1d_{5/2}$ proton configuration is expected to be dominant (77%), as shown by the calculations of Hooper et al.[5]

The core reaction ^{120}Sn$(p, t)^{118}$Sn was also studied[6] in the same experimental conditions to obtain information, via integrated cross section comparison, on the role of the extra unpaired proton. The study of the ^{121}Sb$(p, t)^{119}$Sb, never measured before the present high resolution experiment, is aimed to characterize the low-spin states of ^{119}Sb. The reaction data have been compared with the theoretical predictions supplied by the Quasiparticle-Phonon Model (QPM),[7] which accounts for the interaction between simple and complex configurations of nuclear excitation. Within QPM, phonons of different multipolarities and parities are obtained by solving quasiparticle random phase approximation equations. The single particle spectrum and phonon basis are determined from the calculations on the neighboring even-even nuclear core. On the other hand, the QPM analysis can be easily transformed into the spectator approach by switching off the interaction between different configurations in the model space. The followed procedure is described in details in ref.[4]

2. Experimental Results

The ^{121}Sb$(p, t)^{119}$Sb reaction has been measured using the 21 MeV proton beam delivered by the HVEC MP Tandem accelerator of Munich. The ^{121}Sb isotopic enriched (99.53%) target had a thickness of 102 μg/cm^2 deposited on a 10 μg/cm^2 carbon backing. The beam current intensity was up to 500 nA. The tritons, analyzed with the Munich Q3D spectrograph at 11 angles in different magnetic field settings to reach an excitation energy of 2.874 MeV, were detected in the focal plane of the Q3D magnetic spectrograph by the 1.8 m long focal plane detector for light ions.[8] The energy resolution was 8 keV full width at half maximum, while the uncertainty in our quoted energies is estimated to be 3 keV. Absolute cross sections were estimated with an uncertainty of $\sim$ 15%. Table 1 summarizes the results obtained in the present experiment. The integrated cross sections reported are calculated in the interval $10^o \leq \theta_{lab} \leq 65^o$.

Table 1. Present experiment

E_{exc} (MeV)	L_{tran}	J^π	σ_{int} (μb)	E_{exc} (MeV)	L_{tran}	J^π	σ_{int} (μb)
0.0	0	$5/2^+$	2371± 17	2.282	3	$(1/2-11/2)^-$	19± 2
0.271	4	$(3/2-13/2)^+$	5.0± 0.9	2.294	2	$(1/2-9/2)^+$	3.1± 0.4
0.644	2	$(1/2-9/2)^+$	26± 2	2.322	2+4+6	$(7/2,9/2)^+$	33± 2
0.700	2	$(1/2-9/2)^+$	8.8± 1.2	2.339	3	$(1/2-11/2)^-$	11± 1
1.048	2	$(1/2-9/2)^+$	149± 5	2.380	2+4+6	$(7/2,9/2)^+$	22± 1
1.213	2	$(1/2-9/2)^+$	159± 5	2.403	3	$(1/2-11/2)^-$	23± 1
1.250	2	$(1/2-9/2)^+$	29± 2	2.412	3+5+7	$(9/2,11/2)^-$	11± 1
1.334	2	$(1/2-9/2)^+$	44± 3	2.419	6	$(7/2-17/2)^+$	2.4± 0.4
1.366	5	$(5/2-15/2)^-$	29± 2	2.448	0	$5/2^+$	13± 1
1.413	3	$(1/2-11/2)^-$	3.8± 0.8	2.475	3+5+7	$(9/2,11/2)^-$	8.2± 0.7
1.469	2	$(1/2-9/2)^+$	28± 2	2.490	4	$(3/2-13/2)^+$	5.2± 0.6
1.646	0	$5/2^+$	25± 1	2.514	1	$(3/2-7/2)^-$	17± 1
1.662	2	$(1/2-9/2)^+$	3.5± 0.5	2.527	0	$5/2^+$	33± 2
1.675	4	$(3/2-13/2)^+$	0.8± 0.2	2.539	2	$(1/2-9/2)^+$	23± 1
1.727	2	$(1/2-9/2)^+$	1.7± 0.4	2.554	3	$(1/2-11/2)^-$	20± 1
1.750	2	$(1/2-9/2)^+$	41± 2	2.586	3+5+7	$(9/2,11/2)^-$	13± 1
1.821	2	$(1/2-9/2)^+$	3.0± 0.4	2.622	3+5+7	$(9/2,11/2)^-$	14± 1
1.875	2	$(1/2-9/2)^+$	5.3± 0.6	2.637	2	$(1/2-9/2)^+$	11± 1
1.968	4	$(3/2-13/2)^+$	2.0± 0.4	2.670	3+5+7	$(9/2,11/2)^-$	15± 1
2.019	4	$(3/2-13/2)^+$	2.8± 0.4	2.687	3+5+7	$(9/2,11/2)^-$	7.3± 0.7
2.038	5	$(5/2-15/2)^-$	3.4± 0.5	2.728	2	$(1/2-9/2)^+$	11± 1
2.094	2+4+6	$(7/2,9/2)^+$	14± 1	2.755	4	$(3/2-13/2)^+$	6.1± 0.6
2.114	4	$(3/2-13/2)^+$	12± 1	2.777	3	$(1/2-11/2)^-$	7.6± 0.7
2.130	3	$(1/2-11/2)^-$	53± 2	2.788	3	$(1/2-11/2)^-$	8.2± 0.7
2.138	6	$(7/2-17/2)^+$	3.0± 0.5	2.803	3+5+7	$(9/2,11/2)^-$	23± 1
2.162	3	$(1/2-11/2)^-$	2.1± 0.4	2.815	9	$(13/2-23/2)^-$	12± 1
2.194	2+4+6	$(7/2,9/2)^+$	40± 2	2.829	3	$(1/2-11/2)^-$	5.2± 0.6
2.202	6	$(7/2-17/2)^+$	11± 1	2.849	5	$(5/2-15/2)^-$	23± 1
2.232	3	$(1/2-11/2)^-$	64± 2	2.874	2	$(1/2-9/2)^+$	7.5± 0.7

In general, more than one L-transfer contribute to a given final state in (p, t) reaction from non zero-spin target nucleus. As a consequence the angular distribution will be composed by all the allowed L-transfers, whose contributions must be incoherently added.

When only one L-transfer dominates a given transition amplitude, the accuracy of the spectroscopic information is higher.

The measured differential cross sections display two kinds of shapes: one exhibits relevant angular structure, significant enough to allow different L-transfers to be distinguished; the other, rather featureless, is distinctive of more L-transfer contributions. The DWBA analysis of the experimental angular distributions has been carried out assuming a one-step semimicroscopic dineutron cluster pickup. The calculations have been performed using the computer code TWOFNR[9] in a finite range approximation. The parameters for the proton entrance channel are deduced from a systematic survey of elastic scattering by Perey[10] and for the triton exit channel by Fleming et al.[11] Transferred angular momentum L has been assigned for all the observed transitions by comparing the shapes of the experimental $d\sigma/d\Omega$ with the calculated ones.

The angular distribution of most part of the observed transitions are reproduced assuming only one L-transfer, while eleven transitions display rather featureless angular distributions, distinctive of more L-transfer contributions. Examples of typical analysis are shown in Fig. 1.

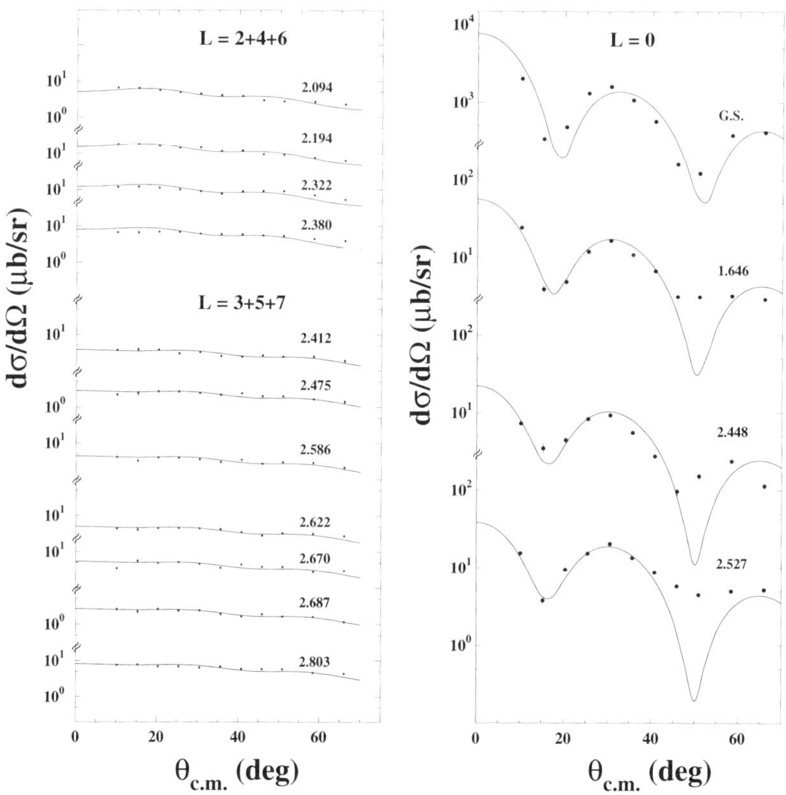

Fig. 1. Example of cluster DWBA calculations.

3. Theoretical analysis

Microscopic calculations of the ^{121}Sb$(p,t)^{119}$Sb reaction integrated cross sections have been performed since it is not possible to unequivocally identify the spin of most of the excited levels of ^{119}Sb from the analysis of the angular distributions. Firstly, calculations are carried out for the neighboring even-even ^{118}Sn nucleus.

QPM calculations for the residual even-even nucleus ^{118}Sn have been performed with a wave function of excited states with a multipolarity J, written as a combination of one- and two-phonon configurations:

$$\Psi^{\nu}(JM) = \left\{ \sum_i R^{\nu}_{Ji}\, Q^{+}_{JMi} + \sum_{\lambda_1 i_1 \lambda_2 i_2} T^{\nu}_{\lambda_1 i_1 \lambda_2 i_2}(J) \frac{[Q^{+}_{\lambda_1 i_1} Q^{+}_{\lambda_2 i_2}]_{JM}}{\sqrt{1+\delta_{\lambda_1 i_1, \lambda_2 i_2}}} \right\} |^{118}\,\mathrm{Sn}\rangle_{g.s.} \quad (1)$$

where $Q^{+}_{\lambda\mu i}$ is a phonon creation operator and $|^{118}\,\mathrm{Sn}\rangle_{g.s.}$ is the ground state wave function of this nucleus. The properties of the phonon excitations have been obtained by solving quasiparticle RPA equations for different natural-parity multipolaries from 0^+ to 7^-. These equations yield phonon's energies (where index $i = 1,\ 2,\ \ldots$ in the phonon operator means its order number) and an internal fermion structure of each phonon. The strength of the residual interaction in the model Hamiltonian has been adjusted to the experimental properties of the collective 2^+_1 and 3^-_1 levels. Coefficients R and T in the wave function (1) and eigen energies of these states have been obtained by diagonalization of the model Hamiltonian in the set of (1).

Assuming one-step mechanism of the reaction, the cross section of the $^{120}\mathrm{Sn}_{g.s.}(\mathrm{p,t})^{118}\mathrm{Sn}_{LMi}$ reaction has the form:

$$\frac{d\sigma}{d\Omega}(^{120}\mathrm{Sn}_{g.s.} \rightarrow ^{118}\mathrm{Sb}_{J\nu}) = \left| \sum_i R^{\nu}_{Ji}\, A_{Ji} \right|^2 \quad (2)$$

where A_{Ji} are the reaction amplitudes. These amplitudes have been extracted from experimental data to reach the best agreement between calculation and experiment for each multipolarity J.

Calculations for the odd-mass ^{119}Sb have been performed with a wave function which contains quasiparticle-, [quasiparticle x one-phonon]-, and [quasiparticle x two-phonon]-configurations:

$$\Psi^{\nu}(JM) = \left\{ C^{\nu}(J)\, \alpha^{+}_{JM} + \sum_{j\lambda i} S^{\nu}_{j\lambda i}(J)\, [\alpha^{+}_j Q^{+}_{\lambda i}]_{JM} + \sum_{j\lambda_1 i_1 \lambda_2 i_2 l} D^{\nu}_{j\lambda_1 i_1 \lambda_2 i_2}(J) \frac{[\alpha^{+}_j [Q^{+}_{\lambda_1 i_1} Q^{+}_{\lambda_2 i_2}]_l]_{JM}}{\sqrt{1+\delta_{\lambda_1 i_1, \lambda_2 i_2}}} \right\} |^{118}\,\mathrm{Sn}\rangle_{g.s.} \quad (3)$$

where α^{+}_{jm} is a quasiparticle creation operator (on a proton mean field level $jm = |nljm>$ and $Q^{+}_{\lambda\mu i}$ is a phonon excitation in the ^{118}Sn core. To obtain the coefficient C, S, and D in (3) we diagonalize the model Hamiltonian on the set of wave functions (3). This diagonalization performed for different J also yields eigen energies of the odd nucleus excited states. The ^{121}Sb$(p,t)^{119}$Sb cross section to the final state $J\nu$ by momentum transfer L, has the form:

$$\frac{d\sigma}{d\Omega}(^{121}\mathrm{Sb}_{g.s.} \rightarrow ^{119}\mathrm{Sb}_{J\nu}) = \Big| C^{\nu}(J)\ \delta_{J,5/2^+}\ A_{g.s.} + \sum_i S^{\nu}_{jLi}(J)\ \delta_{j,5/2^+}\ A_{Li} \Big|^2 \tag{4}$$

where $J = 5/2^+$ is the ground state of ^{119}Sb. Since experimental conditions for the ^{120}Sn$(p,t)^{118}$Sn and ^{121}Sb$(p,t)^{119}$Sb reaction were very close, we have used in (4) amplitudes A from (2), and the amplitude $A_{g.s.}$ corresponds to the transition between the ground states $^{120}\mathrm{Sn}_{g.s.} \rightarrow ^{118}\mathrm{Sn}_{g.s.}$. In fact when the ^{119}Sb is excited, the same set of phonons of the core ^{118}Sn is involved and the unpaired proton of Antimony does not influence the excitation process in a one step transfer. In Fig. 2 the experimental integrated cross sections of the ^{121}Sb$(p,t)^{119}$Sb reaction (top) is compared with the complete calculations (bottom) carried out with the the wave function (3) and with the simplified calculations (middle) in which the unpaired proton is considered as a pure spectator.

For that, we have switched off the term of the residual interaction in the model Hamiltonian which is responsible for mixing between simple and complex configurations. Then the wave function for an excited state has the form of (3) in which only one term C, S, or D is equal to one and the rest are zero. Of course, in this case all multiplets are energy degenerate. The spectator approach reproduces the general features of the experimental distribution, but is not able to describe the splitting of the multiplets. On the contrary the realistic calculations performed with the wave function (3) reasonably well reproduce at higher excitation energy the fragmentation of the (p,t) cross section and the absence of the (p,t) strength above 2.9 MeV.

In the excitation energy region up to 2.597 MeV, besides the ground state, three $5/2^+$ states have been identified because of L=0 transfer. The corresponding experimental integrated cross sections are plotted in Fig.4a. The theoretical predictions for all the four $5/2^+$ states are shown in Fig.4b. The agreement between experiment and theory for these states, both in position and integrated (p,t) cross sections, is very good.

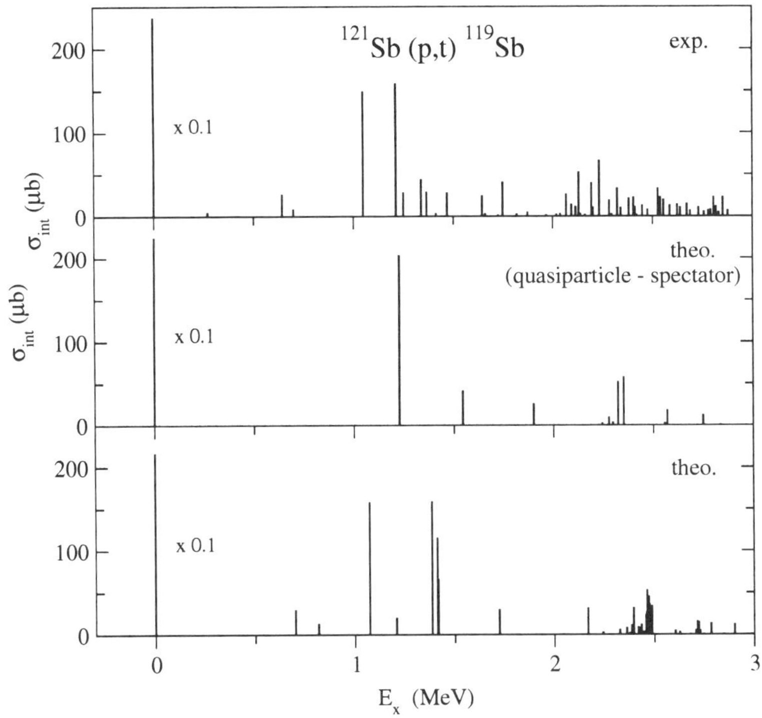

Fig. 2. Comparison between experimental and QPM integrated cross sections.

4. Summary

The high resolution measurement of the (p,t) reaction angular distributions for the transitions to the levels of ^{119}Sb nucleus allows us to confirm or determine the energies of 59 levels, 23 of which have been observed for the first time and to determine the angular momentum transfer values and a well-defined range for the J values. The DWBA analysis has been performed in a finite range approximation, assuming a dineutron cluster pickup mechanism, and using conventional Woods-Saxon potential for the entrance proton and exit triton channel.

The present (p,t) data have been supplemented by microscopic calculations in the framework of QPM, giving a reasonably good description of the experimental fragmentation of the cross sections and the absence of (p,t) strength above 2.9 MeV.

The $5/2^+$ states identified are well reproduced by the theoretical calculations both in energies and integrated cross sections.

Simplified calculations in which the unpaired quasiparticle is considered

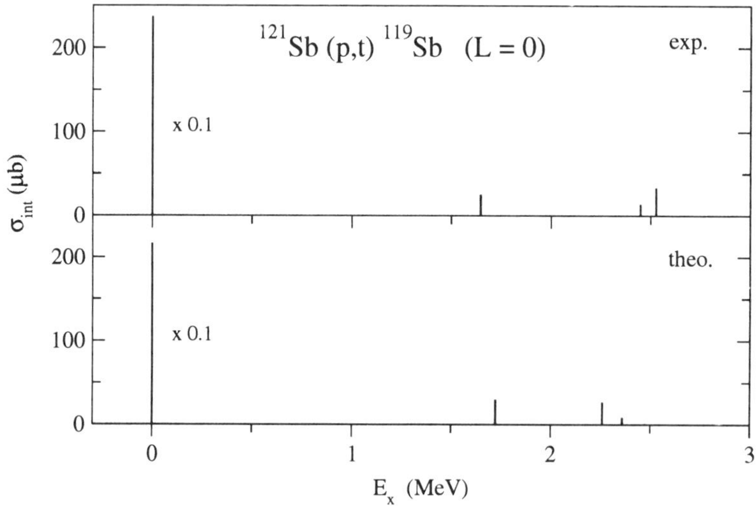

Fig. 3. Experimental and calculated $5/2^+$ states (L_{tran}=0).

as pure spectator are able to reproduce the general features of the (p, t) cross section distribution, but fail to describe fragmentation.

References

1. P. Guazzoni, M. Jaskóla, L. Zetta, J.N. Gu, A. Vitturi, G. Graw, R. Hertenberger, D. Hofer, P. Schiemenz, B. Valnion, U. Atzrott, G. Staudt, *Z. Phys.* **A356**, 381 (1997).
2. P. Guazzoni, M. Jaskóla, L. Zetta, J.N. Gu, A. Vitturi, G. Graw, R. Hertenberger, P. Schiemenz, B. Valnion, U. Atzrott, and G. Staudt, *Eur. Phys.* **A1**, 365 (1998).
3. P. Guazzoni, L. Zetta, B. F. Bayman, A. Covello, A. Gargano, G. Graw, R. Hertenberger, H.-F. Wirth, and M. Jaskóla, *Phys. Rev.* **C72**, 044604 (2005).
4. P. Guazzoni, M. Jaskóla,V. Yu. Ponomarev, L. Zetta, G. Graw, R. Hertenberger, and G. Staudt, *Phys. Rev.* **C62**, 054312 (2000).
5. H. R. Hooper, P. G. Green, H.E. Siefken, G. C. Neilson, W. J. McDonald, D. M. Sheppard, and W. K. Dawson, *Phys. Rev.* **C20**, 2941 (1979).**A310**
6. P. Guazzoni, L. Zetta, T. Faestermann, G. Graw, R. Hertenberger, H.-F. Wirth, and M. Jaskóla, these proceedings.
7. V. G. Soloviev, Theory of Complex Nuclei (Pergamon Oxford, 1976).
8. E. Zanotti, M. Bisenberger, R. Hertenberger, H. Kader, G. Graw, *Nucl. Instrum. Methods Phys. Res.* **A310**, 706 (1991).
9. M. Igarashi, computer code TWOFNR (1977) unpublished.
10. F. G. Perey *Phys. Rev.* **131**, 745 (1963).
11. D. G. Fleming, M. Blann, H. W. Fulbright, and J. A. Robbins, *Nucl. Phys.* **A157**, 1 (1970).

CHIRAL BANDS IN NUCLEI ?

I. HAMAMOTO

Division of Mathematical Physics, LTH, University of Lund,
P.O.Box 118, S-22100 Lund, Sweden
E-mail: ikuko@MATFYS.LTH.SE

G. B. HAGEMANN

The Niels Bohr Institute,
Blegdamsvej 17, DK-2100 Copenhagen Ø, Denmark

The observation of two approximately degenerate $\Delta I = 1$ rotational bands with the same parity has been often taken as a sign of chiral bands. A critical analysis of observed electromagnetic properties of the doublet bands in the nucleus ${}^{134}_{59}\mathrm{Pr}_{75}$, which is so far experimentally best studied, is carried out. It is concluded that chiral pair bands have not yet been identified in nuclei.

Keywords: chiral bands; electromagnetic properties; ${}^{134}\mathrm{Pr}$

1. Introduction

Spontaneous symmetry breaking is an interesting and important phenomenon. In particular, spontaneous formation of handedness or chirality is a subject of general interest in molecular physics, in the characterization of elementary particles, and in optical physics. Chirality in triaxial nuclei is a topic of current interest in nuclear-structure physics, being intensively studied. It is characterized by the presence of three angular-momentum vectors, which are non-coplanar and thereby make it possible to define chirality.[1] In the possible candidates for chiral bands of odd-odd nuclei in the A≈130 region the three angular momenta consist of the angular momentum of collective rotation $\vec{R}$, that of an odd neutron-hole $\vec{j_n}$ and that of an odd proton-particle $\vec{j_p}$, where both the odd proton-particle and odd neutron-hole are in the $h_{11/2}$ shell. The core rotational angular-momentum prefers to pointing to the direction of the intermediate axis of the triaxial shape, in a way similar to the case for hydrodynamical moments of inertia. On the other hand, one particle (hole) in a high-j shell would point to the direction of the shortest (longest) axis, in order to obtain the largest (smallest) overlap with the triaxial core. The odd-odd nucleus, of which the so-called chiral pair bands are best studied so far, including electromagnetic properties, is ^{134}Pr.

Having three non-coplanar vectors and thereby realizing chiral geometry, in the intrinsic system one writes left- and right-handed geometry states with a given I as $\mid IL\rangle$ and $\mid IR\rangle$, respectively, where $\vec{I} = \vec{R} + \vec{j}_p + \vec{j}_n$. Since the total Hamiltonian is invariant under the space reflection, $R \leftrightarrow L$, two chiral-degenerate states in the laboratory system are written as[2]

$$\mid I+\rangle = \frac{1}{\sqrt{2}}(\mid IL\rangle + \mid IR\rangle) \tag{1}$$

$$\mid I-\rangle = \frac{i}{\sqrt{2}}(\mid IL\rangle - \mid IR\rangle) \tag{2}$$

For states with $I \gg 1$ it is expected that

$$\begin{aligned} \langle IL \mid E2 \mid IR\rangle &\approx 0 \\ \langle IL \mid M1 \mid IR\rangle &\approx 0 \end{aligned} \tag{3}$$

where the electric quadrupole and magnetic-dipole operators are denoted by $E2$ and $M1$, respectively. If $\approx$ in (3) is replaced by $=$, then within the pair of chiral bands

$$\begin{aligned} B(EM; I_2+ \to I_1+) &= B(EM; I_2- \to I_1-) \\ B(EM; I_2+ \to I_1-) &= B(EM; I_2- \to I_1+) \end{aligned} \tag{4}$$

where EM expresses either $E2$ or $M1$.

In actual nuclei chiral bands may be realized at best only in a region of medium values of I. This is because around the band head the collective rotational angular momentum $\vec{R}$ is not large enough to gain the total energy by pointing to the intermediate axis, along which the moment of inertia is largest. Namely, in order to construct a given value of I, for small values of I (then, small values of R) the energy of the state in which the vector $\vec{R}$ lies in the plane defined by two vectors, $\vec{j}_p$ and $\vec{j}_n$, can be smaller, depending on the ratios of three moments of inertia of the core. On the other hand, at high spins both $\vec{j}_p$ and $\vec{j}_n$ would align to the direction of the total angular momentum and consequently chiral geometry will be destroyed.

The observation of approximately degenerate $\Delta I = 1$ rotational bands with the same parity has been so far taken as a sign of chiral bands. However, the electromagnetic properties carry more stringent information on the intrinsic structure than energies.[3] In ideal chiral pair bands all corresponding properties such as energies, spin alignments, shapes, electromagnetic properties, etc. must be identical or, in practice, very similar. The observation of the identity is the necessary condition for claiming chiral bands. Furthermore, in the case of chiral pair bands with $j_p = j_n$ and $\gamma = -30°$ one can assign a quantum number to eigenstates and obtains more specific detailed selection rules for electromagnetic transitions.[4]

In the present talk I limit myself to the discussion of the shape of the observed pair bands in ^{134}Pr, in which the observed two bands are approximately degenerate in the spin region of $I = 14-18$. In fact, available observed values of corresponding B(M1) and B(E2) in the two bands are considerably different.[5–8] Using available experimental information, we extract the shape difference of the so-called chiral pair bands.[3]

In Sec. 2 the relevant level scheme of ^{134}Pr, our way of analyzing data, the numerical results obtained, and discussions are presented, while a conclusion is drawn in Sec. 3.

2. Model, Analysis of Data, and Discussions[3]

In Fig. 1 the observed level scheme of ^{134}Pr is shown.[5,9] Bands 1 and 2 are organized so that observed strong ΔI=2, E2 transitions are identified as intra-band transitions. Band 1 is yrast for $I \leq 15$, while the band 2 is yrast for $16 \leq I \leq 20$. Namely, the two bands cross between $I = 15$ and 16. The observed energy difference, $E(I)_2 - E(I)_1$, is 163, 36, -44, -118 and -173 keV for $I = 14$, 15, 16, 17 and 18, respectively. (Or, in the notation used later, $\Delta E_I \equiv E(I)_{ny} - E(I)_y$ is 163, 36, 44, 118 and 173 keV for $I = 14$, 15, 16, 17 and 18, respectively.) Using observed intra- and inter-band E2 transitions for the initial states with $I = 15-18$, namely in the band-crossing region, we will extract the ratio of the intrinsic quadrupole moment of band 1 to that of band 2.

When a particle-rotor model, in which one proton-particle and one neutron-hole in the $h_{11/2}$ shell are coupled to a $\gamma = -30°$ triaxial rotor, is solved adjusting other parameters to be appropriate for odd-odd nuclei in the A=130 region, one obtains an almost ideal chiral pair bands in the region of $I = 14-21$ and the two bands cross around $I = 15-17$. See, for example, Fig. 3 of Ref. 4. In the calculated result the selection rule of electromagnetic transitions in chiral bands[4] works exactly. Therefore, the observation of crossing bands in Fig. 1 is not in conflict with the interpretation of chiral bands. However, we note that in the case of ideal chiral bands there is no interaction between the two bands. Consequently, a given state

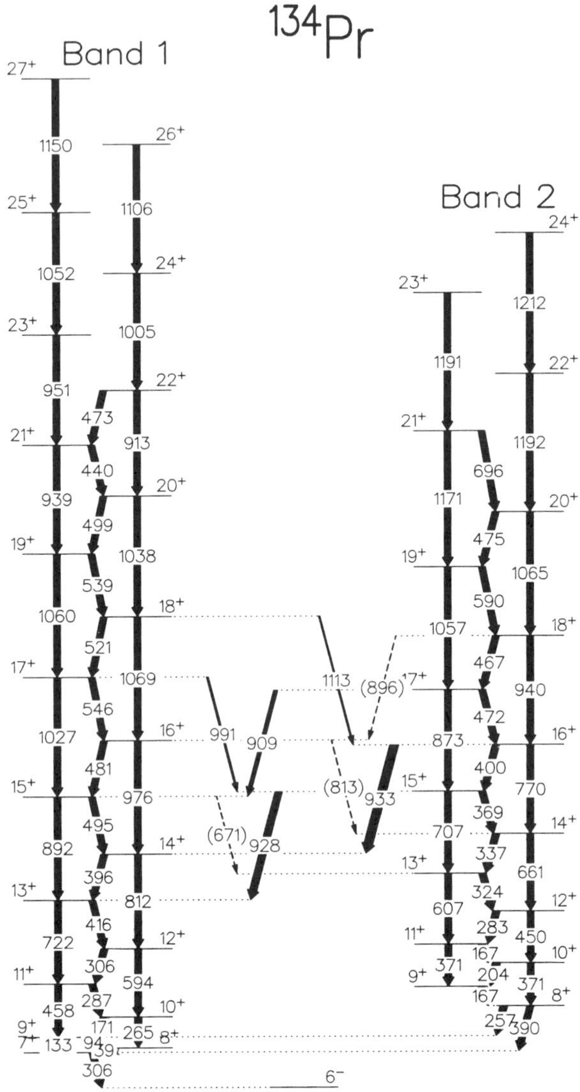

Fig. 1. Observed level scheme of ΔI=1 pair bands in ^{134}Pr. The data are taken from Refs. 9 and 5. A band is defined so that the members are connected by strong ΔI=2, E2 transitions.

cannot decay by both intra and inter ΔI=2, E2 transitions. In contrast, decays of that kind are observed in the I = 15–18 states of ^{134}Pr.

Since in the region of I = 14–18 the members of two bands with a given I lie within 200 keV and no other states with the same I are observed in their neighborhood, we assume that only those two bands (1 and 2) mix with each other. Then, wave functions of the observed states with spins I are written as

$$|y\rangle = \alpha_I|1\rangle + \sqrt{1-\alpha_I^2}|2\rangle \tag{5}$$

$$|ny\rangle = \sqrt{1-\alpha_I^2}|1\rangle - \alpha_I|2\rangle, \tag{6}$$

where $|1\rangle$ and $|2\rangle$ denote the wave functions of the original bands 1 and 2, respectively, while $|y\rangle$ and $|ny\rangle$ denote the observed yrast and non-yrast states, respectively. The relation between the admixed amplitudes α_I, the interaction strength $|V|$ and level spacing ΔE_I can be expressed as

$$\frac{2V}{\Delta E_I} = sin(2 \cdot sin^{-1}(\alpha_I)). \tag{7}$$

The resulting ratios for the transition probabilities between the states with spins I and $I-2$ read:

$$\frac{B[E2, I^{ny} \to (I-2)^{y}]}{B[E2, I^{ny} \to (I-2)^{ny}]} = \left(\frac{Q_{0,1}\alpha_{I-2}\sqrt{1-\alpha_I^2} - Q_{0,2}\alpha_I\sqrt{1-\alpha_{I-2}^2}}{Q_{0,1}\sqrt{1-\alpha_{I-2}^2}\sqrt{1-\alpha_I^2} + Q_{0,2}\alpha_I\alpha_{I-2}} \right)^2 \tag{8}$$

$$\frac{B[E2, I^{y} \to (I-2)^{ny}]}{B[E2, I^{y} \to (I-2)^{y}]} = \left(\frac{Q_{0,1}\alpha_I\sqrt{1-\alpha_{I-2}^2} - Q_{0,2}\alpha_{I-2}\sqrt{1-\alpha_I^2}}{Q_{0,1}\alpha_I\alpha_{I-2} + Q_{0,2}\sqrt{1-\alpha_I^2}\sqrt{1-\alpha_{I-2}^2}} \right)^2 \tag{9}$$

where $Q_{0,1}$ and $Q_{0,2}$ express the E2 transition moments of bands 1 and 2, respectively. Later we use a notation $B(E2)^{y}_{in}$ for $B[E2, I^y \to (I-2)^y]$, etc.

It is noted that the ratio (8) becomes equal to (9) in the limit of either $Q_{0,1}/Q_{0,2} = 1$ (namely, the shape of band 1 is equal to that of band 2) or $\alpha_I = 0$ (namely, the yrast I state is pure $|\ 2\rangle$). In contrast to what is expected for chiral bands with the same shape, the two experimental ratios for the decays of the states 16^{ny} and 16^{y}, 17^{ny} and 17^{y} and 18^{ny} and 18^{y} in ^{134}Pr are 0 ± 1 vs. 1.3 ± 0.3, 0.3 ± 0.1 vs. 0.6 ± 0.1, and 0.22 ± 0.09 vs. < 0.08, respectively. See Table 1. Therefore, already from these differences between observed ratios defined in (8) and (9) it is seen that the intrinsic shape of band 1 is different from that of band 2.

Having measured values of ΔE_I, ΔE_{I-2}, $B(E2, I^{ny} \rightarrow (I-2)^y)/B(E2)^{ny}_{in}$ and $B(E2, I^y \rightarrow (I-2)^{ny})/B(E2)^y_{in}$, one obtains α_I, α_{I-2}, V and $Q_{0,1}/Q_{0,2}$ from Eqs. (7)–(9). In the region of $I = 14$–18 we have eight observed $B(E2, I \rightarrow I-2)$ values and five observed ΔE_I values. Using those thirteen observed quantities, we can determine the following eight quantities: the ratio $Q_{0,1}/Q_{0,2}$, α_I of the five states, V for the odd-I sequence and V for the even-I sequence. We assume that $Q_{0,1}/Q_{0,2}$ is I-independent in the region of $I = 14$–18 and the interaction strength V may be different for the even-I and odd-I sequences.

Table 1. Comparison of experimental and calculated values of branching ratios, $\frac{B(E2, I\rightarrow I-2)_{out}}{B(E2, I\rightarrow I-2)_{in}}$, in the crossing region of the pair-band in ^{134}Pr, using $Q_{0,1}/Q_{0,2} = 2$. Calculated values of $|V|$ and admixed amplitudes α_I are also listed. Experimental data are taken from Ref. 5.

I	$E_{\gamma,out}$ (keV)	$E_{\gamma,in}$ (keV)	B(E2)$_{out}$/B(E2)$_{in}$ Exp.	Calc.	$\|V\|$ (keV)	α_I
17^{ny}	991	1027	0.3±0.1	0.30	13.5	0.114
17^{y}	909	873	0.6±0.1	0.61	13.5	-
15^{y}	671	892	0±1	0.030	13.5	0.905
15^{ny}	928	707	1.1±0.5	0.79	13.5	-
18^{ny}	1113	1069	0.22±0.09	0.16	16.5	0.096
18^{y}	896	940	< 0.08	0.060	16.5	-
16^{ny}	813	976	0±1	0.11	16.5	0.411
16^{y}	933	770	1.3±0.3	1.23	16.5	-

Considering the relatively large ambiguity in the measured $B(E2)_{out}/B(E2)_{in}$ ratios, the possible ranges of V and $Q_{0,1}/Q_{0,2}$ allowed by the experimental data are determined. We find $V \sim 13.5$ and ~ 16.5 keV for the odd and even spin sequences, respectively, with the same range of $Q_{0,1}/Q_{0,2}$ between 1.6 and 2.4. In Table 1 the data are compared with values calculated from these extracted ratios. The calculated values B(E2)$_{out}$/B(E2)$_{in}$ and $|V|$ given in Table 1 are slightly different from those in Table I of Ref. 3 and show the values obtained from an improved analysis after the publication of Ref. 3.

The very small interaction strength $|V| \approx 15$ keV indicates a large difference in the structure of the two bands. For example, the smallness might come from a large shape difference or a difference in some quantum numbers or their chiral character. We note that in the case of ideal chiral bands the interaction strength vanishes.

The value of $Q_{0,1}/Q_{0,2}$ which we obtained clearly indicates that the shapes of the nearly degenerate bands in ^{134}Pr are very different and, thus, they cannot be interpreted as chiral bands. The pronounced difference[7] in the measured values of $B(E2)_{in}$ is in qualitative agreement with this result.

3. Conclusion

Having critically analyzed the observed $\Delta I = 2$, E2 transitions of so-called chiral pair bands in ^{134}Pr, we have obtained a value of 2.0 ± 0.4 for the ratio of the E2 transition moments, $Q_{0,1}$ and $Q_{0,2}$, of the two bands. Thus, there must be a significant difference in the shapes of the two bands. We note that our analysis is not based on the assumption of a rigid shape. The significant difference of the shapes does not agree with what is expected for chiral bands, in spite of the near degeneracy of the observed levels with $13 < I < 19$. It is therefore questionable if and to what extent a reminiscence of the chiral geometry is present in the ^{134}Pr data. We come to conclude that chiral bands in nuclei have not yet been identified.

References

1. S. Frauendorf and J. Meng, Nucl. Phys. **A617**, 131 (1997).
2. K. Starosta, T. Koike, C. J. Chiara, D. B. Fossan and D. R. LaFosse, Nucl. Phys. **A682**, 375c (2001).
3. C. M. Petrache, G. B. Hagemann, I. Hamamoto and K. Starosta, Phys. Rev. Lett. **96**, 112502 (2006).
4. T. Koike, K. Starosta and I. Hamamoto, Phys. Rev. Lett. **93**, 172502 (2004).
5. K. Starosta *et al.* (GS2K009 Collaboration), AIP Conf. Ptoc. no. 610 (AIP, New York, 2002) p.815.
6. T. Koike *et al.*, Phys. Rev. C **67**, 044319 (2003).
7. D. Tonev *et al.*, Phys. Rev. Lett. **96**, 052501 (2006).
8. E. Grodner *et al.*, Phys. Rev. Lett. **97**, 172501 (2006).
9. C. M. Petrache *et al.*, Nucl. Phys. **A597**, 106 (1996).

EVEN- AND ODD-PARITY BANDS IN ^{108,110,112}Ru AND ODD-PARITY DOUBLETS IN ^{106}Mo

J.H. HAMILTON*[1], S.J. ZHU[2], Y.X. LUO[1,3], J.O. RASMUSSEN[3], A.V. RAMAYYA[1], J.K. HWANG[1], X.L. CHE[2], Z. JIANG[2], C. GOODIN[1], P.M. GORE[1], E.F. JONES[1], K. LI[1], S. FRAUENDORF[4], V. DIMITROV[4], J.Y. ZHANG[5], I. STEFANESCU[6], A. GELBERG[7], J. JOLIE[7], P. VAN ISACKER[8], P. VON BRENTANO[7], J.L. WOOD[9], M.A. STOYER[10], R. DONANGELO[11], J.D. COLE[12], N.J. STONE[5,13], J. STONE[5,14]

[1] *Department of Physics, Vanderbilt University, Nashville, TN 37235, USA*
[2] *Physics Department, Tsinghua University, Beijing 100084, Peoples Republic of China*
[3] *Lawrence Berkeley National Laboratory, Berkeley, CA 24720, USA*
[4] *Physics Department, Notre Dame University, Notre Dame, IN 46556, USA*
[5] *Department of Physics, University of Tennessee, Knoxville, Tennessee 37996, USA*
[6] *PKU Leuven, Instituuut voor Kern-en Stralingsfysica, B-3001 Leuven, Belgium*
[7] *University of Cologne, Institute of Kernphysics, Cologne, D-50937, Germany*
[8] *Grand Accelerateur National d'Ions Lourds, CEA/DSM-CNRS/IN2P3, BP 55027 F-14076 Caen Cedex 5, France*
[9] *School of Physics, Georgia Institute of Technology, Atlanta, GA 30332, USA*
[10] *Lawrence Livermore National Laboratory, Livermore, CA 94550, USA*
[11] *Universidade Federal do Rio de Janeiro, CP 68528, RG Brazil*
[12] *Idaho National Laboratory, Idaho Falls, ID 83415, USA*
[13] *Department of Physics, Oxford University, Oxford OX1 3PU, United Kingdom*
[14] *Department of Chemistry and Biochemistry, University of Maryland, College Park, Maryland 20742, USA*
* *E-mail: j.h.hamilton@vanderbilt.edu*

The band structures of ^{108,110,112}Ru have been further investigated in $\gamma-\gamma-\gamma$ coincidence studies of prompt γ rays emitted in the spontaneous fission of ^{252}Cf. New features of the ground and γ vibrational bands are found. Calculations in IBM1 and IBM1+V3 with three body terms were carried out and compared with our data. The comparisons show ^{108}Ru is best fitted as a γ-soft nucleus by IBM1 and ^{110,112}Ru are more like rigid triaxial rotors. New ΔI=1 negative parity doublet bands are found in ^{108,110,112}Ru. There is a change in level spacings in the non-yrast member of the doublet in ^{108}Ru that is not seen in ^{110,112}Ru. This difference is thought to arise from the γ-soft nature of ^{108}Ru. The agreement of our data with the expectations for chiral vibrational bands indicates ^{106}Mo and ^{110,112}Ru are strong candidates for chiral doublets.

Keywords: ^{252}Cf, Gammasphere, Chiral vibrational doublet bands

1. Introduction

The neutron rich A=98-112 region has been a rich source of nuclear structure information for many years [1,2,3]. The rapidly changing structure from spherical to super-deformed ground states was one of the early surprise as described in Refs. 1 and 2. Our recent work on odd Z nuclei in this region has shown an evolution from super-deformed axially symmetric ^{99}Y to maximum triaxiality in ^{111,113}Rh with $\gamma \approx 28^o$ [4-7]. We have recently experimentally and theoretically investigated ^{108,110,112}Ru [8,9,10]. Comparisons of our IBM calculations with our experimental results for the even parity ground and quasi-γ vibrational bands indicate a change from a γ-soft type nucleus in ^{108}Ru to a more rigid maximum triaxiality in ^{110,112}Ru [8]. We have found in ^{108,110,112}Ru pairs of ΔI=1 negative parity doublet bands [10]. Our angular correlation data establish spins for several of the band heads and the dipole multipolearity of the depopulating transitions to support their assignments as negative parity bands. The life times of the band heads were measured to be less than 1 ns to indicate K is not a good quantum number. These bands and our earlier reported doublets in ^{106}Mo [11] have all the expected chracteristics of chiral bands as proposed by Frauendorf and collaborators [12,13,14]. One of the doublet bands in ^{108}Ru exhibits an energy staggering that can be understood as a pertubation brought on by the γ softness as was used to explain the energy pattern of the doublet bands in ^{106}Ag recently [15]. All the evidences for chiral structure will be presented. The detail studies of these full ΔI=1 bands were made possible by our high statistics data set, 5.7x10^{11} triple and higher folded coincidences, taken with a 62 μCi ^{252}Cf source in Gammasphere (see Refs. 4 and 5 for futher experimental details).

2. Even parity bands in ^{108,110,112}Ru

Our new experimental results for ^{110}Ru revealed unusual features with the ground and quasi-γ bands both have nearly degenerate in energy side bands that feed into them around 10^+. This behavior is not seen in other nuclei in the region and the origin of these extra bands is not known. The quasi-γ bands show an unexpected change in the odd-even spin energy staggering between ^{108}Ru and ^{100,112}Ru as seen Fig. 1. IBM1 calculations were carried out for all three nuclei. The even-even Ru nuclei can be described as transitional, situated between the U(5) (spherical vibration) and SO(6) (γ-unstable rotor) symmetries of the IBM1. First a Hamiltonian with only one and two body terms was used. Excitation energies and B(E2) ratios

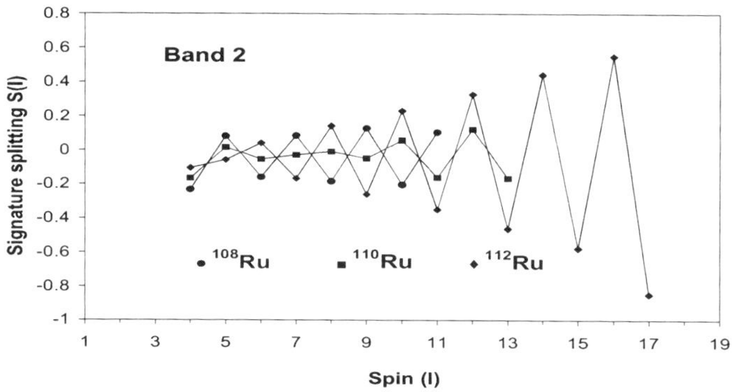

Fig. 1. Signature splitting S(I) of band (2) in ^{110,112}Ru

of the gamma transitions were calculated. Careful attention was given to extracting the experimental branching ratios by using double gated spectra.

The agreement of the calculated and experimental excitation energies and B(E2) ratios was quite satisfactory for ^{108}Ru in this first calculation. The odd-even spin energy band staggering in the quasi-γ bands in ^{110,112}Ru could not be fitted in these calculations. The pattern is like that of a rigid triaxial rotor in ^{110,112}Ru but not that of a γ soft SU(6) nucleus. This descrepancy and better fits to the branching ratios and ground band energies, were obtained by including three body terms in the Hamiltonian denoted IBM1+V3. This produces an energy surface with a triaxial minimum. The presence of the strong triaxiality in these two nuclei is supported by our work on ^{111}Tc and ^{111,113}Ru [4,7] where near maximum triaxiality is found.

3. Negative parity doublet bands - Chiral doublets

We have identified extended negative parity $\Delta I = 1$, doublet bands in ^{106}Mo [11] and ^{108,110,112}Ru [10]. From our $\gamma - \gamma(\theta)$ data, we have determined the spins of the one member of both bands in ^{110}Ru and of one band in ^{106}Mo and ^{108,112}Ru as well as established the dipole nature of the depopulating electromagnetic transitions. The latter support the negative parity assignments. All of the band heads are measured to have lifetimes less than 1 ns so these cannot be high K axial rotor bands. In contrast to ^{110,112}Ru, the non-yrast band in ^{108}Ru exhibits an odd-even spin staggering pattern not seen in its partner nor in any of these bands in ^{110,112}Ru.

Frauendorf and his collabarator [12,13,14] proposed that nuclei could have chiral symmetry doublet bands related to angular momenta. In well deformed triaxial nuclei, when there are substantial components of the angular momentum along all three axis, chiral doublet bands can occur. Chiral symmetry breaking in nuclei gives rise to two sets of nearly degenerate in energy for the same spin, ΔI=1 bands of the same parity. Such can occur in a well deformed triaxial nucleus when one has a high-j proton (neutron) particle with its angular momentum aligned along the short axis, a high-j neutron (proton) hole with its angular momentum aligned along the long axis, and the rotational angular momentum aligned along the intermediate axis so the axis of rotation is not along any of these three axises.

One of the first reports of chiral bands was in ^{134}Pr as discussed in Ref. 14. A number of cases were reported in this region and then around A=105 for example in ^{104}Rh [16] and ^{106}Rh [17]. It was reported that "the best chiral properties observed to date "are in ^{104}Rh [16]. These results were followed by the first report of chiral doublet bands in an even-even nuclei ^{106}Mo [11] with a vibrational character.

Then Petrache et al. [18] pointed out the possible misinterpretation of nearly degenerate pairs of bands as chiral partners in nuclei for example in ^{134}Pr using new lifetime measurements [19]. The degeneracy of the 15^{+} and 16+ states are related to multiple band crossings [18]. More important, they emphasize that the ratio of 2.0(4) for the E2 transition moments for the two bands indicates they have different intrinsic structures since the ratio should be one if the two sets of bands are chiral. They also note that the present data for the two sets of bands in ^{136}Pm are not in favor of their reported chiral character. They emphasize that chiral doublet bands should have identical or very similar energies, spin alignments, shapes and electromagnetic transition probabilities.

Recently Joshi et al [15] reported on the effect of γ softness on the stability of the chiral symmetry in ^{106}Ag. Two strongly coupled negative parity bands which cross each other at I$\approx$14 were observed in ^{106}Ag. They suggest that the quite different energy spacings for the levels in the two bands is related to differences in shapes, the yrast one having a triaxial shape while its partner may be explained in terms of an axial symmetric shape. They suggest that "In such γ-soft nuclei the one phonon chiral vibrational state which is responsible for the excited band, may generate a transformation from a triaxial shape to an axial shape "[15]. By looking at the kinematic moments of inertia and quasiparticle alignments as a function of spin for doublet bands in odd-odd nuclei in A≈100 region, they propose that the bands ^{106}Rh are the best candidate to have a chiral structure.

We have compared our ΔI=1 doublet bands for ^{106}Mo and ^{108,110,112}Ru with those of ^{106}Ag[15], ^{104,106}Rh [16,17] where chiral structures are reported. We have already noted the change in structure from γ-softness in ^{108}Ru to more maximum triaxiality in ^{110,112}Ru and their odd Z neighbors.

Since the two quasiproton states lie at higher energy than the two quasi neutron states, our new negative parity bands are interpreted as two quasi neutron excitations. The lowest configuration corresponds to the excitation of a neutron from the highest $h_{11/2}$ level to the low-lying mixed $d_{5/2}$-$g_{7/2}$ levels. Our 3D-Tilted Axis Cranking (TAC) calculations yielded $\beta_2 = 0.31$ and $\gamma = 31^o$ for ^{106}Mo. While the TAC calculations give a $d_{5/2}$ - $g_{7/2}$ neutron hole strongly aligned with the long axis and a $h_{11/2}$ neutron lying in the short intermediate plane, the microscopic TAC calculations cannot be reduced to the simple picture for odd-odd nuclei. Chirality comes about from the interplay of the neutrons in the open shell.

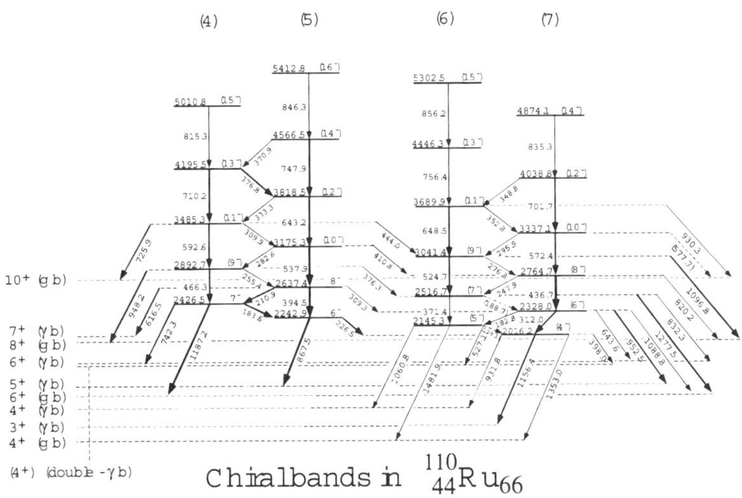

Fig. 2. Proposed chiral bands in ^{110}Ru

Table 1. Branching ratios of the in band E2 cross over and the related M1(+E2) transitions in doublet bands 4-5 and 6-7 in ^{110,112}Ru and band 4-5 in ^{106}Mo.

	^{110}Ru		^{112}Ru		^{106}Mo	
spin	4-5	6-7	4-5	6-7	4	5
13	>4.4(6)					
12	9.2(11)	6.4(9)	>4.0(6)	5.1(8)	>6.0	>4.8
11	5.9(6)	1.14(15)	7.0(11)	5.6(9)	4.7	>3.9
10	6.9(6)	4.9(5)	5.7(8)	7.0(11)	6.8	8.5
9	3.3(3)	7.9(7)	10.4(17)	8.3(13)	8.3	5.8
8	2.6(3)	3.4(3)		4.2(4)	6.4	2.7
7					3.2	1.3

Another test for these bands being chiral vibrational doublets or being accidentally degenerate from the coupling of say an $h_{11/2}$ neutron to two different neutron bands in ^{105}Mo, and ^{109,111}Ru was carried out. The ratio of the E2 to M1 + (E2) strengths within the two sets of doublet bands in ^{106}Mo, ^{110}Ru and ^{112}Ru were carefully extracted, as given in Table 1. These electromagnetic transition intensity ratios for each spin state are in very good agreement to indicate they have very similar structures as required for chiral doublets.

Calculations were carried out for various quasi-particle configurations with negative parity in ^{106}Mo, and ^{110,112}Ru for both axial shapes and $\gamma=30^o$. In all cases, the B(E2)/B(M1) ratios of the two lowest bands differ typically by one order of magnitude. The TAC calculations are consistent with the findings in the framework of the triaxial rotor plus two quasi-particle model. The clear disagreement of the B(E2)/B(M1) ratios based on various quasi-particle configurations with the experimental data is strong evidence these doublet bands do not arise from the couplings of different quasi-particle configurations which are just accidentally degenerate.

Also the parameter S(I) = [E(I)-E(I-1)]/2I should be constant with increasing spin and be equal for the two doublet if they are chiral doublet [16]. We compare in Fig. 3 S(I) our ^{106}Mo and ^{110}Ru data with that for ^{106}Rh said to be the best candidate for chiral doublets. Both our cases are more constant with spin and are comparable or more equal than for the two bands in ^{106}Rh and ^{104}Rh (not shown).

Finally states of the same spin should be degenerate in energy. In Fig. 4 are shown the same spin level energy differences for ^{106}Mo, ^{110,112}Ru and ^{104,106}Rh. Note our three cases have the most nearly degenerate energies, considerably smaller energy differences than those of ^{106}Rh.

The ^{106}Mo, ^{110,112}Ru ΔI=1 doublet bands have equal electromagnetic intensity ratios, identical and constant with spin S(I) values and are the most nearly degenerate in energy of any of the proposed chiral bands. Thus since they have all the properties of good chiral vibrational bands, we propose that these ΔI=1 doublet bands in ^{106}Mo and ^{110,112}Ru are strong candidates for chiral vibrational bands.

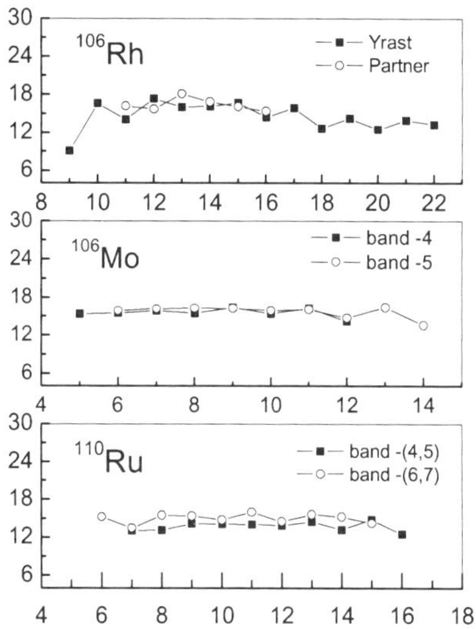

Fig. 3. S(I) for ^{106}Rh, ^{106}Mo and ^{110}Ru.

4. Acknowledgments

The work at Vanderbilt University, Lawrence Berkeley National Laboratory, Lawrence Livermore National Laboratory, and Idaho National Engineering and Environmental Laboratory are supported by U.S. Department of Energy under Grant No. DE-FG05-88ER40407 and Contract Nos. W-7405-ENG48, DE-AC03-76SF00098, and DE-AC07-76ID01570.The work at Tsinghua University was supported by the NNSFC under Grant No. 10575057 and 10375032.

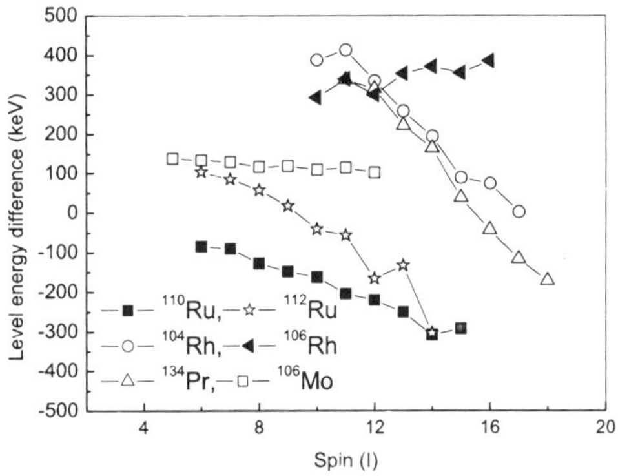

Fig. 4. Energy differences between states of the same spin in the chiral doublets for ^{106}Mo, ^{110,112}Ru, ^{104,106}Rh and ^{134}Pr.

References

1. J.H. Hamilton et al., J. Phys. **G10**, L87 (1984).
2. J.H. Hamilton, *Treatise on Heavy Ion Science* **Vol. 8**, ed. by Allan Bromley, (Plenum Press, New York, 1989) p. 2.
3. J.H. Hamilton et al., Prog. Part. Nucl. Phys. **35**, 635 (1995).
4. Y.X. Luo et al., Phys. Rev. **C69**, 024315 (2004).
5. Y.X. Luo et al., Phys. Rev. **C70**, 044310 (2004).
6. Y.X. Luo et al., J. Phys. **G31**, 1303 (2005).
7. Y.X. Luo et al., Phys. Rev. **C74**, 024308 (2006).
8. I. Stefanescu et al., Submitted to Elsevier Science (2007). Phys. Rev. **C55**, 1146 (1997).
9. S.J. Zhu et al.. Eur. Phys. J. **A**, Submitted (2007).
10. Y.X. Luo et al., To be submitted (2007).
11. S.J. Zhu et al., Eur. Phys. J. **A25**, 459 Suppl.1 (2005).
12. S. Frauendorf, Rev. Mod. Phys. **73**, 463 (2001).
13. V.I. Dimitrov et al., Phys. Rev. Lett. **84**, 5732 (2000).
14. V.I. Dimitrov and S. Frauendorf, Proc. *Third Int. Conf. Fission and Properties of Neutron-rich Nuclei* ed. by J.H. Hamilton et al. (World Scientific, Singpore, 2003) p. 93.
15. P. Joshi et al., Phys. Rev. Lett. **98**, 102501 (2007).
16. C. Varman et al., Phys. Rev. Lett **92**, 032501 (2004).
17. P. Joshi et al., Phys. Lett. **B595**, 135 (2004).
18. C.M. Petrache et al., Phys. Rev. Lett. **96**, 112502 (2006).
19. D. Toner et al., Phys. Rev. Lett. **96**, 052501 (2006).

IDENTIFICATION OF LEVELS IN ^{144}Cs

E. F. JONES, P. M. GORE, Y. X. LUO[1,2,4], J. H. HAMILTON[1], A. V. RAMAYYA[1], J. K. HWANG[1], H. L. CROWELL[1], K. LI[1], C. T. GOODIN[1], J. O. RASMUSSEN[4], AND S. J. ZHU[1,2,3]

[1]*Department of Physics, Vanderbilt University, Nashville, Tennessee 37235 USA*
[2]*Joint Institute for Heavy Ion Research, Oak Ridge, TN 37830 USA*
[3]*Department of Physics, Tsinghua University, Beijing 100084, Peoples Republic of China*
[4]*Lawrence Berkeley National Laboratory, Berkeley, California 94720 USA*

From the analysis of γ-γ-γ coincidence data taken with Gammasphere of the prompt γ rays in the spontaneous fission of ^{252}Cf, a cascade of six transitions, at 108.0, 115.1, 263.8, 404.8, 535.2, and tentatively 679.5 keV, was identified in ^{144}Cs for the first time. The transitions were assigned to a cascade with these energies in ^{144}Cs from their relative intensities and by identifying their coincidences with the known transitions in ^{105}Tc and ^{106}Tc, the 3n and 2n fission partners of ^{144}Cs, and comparing the Tc intensities to the respective Tc yield tables. The energy levels in ^{138,140,142}Cs with N = 83, 85, and 87, respectively, just above the spherical closed shell at N = 82, each have first excited states around 10 keV and second excited states between 16 and 65 keV, and no rotational-type bands. These states are associated with single particle states in the spherical region. With N = 89 in ^{144}Cs, one has crossed the region between N = 88 and 90 where there is a rather sudden change from spherical nuclei with N $\leq$ 88 to significant deformation in N $\geq$ 90 nuclei. The levels in ^{144}Cs look like a rotational band in a more well-deformed nucleus.

Keywords: 144Cs, nuclear structure

1. Introduction

We have used our γ-γ-γ data (5.7x10^{11} triples and higher folds) from the spontaneous fission of ^{252}Cf taken with Gammasphere to identify transitions in ^{144}Cs.

Previous studies[1] have found that ^{138,140,142}Cs with N = 83, 85, and 87, respectively, just above the spherical closed shell at N = 82, each have first excited states around 10 keV and second excited states between 16 and 65 keV. These states are associated with single-particle states in the spherical region. In Figure 1, we see the level scheme of ^{138}Cs with a 10.86 keV first excited state. (figure from NNDC[1]). Figure 2 depicts the level scheme of

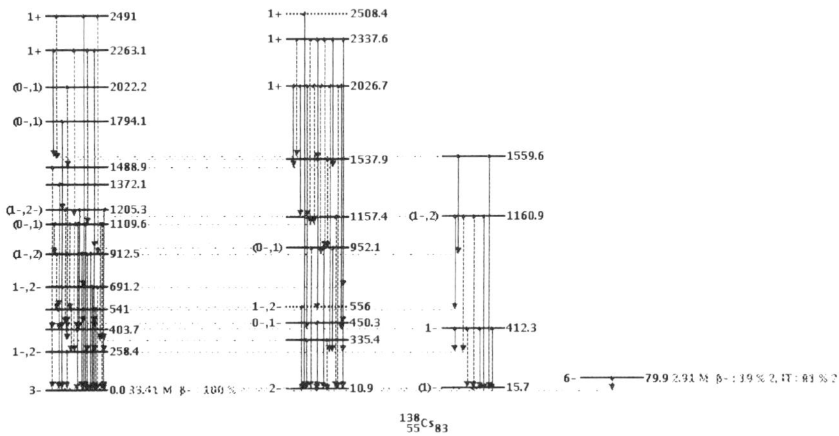

Fig. 1. Level scheme of ^{138}Cs (from NNDC[1]).

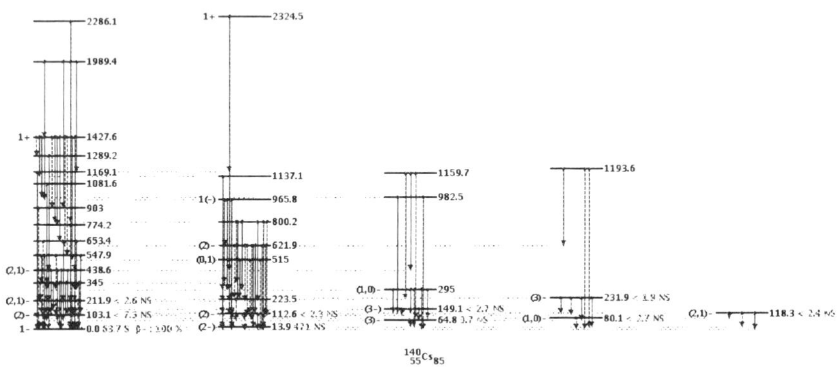

Fig. 2. Level scheme of ^{140}Cs (from NNDC[1])

^{140}Cs where there is seen a 13.91 keV first excited state. Figure 3 shows the level scheme of ^{142}Cs where there is seen a 12.94 keV first excited state.

In Figure 4 are seen the 2^+ level energies of even-even nuclei in this region. We can see that between N = 88 and 90 there is a sudden onset

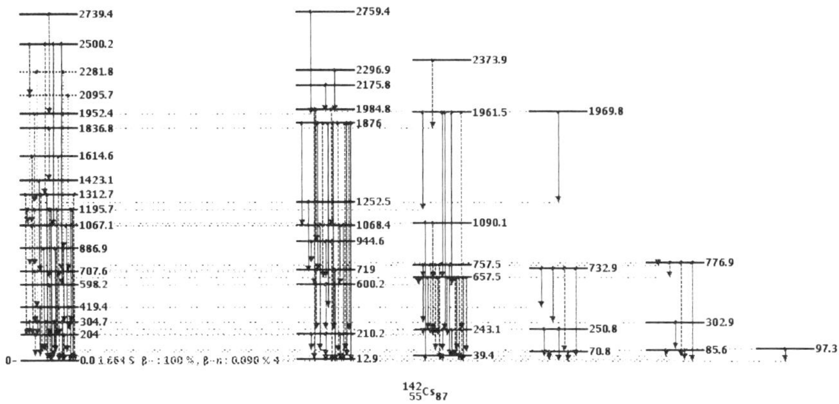

Fig. 3. Level scheme of ^{142}Cs (from NNDC[1]).

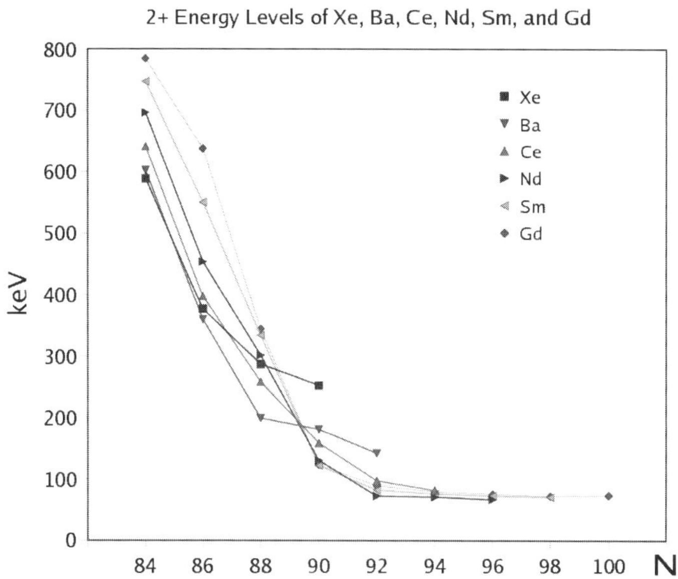

Fig. 4. The 2^+ level energies of even-even nuclei between N = 84 and 100.

from spherical to significant deformation. With N=89 in ^{144}Cs, one has crossed the region between N = 88 and 90 where there is a sudden onset from spherical to significant deformation in N = 90 nuclei.

In a nearby isotone of ^{144}Cs, the known levels in ^{146}La are built on a high-spin isomer[2] (Figure 5).

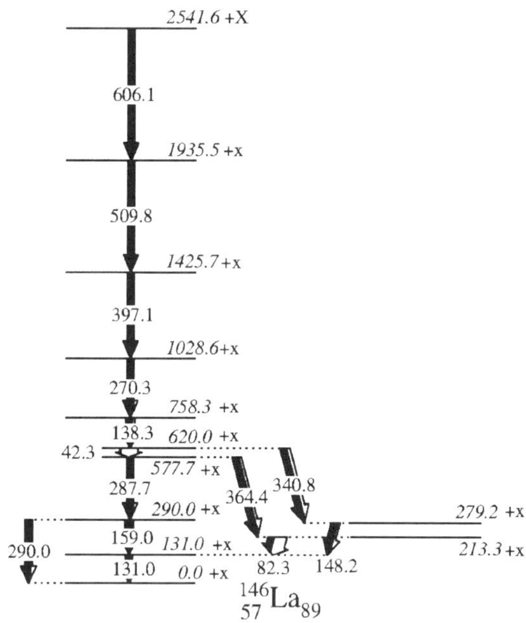

Fig. 5. The known levels in ^{146}La.

2. Results and Analysis

We identified the ^{144}Cs peaks by gating on the fission partners. For example, Figure 6 displays a double gate on the known 85.5 keV peak of the 3n partner, ^{105}Tc, and the proposed 263.5 keV peak of ^{144}Cs. We see peaks in the 3n partner ^{105}Tc as well as other peaks in ^{144}Cs. The spectrum in Figure 7 shows a double gate on the proposed 108.0 keV and 263.8 keV transitions in ^{144}Cs. We can see transitions in the 2n partner ^{106}Tc and the 3n partner ^{105}Tc as well as other transitions in ^{144}Cs. Based on their relative intensities and by identifying their coincidences with the known transitions in ^{105}Tc and ^{106}Tc, we identied a cascade of six transitions, at 108.0, 115.1, 263.8, 404.8, 535.2, and tentatively 679.5 keV in ^{144}Cs. This appears to be

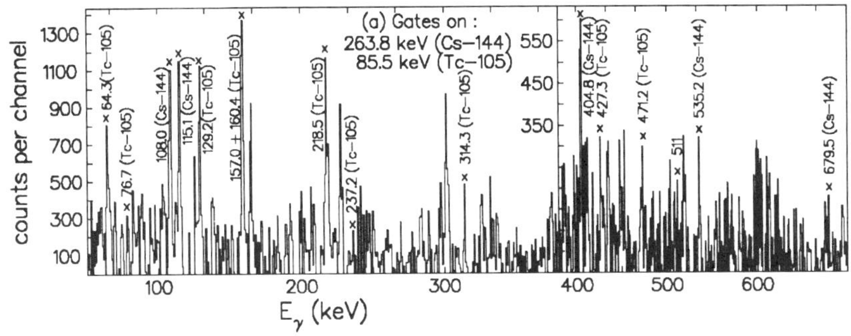

Fig. 6. Double gate on the known 85.5 keV peak of the 3n partner, ^{105}Tc, and the proposed 263.5 keV peak of ^{144}Cs.

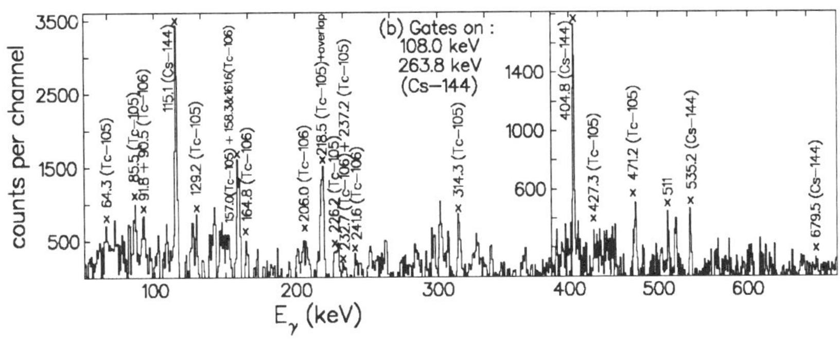

Fig. 7. Double gate on the proposed 108.0 keV and 263.8 keV transitions in ^{144}Cs.

a well-deformed rotational nucleus. Our proposed level scheme is shown in Figure 8.

Previous studies[3] identified a ground state and a high spin isomer in ^{144}Cs. The levels we see in ^{144}Cs are likely a deformed band built on the high-spin isomer, since spontaneous fission populates high-spin states.

3. Summary

We have identified levels in ^{144}Cs. There is a rather sudden change from spherical nuclei with N $\leq$ 88 to signicant deformation in N $\geq$ 90 nuclei. With ^{144}Cs, where N = 89, one has crossed the region between N = 88 and 90. The levels in ^{144}Cs look like a rotational band in a more well-deformed

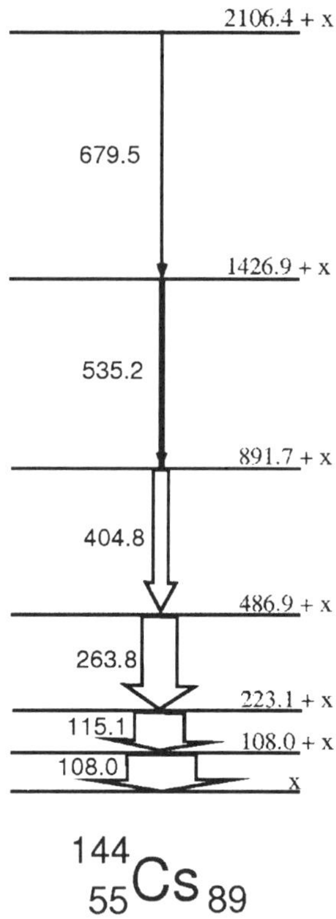

Fig. 8. Proposed level scheme of ^{144}Cs.

nucleus. Because spontaneous fission populates high-spin states, these ^{144}Cs levels are likely a deformed band built on the high-spin isomer.

4. Acknowledgements

The authors are indebted for the use of ^{252}Cf to the office of Basic Energy Sciences, U.S.DOE, through the transplutonium element production facilities at ORNL and acknowledge the essential help of I. Ahmad, J. Greene, and R. V. F. Janssens in preparing and lending the ^{252}Cf source we used in the year 2000 runs. Work at VU, INEEL, LBNL, LLNL, MSU, and ANL was supported by U.S. DOE grants and contracts DE-FG05-88ER40407, DE-AC07-76ID01570, DE-AC03-76SF00098, W-7405-ENG-48, DE-FG05-95ER40939, and W-31-109-ENG-38; Tsinghua by Nat'l. Nat. Sci. Found. of China and Sci. Found. for Nucl. Ind. The JIHIR is supported by U.TN,

VU, ORNL, and U.S.DOE.

References

1. National Nuclear Data Center, Brookhaven National Laboratory, 2007: http://www.nndc.bnl.gov/nudat2/.
2. J.K.Hwang, A.V.Ramayya, J.Gilat, J.H.Hamilton, L.K.Peker, J.O.Rasmussen, J.Kormicki, T.N.Ginter, B.R.S.Babu, C.J.Beyer, E.F.Jones, R.Donangelo, S.J.Zhu, H.C.Griffin, G.M.Ter-Akopyan, Yu.Ts.Oganessian, A.V.Daniel, W.C.Ma, P.G.Varmette, J.D.Cole, R.Aryaeinejad, M.W.Drigert, M.A.Stoyer. Rotational Bands in $^{101-103}$Nb and ^{98,100}Y Nuclei and Identification of Yrast Bands in ^{146}La and ^{149}Pr. *Phys.Rev. C* **58** 3252 (1998).
3. E. Monnand, J. Blachot, F. Schussler, K. Jung, K. Wunsch, B. Fogelberg, J. Feenstra, and J. van Klinken. γ Transitions in the Decay of ^{144}Cs. Priv.Comm. with NNDC (July 1978).

NUCLEAR STRUCTURE AT EXTREME CONDITIONS THROUGH GAMMA SPECTROSCOPY MEASUREMENTS

S. LEONI

University of Milano, Department of Physics and INFN sez. Milano
Via Celoria 16, 20133 Milano, Italy

The γ-decay of the excited nucleus is experimentally investigated in order to obtain information on the nuclear structure properties at extreme conditions of temperature and angular momentum. First, the order-to-chaos transition is here studied in the warm rotating ^{163}Er nucleus, in terms of vanishing of selection rules on the K quantum number. Secondly, the properties of the excited superdeformed bands in ^{151}Tb and ^{196}Pb nuclei are investigated, mostly in connection with their decay-out into the low deformation states. In both cases, comparisons with cranked shell model calculations are presented.

1. Introduction

The collective response of the atomic nucleus under extreme conditions of temperature and angular momentum can be experimentally investigated through the γ-decay of excited nuclei formed by nuclear reactions between heavy ions [1]. In particular, the temperature degree of freedom allows one to study the order-to-chaos transition and the persistence of collectivity and damping mechanisms, while the angular momentum induces changes in pairing correlations and in the nuclear shape [2].

Fast rotating nuclei formed at the highest values of angular momentum and excitation energies de-excites emitting long sequences of γ transitions, ending up in discrete regular rotational bands when the nuclear temperature T of the system is almost zero. Therefore, by detecting the largest number of emitted γ-rays, nuclear structure properties can be investigated as a function of angular momentum and excitation energy. This can be achieved by the use of high-efficiency Ge-array spectrometers (such as EUROBALL [3] and AGATA [4] in the near future), consisting of more than 100 Ge crystals in 4π geometry around the reaction center, usually combined with other types of detectors.

By constructing γ–γ spectra from the measured E_γ energies of high-fold γ cascades, one can distinguish between the contribution from the COLD region of regular decay close to the yrast line ("ridges") and from the WARM region above ≈ 1 MeV internal excitation energy ("valley"). This is where the nuclear rotation becomes damped and the rotational decay is fragmented over a large number of states with an energy spread $\Gamma_{rot} \approx 200$ keV, as a consequence of the rapidly increasing level density and of the presence of a residual two-body interaction mixing the nuclear states [1].

In this paper we discuss results from high statistics EUROBALL experiments on the normal deformed (ND) ^{163}Er nucleus and on the superdeformed (SD) ^{151}Tb and ^{196}Pb nuclei. In the ^{163}Er case we focus on the transition between order and chaos by studying the validity of the selection rules associated with the K quantum number, as a function of internal energy above yrast. In the case of the superdeformed systems we study instead the properties of the discrete excited bands, in particular in connection with the tunneling through the potential energy barrier between the ND and SD well. In all cases the experimental quasi-continuum γ–γ coincidence spectra are analyzed by statistical fluctuations and spectral shape analysis methods and the results are compared with cranked-shell model calculations at finite temperature for the specific nuclei.

1.1. *The order-to-chaos transition in the warm rotating nucleus*

It has been shown, both experimentally and theoretically, that the atomic nucleus displays properties typical of an ordered system at temperature T = 0 [5], and of a chaotic systems at the compound nucleus level [6]. Therefore, the γ-decay of the nucleus at high spins and moderate excitation energy offers the possibility to study the transition between order and chaos in a finite quantum system. This can be done by investigating the gradual vanishing (at T≠0) of the selection rules associated with the good quantum numbers at T = 0, such as K (the projection of the angular momentum on the symmetry axis). In this respect, the deformed nucleus ^{163}Er is an ideal case, being characterized by a number of rotational bands having low-K (K=5/2) and high-K (K=19/2) values [7].

The nucleus ^{163}Er has been populated by the reaction ^{18}O + ^{150}Nd, at E_{beam} = 87, 93 MeV, and its subsequent γ-decay has been detected using the EUROBALL array at the IReS Laboratory (France). The ^{150}Nd target was made of a stack of two thin foils for a total thickness of 740 $\mu g/cm^2$ and the corresponding maximum angular momentum reached in the reaction has been calculated to be 40 and 45 $\hbar$, for the two different bombarding energies. Figure 1 shows examples of 60 keV wide projections, perpendicular to the $E_{\gamma 1} = E_{\gamma 2}$ diagonal of experimental matrices of ^{163}Er, gated by transitions among states with low-K (K=5/2) and high-K (K=19/2) quantum numbers [8]. The spectra, showing the typical ridge-valley structure resulting from the γ-decay of a deformed rotating nucleus, have been first analyzed by statistical analysis methods [9,10].

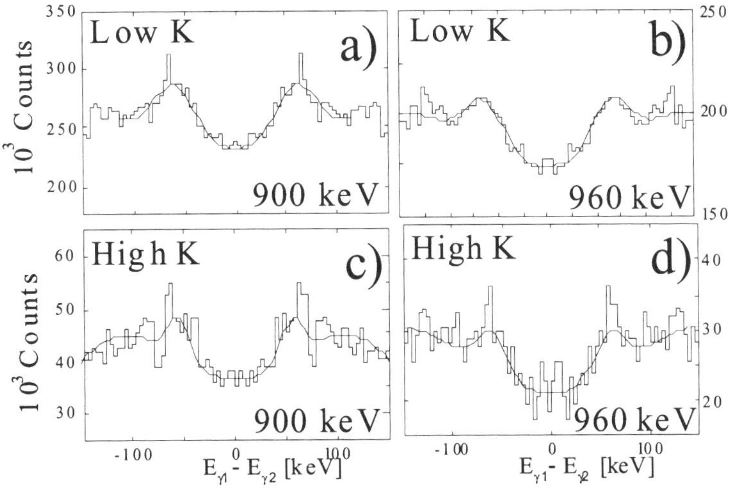

Figure 1. 60 keV wide projections on the $E_{\gamma 1}$-$E_{\gamma 2}$ axis of experimental matrices of ^{163}Er, at the average transition energies E_γ = 900 and 960 keV. Panels a) and b) (c) and d) show spectra obtained from γ–γ matrices gated by low-K (high-K) configurations of ^{163}Er. The smooth curves represent the interpolation of the data by a two-component spectral function [8].

The analysis of the fluctuations of the events collected in the γ–γ spectra allows to estimate the number of bands (named paths, N_{path}) both in the ridge and valley region, through the relation $N_{path} = N_{eve}/(\mu_2/\mu_1-1)$, being N_{eve}, μ_1 and μ_2 the number of events and the first and second moment of the distribution of counts in a given sector [9]. It is found that a total number of ~40 discrete excited bands populates the ridge structures of ^{163}Er, half of which of high-K nature. On the contrary, many more bands, of the order of 10^3-10^5, are found to populate the valley region, with large differences between low-K and high-K states, being the latter ~10 times fewer, as shown in the left panel of figure 2. This suggests that the K-quantum number is at least partially conserved up to moderate excitation energies, of the order of ~1.5 MeV above yrast (corresponding to transition energies E_γ ~1.1 MeV in figure 2), where the rotational motion is damped. At higher excitation energies (namely for E_γ > 1.1 MeV) more similar number of bands are obtained for low-K and high-K gated spectra, pointing to a vanishing of selection rules on K and to the onset of a chaotic regime, in which quantum numbers and selection rules loose their meaning. This result is supported by the analysis of the correlations between spectra gated by different K-states, which can be evaluated in terms of covariance of the spectral fluctuations [10]. The right panel of figure 2 gives the correlation coefficient r obtained from the analysis of the valley region between low-K and high-K spectra. r is obtained from the variance μ_2 and covariance $\mu_{2,cov}$ of the events distributions of the 2 spectra, as discussed in ref. [10] and assumes values between 0 (for completely different event distributions, as in the ordered regime) and 1 (for identical spectra, as in the chaotic limit). The experimental values of r shown in figure 2 gradually approach 1 at the highest transition energy values, indicating strong similarities/correlation between the high-K and low-K gated distributions, as expected in a chaotic regime [11].

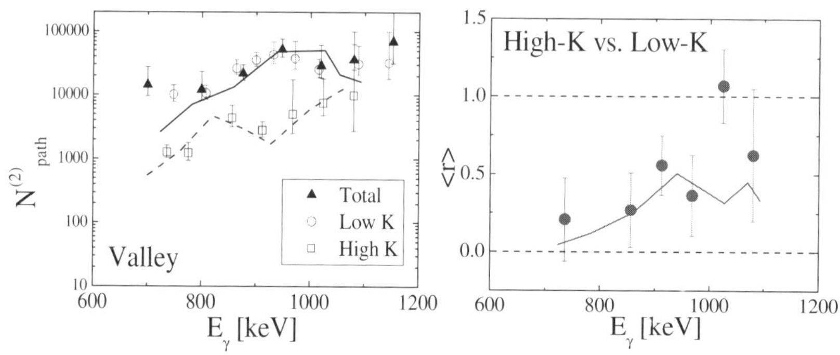

Figure 2. Results of the statistical analysis of the $E_{\gamma 1} = E_{\gamma 2}$ valley region of γ–γ matrices of ^{163}Er. The left panel shows the number of strongly interacting bands (indicated by $N^{(2)}_{path}$), obtained from the fluctuation analysis of experimental spectra gated by Low-K and High-K structures. The lines give the corresponding prediction from cranked-shell model calculation at finite temperature. The right panel gives the results from the experimental analysis of the covariance between Low-K and High-K spectra. The solid line is the cranked-shell model prediction, while the dashed lines represent the values of the r correlation coefficient in the two opposite limits of ordered (r=0) and chaotic regime (r=1) [11].

The experimental results are well reproduced by the shell model of ref. [12] which combines a cranked mean-field and a residual two-body interaction, together with a terms taking into account the angular momentum carried by the K-quantum number. According to the model, K-mixing is induced by the interplay of the Coriolis and residual interaction, and it is found to gradually increase until a complete violation of the K-quantum number is reached around 2-2.5 MeV of internal energy, in good agreement we the experimental findings.

The spectral shape of the ridge-valley structure of the high-K and low-K gated matrices of ^{163}Er has also been studied in order to extract the rotational damping with Γ_{rot} [13]. Figure 3 shows by symbols the values obtained for the rotational damping width, by interpolating the ridge-valley event distribution by a two component function (smooth curves in figure 1). The damping width is found to depend on the K value, being ~ 200 keV for low-K and ~ 150 keV for high-K states, in agreement with the calculation of the model [12], giving further support to the conservation of K up to moderate excitation energies [8,].

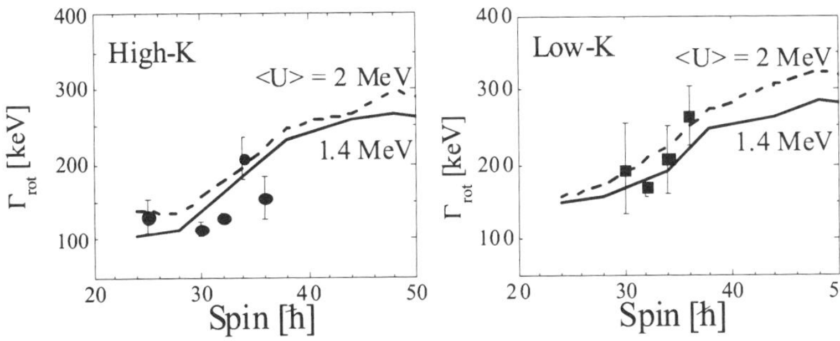

Figure 3. Experimental values of the rotational damping width Γ_{rot}, as extracted from the spectral shape analysis of low-K (right) and high-K (left) spectra of ^{163}Er [13]. Predictions from cranked shell model calculations [9] for average excitation energies of 1.4 and 2 MeV are shown by solid and dashed lines, respectively [7].

1.2. *Quantum tunneling of the excited superdeformed band*

The study of the properties of the COLD atomic nucleus as a function of angular momentum offers opportunities to learn about the phase transition between normal and superfluid systems [1,2]. This topic has been investigated in a number of nuclei, mostly in connection with the sudden decay-out from superdeformed (SD) to normal deformed (ND) structures.

The depopulation of the SD states is understood in terms of coupling to the compound ND states, which coexist over a wide spin range. The decay between different configurations is hindered by the potential energy barrier separating the normal and superdeformed configurations and a tunnelling through it becomes suddenly possible over few units of spins when the nucleus undergoes an abrupt transition from normal to superfluid. This decay mechanism has been established for the discrete SD bands at T=0, and it can also be used to describe the decay-out of the excited states (T≠0). In this case, the tunneling model has to be coupled to cranked shell model calculations including also a two-body residual interaction, needed to properly describe the warm rotational decay [14]. Experimentally, only few cases have been studied in details [15-17], since selective data obtained by gating on transitions collecting only few percent of the decay flow are needed with high statistics.

In this paper we present results from two different EUROBALL experiments, both performed at the IReS Laboratory (France). The first one aimed at the population of ^{151}Tb by means of the reaction ^{27}Al, at 155 MeV, on a thin ^{130}Te target [18]; the second experiment employed the reaction ^{30}Si at 150 MeV on a thin target of ^{170}Er, leading to ^{196}Pb after 4 neutron evaporation. The first experiment was performed using the EUROBALL array in the standard configuration, while for the second experiment, the low efficiency Ge detectors in the forward hemisphere were substituted by 8 large volume BaF_2 scintillators to measure high energy γ-rays from the decay of the giant dipole resonance (GDR). In both experiments an InnerBall of BGO detectors was also used to determine the γ multiplicity of each single event by measuring the number of fired detectors.

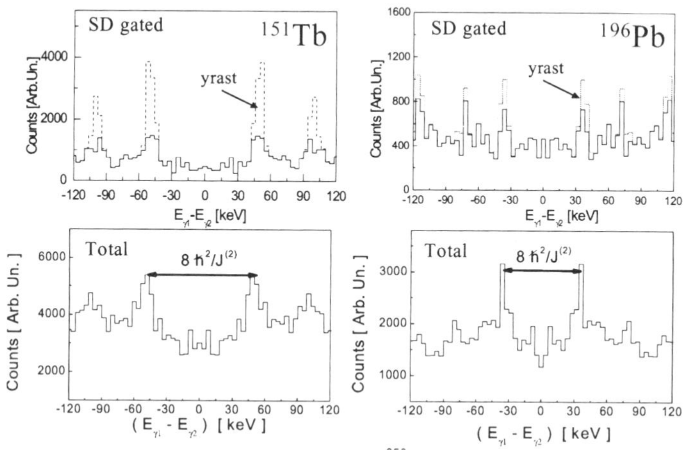

Figure 4. Wide projections on the $E_{\gamma1}$-$E_{\gamma2}$ axis of experimental matrices of ^{151}Tb (left column) and ^{196}Pb (right column). In the bottom spectra, obtained from the *Total* matrices, the arrow indicate the separation between the two most inner ridges, corresponding to $2 \times 4\hbar^2/J^{(2)} \sim 100$ keV and 70 keV, respectively. The top spectra are instead gated on transitions of the SD yrast band, which contributes to the spectrum with the intensity shown by the dotted lines. The cuts for ^{151}Tb and ^{196}Pb are performed at the average energy $\langle E_\gamma \rangle = 1228$ and 532 keV, corresponding to the spin values 52 and 26 $\hbar$, respectively.

The data have been sorted into a number of γ–γ matrices in coincidence with the isotope of interest (^{196}Pb or ^{151}Tb), named *Total,* and in coincidence with the SD yrast band of each nucleus (named *SD*), with a condition on high-fold events (F≥20) to better focus on high-multiplicity cascades. In both sets of spectra cuts perpendicular to the main diagonal reveal the existence of ridge structures extending over a large range of γ transition energies (corresponding to the spin regions 40 to 60 $\hbar$ and 10 to 40 $\hbar$ for ^{151}Tb and ^{196}Pb), with a spacing between the two most inner ridges equal to $4\times\hbar^2/J^{(2)}$, being $J^{(2)}$ the moment of inertia of the SD yrast band in each nucleus. Examples of such cuts are given in figure 4.

In both nuclei, the ridge structures are found to be populated by several unresolved discrete transitions, as confirmed in first place by the evaluation of their intensities (up to 3 times larger than the SD yrast one) and their widths (almost 4 times wider than that of the discrete bands). In order to provide information on the dynamics of the γ-decay in the SD well and on the coupling with the ND states, the number of bands populating the SD ridge structures have been evaluated, making use of the fluctuation analysis technique [9]. The results of the fluctuation analysis are shown in figure 5 for ^{151}Tb (left panel) and ^{196}Pb (right panel). In both cases a rather large number of SD rotational bands (up to 40 in ^{151}Tb and more than 50 in ^{196}Pb) are found to populate the SD ridges (filled squares), half of which directly feeding into the SD yrast band (open circles) [18].

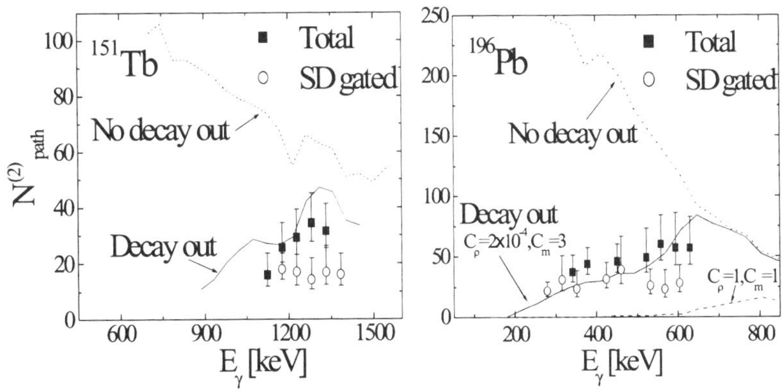

Figure 5. The experimental number of discrete excited bands (symbols) obtained from the fluctuation analysis of the SD ridge structures of ^{151}Tb [18] and ^{196}Pb. The filled (open) symbols refer to the analysis of the Total (SD gated) matrix. In both panels the dashed lines correspond to the model of ref. [14], not taking into account the decay-out towards the ND states. The full lines show instead the predictions including also a probability to decay into the ND well. In the case of ^{196}Pb, good agreement with the data is obtained renormalizing by the factor $C_\rho = 2\times10^{-4}$ and $C_m = 3$ the ND level density and the inertial mass, respectively, which are the main quantities entering into the calculation of the tunneling probability through the potential barrier between the ND and SD wells.

The interpretation of the results of the fluctuation analysis of the SD ridges in both nuclei has been based on the cranked shell model calculations discussed in ref. [14], successfully employed in the case of the SD nucleus ^{143}Eu [15]. As shown in both panels by dotted lines, the model largely overestimates the experimental data in both nuclei, especially at low transition energies, where the experimental number of decay paths decreases as a consequence of the decay out towards ND structures. A better agreement between theory and experiment is obtained once we take into account, for each SD excited band, a probability P_{out} to decay towards ND states. In the calculation P_{out} depends significantly on the level density ρ_{ND} of the ND states and on the inertial mass M_0 of the nucleus, especially in the heaviest mass region A = 190. In this case, a renormalization of these two quantities by the constant $C_\rho = 2\times10^{-4}$ and $C_m = 3$ is needed to reproduce both the decay-out spin values and the number of excited bands, as shown by solid line in figure 5. One can than conclude that the decay-out mechanism modeled as in the case of the SD yrast states is an essential ingredient to interpret the experimental results, providing additional information on the key parameter governing the decay-out process.

Acknowledgments

This work has been partially supported by EU (contract No. EUROVIV: HPRI-CT-1999-00078) and by the Italian Institute of Nuclear Physics (INFN).

References

1. A. Bracco and S. Leoni, *Rep. Prog. Phys.* **65**, 299(2002).
2. A. Bracco et al., *Cont. Phys.* **37**, 183(1996).
3. J. Simpson *Z. Phys.* **A358,** 139 (1997).
4. AGATA: *Technical Proposal for Advanced Gamma Tracking Array for the European Gamma Spectroscopy Community,* J. Gerl, W. Korten (2001).
5. J.D. Garrett et al, *Phys. Lett.* **B392,** 24 (1997).
6. R.U. Haq, A. Pandy and O. Bohigas, *Phys. Rev. Lett.* **48,** 1086 (1982).
7. G.B. Hagemann et al., *Nucl. Phys.* **A618**, 199(1997).
8. S. Leoni et al., *Phys. Rev.* **C72**, 034307(2005).
9. T. Døssing et al., *Phys. Rep.* **268**, 1(1996).
10. S. Leoni et al., *Nuc. Phys.* **A671**, 71(2000).
11. G. B. Benzoni et al., *Phys. Lett.* **B615,**160 (2005).
12. M. Matsuo et al., *Nucl. Phys.* **A736**, 223(2004).
13. S. Leoni et al., *Phys. Rev. Lett.* **93**, 022501(2004).
14. K. Yoshida, M. Matsuo and Y.R. Shimizu, *Nuc. Phys.* **A696**,85(2001).
15. S. Leoni et al., *Phys. Lett.* **B498**,137(2001).
16. A. Lopez-Martens et al., *Phys. Rev. Lett.* **77**, 1707(1996).
17. T. Lauritzen et al., *Phys. Rev.* **C75**, 064309(2007).
18. G. Benzoni et al., *Phys. Rev.* **C75**,047301(2007).

SELECTION RULES FOR THE INTRA AND INTERBAND TRANSITIONS AND QUANTUM NUMBERS FOR THE TRIAXIAL ROTOR IN ODD-A NUCLEI

KAZUKO SUGAWARA-TANABE

Otsuma Women's University,
Tama, Tokyo 206-8540, Japan

KOSAI TANABE

Department of Physics, Saitama University,
Sakura-Ku, Saitama 338-8570, Japan

An application of Holstein-Primakoff boson expansion to both single-particle and total angular momenta provides an algebraic solution for the particle-rotor model with one high-j nucleon coupled to a triaxially deformed core, and leads two kinds of quantum numbers to classify the rotational bands. The selection rules for the intraband and interband transitions are derived referring to these quantum numbers. The variable moments of inertia are preferable to fit the experimental data both for the positive and negative parity bands in Lu isotopes.

Keywords: TSD bands in Lu isotopes; Selection rules; Top-on-top mechanism.

1. Introduction

We have demonstrated that the top-on-top mechanism[1] can explain not only the energy levels but also the transition rates of the triaxial strongly deformed (TSD) bands in odd-A Lu isotopes. It is based on the particle-rotor model with one valence nucleon in a high-j orbital coupled to a triaxially deformed even-even core. We extended the Holstein-Primakoff (HP) transformation in even nuclei[2] to the case of odd-A nuclei by introducing two kinds of bosons for the total angular momentum $\vec{I}$ and the single-particle angular momentum $\vec{j}$. We take into account the invariance of the nuclear states under Bohr symmetry group,[3,4] which imposes the restriction on the quantum numbers. We can identify the nature of each band referring to two kinds of quantum numbers which indicate the precessions of $\vec{I}$ and $\vec{j}$, where the precession of the core angular momentum $\vec{R} = \vec{I} - \vec{j}$ correlates with that of $\vec{j}$. Such an interplay between two tops with $\vec{R}$ and $\vec{j}$ is called the "top-on-top mechanism". Our attempt in this paper is to reproduce detailed behavior of experimental energy levels in reference to $E^* - aI(I+1)$ with a=0.0075MeV. We take into account the decrease of the pairing correlation in the neutron core by increasing the moments of inertia as functions of I, and apply it to both positive and negative parity bands in Lu isotopes.[5–7] We also extend our algebraic method to the electric transition rates among various bands near the yrast to derive the selection rules. In Sec. 2, we review our algebraic formula[1] briefly. In Sec. 3, the selection rules are derived from the approximate expressions of the matrix elements for the interband and intraband transitions, and the theoretical results with variable moments of inertia method are compared with experimental data of energy levels. In Sec. 4, the paper is concluded.

2. Formalism

The particle-rotor Hamiltonian is given by $H = H_{\rm rot} + H_{\rm sp}$, with

$$
\begin{aligned}
H_{\rm rot} &= \sum_{k=x,y,z} A_k (I_k - j_k)^2, \\
H_{\rm sp} &= \frac{V}{j(j+1)}[\cos\gamma(3j_z^2 - \vec{j}^2) - \sqrt{3}\sin\gamma(j_x^2 - j_y^2)],
\end{aligned}
\tag{1}
$$

where $\vec{I}$ is the total angular momentum, $\vec{j}$ the single-particle angular momentum and $A_k = 1/(2\mathcal{J}_k)$ ($k = 1, 2, 3$ or x, y, z). We adopt the rigid-body model in Lund convention,

$$
\mathcal{J}_k = \frac{\mathcal{J}_0}{1 + (\frac{5}{16\pi})^{1/2}\beta_2}[1 - (\frac{5}{4\pi})^{1/2}\beta_2\cos(\gamma + \frac{2}{3}\pi k)], \tag{2}
$$

where β_2 and γ are the deformation parameters describing ellipsoidal shape of the rotor. Note that the maximum moment of inertia is about x-axis and the relation $\mathcal{J}_x \geq \mathcal{J}_y \geq \mathcal{J}_z$ holds in the range of $0 \leq \gamma \leq \pi/3$. The sign of γ in $H_{\rm sp}$ is consistent with the largest $\mathcal{J}_x$. We do not use hydrodynamical moments of inertia, as it is not suitable to simulate the experimental data which has been shown in Ref. 1.

We pay special attention to the symmetry properties of the nuclear Hamiltonian (D_2 symmetry) and the nuclear state (Bohr symmetry). The effect of an operation relabeling three axes is compensated by a proper change of deformation parameters (β_2, γ). These operations, each of which leaves a nuclear state invariant, compose Bohr symmetry group.[3,4] Now we consider the case where x-axis is chosen as a quantization axis. A complete set of the physical states which is invariant under Bohr symmetry is given by,

$$\{\sqrt{\frac{2I+1}{16\pi^2}}[\mathcal{D}^I_{MK}(\theta_i)\phi^j_\Omega+(-1)^{I-j}\mathcal{D}^I_{M-K}(\theta_i)\phi^j_{-\Omega}];\ |K-\Omega|=\text{even},\ \Omega>0\},\tag{3}$$

where K and Ω denote an eigenvalue of I_x and j_x, respectively. The wave function ϕ^j_Ω stands for spherical bases for the single-particle state, and $\mathcal{D}^I_{MK}(\theta_i)$ Wigner $\mathcal{D}$-function. Since the magnitude R of the rotor angular momentum $\vec{R}=\vec{I}+(-\vec{j})$ is given by $R=|I-j|,|I-j|+1,\cdots,I+j-1$, or $I+j$, an integer $n_{\beta'}$ defined by $R=I-j+n_{\beta'}$ takes $n_{\beta'}=0,1,2,\cdots,2j-1$, or $2j$. As R_x runs from R to $-R$, and $R_x=I_x-j_x=K-\Omega=$ even, an integer $n_{\alpha'}$ defined by the relation $R_x=R-n_{\alpha'}$ takes

$$\begin{aligned} n_{\alpha'}&=0,2,4,\cdots,\text{or}\ \ 2R, &&\text{for}\quad R=\text{even},\\ n_{\alpha'}&=1,3,5,\cdots,\text{or}\ \ 2R-1, &&\text{for}\quad R=\text{odd}. \end{aligned}\tag{4}$$

Physical states are realized for a set of non-negative integers $n_{\alpha'}$ and $n_{\beta'}$, which are related to the magnitude of rotor angular momentum R and its x-component R_x through the relations $R=I-j+n_{\beta'}$ and $R_x=I-j+n_{\beta'}-n_{\alpha'}$ by the Bohr symmetry rule.

It has been shown by the present authors[2] thirty-five years ago that the inclusion of higher order terms in Holstein-Primakoff (HP) boson expansion is necessary to reproduce the rotational spectra of the triaxially deformed rotor of an even nucleus. In Ref. 1, we found that this is also true for the odd-A nucleus in association with Bohr symmetry of the bosonized Hamiltonian. We choose diagonal forms for the components I_x and j_x in the HP boson representation as follows:

$$I_+=I_-^\dagger=I_y+iI_z=-\hat{a}^\dagger\sqrt{2I-\hat{n}_a},\ I_x=I-\hat{n}_a\ \text{with}\,\hat{n}_a=\hat{a}^\dagger\hat{a};\tag{5}$$

$$j_+=j_-^\dagger=j_y+ij_z=\sqrt{2j-\hat{n}_b}\hat{b},\ j_x=j-\hat{n}_b\ \text{with}\,\hat{n}_b=\hat{b}^\dagger\hat{b}.\tag{6}$$

Using these representations, we rewrite the Hamiltonian in terms of two kinds of boson operators, $\hat{a}$ and $\hat{b}$. We have proved that the invariance of the Hamiltonian under Bohr symmetry is achieved in the expansion of $\sqrt{2I-\hat{n}_a}$ and $\sqrt{2j-\hat{n}_b}$ into series in $\hat{n}_a/(2I)$ and $\hat{n}_b/(2j)$ up to the next to leading order.[1] We arrive at an approximate Hamiltonian written in terms of two kinds of HP bosons,

$$H_{\mathrm{B}}\cong H_0+H_2+H_4,\tag{7}$$

where H_0 denotes a constant which collects all the terms independent of boson operators, H_2 the bilinear forms of boson operators, and H_4 the fourth order terms. Diagonalization of H_2 is attained by the unitary transformation (or the boson Bogoliubov transformation) connecting boson operators $(\hat{a}, \hat{b}, \hat{a}^\dagger, \hat{b}^\dagger)$ to quasiboson operators $(\alpha, \beta, \alpha^\dagger, \beta^\dagger)$. Finally, H_2 is diagonalized as

$$H_2 = 2\omega_\alpha(\hat{n}_\alpha + 1/2) + 2\omega_\beta(\hat{n}_\beta + 1/2), \tag{8}$$

where eigenvalues ω_α and ω_β are obtained by solving a fourth order algebraic equation, and $\hat{n}_\alpha = \alpha^\dagger\alpha$ and $\hat{n}_\beta = \beta^\dagger\beta$. In order to take account of higher order contributions, we apply the boson transformation to H_4, and retain only diagonal terms which are expressed in terms of $\hat{n}_\alpha$ and $\hat{n}_\beta$. Thus, the particle-rotor Hamiltonian is approximately expressed in terms of two kinds of quantum numbers as

$$\begin{aligned} H_\mathrm{B} = H_0 + \omega_\alpha + \omega_\beta + C_0 + (2\omega_\alpha + C_\alpha)\hat{n}_\alpha \\ +(2\omega_\beta + C_\beta)\hat{n}_\beta + C_{\alpha\alpha}\hat{n}_\alpha^2 + C_{\beta\beta}\hat{n}_\beta^2 + C_{\alpha\beta}\hat{n}_\alpha\hat{n}_\beta. \end{aligned} \tag{9}$$

In order to clarify physical meaning of two quantum numbers, n_α and n_β, which are the eigenvalues of $\hat{n}_\alpha$ and $\hat{n}_\beta$, we consider the pure rotor case, i.e., $V = 0$ in Eq. (1), and the rotational energy in the symmetric limit of $A_y = A_z$ goes to well-known expression

$$E_\mathrm{rot} = A_z R(R+1) - (A_z - A_x)(R - n_\alpha)^2. \tag{10}$$

The eigenvalue R can be regarded as an effective magnitude of the rotor angular momentum, and $R - n_\alpha$ as its x-component R_x. It turns out that these n_α and n_β are the same integers $n_{\alpha'}$ and $n_{\beta'}$ as defined in Eq. (4). This allows us to interpret the quantum number n_α as the "precession" of $\vec{R}$ (so-called the "wobbling"[4]) because of $R_x = R - n_\alpha$, and the quantum number n_β as the "precession" of $\vec{j}$ about the intrinsic x-axis because of Eq. (6). The precession of $\vec{R}$ and/or $\vec{j}$ gives rise to the precession of the total angular momentum $\vec{I}(= \vec{R} + \vec{j})$. Due to the mixing of bosons $\hat{a}$ and $\hat{b}$, the physical contents of n_α and n_β change, but they keep the same values as in the symmetric limit whole through the adiabatic change of interaction parameter V and deformation parameters β_2 and γ. Thus, the rotational bands can be classified in terms of a pair of quantum numbers (n_α, n_β) according to the restriction imposed by the Bohr symmetry rule.

3. The Selection Rule of $E2$ Transition Rates and the Variable Moments of Inertia

Exact diagonalization of H yields an eigensolution belonging to λ-th eigenvalue E^λ, which is expressed by the complete set of wavefunctions which satisfy Bohr symmetry. The $E2$ transition operator is given by

$$\mathcal{M}(E2,\mu) = \sqrt{\frac{5}{16\pi}} e[Q_0 \mathcal{D}^2_{\mu 0} + Q_2(\mathcal{D}^2_{\mu 2} + \mathcal{D}^2_{\mu -2})], \tag{11}$$

where Q_0 and Q_2 are components of the intrinsic quadrupole moment. The deformation parameter γ is related to the ratio of Q_2 to Q_0 by

Table 1. The approximate B(E2) values without common kinematical factor for the $\Delta I = 2$ transition from initial I to final $I-2$ states among the bands with (n_α, n_β). Forbidden transition is denoted by $f.$, and higher order contribution by $h.o.$.

final \ initial	(0,0)	(1,0)	(2,0)	(3,0)
(0,0)	$(Q_2' G^2_{0000})^2$	$f.$	$\frac{3}{I^2}(Q_0' G_{0000} G_{2020})^2$	$f.$
(1,0)	$f.$	$(Q_2' G^2_{1010})^2$	$f.$	$\frac{9}{I^2}(Q_0' G_{3030} G_{1010})^2$
(2,0)	$h.o.$	$f.$	$(Q_2' G^2_{2020})^2$	$f.$
(3,0)	$f.$	$h.o.$	$f.$	$(Q_2' G^2_{3030})^2$

$Q_2/Q_0 = -\tan\gamma/\sqrt{2}$. While the $B(E2)$ values for the intraband and interband transitions can be calculated directly from the eigenfunctions of H, it is useful to calculate them by the algebraic method in order to find the selection rules for the transitions among the bands specified by different quantum numbers. For this purpose, we need the overlaps between two boson Fock spaces, i.e., the one is generated by quasibosons $(\alpha^\dagger, \beta^\dagger)$ on the quasivacuum $|0\rangle_\alpha$ and the other by HP bosons $(\hat{a}^\dagger, \hat{b}^\dagger)$ on the vacuum $|0\rangle_a$. Such overlaps are calculated by applying the extended form of the generalized Wick theorem. The overlap between $|n_a n_b, Ij\rangle$ and $|n_\alpha n_\beta, Ij\rangle$ is

$$G^{Ij}_{n_a, n_b; n_\alpha, n_\beta} \equiv \frac{{}_a\langle 0|\hat{a}^{n_a}\hat{b}^{n_b}(\alpha^\dagger)^{n_\alpha}(\beta^\dagger)^{n_\beta}|0\rangle_\alpha}{(n_a! n_b! n_\alpha! n_\beta!)^{1/2}} \tag{12}$$

In this expression $n_a(=I-K)$ and $n_b(=j-\Omega)$ stand for the eigenvalues of $\hat{n}_a$ and $\hat{n}_b$, respectively. Some simple examples of $G^{Ij}_{n_a, n_b; n_\alpha, n_\beta}$ are listed in Appendix C of Ref. 1. Associated with the change of quantization axis from z- to x-axis, the components of quadrupole moment must be transformed to $Q_0 \to Q_0' = -Q_0/2 + \sqrt{\frac{3}{2}}Q_2$ and $Q_2 \to Q_2' = (\sqrt{\frac{3}{2}}Q_0 + Q_2)/2$. Therefore, Q_0 and Q_2 in Eq. (11) must be replaced by Q_0' and Q_2', respectively. An approximation collecting only a few lowest order terms of $G^{Ij}_{n_a, n_b; n_\alpha, n_\beta}$ is enough to derive selection rules and to estimate the order of magnitude of the transition matrix elements. We employ an asymptotic estimation by assuming that I is large enough and the difference of the I-dependence of

$G^{Ij}_{n_a,n_b;n_\alpha,n_\beta}$ between the initial and the final states is negligible. We drop indices I and j from $G^{Ij}_{n_a,n_b;n_\alpha,n_\beta}$ and employ its abbreviation as $G_{n_a n_b n_\alpha n_\beta}$. We employ an approximation retaining only terms with $G_{n_a n_b n_\alpha n_\beta}$ whose suffices satisfy $n_a + n_b - n_\alpha - n_\beta = 0$ ("$\Delta n = 0$"). The approximate $B(E2)$ values are tabulated for $\Delta I = 2$ transitions in Table 1 and $\Delta I = 1$ transitions in Table 2, where a common kinematical factor $5e^2/(16\pi)$ is abbreviated. In both Tables, "*f.*" denotes forbidden transition, while "*h.o.*"

Table 2. The approximate B(E2) values without common kinematical factor for the $\Delta I = 1$ transition from initial I to final $I-1$ states among the bands with (n_α, n_β). The symbols *f.* and *h.o.* are the same as defined in Table 1, and the line – indicates no transition.

final \ initial	(0,0)	(1,0)	(2,0)	(3,0)
(0,0)	—	$\frac{3}{I}(Q'_0 G_{0000} G_{1010})^2$	*f.*	$\frac{3}{I^3}(Q'_2 G_{0000} G_{3030})^2$
(1,0)	$\frac{2}{I}(Q'_2 G_{0000} G_{1010})^2$	—	$\frac{6}{I}(Q'_0 G_{2020} G_{1010})^2$	*f.*
(2,0)	*f.*	$\frac{4}{I}(Q'_2 G_{2020} G_{1010})^2$	—	$\frac{9}{I}(Q'_0 G_{2020} G_{3030})^2$
(3,0)	*h.o.*	*f.*	$\frac{6}{I}(Q'_2 G_{3030} G_{2020})^2$	—

denotes higher order contribution, taking $n_a + n_b - n_\alpha - n_\beta = 2$ or 4. The symbols with – in the diagonal places in Table 2 for $B(E2)$ with $\Delta I = 1$ denote that these transitions are hindered. These approximated transition rates lead useful estimation. In ^{163}Lu, the transitions from TSD3 ($n_\alpha = 2, n_\beta = 0$) to TSD1 $(0,0)$ and to TSD2 $(1,0)$ are measured.[6] The $\Delta I = 2$ transition value of $B(E2)_{\text{out}}/B(E2)_{\text{in}}$ is around 0.02 from initial $I = 37/2 \sim 57/2$. The $\Delta I = 1$ transition value of $B(E2)_{\text{out}}/B(E2)_{\text{in}}$ is 0.51 ± 0.13 at $I = 45/2$. As for the transitions between TSD2 $(1,0)$ and TSD1 $(0,0)$, $\Delta I = 1$ transition value of $B(E2)_{\text{out}}/B(E2)_{\text{in}}$ is around 0.2, which is well explained in Ref. 1. From Table 1 and 2, the following relations are derived.

$$\frac{B(E2; I, 20 \to I-2, 00)}{B(E2; I, 20 \to I-2, 20)} \sim \frac{3}{I^2}\left(\frac{Q'_0 G_{0000}}{Q'_2 G_{2020}}\right)^2, \tag{13}$$

$$\frac{B(E2; I, 10 \to I-1, 00)}{B(E2; I, 10 \to I-2, 10)} \sim \frac{3}{I}\left(\frac{Q'_0 G_{0000}}{Q'_2 G_{1010}}\right)^2, \tag{14}$$

$$\frac{B(E2; I, 20 \to I-1, 10)}{B(E2; I, 20 \to I-2, 20)} \sim \frac{6}{I}\left(\frac{Q'_0 G_{1010}}{Q'_2 G_{2020}}\right)^2. \tag{15}$$

Hence, if the quantity in (14) is 0.2, we get $(13) = (G_{1010}/G_{2020})^2/I \times (14) \sim 0.2/I \sim 0.01$. Similarly, $(15) = 2G^4_{1010}/(G^2_{0000}G^2_{2020}) \sim 2 \times (14) \sim 0.4$.

So far, there is no theoretical work which reproduces the excitation energy relative to a reference, $E^* - aI(I+1)$ with $a = 0.0075\text{MeV}^{-1}$. As there seems to remain the Coriolis Anti-Pairing effect (CAP) especially in the neutron core, we include the effect of decreasing pairing on the moments of inertia as an I-dependence of the moments of inertia. We adopt the formula

$$\mathcal{J}_0 \to \mathcal{J}_0 \frac{I - 0.69}{I + 23.5}. \tag{16}$$

We call this as variable moments of inertia. For positive parity levels,

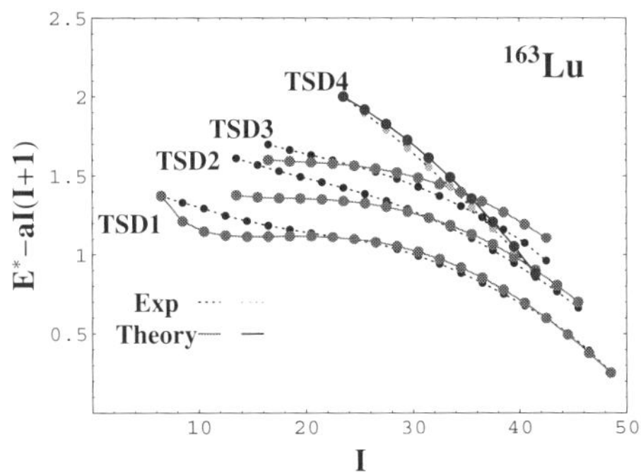

Fig. 1. The comparison between the experimental and the theoretical energy levels, $E^* - aI(I+1)$ as functions of angular momentum I for ^{163}Lu. The vertical axis is in unit of MeV. The experimental data are from Refs. 5 and 6.

$\pi i_{13/2}$ is assumed for the single-particle orbital, while for negative parity levels, $\pi j_{15/2}$ is assumed. The value of $\mathcal{J}_0$ is 77.6 for $\pi i_{13/2}$ and 81.6 for $\pi j_{15/2}$, and only the bandhead energy of $(0,0)$ band is adjusted to the experimental TSD1 bandhead energy in ^{163}Lu, and is kept the same value over the isotopes. The value of $\gamma = 17°$ and $V = 2.6$MeV are fixed. In Fig. 1 and 2, we show the comparison between experimental and theoretical energy levels of $E^* - aI(I+1)$ curves (red circles connected by solid lines for $i_{13/2}$ and blue circles for $j_{15/2}$ connected by solid lines) with experimental values (black small circles for $i_{13/2}$ and green small circles for $j_{15/2}$ connected by dashed lines). Quite good fit to the experimental data are also obtained for ^{161}Lu and ^{167}Lu with the same set of parameters.

4. Conclusion

Our algebraic method leads two kinds of quantum numbers n_α describing the precession of core angular momentum, and n_β that of the single-particle

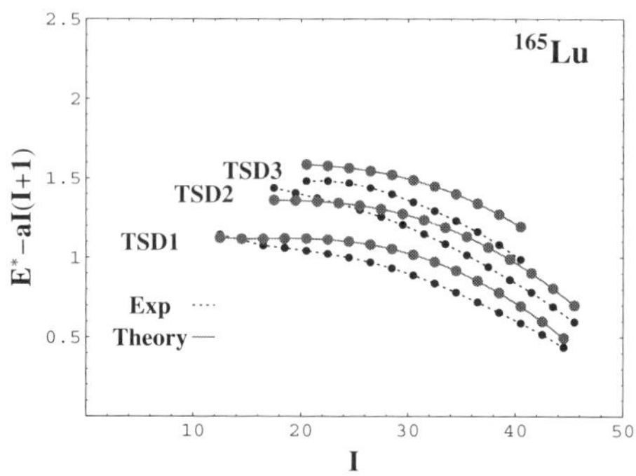

Fig. 2. The comparison between the experimental and the theoretical energy levels, $E^* - aI(I+1)$ as functions of angular momentum I for ^{165}Lu. The vertical axis is in unit of MeV. The experimental data are from Ref. 7.

angular momentum. Each rotational band is characterized by a set (n_α, n_β) whose values are restricted by Bohr symmetry rule depending on the total angular momentum I. We have accounted for the pairing effect on the neutron core with the variable moments of inertia method. Then, the theoretical excitation energy relative to a reference fit to the experimental values quite well. The B(E2) values are derived by the overlap integral between the wave functions with $\gamma = 0$ and with $\gamma \neq 0$. In the lowest order apprximation taking only the contributions from $n_a + n_b - n_\alpha - n_\beta = 0$ in the overlap $G^{Ij}_{n_a,n_b;n_\alpha,n_\beta}$ gives selection rules, which can explain the experimental data in ^{163}Lu among TSD1 (0,0), TSD2 (1,0) and TSD3 (2,0) around I ~20.

References

1. K. Tanabe and K. Sugawara-Tanabe, Phys. Rev. C **73**, 034305 (2006).
2. K. Tanabe and K. Sugawara-Tanabe, Phys. Lett. **B34**, 575 (1971).
3. A. Bohr, Mat. Fys. Medd. Dan. Vid. Selsk. **26**, no.14 (1952).
4. A. Bohr and B. R. Mottelson, *Nuclear Structure* (Benjamin, Reading, MA, 1975), Vol. II, Chap. 4.
5. S. W. Ødegård *et al.*, Phys. Rev. Lett. **86**, 5866 (2001).
6. D. R. Jensen *et al.*, Phys. Rev. Lett. **89**, 142503 (2002).
7. G. Schönwaßer *et al.*, Phys. Lett. B **552**, 9 (2003).

COLLECTIVE BANDS IN SUPERDEFORMED NUCLEI

J. KVASIL[1], N. LO IUDICE[2], F. ANDREOZZI[2], A. PORRINO[2], F. KNAPP[1]

[1] *Institute of Particle and Nuclear Physics, Charles University, V.Holešovičkách 2, CZ-18000 Praha 8, Czech Republic*

[2] *Dipartimento di Scienze Fisiche, Universitá di Napoli "Federico II" and Istituto Nazionale di Fisica Nucleare, Monte S Angelo, Via Cintia I-80126 Napoli, Italy*

The collective features of superdeformed nuclei have been investigated in self-consistent cranked Nilsson plus quasiparticle random-phase approximation. The analysis supports the octupole collectivity of some low-lying excited bands observed recently and, moreover, shows that the onset of supederformation changes considerably the $E1$ and $E2$ strength distribution and greatly enhances the collectivity of the low-lying scissors mode.

Keywords: Fast rotation; Superdeformation; Collective bands.

1. Introduction

The discovery of superdeformed (SD) rotational bands[1] has stimulated intensive experimental and theoretical investigations which have greatly contributed to the understanding of nuclear structure under extreme conditions of fast rotation and large deformation.

Most of the properties of SD nuclei can be explained in terms of single particle motion.[2–8] On the other hand, the octupole degrees of freedom are expected to play an important role, because of the presence of $\Delta l = 3$ intruder states in the region of the SD Fermi surface.

Indeed, the presence of octupole fluctuations around SD minima could account for the deviations of empirical dynamical moments of inertia form the expected yrast-band value.[9,10]

A more direct evidence in favor of octupole collectivity in SD nuclei was provided by the strong $E1$ transitions connecting the excited SD to the yrast SD band measured in several experiments.[11]

In a recent work,[12,13] we have adopted a cranked Nilsson plus quasiparticle random-phase approximation (QRPA) to investigate how fast rotation and the onset of superdeformation affect the nuclear collective properties.

We have focused our attention on the octupole mode first and, then, on other collective modes, well established at normal deformation and low rotational frequency.

We discuss here how superdeformation affects the low-lying scissors $M1$ mode.[14–16] The strong impact of superdeformation on such a mode was established long ago.[17] That fast rotation affects dramatically the orbital $M1$ excitations was pointed out recently.[18]

2. RPA in the rotating frame

The Hamiltonian adopted here is

$$H_\Omega = H_0(\Omega) + V_{pair} + V_{FF}. \tag{1}$$

$H_0(\Omega)$ is a cranked one-body term

$$H_0(\Omega) = H_0 - \sum_{\tau=n,p} \lambda_\tau N_\tau - \hbar\Omega I_1, \tag{2}$$

where H_0 is a modified triaxial harmonic oscillator (HO) Nilsson Hamiltonian plus a local Galilean invariance restoring piece of the form given in Refs.,[10,18] the second piece provides a constraint in the neutron and proton numbers, the third is the cranking term.

The equilibrium deformation may be determined by minimizing at each Ω the expectation value of the above one-body Hamiltonian with respect to the HO frequencies ω_i, also dependent on Ω, under the volume conserving constraint $\omega_1\omega_2\omega_3 = \omega_0^3$.

As regard to the two-body potential, V_{pair} is a proton-proton and neutron-neutron monopole pairing term, while V_{FF} a sum of separable potentials. The sum includes quadrupole-quadrupole plus monopole-monopole plus spin-spin separable potentials, in the positive parity sector, and dipole-dipole plus octupole-octupole interactions, effective in the negative parity subspace.

All multipole and spin-multipole fields have good isospin T and signature $r = \pm$ $(\alpha = 0, 1)$ and are expressed in terms of doubly stretched coordinates $x_i'' = (\omega_i/\omega_0)\, x_i$[19] so as to ensure the separation of the pure rotational mode from the intrinsic excitations.

We express the Hamiltonian (1) in terms of quasi-particle operators and plug the transformed Hamiltonian into the RPA equations of motion

$$[H_\Omega, P_\nu] = i\,\hbar\omega_\nu^2\, X_\nu, \quad [H_\Omega, X_\nu] = -i\,\hbar\, P_\nu, \quad [X_\nu, P_{\nu'}] = i\hbar\delta_{\nu\,\nu'}, \tag{3}$$

where X_ν, P_ν are, respectively, the collective coordinates and their conjugate momenta.

The RPA equations are solved under the symmetry constraints

$$[H_\Omega\,,\,N_{\tau=n,p}\,]_{RPA} \;=\; 0, \quad [H_\Omega\,,\,I_1\,]_{RPA} = 0 \tag{4}$$

and

$$[H_\Omega(-), \Gamma^\dagger] = \Omega\Gamma^\dagger, \tag{5}$$

where $\Gamma^\dagger = (I_2 + iI_3)/\sqrt{2\langle I_1\rangle}$ and $\Gamma = (\Gamma^\dagger)^\dagger = (I_2 - iI_3)/\sqrt{2\langle I_1\rangle}$ fulfill the commutation relation $[\Gamma, \Gamma^\dagger] = 1$. Eqs. (4) yield two Goldstone modes, one associated to the violation of the particle number operator, the other to the breaking of spherical symmetry. Eq. (5), on the other hand, yields a negative signature redundant solution of energy $\omega_\lambda = \Omega$, arising from the symmetries broken by the external rotational field (the cranking term). Eqs. (4) and (5) ensure the separation of the spurious or redundant solutions from the intrinsic ones.

We have adopted the method developed in Ref.[20] to compute the strength function

$$S_{X\lambda}(E) \;=\; \sum_{\nu\, I'} B(X\lambda,\, I \;\rightarrow\; I',\, \nu)\, \delta(E - \hbar\omega_\nu), \tag{6}$$

where $X = E, M, \nu$ labels all the excited states with a given I'. The reduced strengths $B(X\lambda,\, I \;\rightarrow\; I',\, \nu)$ were computed in the limits of zero and high angular frequencies.

The mentioned method avoids the explicit determination of RPA eigenvalues and eigenfunctions and yields the n-th moments as

$$m_n(X\lambda) \;=\; \int_0^\infty \; E^n\, S_{X\lambda}(E)\, dE. \tag{7}$$

The $m_0(X\lambda)$ and $m_1(X\lambda)$ moments give, respectively, the energy unweighted and weighted summed strengths.

3. Calculations and results

We determined the equilibrium deformations at each angular velocity by minimizing at each Ω the BCS expectation value of the cranked Nilsson Hamiltonian (2) plus the pairing potential. For all nuclei under investigation and all rotational frequencies, we obtained two minima, both axially symmetric ($\gamma = 0$). One of them is the SD minimum and is insensitive to Ω. The other falls at low deformation and is weakly dependent on Ω. In all Hg isotopes, the ND minima are deeper than the SD ones at all frequencies. Consequently, the SD band is never yrast and, therefore, decays to the ND

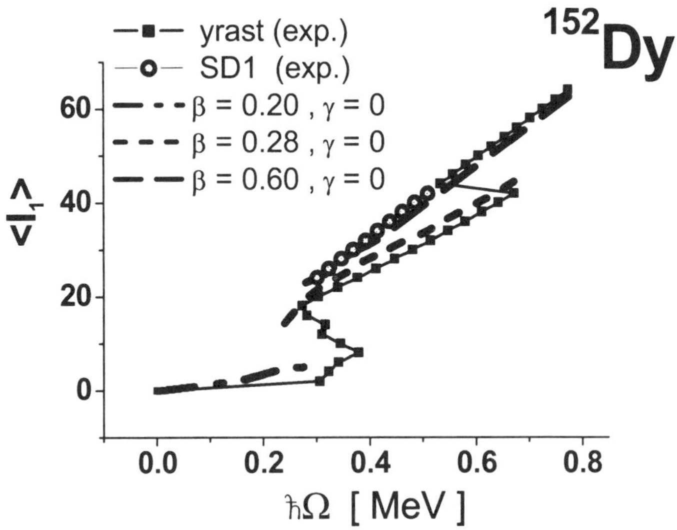

Fig. 1. Angular momenta versus rotational frequency.

states. In ^{152}Dy, instead, the SD band becomes yrast starting from $\Omega \sim 0.6$ MeV.

The validity of our treatment is provided by its success in describing the angular momentum expectation value, $< I_x >_\Omega = < \Omega | I_x | \Omega >$, along the yrast line. As shown in Fig. 1, the calculation reproduces fairly well the experimental angular momenta. The plot shows clearly a close connection between β discontinuities and band crossings as well as the onset of first and second backbending.

Because of the combined effect of fast rotation and superdeformation, collective octupole bands intrude in the low energy domain and induce pronounced fluctuations in the dynamical moment of inertia, which cannot be reproduced by the Harris formula adopted to fit the moment of inertia of the yrast SD band (Fig. 2). The connection between level crossings and fluctuations has been shown by our RPA calculation[12,13] in substantial agreement with Refs.[9,10]

A more direct test of the collective character of these low-lying octupole SD bands has been provided by the measurements of strong $E1$ transitions connecting those bands to the yrast band. As shown in table 1, our calculation reproduces remarkably well the experimental strengths.

Fast rotation together with shape transition have a deep impact on other electric and magnetic responses. We discuss here the behavior of

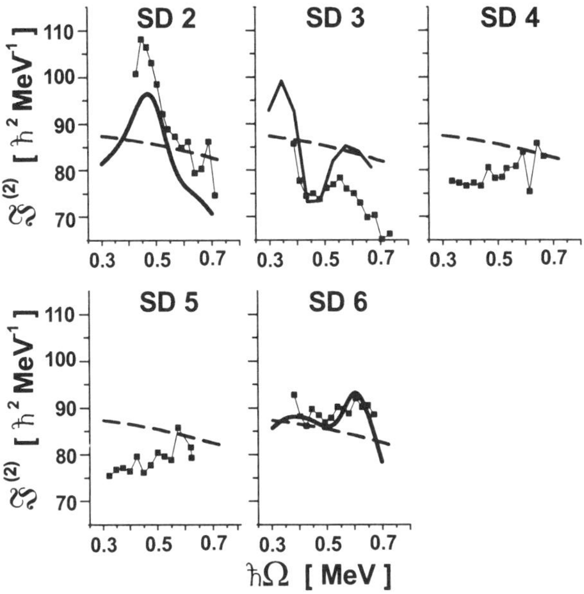

Fig. 2. Dynamical moment of inertia versus rotational frequency. The dashed line corresponds to the Harris fit of the yrast SD band

Table 1. Strengths of the $E1$ transitions from the $SD6$ to the yrast SD band in ^{152}Dy. The experimental data are taken from.[11]

$J_i^\pi \rightarrow J_f^\pi$	E_γ (keV)	$B(E1)_{exp}$ (W.u.)	$B(E1)_{th}$ (W.u.) (W.u.)
$33^- \rightarrow 32^+$	1676	2.2×10^{-4}	2.86×10^{-4}
$35^- \rightarrow 34^+$	1696	3.8×10^{-4}	1.86×10^{-4}
$37^- \rightarrow 36^+$	1715	4.5×10^{-4}	1.94×10^{-4}
$39^- \rightarrow 38^+$	1734	3.9×10^{-4}	1.74×10^{-4}
$41^- \rightarrow 40^+$	1751	4.9×10^{-4}	3.69×10^{-4}

the $M1$ response which looks quite intriguing. Most of the $M1$ strength remains concentrated in the energy range $2 \div 10$ MeV at all frequencies and deformations. At low rotational frequency and normal deformation ($\beta = 0.2$), most of the strength is due to spin excitations. Only in the low-energy tail $(2 \div 4)$ MeV, the orbital contribution is comparable to the one due to spin.

The onset of superdeformation enhances strongly the orbital strength, which gets strongly peaked around 6 MeV, and has a damping and spreading effect on spin transitions, which get scattered all along the $2 \div 10$ MeV interval. The low-lying orbital strength amounts to $\sim 20\,\mu_N^2$. Orbital and spin amplitudes interfere constructively, yielding a total $M1$ strength of $\sim 30\mu_N^2$.

A high energy peak, which can hardly be noticed at normal deformation, is well noticeable in the SD phase. It is rather broad and carries a $M1$ strength of only $\sim 5\,\mu_N^2$, much smaller than the the one collected by the low-energy excitations.

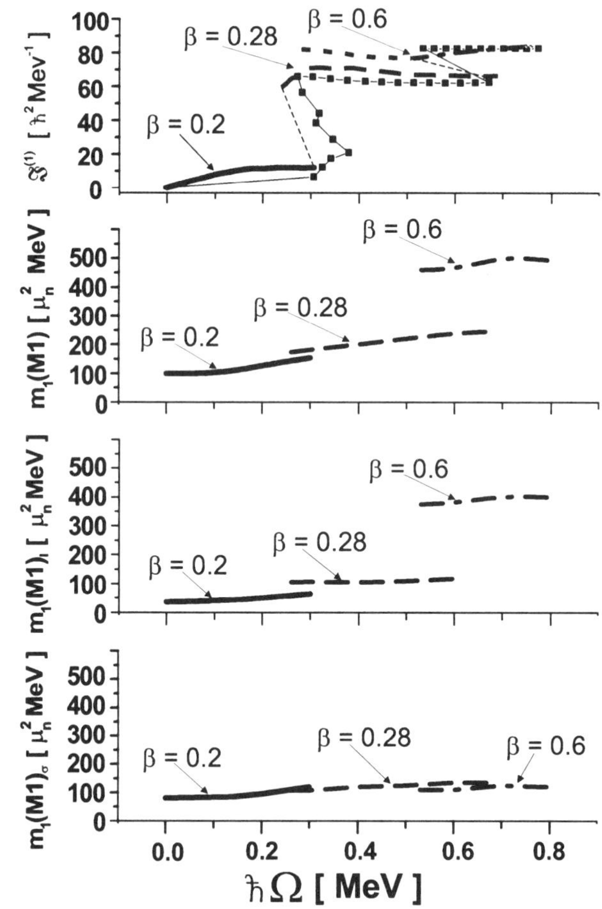

Fig. 3. The yrast line kinematical moment of inertia (upper panel) versus the total (second panel), orbital (third panel), and spin (bottom panel) $m_1(M1)$ moments.

It is possible to identify one of the mechanisms responsible for such a large enhancement by comparing the Ω behavior of the orbital and total $m_1(M1)$ moments with the corresponding evolution of the kinematical moment of inertia $\Im = I/\Omega$. As shown in Fig. 3, the strikingly similar behavior of the orbital $m_1(M1)$ and the moment of inertia shows that the two quantities are closely correlated at all rotational frequencies.

The close correlation between scissors $M1$ strength and moment of inertia was already established at zero frequency. Indeed, we have for the scissors $M1$ NEWSR[16] and EWSR,[21] respectively

$$m_0^{(sc)}(M1) \propto \Im\omega,$$
$$m_1^{(sc)}(M1) \propto \Im\omega^2. \tag{8}$$

4. Conclusions

A basically selfconsistent cranked Nilsson plus QRPA approach has proved to be successful in describing the properties of low-lying SD excited bands.

The approach reproduces fairly well the fluctuating behavior of the dynamical moment of inertia versus deformation and angular velocity, thereby confirming the conclusion drawn in Refs.[9,10] about the octupole character of the negative parity excited SD bands near the yrast line.

It accounts quantitatively well for the strong $E1$ transitions connecting these excited SD band to the yrast SD band. The deexcitation of these SD bands via strong $E1$ transitions is a more direct test of their octupole character.

Our analysis shows that the onset of superdeformation has a strong impact on the other electric collective modes and, to a much higher degree, on the orbital $M1$ response. This is dramatically enhanced over the spin around 6 MeV above the yrast line and confers to these low-lying $M1$ transitions the typical features of the scissors mode. Being the lowest in energy, apart from the octupole excitations, such a mode should have good changes of being detected in γ cascade processes, thanks to the very effective modern methods of analysis of the decay spectra.

Acknowledgments

Work was partly supported by the Czech Ministry of Education contract No. VZ MSM 0021620859 and by the Italian Ministero dell'Istruzione, Universitá and Ricerca (MIUR).

References

1. P. J. Twin *et al.*, Phys. Rev. Lett. **57**, 811 (1986).
2. W. Nazarewicz, R. Wyss, A. Johnson, Nucl. Phys. **A503**, 285 (1989).
3. Y.R. Shimizu, E. Viggezi, R.A. Broglia, Nucl. Phys. **A509**, 80 (1990).
4. T. Bengtsson, I. Ragnarson, and S. Aberg, Phys. Lett. **B 208**, 39 (1988).
5. R.V.F. Janssens and T.L. Khoo, Annu. Nucl. Part. Sci. **41**, 321 (1991).
6. W. Satula, J. Dobaczewski, J. Dudek, and W. Nazarewicz, Phys. Rev. Lett. **77**, 5182 (1996).
7. L.B. Karlsson, I. Ragnarsson, S. Aberg, Nucl. Phys. **A639**, 654 (1998).
8. S.T. Clark *et al.*, Phys. Rev. Lett. **87**, 172503 (2001).
9. T. Nakatsukasa, K. Matsuyanagi, S. Mizutori, W. Nazarewicz, Phys. Lett. **B343**, 19 (1995).
10. T. Nakatsukasa, K. Matsuyanagi, S. Mizutori, Y.R. Shimizu, Phys. Rev.C **53**, 2213 (1996).
11. T. Lauritsen *et al.*, Phys. Rev. Lett. **89**, 282501 (2002) and references therein.
12. J. Kvasil, N. Lo Iudice, F. Andreozzi, F. Knapp, A. Porrino, Phys. Rev. C **73**, 034302 (2006).
13. J. Kvasil, N. Lo Iudice, F. Andreozzi, F. Knapp, A. Porrino, Phys. Rev. C **75**, 034306 (2007).
14. N. Lo Iudice and F. Palumbo, Phys. Rev. Lett. **41**, 1532 (1978).
15. D. Bohle, A. Richter, W. Steffen, A.E.L. Dieperink, N. Lo Iudice, F. Palumbo, and O. Scholten, Phys. Lett. B **137**, 27 (1984).
16. For an exhaustive list of reference N. Lo Iudice, Rivista Nuovo Cimento **9**, 1 (2000).
17. I. Hamamoto, W. Nazarewicz, Phys. Lett. **B297**, 25-30 (1992).
18. J. Kvasil, N. Lo Iudice, R.G. Nazmitdinov, A. Porrino, F. Knapp, Phys. Rev. C **69**, 064308 (2004).
19. T. Kishimoto, J. M. Moss, D.H. Youngblood, J.D. Bronson, C.M. Rozsa, D.R. Brown, and A.D. Bacher, Phys. Rev. Lett. **35**, 552 (1975).
20. J. Kvasil, N. Lo Iudice, V.O. Nesterenko, and M. Kopal, Phys. Rev. C **58**, 209 (1998).
21. N. Lo Iudice, Phys. Rev. C **57**, 1246 (1998).

NUCLEAR STRUCTURE CALCULATIONS IN A LARGE DOMAIN OF EXCITATION ENERGIES

CH. STOYANOV*, D. TARPANOV

Institute for Nuclear Research and Nuclear Energies, BAS, Sofia 1734, Bulgaria

**E-mail: stoyanov@inrne.bas.bg*

A microscopic nuclear model (*Quasiparticle – phonon Model*) is used to calculate the structure of excited states in large domain of excitation energy. The new quantity *Invariant Correlational Entropy* is introduced to estimate the complexity of high-lying single particle mode. The properties of low-lying mixed symmetry states are calculated in nuclei near shell closure. The low-lying part of $E1$ distribution, known as *Pygmy* resonance is also calculated.

1. Introduction

Recently, new experimental results reveal some details of nuclear structure at diverse excitation energy that were unknown before. New data are available for the tail of giant dipole resonance. The new structure, called Pygmy resonance is investigated via α and γ scattering experiments. Data are collected for mixed symmetry states in nearly spherical nuclei. These achievements are challenge for nuclear models. It is necessary to describe the properties of excited states by means of models enabling to reproduce the changing of the structure with excitation energy. The presented paper is an attempt to reveal the opportunities of the microscopic model, known as "*Quasiparticle – phonon model*" (QPM) [1] and to apply the model to the new structure phenomena as statistical properties of high-lying single particle states, mixed symmetry states and Pygmy resonance.

The intrinsic QPM Hamiltonian having the following structure

$$H = H_{sp} + V_{pair} + V_M^{ph} + V_{SM}^{ph} + V_M^{pp} \,. \tag{1}$$

H_{sp} is the one-body Hamiltonian, V_{pair} the monopole pairing, V_M^{ph} and V_{SM}^{ph} are respectively sums of particle-hole separable multipole and spin-multipole interactions, and V_M^{pp} is the sum of particle-particle multipole pairing potentials.

Following the QPM procedure, the Hamiltonian is transformed into a multiphonon one of the form

$$H_{QPM} = \sum_{i\mu} \omega_{i\lambda} Q^{\dagger}_{i\lambda\mu} Q_{i\lambda\mu} + H_{vq}, \tag{2}$$

where the first term is the unperturbed phonon Hamiltonian and H_{vq} is a phonon-coupling part whose exact expression can be found in Ref. [1]. Both terms are expressed in terms of the RPA phonon operators

$$Q^{\dagger}_{i\lambda\mu} = \frac{1}{2}\sum_{jj'} \left\{ \psi^{i\lambda}_{jj'} [\alpha^{\dagger}_{j}\alpha^{\dagger}_{j'}]_{\lambda\mu} - (-1)^{\lambda-\mu} \varphi^{i\lambda}_{jj'} [\alpha_{j'}\alpha_{j}]_{\lambda-\mu} \right\}_{\tau} \tag{3}$$

of multipolarity $\lambda\mu$ and energy $\omega_{i\lambda}$, where $\alpha^{\dagger}_{jm}(\alpha_{jm})$ are quasiparticle operators obtained from the corresponding particle operators through a Bogoliubov transformation.

Recently, the QPM approach was extended in the case of non-separable Skyrme interaction. Details about this application are given in Ref. [2]

2. The correlation entropy as a measure of the complexity of high-lying single particle mode

The complexity of stationary many-body states can be quantified with the aid of such characteristic as information entropy [3]. The quantity smoothly changes with excitation energy revealing strong mixing of original states and loss of simple quantum numbers, typical features of quantum chaos. An entropy-like quantity, called *invariant correlational entropy* (ICE), was suggested in Ref. [4] as a measure of complexity of the states. This quantity is by construction invariant with respect to the basis transformations.

An example of complex state is the wave function of the high-lying single-particle mode. At higher excitation energy, a large number of nuclear modes influence the damping of the quasiparticle motion [5]. It is shown in Ref. [6] that ICE could be used to estimate the complexity of the wave function. The ICE method presumes that Hamiltonian $H(\lambda)$ of a system depends on a random parameter λ, member of an ensemble characterized by the normalized distribution function $P(\lambda)$, $\int d\lambda P(\lambda) = 1$.

The ICE is defined as $S^{\alpha} = -\mathrm{Tr}\{\varrho^{\alpha} \ln(\varrho^{\alpha})\}$, where ϱ^{α} is the density matrix of the state $|\alpha\rangle$ averaged over the ensemble. In the basis $|k\rangle$ prior to the averaging the construction of this matrix has to be done for a given value of λ as $\varrho^{\alpha}_{kk'}(\lambda) = C^{\alpha}_{k}(\lambda) C^{\alpha *}_{k'}(\lambda)$, and then average over the ensemble, $\varrho^{\alpha}_{kk'} = \int d\lambda\, P(\lambda) \varrho^{\alpha}_{kk'}(\lambda)$. The ICE reflects the correlations between the wave function components. The value S^α for a given state typically increases with the complexity of the state.

Within QPM the properties of the (A+1)-nucleus can be described in terms of the quasiparticle states $\alpha_a^\dagger \mid 0\rangle$, quasiparticle-plus-phonon states $[\alpha_a^\dagger \otimes Q_\nu^\dagger] \mid 0\rangle$ and quasiparticle-plus-two-phonon states $[\alpha_a^\dagger \otimes [Q_\mu^\dagger \otimes Q_\nu^\dagger]] \mid 0\rangle$. The following wave functions describe in the QPM the ground and excited states of the odd nucleus [5]:

$$\Psi_{JM}^\alpha = C_J^\alpha \{\alpha_{JM}^\dagger + \sum_{a\nu} D_a^\nu \left[\alpha_a^\dagger Q_\nu^\dagger\right]_{JM} + \sum_{a\nu\nu'} F_a^{\nu\nu'} \left[\alpha_a^\dagger \left[Q_\nu^\dagger Q_{\nu'}^\dagger\right]_I\right]_{JM}\}\Psi_0 \ . \tag{4}$$

For studying the ICE, the single-particle $1k_{17/2}$ state in ^{209}Pb was selected. This orbital is quasi-bound in the Woods-Saxon potential being located at 4.88 MeV. Because of its high energy, the state is surrounded by many quasiparticle-plus-phonon and quasiparticle-plus-two-phonon states. The distribution of the single-particle strength of the $1k_{17/2}$-state is shown on Fig. 1 .

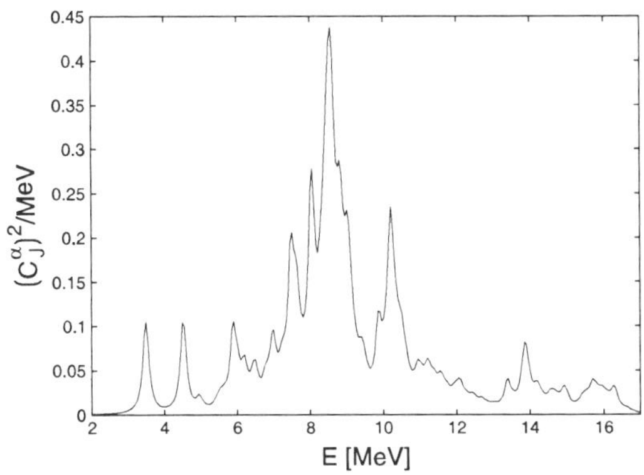

Figure 1. Distribution of the single-particle strength of the state $1k_{17/2}$ in ^{209}Pb.

The correlation entropy S^α connected with the excited states was calculated using Gaussian distribution functions. To test the sensitivity of the results, a new parameter - k was introduced. Multiplying the matrix elements of H_{QPM}, Eq. (2), by this number it is possible to vary the overall strength of the particle-core coupling. The absolute value of entropy, $F(N) = \sum_{\alpha=1}^N S^\alpha$, depends on the value of k, i.e. $F(N) \Rightarrow F(N,k)$. The function $F(N, k)$ is shown on Fig. 2 revealing a pronounced maximum at $k_c = 1.6$.

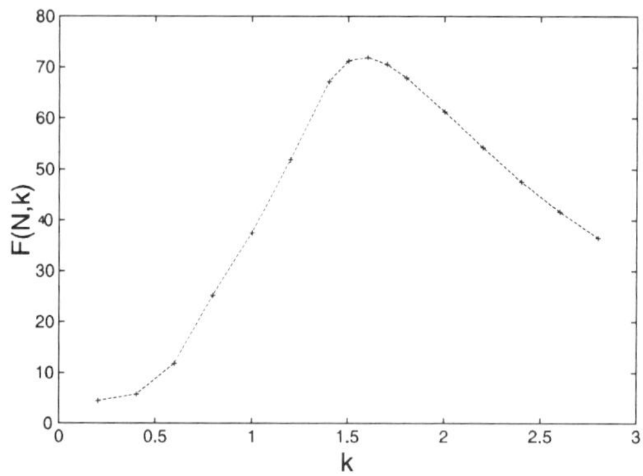

Figure 2. Dependence of the integral correlation entropy on the strength of particle-core coupling.

According to the value of $F(N, k)$ (Fig. 2) it is useful to separate three main domains of parameter k.

For small values of k (k less than 0.5 - 0.6), the mixing of states is suppressed and the ICE decreases rapidly. This case corresponds to the weak particle-core coupling and relatively simple wave function. The strength of single-particle states is distributed in a narrow vicinity of its unperturbed energy.

In the region of $k > 0.6$ and approximately up to $k = k_c = 1.6$ the correlation rapidly increases. This indicates the importance of more complex components and their influence on the damping process.

The third domain includes the values of k larger than 1.6. As it is seen from Fig. 2 ICE decreases rapidly for $k \geq 1.6$. This case correspons to *doubling phase transition.* The single-particle strength is split into two main pieces repelled to low and high excitation energies. The question remains whether such a phenomenon, known in quantum optics, could be observed in real nuclear spectra.

In conclusion, the introduced new quantity – invariant correlational entropy – was used to estimate the growth of complexity as a result of admixture of many new components to the wave function of a quasiparticle. As a function of the overall strength of quasiparticle-phonon interaction, the ICE increases and reaches a pronounced maximum. In this region the system would undergo the quantum doubling transition. The ICE properly reflects this transformation.

3. Microscopic Description of mixed-symmetry states in nearly spherical nuclei

Magnetic dipole excitations are important mode in low-lying part of nucleus spectrum. The existence of the mode was predicted [7] and discovered through inelastic electron scattering experiments [8] in deformed nuclei as scissors mode. An important feature of the scissors mode is its isovector character. States of isovector nature were first considered in a geometrical model as proton-neutron surface vibration [9]. Low-lying isovector excitations are predicted in the algebraic IBM-2 as mixed symmetry states with respect to the exchange between proton and neutron bosons [10]. In spherical nuclei, the lowest mixed symmetry state has $J^\pi = 2^+$ and can be excited from the ground state via weak $E2$ transitions. Its signature, however, is its strong $M1$ decay to the lowest isoscalar $J^\pi = 2^+$ state.

Recently unambiguous evidence in favor of mixed symmetry states was provided by an experiment which combined photon scattering with a $\gamma\gamma$-coincidence analysis of the transitions following β decay of ^{94}Tc to ^{94}Mo [11]. Theoretical calculations were performed in a shell model [12], and within the QPM [13].

Within QPM the Hamiltonian H_{QPM}, Eq.(2) has to be diagonalized in multiphonon space [14]

$$\Psi_\nu(JM) = \sum_i R_i Q_i^\dagger + \sum_{i_1 i_2} P_{i_2}^{i_1} \left[Q_{i_1}^\dagger \otimes Q_{i_2}^\dagger\right]_{JM} + \sum_{\substack{i_1 i_2 \\ i_3 I}} T_{i_3 I}^{i_1 i_2} \left[\left[Q_{i_1}^\dagger \otimes Q_{i_2}^\dagger\right]_I \otimes Q_{i_3}^\dagger\right]_{JM} \Psi_0$$

The calculated structure of excited states in ^{94}Mo is shown in Table 1. It is seen pronounced phonon contribution to the structure which leads to sharper selection rules of transitions. In the case of $E2$ reduced transition probabilities (Table 2) there is a strong one-phonon exchange transition between isoscalar or isovector states. Equally satisfactory is the scheme of the $M1$ transitions (Table 3).

According to the findings, the building blocks of QPM multi-phonon low-lying states in nuclei near shell closure are the first (isoscalar) and second (isovector) $[2^+]_{RPA}$ states. The resulting low-lying QPM states can be classified into two groups, composed respectively of isoscalar and isovector states. All these states have a single dominant component with a given number of phonons. The appreciable $E2$ strengths is obtained only for transitions connecting states differing by one-phonon. They are very large when the states involved in the $E2$ transition are isoscalar, large for transitions between isovector states, the $M1$ operator couples strongly only states of different isospin with an equal number of phonons. The picture seems to be a general feature of nuclei near shell closure and is consistent with the IBM scheme. The isoscalar and isovector states correspond to fully symmetric and mixed symmetry IBM states.

4. Pygmy Resonances

In the recent experiments performed at GSI and TU-Darmstadt [15] , the distribution of electric dipole strength in $^{130,132}Sn$ was measured. Together with some theoretical results [16], it gives a hint for the existence of low-lying dipole mode referred as "pygmy" dipole resonance (PDR). These excited states are known to be slightly collective. The tin isotopic chain ($^{100-132}Sn$) is suitable to study the dependence of PDR on neutron excess. The pre-

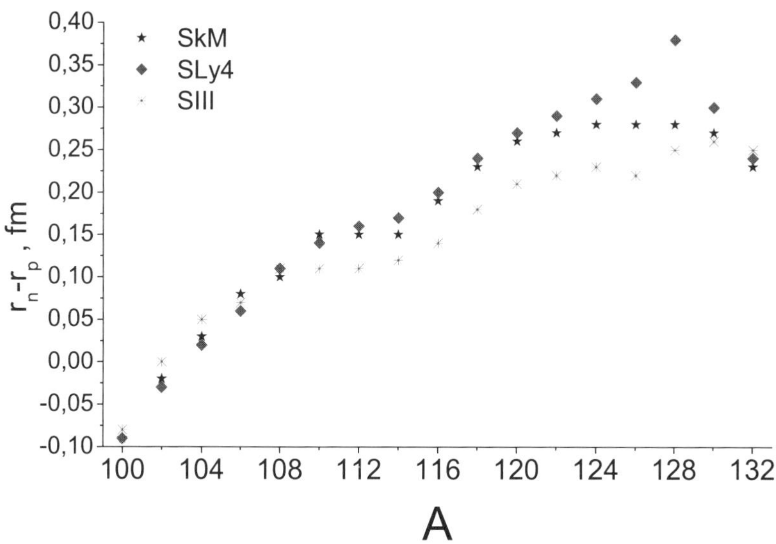

Figure 3. The value of $r_n - r_p$ in $^{110-130}Sn$ calculated by means three Skyrme parametrizations – $SIII(*), SkM^*(\star), SLy4(\blacklozenge)$.

sented distribution of $E1$ strength is calculated by means of QRPA [17]. The residual particle-hole interaction is of Skyrme type. Three Skyrme parametrizations are used in the calculations $-SIII, SkM^*$ and $SLy4$.

On Fig. 3 the dependence of the neutron and proton radii on the mass number is shown. One can trace the evolution of the relative differences of the neutron and proton root mean square radii, with the mass number. The last could be associated with the "neutron skin thickness" .

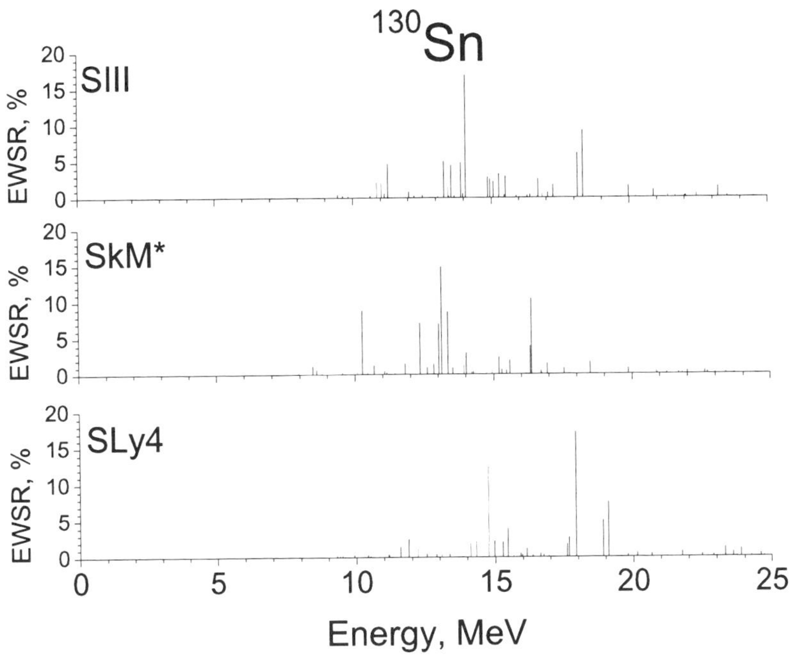

Figure 4. $E1$ strength distribution in ^{130}Sn calculated by means three Skyrme parametrizations $-$ $SIII, SkM^*, SLy4$.

On Fig. 4 the results for the distribution of $E1-$transition strength for ^{130}Sn are presented. It is seen that the distribution depends strongly on the type of Skyrme interaction used. The parameterizations $Sly4$ and $SIII$ reveal larger isovector interaction strength and greater part of the

$E1$ transition strength is shifted upward in comparison with SkM^*. For example, the giant dipole resonance (GDR) in ^{130}Sn lay at $14.20 MeV$ when SkM^* is used, for $SLy4$ it is at $16.30 MeV$, and for $SIII$ it is at $14.85 MeV$. This reflects on the behavior of the low-lying tail of the distribution. The results for SkM^* are quite low in comparison to the experimental data published in [15], and the results obtained with $SLy4$ are quite higher.

5. Conclusions

It is shown that the microscopic nuclear structure model (QPM) could be used to calculate the properties of excited states in large domain of excitation energy. The calculations cover large range of energy including low-lying mixed symmetry states, intermediate energy Pygmy structures and high-lying single-particle states. More details about these application of QPM could be found in Refs. [6,13,17].

References

1. V.G. Soloviev, *Theory of atomic nuclei : Quasiparticles and Phonons* (Institute of Physics Publishing, Bristol and Philadelphia, 1992).
2. N. Van Giai et al. Phys. Rev. C, **57**,1204 (1998).
3. V. Zelevinsky et al., Phys. Lett. B **350**, 141 (1995).
4. V.V. Sokolov, B. A. Brown, and V. Zelevinsky, Phys. Rev. E **58**, 56 (1998).
5. S. Gales, Ch. Stoyanov, and A. Vdovin, Phys. Rep., **166**, 125 (1988).
6. Ch. Stoyanov and V. Zelevinsky, Phys. Rev. C **70**, 014302 (2004).
7. N. Lo Iudice and F. Palumbo, Phys. Rev. Lett. **41**, 1532 (1978).
8. D. Bohle et al., Phys. Lett. **B137**, 27 (1984).
9. A. Faessler et al., Phys. Lett. **B 166**, 367 (1986); J. Phys. G, **13**, 337 (1987).
10. T. Otsuka, A. Arima, and Iachello, Nucl. Phys. **A309**, 1 (1978).
11. N. Pietralla et al., C Phys. Rev. Lett. **83**, 1303 (1999).
12. A.F. Lisetskiy et al., Nucl. Phys. **A 677**, 1000 (2000).
13. N. Lo Iudice and Ch. Stoyanov, Phys. Rev. C **65**, 064304 (2002).
14. V. Yu. Ponomarev et al., Nucl. Phys. **A 635**, 470 (1998).
15. P. Adrich et al., Phys. Rev. Lett. **95**, 132501 (2005); S. Volz et al., Nucl. Phys. A, **779**, 1 (2006); see also Nucl. Phys. A **788** (2007).
16. N. Tsoneva et al., Phys. Lett. B, **586**, 213 (2004); N. Paar et al., Phys. Lett B, **606**, 288 (2005); D. Sarchi et al., Phys. Lett. B, **6001**, 27 (2004)
17. D. Tarpanov et al.,Phys.At. Nucl., **70**, 1402 (2007)

Table 1. Energy and phonon structure of selected low-lying excited states in ^{94}Mo. Only the dominant components are presented.

	State J^{π}	E (keV) EXP	QPM	Structure,%
IS	$2^+_{1,is}$	871	860	$93\%[2^+_{is}]_{RPA}$
	$2^+_{2,is}$	1864	1750	$82\%[2^+_{is} \otimes 2^+_{is}]_{RPA}$
	$4^+_{1,is}$	1573	1733	$82\%[2^+_{is} \otimes 2^+_{is}]_{RPA}$
IV	$1^+_{1,iv}$	3129	2880	$90\%[2^+_{is} \otimes 2^+_{iv}]_{RPA}$
	$2^+_{1,iv}$	2067	1940	$95\%[2^+_{iv}]_{RPA}$
	$2^+_{2,iv}$	2393	2730	$27\%[2^+_{is} \otimes 2^+_{iv}]_{RPA}$
	$2^+_{3,iv}$	2740	3014	$59\%[2^+_{is} \otimes 2^+_{iv}]_{RPA}$
	$4^+_{1,iv}$		3120	$64\%[2^+_{is} \otimes 2^+_{iv}]_{RPA}$
	$3^+_{1,iv}$	2965	2940	$87\%[2^+_{is} \otimes 2^+_{iv}]_{RPA}$

Table 2. $E2$ transitions connecting some excited states in ^{94}Mo calculated in QPM. The experimental data are taken form Ref. [11]

	$B(E2; J_i \to J_f)(e^2fm^4)$	EXP	QPM	IBM-2
$\Delta T = 0$	$B(E2; g.s \to 2^+_{1,is})$	2030(40)	1978	2333
	$B(E2; g.s \to 2^+_{2,is})$	32(7)	35	0
	$B(E2; 2^+_{2,is} \to 2^+_{1,is})$	720(260)	673	592
	$B(E2; 4^+_{1,is} \to 2^+_{1,is})$	670(100)	661	592
$\Delta T = 0$	$B(E2; 2^+_{2,iv} \to 2^+_{1,iv})$		127	
	$B(E2; 2^+_{3,iv} \to 2^+_{1,iv})$		266	
	$B(E2; 1^+_{1,iv} \to 2^+_{1,iv})$	< 690	374	556
	$B(E2; 3^+_{1,iv} \to 2^+_{1,iv})$	$250\ ^{+310}_{-210}$	368	582
		$(1.5^{+1.2}_{-0.6}) \times 10^3$		
	$B(E2; 4^+_{1,iv} \to 2^+_{1,iv})$		274	

Table 3. $M1$ transitions connecting some excited states in ^{94}Mo calculated in QPM. The experimental data are taken from Ref.[11]

	$B(M1; J_i \to J_f)(\mu_N^2)$	EXP	QPM $0.7g^s_{free}$	IBM-2
$\Delta T = 1$	$B(M1; 1^+_{1,iv} \to 2^+_{2,is})$	0.43(5)	0.75	0.36
	$B(M1; 2^+_{1,iv} \to 2^+_{1,is})$	0.48(6)	0.72	0.30
	$B(M1; 2^+_{2,iv} \to 2^+_{2,is})$		0.10	
	$B(M1; 2^+_{3,iv} \to 2^+_{2,is})$	0.35(11)	0.24	0.1
	$B(M1; 3^+_{1,iv} \to 2^+_{2,is})$	$0.24^{+0.14}_{-0.07}$	0.34	0.18
	$B(M1; 3^+_{1,iv} \to 4^+_{1,is})$	$0.074^{+0.044}_{-0.019}$	0.26	0.13
	$B(M1; 4^+_{1,iv} \to 4^+_{1,is})$	0.8(2)	0.75	

TIME-DEPENDENT DENSITY FUNCTIONAL THEORY WITH SKYRME FORCES: DESCRIPTION OF GIANT RESONANCES IN DEFORMED NUCLEI

J. KVASIL[1], W. KLEINIG[2], V. O. NESTERENKO [2], P.-G. REINHARD [3] and P. VESELY [4]

[1] *Institute of Particle and Nuclear Physics, Charles University, CZ-18000 Praha 8, Czech Republic*
* *E-mail: kvasil@ipnp.troja.mff.cuni.cz, vesely@ipnp.troja.mff.cuni.cz*

[2] *Bogoliubov Laboratory of Theoretical Physics, Joint Institute for Nuclear Research, Dubna, Moscow region, 141980, Russia*
* *E-mail: kleinig@theor.jinr.ru, nester@theor.jinr.ru*

[3] *Institut für Theoretische Physik, Universität Erlangen, D-91058, Erlangen, Germany* * *E-mail: Paul-Gerhard.Reinhard@theorie2.physik.uni-erlangen.de*

The isovector giant dipole resonance (GDR) in deformed nuclei ^{154}Sm and ^{238}U is described within the self-consistent separable RPA (random-phase-approximation) with Skyrme forces SkT6, SkM*, SLy6 and SkI3. We examine dependence of the description on the Skyrme force and gross-structure effects, as well as influence of the Coulomb interaction, multipole mixing, and time-odd densities. The results are analyzed in terms of isovector characteristics of the nuclear matter. It is shown that GDR in deformed nuclei can serve as a robust test for Skyrme forces with different effective masses. The best description of the resonance is obtained for the SLy6 force.

Keywords: Style file; LaTeX; Proceedings; World Scientific Publishing.

1. Introduction

Skyrme forces[1–3] are widely used for description of diverse properties of atomic nuclei (see, for recent reviews[4,5]). In particular, there is increasing interest on the description of the dynamics of deformed nuclei, particularly exotic ones.[6–8] However, even the giant resonances (GR) in stable nuclei pose interesting, yet unsolved, problems.[4,5,9] It is still a challenge to describe the isoscalar E2 and the isovector E1 GR with one and the same Skyrme force. Also, one can still work out more relations between properties of the GR and characteristics of nuclear matter.[9]

All these points certainly devote a deep and systematic analysis. At the same time the treatment of excitations in deformed nuclei within self-consistent models is involved and time consuming. Even after development of a new generation of efficient RPA schemes,[6,7] the calculations for deformed rare-earth and actinide nuclei need a great effort. In this connection, we have developed a self-consistent separable RPA (SRPA) method[10–14] allowing to minimize the computational time while keeping the high accuracy. SRPA has been already successfully used for both spherical[11] and deformed[13–15] nuclei. In particular, the systematic exploration of the giant dipole resonance (GDR) with various Skyrme forces was commenced in.[14,15]

In this paper we will continue the analysis. Together with the problems discussed earlier,[14,15] we will also consider dependence of the GDR on the spin-orbital and Coulomb interaction as well as mixing of multipole forces, induced by nuclear deformation.

2. Calculation scheme and Skyrme forces

We use the functional with the Skyrme energy density

$$\begin{aligned}\mathcal{H}_{\text{Sk}} = &\frac{b_0}{2}\rho^2 - \frac{b_0'}{2}\sum_s \rho_s^2 - \frac{b_2}{2}\rho(\Delta\rho) + \frac{b_2'}{2}\sum_s \rho_s(\Delta\rho_s) \\ &+ \frac{b_3}{3}\rho^{\alpha+2} - \frac{b_3'}{3}\rho^{\alpha}\sum_s \rho_s^2 + b_1(\rho\tau - \vec{j}^2) - b_1'\sum_s(\rho_s\tau_s - \vec{j}_s^2) \\ &- b_4\left(\rho(\vec{\nabla}\vec{\Im}) + \vec{\sigma}\cdot(\vec{\nabla}\times\vec{j})\right) - b_4'\sum_s\left(\rho_s(\vec{\nabla}\vec{\Im}_s) + \vec{\sigma}_s\cdot(\vec{\nabla}\times\vec{j}_s)\right) .\end{aligned} \tag{1}$$

It includes contributions from both time-even (nucleon ρ_s, kinetic energy τ_s, and spin-orbital $\Im_s$) and time-odd (current j_s and spin σ_s) densities, where s denotes protons and neutrons. The total densities (like $j = j_p + j_n$) are given in (1) without the index. As was shown in refs.,[2,3] the time-odd densities naturally belong to the Skyrme functional if it involves all the possible bilinear combinations of the nucleon and spin densities together with their derivatives up to the second order. The time-odd densities enter the functional only in specific combinations, as a complement to the time-even ones, so as to keep Galilean and gauge invariance of Skyrme forces.

The dipole response involves contributions from both time-even and time-odd densities as well as from the pairing densities χ_s.[13,14] The contributions of the time-odd densities are driven by the variations

$$\frac{\delta^2 E}{\delta \vec{j}_{s_1} \delta \vec{j}_s}, \quad \frac{\delta^2 E}{\delta \vec{\sigma}_{s_1} \delta \vec{j}_s}, \quad \frac{\delta^2 E}{\delta \vec{j}_{s_1} \delta \vec{\sigma}_s} \tag{2}$$

of the Skyrme functional terms with coefficients b_1, b_1', b_4, b_4'. In addition to previous studies,[14,15] the contribution of the Coulomb interaction and particle-particle pairing channel are taken into account. The code allows to mix multipole forces with the same K^π (projection of the moment to the symmetry z-axis and space parity). Due to *factorization* of the residual interaction, SRPA avoids diagonalization of high-rank RPA matrices and thus drastically simplifies the calculations. This point is crucial for the case of heavy deformed nuclei with their impressive configuration space.

The dipole response is computed as the photo-absorption energy-weighted strength function

$$S(E\lambda\mu; \omega) = \sum_\nu \omega_\nu M^2_{\lambda\mu\nu} \zeta(\omega - \omega_\nu) \tag{3}$$

where $\zeta(\omega-\omega_\nu) = \Delta/[2\pi((\omega-\omega_\nu)^2+(\Delta/2)^2)]$ is the Lorentz weight with the averaging parameter Δ, $M_{\lambda\mu\nu}$ is the matrix element of $E\lambda\mu$ transition from the ground state to the RPA state $|\nu>$, ω_ν is the RPA eigen-energy. By using the SRPA technique,[13,14] we directly compute the strength function with the Lorentz weight, which dramatically reduces the computation time.

The calculations are mainly performed with 2 generating operators

$$\hat{Q}_{\lambda\mu k}(\vec{r}) = f_{\lambda k}(r)\Phi_{\lambda\mu}(\Omega) = r^{\lambda+2(k-1)}(Y_{\lambda\mu}(\Omega) + Y^+_{\lambda\mu}(\Omega)). \tag{4}$$

The operator with $k = 1$ generates the forces mainly localized at the nuclear surface while the term with $k = 2$ allows to touch, at least partly, the nuclear interior. This simple set seems to be a good compromise for the calculations of the giant resonances.[11,14,15] Sometimes, more input operators are used, e.g. with spherical Bessel functions $f_{\lambda k}(r) = j_\lambda(q_k r)$[11] to improve the precision or with $\lambda \to \lambda + 2$ to take into account the multipole mixing. This will be done below for the results exhibited in Fig. 1.

The calculations were performed for axially deformed nuclei ^{154}Sm and ^{238}U. The Skyrme forces SkT6,[16] SkM*,[17] SLy6[18] and SkI3[19] were exploited. Though these forces were fitted with a different bias, they all provide a good overall description of nuclear bulk properties and suitable for deformed nuclei.[4] For our aim it is important that these forces cover different values of nuclear matter characteristics (see Table 1).

The calculations employ a cylindrical coordinate-space grid with the step 0.7 fm. The pairing is treated with delta pairing force at the BCS

Table 1. Nuclear matter and deformation properties for the Skyrme forces under consideration. The table represents the isoscalar effective mass m_0^*/m, symmetry energy $a_{\rm sym}$, density dependence of symmetry energy $a'_{\rm sym} == \partial a_{\rm sym}/\partial_\rho$, sum rule enhancement factor κ, isovector effective mass $m_1^*/m = 1/(1+\kappa)$, and quadrupole moments Q_2 in ^{154}Sm and ^{238}U. The experimental values of Q_2 are taken from.[20,21]

Forces	m_0^*/m	$a_{\rm sym}$ [MeV]	$a'_{\rm sym}$ [MeV fm^3]	κ	m_1^*/m	Q_2 [b] ^{154}Sm	Q_2 [b] ^{238}U
SkT6	1.00	30.0	63	0.001	1.00	6.8	11.1
SkM*	0.79	30.0	95	0.531	0.65	6.8	11.1
SLy6	0.69	32.0	100	0.250	0.80	6.8	11.0
SkI3	0.58	34.8	212	0.246	0.80	6.8	11.0
exp.						6.6	11.1

level.[23] The ground state quadrupole deformation is determined by minimizing the total energy. As is seen from Table 1, all the forces equally well reproduce quadrupole moments of ^{154}Sm and ^{238}U, though with a modest systematic overestimation for ^{154}Sm. This overestimation does not influence the results presented below.

3. Results and discussion

Figure 1 depicts dependence of GDR on some particular contributions to the residual interaction: spin-orbital, Coulomb and multipole mixing. In the later case, the operator $\hat{Q}_3(\vec{r}) = r^3(Y_{3\mu}(\Omega) + Y^*_{3\mu}(\Omega))$ was added to the set (4) so as to take into account the deformation induced mixing of dipole and octupole forces. The results are presented for the force Sly6 which, following our previous exploration,[14,15] seems to be the best for description of GDR in heavy deformed nuclei.

As is seen from the figure, the contributions of spin-orbital and Coulomb are rather weak and multipole mixing is negligible. The same results are obtained with other three Skyrme forces. It worth noting that, following our preliminary calculations, the spin-orbital and Coulomb contributions are more essential for lowest quadrupole and octupole vibrational states.

Figure 2 (left and middle columns) exhibits description of the GDR with different Skyrme forces. Both strengths with and without residual interaction are presented so as to demonstrate the collective shifts. It is seen that E1 unperturbed strengths in the figures exhibit a systematic shift to higher energy from SkT6 to SkI3. This can be explained by the effect of isoscalar effective mass m_0^*/m which results in stretching the single-particle spectra: the less m_0^*/m, the more dilute are the spectra. Simultaneously,

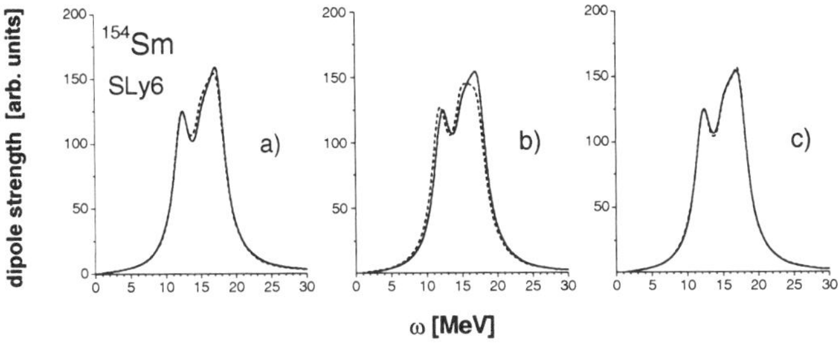

Fig. 1. The isovector dipole strength (E1 giant resonance) in ^{154}Sm calculated with the forces SLy6 for the cases: a) with (solid curve) and without (dash curve) spin-orbital contribution to the residual interaction; b) with (solid curve) and without (dash curve) Coulomb contribution to the residual interaction; c) with (solid curve) and without (dash curve) mixing of the dipole and octupole forces in the residual interaction.

we see the corresponding evolution of the collective energy shifts which decrease from SkT6 to SkI3. The net result is that the final GDR energies become in general rather close to each other for different forces.

Nevertheless, there still remain considerable deviations in the GDR description. The best results is obtained for Sly6 while SkT6 and SkI3 downshift the GDR energy. The force SkM* essentially overestimates the GDR energy which can be explained by exceptionally large value of sum rule enhancement factor κ (and hence too low value of the isovector effective mass m_1^*/m) for this force, see Table 1. Moreover, for SkM* case we get unrealistically high right GDR shoulder and the subsequent overestimation of the resonance width. The effect becomes weaker for SLy6 (thus giving the best description of GDR) and vanishes for SkI3 (already with underestimation of the resonance width). The appearance of the right shoulder for some Skyrme forces, preferably those with a large effective mass, was also noted in the calculations for deformed rare-earth and actinide nuclei within the full (non-separable) Skyrme RPA[8] and in the SRPA calculations for ^{208}Pb.[11] This effect seems to be universal for GDR in heavy nuclei, independently on their shape.

The right shoulder effect is not caused by the deformation splitting but rather is a consequence of the strength fragmentation and excessive collective shift. Indeed, it is seen that the *unperturbed* dipole strength in SkM* also has some kind of a right shoulder. When the collective strength is placed at the shoulder, it is immersed into the sea of two-quasi-particle states and is strongly fragmented. However, if the collective shift is too large, then the strength is pushed beyond the shoulder, i.e. to the region

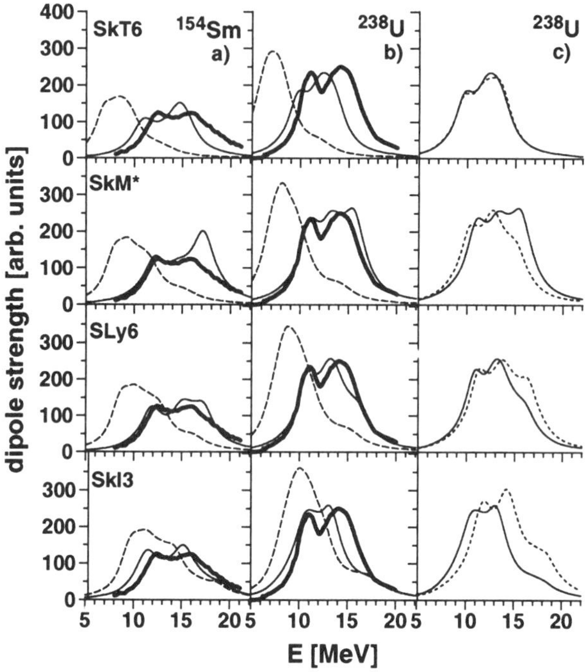

Fig. 2. The isovector E1 giant resonance in ^{154}Sm and ^{238}U, calculated with the Skyrme forces SkT6, SkM*, SLy6 and SkI3. a)-b): The calculated strength (solid curve) is compared with the experimental data[24,25] (triangles). The quasiparticle (unperturbed) strength is denoted by the dotted curve. c): The calculated strength with (solid curve) and without (dash curve) the current contribution.

of weak fragmentation. Then the strength is collected into narrow peak as in the case of SkM*. Other forces either produce less collective shifts (Sly6, SkI3) or have no the right shoulder but rather the smooth tail (Skt6). As a result, they do not exhibit the artificial right-shoulder or three-peak structure or do this in much less extent. So, the right flank of the GDR can serve as a robust indicator of how reasonable is the collective shift.

Figure 2 demonstrates that the GDR energy does not rise with $a_{\rm sym}$ as it might be intuitively expected from the macroscopic arguments.[26] For example, the GDR energy is higher for SLy6 ($a_{\rm sym}$=30.0) than for SkI3 ($a_{\rm sym}$=34.8). As was discussed in,[9] this effect can be explained by density dependence of $a_{\rm sym}$. In other words, we should take into account the relation of GDR properties with $a'_{\rm sym} = \partial a_{\rm sym}/\partial_\rho$ as well.

The right plots of Fig. 2 demonstrate influence of the time-odd densities on the GDR. The calculations show that only the current-current contribution in (2) is essential while the contribution connected with the spin density is negligible. So, we display only effect of the current density. As is seen, the main effect takes place again at the right flank of the GDR. The effect is not systematic. We observe no shift for SkT6, the upshift for SkM* and the down-shifts for SLy6 and SkI3. Again one may note the correlation with the effective masses. For example, SkT6 the effective masses $m_1^*/m = m_0^*/m = 1$ and hence no time-odd contributions. Instead, the specific SkM* case can be connected with very small m_1^*/m for this force. The correlation with the effective mass is quite natural since current density enters the term of the Skyrme functional $\sim b_1, b_1'$ just responsible for generation of the effective masses.

4. Conclusions

Dependence of description of giant dipole resonance (GDR) on various Skyrme forces was considered within the SRPA model. We analyzed the gross-structure effects, relations with nuclear matter characteristics, influence of the spin-orbital, Coulomb and multipole-mixing contributions, impact of the current time-odd density. The best description was obtained for the force Sly6. Altogether, it was demonstrated that GDR in heavy deformed nuclei can serve as a robust test for selection of the Skyrme parameters and related nuclear matter values.

5. Acknowledgments

This work is a part of the research plan MSM 0021620859 supported by the Ministry of Education of the Czech republic. It was partly funded by Czech grant agency (grant No. 202/06/0363), grant agency of Charles University in Prague (grant No. 222/2006/B-FYZ/MFF). We also thank also the support from DFG grant RE 322/11-1 and Heisenberg-Landau (Germany-BLTP JINR) grants for 2006 and 2007 years. W.K. and P.-G.R. are grateful for the BMBF support under contracts 06 DD 139D and 06 ER 808.

References

1. T.H.R. Skyrme, *Phil. Mag.* **1**, 1043 (1956); D. Vauterin, D.M. Brink, *Phys. Rev. C* **5**, 626 (1972).
2. Y.M. Engel, D.M. Brink, K. Goeke, S. J. Krieger, and D. Vauterin, *Nucl. Phys. A* **249**, 215 (1975).
3. J. Dobaczewski and J. Dudek, *Phys. Rev. C* **52**, 1827 (1995).

4. M. Bender, P. -H. Heenen, and P.-G. Reinhard, *Rev. Mod. Rhys.* **75**, 121 (2003).
5. J. R. Stone and P. -G. Reinhard, *Prog. Part. Nucl. Phys.* **58**, 587 (2007).
6. M. V. Stoitsov, J. Dobaczewski, W. Nazarewitcz, S. Pittel, and D. J. Dean, *Phys. Rev. C* **68**, 054312 (2003).
7. A. Obertelli, S. Peru, J. -P. Delaroche, A. Gillibert, M. Girod, and H. Goutte, *Phys. Rev. C* **71**, 024304 (2005).
8. J. A. Maruhn, P. -G. Reinhard, P. D. Stevenson, J. Rikovska Stone, and M. R. Strayer, *Phys. Rev. C* **71**, 064328 (2005).
9. P.-G. Reinhard, *Nucl. Phys. A* **649**, 305c (1999).
10. J. Kvasil, V.O. Nesterenko, and P.-G. Reinhard, in Proceed. of 7th Inter. Spring Seminar on Nuclear Physics, Miori, Italy, 2001, edited by A.Covello, World Scient. Publ., p.437, 2002; arXiv nucl-th/00109048.
11. V.O. Nesterenko, J. Kvasil, and P.-G. Reinhard, *Phys. Rev. C* **66**, 044307 (2002).
12. V.O. Nesterenko, J. Kvasil and P.-G. Reinhard, *Progress in Theoretical Chemistry and Physics*, **15** 127 (2006); ArXiv: physics/0512060.
13. V.O. Nesterenko, J. Kvasil, W. Kleinig, P.-G. Reinhard, and D. S. Dolci, arXiv: nucl-th/0512045.
14. V.O. Nesterenko, W. Kleinig, J. Kvasil, P. Vesely, P.-G. Reinhard, and D. S. Dolci, *Phys. Rev. C* **74**, 054306 (2006).
15. V.O. Nesterenko, W. Kleinig, J. Kvasil, P. Vesely, and P.-G. Reinhard, *Int. J. Mod. Phys. E* **16**, 624 (2007).
16. F. Tondeur, M. Brack, M. Farine, and J.M. Pearson, *Nucl. Phys. A* **420**, 297 (1984).
17. J. Bartel, P. Quentin, M. Brack, C. Guet, and H.-B. Håakansson, *Nucl. Phys. A* **386**, 79 (1982).
18. E. Chabanat, P. Bonche, P. Haensel, J. Meyer, and R. Schaeffer, *Nucl. Phys. A* **643**, 441(E) (1998).
19. P.-G. Reinhard and H. Flocard, *Nucl. Phys. A* **584**, 467 (1995).
20. S. Raman, *At. Data and Nucl. Data Tables* **36**, 1 (1987).
21. A.S Goldhaber and G.S. Goldhaber, *Phys. Rev. C* **17**, 1171 (1978).
22. P.-G. Reinhard, *Ann. Phys.* (Leipzig) **1**, 632 (1992).
23. M. Bender, K. Rutz, P.-G. Reinhard, and J.A. Maruhn, *Eur. Phys. J. A*, **8** 59 (2000).
24. JANIS database: l0025.029-0 (Sac1971).
25. S.S. Dietrich and B.L. Bergman, *At. Data Nucl. Data Tables*, **38**, 199 (1998); IAEA Photonuclear Data.
26. G.F. Bertsch and R.A. Broglia, *Oscillations in Finite Quantum Systems* (Cambridge University Press, Capbridge, 1994).

FINITE RANK APPROXIMATION FOR SKYRME FORCES AND EFFECTS OF THE PARTICLE-PARTICLE CHANNEL

A.P. SEVERYUKHIN, V.V. VORONOV*

Bogoliubov Laboratory of Theoretical Physics, JINR,
141980 Dubna, Moscow region, Russia
**E-mail: voronov@theor.jinr.ru*

NGUYEN VAN GIAI

Institut de Physique Nucléaire, CNRS-IN2P3,
Université Paris-Sud, F-91406 Orsay Cedex, France

A finite rank separable approximation for the QRPA calculations with Skyrme interactions is generalized to take into account the particle-particle residual interaction. As an illustration of the method properties of the first quadrupole states in some Te isotopes are considered.

Keywords: QRPA; particle-particle channel; Tellurium.

1. Introduction

The new spectroscopic studies of the exotic nuclei stimulate a development of the nuclear models to describe properties of nuclei away from the stability lines. One of the standard tools for nuclear structure studies is the quasiparticle random phase approximation (QRPA) with the self-consistent mean-field derived by making use of the effective nucleon-nucleon interaction.[1,2] When the residual interaction is separable,[3] the QRPA problem can be easily solved no matter how many two-quasiparticle configurations are involved. Starting from an effective interaction of the Skyrme type, a finite rank separable approximation was proposed[4] for the particle-hole (p-h) residual interaction. Such an approach allows one to perform structure calculations in very large particle-hole spaces. Thus, the self-consistent mean field can be calculated within the Hartree-Fock method with the original Skyrme interaction whereas the RPA solutions would be obtained with the finite rank approximation to the p-h matrix elements. This approach was extended to include the pairing correlations within the BCS approximation.[5] Recently, we generalized our approach to take into account a coupling between the one- and two-phonon components of wave functions.[6]

Here, we present an extension of our approach by taking into account the particle-particle (p-p) residual interaction. This extension is described in detail elsewhere.[7] As an illustration of the method properties of the lowest quadrupole states in some Te isotopes are considered.

2. The method

The starting point of the method is the HF-BCS calculation[8] of the ground states, where spherical symmetry is imposed on the quasiparticle wave functions. The continuous part of the single-particle spectrum is discretized by diagonalizing the HF hamiltonian on a harmonic oscillator basis.[9] We work in the quasiparticle representation defined by the canonical Bogoliubov transformation:

$$a^{+}_{jm} = u_j \alpha^{+}_{jm} + (-1)^{j-m} v_j \alpha_{j-m}, \tag{1}$$

where jm denote the quantum numbers $nljm$. The hamiltonian includes the Skyrme interaction[10] in the p-h channel and the surface peaked density-dependent zero-range force

$$V_{pair}(\mathbf{r}_1, \mathbf{r}_2) = V_0 \left(1 - \frac{\rho(r_1)}{\rho_c}\right) \delta(\mathbf{r}_1 - \mathbf{r}_2) \tag{2}$$

in the particle-particle channel. The parameters of the p-p interaction (2) are fixed to reproduce the odd-even mass difference of neighboring nuclei.

The residual interaction in the p-h channel V^{ph}_{res} and in the p-p channel V^{pp}_{res} can be obtained as the second derivative of the energy density functional with respect to the particle density ρ and the pair density $\tilde{\rho}$, respectively. Following our previous paper[4] we simplify V^{ph}_{res} by approximating it by its Landau-Migdal form. For Skyrme interactions all Landau parameters with $l > 1$ are zero. We keep only the $l = 0$ terms in V^{ph}_{res} and the expressions for $F^{ph}_0, G^{ph}_0, F^{ph'}_0, G^{ph'}_0$ in terms of the Skyrme force parameters can be found in Ref.[11] Our calculations[5,12] show that, for analysis of the normal parity states one can neglect the G^{ph}_0 and $G^{ph'}_0$ terms of the residual interaction V^{ph}_{res}. Therefore we can write the residual interaction in the following form:

$$V^{a}_{res}(\mathbf{r}_1, \mathbf{r}_2) = N_0^{-1} [F^{a}_0(r_1) + F^{a'}_0(r_1)(\tau_1 \cdot \tau_2)] \delta(\mathbf{r}_1 - \mathbf{r}_2), \tag{3}$$

where a is the channel index $a = \{ph, pp\}$; σ_i and τ_i are the spin and isospin operators, and $N_0 = 2k_F m^*/\pi^2\hbar^2$ with k_F and m^* standing for the Fermi momentum and nucleon effective mass.

The two-body matrix element

$$V_{1234} = \int \phi_1^*(\mathbf{r}_1)\phi_2^*(\mathbf{r}_2)V_{res}(\mathbf{r}_1,\mathbf{r}_2)\phi_3(\mathbf{r}_1)\phi_4(\mathbf{r}_2)\mathbf{dr}_1\mathbf{dr}_2 \tag{4}$$

can be written as:

$$V^{ph}_{j_1m_1j_2m_2j_3m_3j_4m_4} = \frac{1}{(2J+1)}\sum_{JM}\langle j_1m_1j_3-m_3 \mid J-M\rangle \times\langle j_2m_2j_4-m_4 \mid JM\rangle(-1)^{J-M+j_3-m_3+j_4-m_4}V^{ph(J)}_{j_1j_2j_3j_4} \tag{5}$$

in the p-h channel and

$$V^{pp}_{j_1m_1j_2m_2j_3m_3j_4m_4} = \sum_{JM}\langle j_1m_1j_2m_2 \mid JM\rangle\langle j_3m_3j_4m_4 \mid JM\rangle V^{pp(J)}_{j_1j_2j_3j_4} \tag{6}$$

in the p-p channel, where

$$V^{ph(J)}_{j_1j_2j_3j_4} = \langle j_1||i^JY_J||j_3\rangle\langle j_2||i^JY_J||j_4\rangle I^{ph}(j_1j_2j_3j_4), \tag{7}$$

$$V^{pp(J)}_{j_1j_2j_3j_4} = \sum_\lambda \left\{\begin{matrix} j_4 & j_3 & J \\ j_1 & j_2 & \lambda\end{matrix}\right\}(-1)^{j_2+j_3} \times\langle j_1||i^JY_J||j_3\rangle\langle j_2||i^JY_J||j_4\rangle I^{pp}(j_1j_2j_3j_4). \tag{8}$$

In the above expressions, $\langle j_1||i^JY_J||j_3\rangle$ is the reduced matrix element of the spherical harmonics $Y_{J\mu}$ and $I^a(j_1j_2j_3j_4)$ is the radial integral:

$$I^a(j_1j_2j_3j_4) = N_0^{-1}\int_0^\infty (F_0^a(r) + F_0^{a'}(r)\tau_1\cdot\tau_2) \times u_{j_1}(r)u_{j_2}(r)u_{j_3}(r)u_{j_4}(r)\frac{dr}{r^2}, \tag{9}$$

where $u(r)$ is the radial part of the single-particle wave function. The radial integrals can be calculated accurately by choosing a large enough cutoff radius R and using a N-point integration Gauss formula with abscissas r_k and weights w_k. As it is shown in Refs.,[4,5,7] the residual interaction can be reduced to a finite rank separable form:

$$\hat{V}_{res} = -\frac{1}{2}\sum_a\sum_{\lambda\mu}\sum_{k=1}^{N}\sum_{q=\pm1}\sum_\tau\left(\kappa_0^{(a,k)} + q\kappa_1^{(a,k)}\right) \times :M^{(a,k)+}_{\lambda\mu}(\tau)M^{(a,k)}_{\lambda\mu}(q\tau):. \tag{10}$$

We sum over the proton(p) and neutron(n) indexes and the notation $\{\tau = (n,p)\}$ is used. A change $\tau\leftrightarrow-\tau$ means a change $p\leftrightarrow n$. $\kappa^{(ph,k)}$ ($\kappa^{(pp,k)}$)

are the multipole interaction strengths in the p-h (p-p) channel and they can be expressed as:

$$\begin{pmatrix} \kappa_0^{(a,k)} \\ \kappa_1^{(a,k)} \end{pmatrix} = -N_0^{-1} \frac{Rw_k}{2r_k^2} \begin{pmatrix} F_0^a(r_k) \\ F_0^{a'}(r_k) \end{pmatrix}. \tag{11}$$

The multipole operators entering the normal products in Eq.(10) are defined as follows:

$$M_{\lambda\mu}^{(ph,k)+}(\tau) = \frac{1}{\sqrt{2\lambda+1}} \sum_{jj'mm'}^{\tau} (-1)^{j+m} \langle jmj' - m' \mid \lambda\mu\rangle f_{j'j}^{(\lambda,k)} a_{jm}^{+} a_{j'm'}, \tag{12}$$

$$M_{\lambda\mu}^{(pp,k)+}(\tau) = \frac{(-1)^{\lambda-\mu}}{\sqrt{2\lambda+1}} \sum_{jj'mm'}^{\tau} \langle jmj'm' \mid \lambda\mu\rangle f_{jj'}^{(\lambda,k)} a_{jm}^{+} a_{j'm'}^{+}, \tag{13}$$

where $f_{j_1 j_2}^{(\lambda,k)}$ are the single-particle matrix elements of the multipole operators:

$$f_{j_1 j_2}^{(\lambda k)} = u_{j_1}(r_k) u_{j_2}(r_k) \langle j_1 || i^{\lambda} Y_{\lambda} || j_2 \rangle. \tag{14}$$

The residual interaction (10) is represented in terms of bifermion quasiparticle operators and their conjugates:

$$B(jj^{'};\lambda\mu) = \sum_{mm'} (-1)^{j'+m'} \langle jmj'm' \mid \lambda\mu\rangle \alpha_{jm}^{+} \alpha_{j'-m'}, \tag{15}$$

$$A^{+}(jj^{'};\lambda\mu) = \sum_{mm'} \langle jmj'm' \mid \lambda\mu\rangle \alpha_{jm}^{+} \alpha_{j'm'}^{+}. \tag{16}$$

It is necessary to emphasize that the hamiltonian of our method has the same form as the hamiltonian of the well-known quasiparticle-phonon model,[3] but the single-quasiparticle spectrum and the parameters of the residual interaction are calculated by the Skyrme forces.

We introduce the phonon creation operators

$$Q_{\lambda\mu i}^{+} = \frac{1}{2} \sum_{jj'} \left(X_{jj'}^{\lambda i} A^{+}(jj^{'};\lambda\mu) - (-1)^{\lambda-\mu} Y_{jj'}^{\lambda i} A(jj^{'};\lambda-\mu) \right), \tag{17}$$

where the index λ denotes total angular momentum and μ is its z-projection in the laboratory system. One assumes that the ground state is the QRPA phonon vacuum $\mid 0\rangle$. We define the excited states for this approximation by $Q^{+}_{\lambda\mu i} \mid 0\rangle$. Making use of the linearized equation-of-motion approach one can get the QRPA equations:[8]

$$\begin{pmatrix} \mathcal{A} & \mathcal{B} \\ -\mathcal{B} & -\mathcal{A} \end{pmatrix} \begin{pmatrix} X \\ Y \end{pmatrix} = w \begin{pmatrix} X \\ Y \end{pmatrix}. \tag{18}$$

In QRPA problems there appear two types of interaction matrix elements, the $\mathcal{A}^{(\lambda)}_{(j_1 j_1')(j_2 j_2')}$ matrix related to forward-going graphs and the $\mathcal{B}^{(\lambda)}_{(j_1 j_1')(j_2 j_2')}$ matrix related to backward-going graphs. Solutions of this set of linear equations yield the eigen energies and the amplitudes X, Y of the excited states. For our case expressions for $\mathcal{A}, \mathcal{B}$ and X, Y are given in Ref.[7] The dimension of the matrices $\mathcal{A}, \mathcal{B}$ is the space size of the two-quasiparticle configurations. Using the finite rank approximation the matrix dimensions never exceed $6N \times 6N$ independently of the configuration space size. If we omit terms of the residual interaction in the p-p channel then the matrix dimension is reduced by a factor 3.[4,5] So this approach enables one to reduce remarkably the dimensions of the matrices that must be inverted to perform structure calculations in very large configuration spaces.

3. Results of calculations

As an application of the method we investigate the 2^+_1 state energies and transition probabilities B(E2) in the $^{126-136}$Te isotopes. The results of our calculations and the experimental data[13,14] are displayed in Fig. 1. As it is seen from Fig. 1 there is a remarkable decrease of the 2^+_1 energies due to an inclusion of the p-p channel. At the same time the $B(E2)$-values don't change practically. It means that a collectivity of the 2^+_1 states is reduced. Our calculations are in a reasonable agreement with the available experimental data. It is worth to mention that the anharmonical effects are strong for the light Te isotopes and the QRPA is not very good in such a case. The $B(E2)$-value in the neutron-rich isotope ^{136}Te is only slightly larger than one for ^{134}Te, in contrast to the trend of Ce, Ba and Xe isotopes.[13,14] Such a behaviour of $B(E2)$ is related with the proportion between the QRPA amplitudes for neutrons and protons in Te isotopes.[7] It is worth to mention that the first prediction of the anomalous behaviour of 2^+ excitations around ^{132}Sn based on the QRPA calculations with a separable quadrupole-plus-pairing hamiltonian has been done in Ref.[15]

4. Conclusions

A finite rank separable approximation for the QRPA calculations with Skyrme interactions that was proposed in our previous work is extended

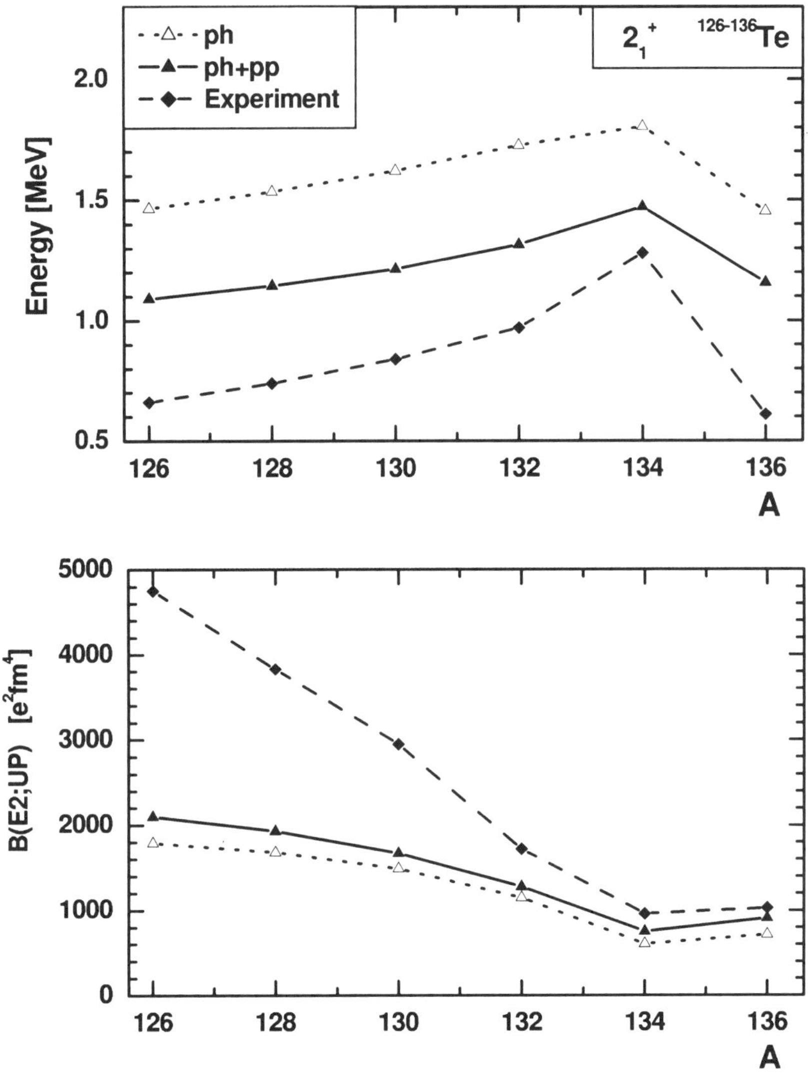

Fig. 1. Energies and transition probabilities for neutron-rich Te isotopes.

to take into account the particle-particle residual interaction. This approximation enables one to reduce remarkably the dimensions of the matrices that must be inverted to perform structure calculations in very large configuration spaces. As an illustration of the method we have studied the energies and transition probabilities of the 2_1^+ states in the $^{126-136}$Te isotopes.

Our results are in a reasonable agreement with available experimental data. An inclusion of the quadrupole p-p interaction results in a reduction of a collectivity and may be more important for nuclei far from closed shells.

Acknowledgments

A.P.S. and V.V.V. thank the hospitality of IPN-Orsay where the part of this work was done. This work is partly supported by the IN2P3-JINR agreement.

References

1. D. Vautherin and D. M. Brink, *Phys. Rev. C* **5**, 626 (1972).
2. D. Gogny, in *Nuclear Self-consistent Fields*, eds. G. Ripka and M. Porneuf (North-Holland, Amsterdam, 1975).
3. V. G. Soloviev, *Theory of Atomic Nuclei: Quasiparticles and Phonons* (Institute of Physics, Bristol and Philadelphia, 1992).
4. Nguyen Van Giai, Ch. Stoyanov and V. V. Voronov, *Phys. Rev. C* **57**, 1204 (1998).
5. A. P. Severyukhin, Ch. Stoyanov, V. V. Voronov and Nguyen Van Giai, *Phys. Rev. C* **66**, 034304 (2002).
6. A. P. Severyukhin, V. V. Voronov and Nguyen Van Giai, *Eur. Phys. J. A* **22**, 397 (2004).
7. A. P. Severyukhin, V. V. Voronov and Nguyen Van Giai, submitted to *Phys. Rev. C*.
8. P. Ring and P. Schuck, *The Nuclear Many Body Problem* (Springer, Berlin, 1980).
9. J. P. Blaizot and D. Gogny, *Nucl. Phys. A* **284**, 429 (1977).
10. E. Chabanat , P. Bonche, P. Haensel, J. Meyer and R. Schaeffer, *Nucl. Phys. A* **635**, 231 (1998).
11. Nguyen Van Giai and H. Sagawa, *Phys. Lett. B* **106**, 379 (1981).
12. A. P. Severyukhin, V. V. Voronov, Ch. Stoyanov and Nguyen Van Giai, *Physics of Atomic Nuclei* **66**, 1434 (2003).
13. S. Raman, C.W. Nestor Jr. and P. Tikkanen, *At. Data and Nucl. Data Tables* **78**, 1 (2001).
14. D. C. Radford et al., *Phys. Rev. Lett.* **88**, 222501 (2002).
15. J. Terasaki, J. Engel, W. Nazarewicz and M. Stoitsov, *Phys. Rev. C* **66**, 054313 (2002).

LOW-LYING 0^+ STATES IN DEFORMED NUCLEI

A. V. SUSHKOV[1], N. LO IUDICE[2], AND N. YU. SHIRIKOVA[1]

[1]*Bogoliubov Laboratory of Theoretical Physics, Joint Institute for Nuclear Research, 141980 Dubna, Russia*

[2]*Bogoliubov Laboratory of Theoretical Physics, Joint Institute for Nuclear Research, 141980 Dubna, Russia*
E-mail: Sushkov@jinr.thsun1.ru

Dipartimento di Scienze Fisiche, Universitá di Napoli "Federico II" and Istituto Nazionale di Fisica Nucleare, Monte S Angelo, Via Cintia I-80126 Napoli, Italy

The low-lying excited 0^+ states populated in large abundance in some deformed nuclei through two-nucleon transfer reaction experiments are investigated within a microscopic multiphonon framework. The approach provides an exhaustive description of the properties of those states.

Keywords: Zeroplus; Deformation; Two-nucleon transfer.

1. Introduction

A large number of low-lying excited 0^+ levels in several nuclei of the rare-earth and actinide regions have been populated in recent high resolution (p, t) experiments.[1–6]

Such an unexpected abundance has stimulated theoretical investigations aiming at establishing the nature of those states. The first theoretical study was carried out within the interacting boson model (IBM).[7] It was found that only by enlarging the boson space so as to include the (pf) in addition to the (sd) bosons, it is possible to account for a large fraction of the detected 0^+ states in 158Gd. The $(sdpf)$ IBM analysis points out the importance of the octupole degrees of freedom. In fact, several 0^+ levels were associated to collective octupole two-phonon IBM states.

The microscopic investigations have been carried out either within a projected shell model (PSM)[8] or by adopting[9,10] the quasiparticle-phonon model (QPM).[11] The PSM calculation is in principle exact but is forcefully confined within a severely truncated shell model space which may exclude important configurations. The QPM is based on the quasi-boson approximation but is able to include an arbitrarily large number of configurations. We have adopted this approach to perform a systematic investigation of the 0^+ states in all nuclei explored experimentally.

2. The QPM study

In the QPM, one adopts a Hamiltonian of the form

$$H = H_0 + V_{pair} + V^{(P)} + V^{(M)}. \tag{1}$$

The one-body piece H_0 is composed of a kinetic term plus an axially deformed Woods-Saxon potential, while the two-body potential consists of a monopole pairing V_{pair} and of other two main pieces acting in the particle-particle $V^{(P)}$ and particle-hole $V^{(M)}$ channels respectively. Both terms, $V^{(P)}$ and $V^{(M)}$, are sums of several separable potentials of different multipolarity λ.

In the QPM one expresses the Hamiltonian in terms of quasi-particle creation ($\alpha_q^\dagger$) and annihilation (α_q) operators by means of the Bogoliubov canonical transformation and, then, in terms of the RPA phonon operators

$$Q_{i\nu}^\dagger = \frac{1}{2}\sum_{q_1 q_2}\left(\psi_{q_1 q_2}^{i\nu} A_{q_1 q_2}^\dagger - \phi_{q_1 q_2}^{i\nu} A_{q_1 q_2}\right) \tag{2}$$

where $A_{q_1 q_2}^\dagger = \alpha_{q_1}^\dagger \alpha_{\bar{q}_2}^\dagger$.

This procedure yields the interacting phonon Hamiltonian

$$H_{QPM} = \sum_{v_i} \omega_{v_i} Q_{v_i}^\dagger Q_{v_i} + H_{vq}, \tag{3}$$

where $v_i = \{i\,\lambda_i \mu_i\}$. The first term is an unperturbed Boson Hamiltonian diagonal in the basis of the RPA phonon states $\mid v_i\rangle = Q_{v_i}^\dagger \mid 0\rangle$ of energies ω_{v_i}. These are coupled by the term H_{vq}.

The interacting phonon Hamiltonian is put in diagonal form through the variational principle with the trial wave function

$$\Psi_{nK^\pi=0^+} = \sum_i C_i(n) \mid i, K^\pi = 0^+\rangle + \sum_{v_1 v_2} C_{v_1 v_2}(n) \mid \Big[v_1 \otimes v_2\Big]_0^+\rangle, \tag{4}$$

where $\mid i, K^\pi = 0^+\rangle = Q_{i0^+}^\dagger \mid 0\rangle$ and $\mid \Big[v_1 \otimes v_2\Big]_0^+\rangle = \Big[Q_{v_1}^\dagger \otimes Q_{v_2}^\dagger\Big]_0^+ \mid 0\rangle$. The two-phonon basis contains phonons of different multipolarity, including the octupole ones.

We have used this wave function to compute the (p,t) normalized spectroscopic factors

$$\mathcal{S}_n(p,t) = \left(\frac{\Gamma_n(p,t)}{\Gamma_0(p,t)}\right)^2 , \tag{5}$$

where

$$\Gamma_n(p,t) = \langle 0_n^+, N-2 \mid \sum_q^{(\tau=n)} a_q a_{\bar{q}} \mid 0_0^+, N\rangle = \sum_i C_i^{(n)} \Gamma_i(p,t) \tag{6}$$

are the QPM (p,t) transfer amplitudes. The $\Gamma_i(p,t)$ include the amplitudes of the *allowed* transitions involving the one-phonon components $\mid i, K^\pi = 0^+\rangle$ of the QPM state (4) as well as the *forbidden* transitions from the ground to the two-phonon components $\mid [v_1\, v_2]_{0^+}\rangle$. The two-phonon contribution, however, came out to be negligible, so that the full strength is carried by the one-phonon allowed transitions, whose amplitudes are given by

$$\Gamma_i(p,t) \simeq -2 \sum_{qq'}^{\tau=n} \left[\psi^i_{qq'} v_q v_{q'} - \phi^i_{qq'} u_q u_{q'}\right] , \tag{7}$$

where u_q and v_q are the occupation number coefficients of the Bogoliubov transformation.

2.1. *Numerical results*

The QPM calculation not only accounts for more than all 0^+ levels in all nuclei but reproduces fairly well all the energy levels. For the sake of illustration, we discuss few typical cases. One of the most interesting cases is represented by 168Er, where about 25 0^+ and 63 2^+ have been populated through the (p,t) experiments.[3] The calculation yields an even larger number of levels in both cases. As shown in Fig. 1, the QPM 0^+ levels below 4 Mev are about 40. Almost all the 0^+ in excess fall in the range $\sim 3.2 \div 3.5$ MeV and carry a negligible (p,t) strength (second panel).

Almost all the 0^+ excitations below 3 MeV are described by one-phonon states which, in some cases, are rather fragmented. Only one, in this energy range, is a two-phonon state built out of a phonon with high multipolarity. The weight of the two-phonon components becomes appreciable just above 3 MeV and increases more and more with the energy. The octupole phonons do not play any special role, except for few states above 3 MeV, where they are dominant. The ^{168}Er picture inferred from these results is quite different

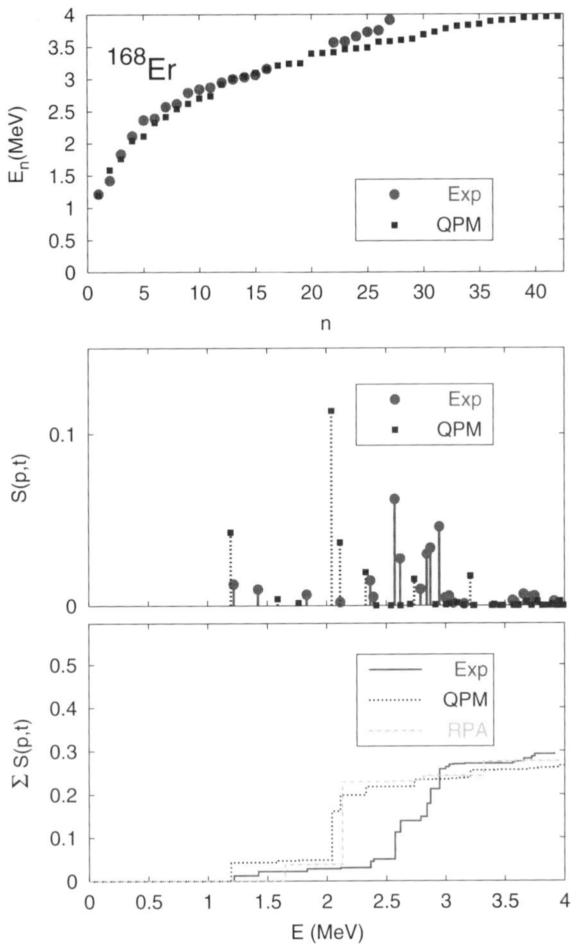

Fig. 1. QPM versus experimental data in ^{168}Er. These include the energies of the 0^+ states (top panel), the energy distribution of the (p, t) spectroscopic factors (middle) and their running sum (bottom).

from the one found for ^{158}Gd, where several 0^+ levels are associated to states with dominant two-phonon components, some of octupole nature.[9]

The information about the e.m. properties of these 0^+ is scarce. The only known strength is the one measured for the $E2$ decay transition from the 0_1^+ bandhead to the 2^+ of the ground band[12] and came out to be small, $B^{(exp)}(E2; 0_1^+ \to 2_0^+) = 0.08(1)$ W.u. The QPM corresponding value is quite close, $B^{(QPM)}(E2; 0_1^+ \to 2_0^+) = 0.11$ W.u.. According to the

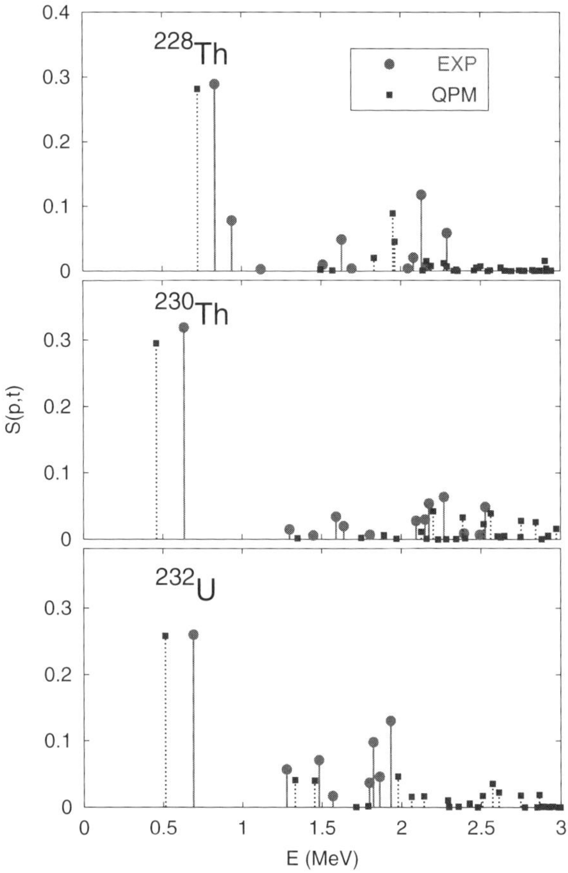

Fig. 2. Experimental versus QPM energy distributions of (p, t) spectroscopic factors in the actinides. The experimental data are taken from.[2]

QPM calculation, the $E2$ decay strengths of the other 0^+ states are even smaller. Also the normalized monopole transitions are generally weak. The strength collected by the bandhead 0_1^+ is close to the experimental value $10^3\,\rho^2_{exp}(E0; 0_1^+ \to 0^+) = 0.8(8)$.[13–15] We find therefore an almost complete lack of quadrupole collectivity in all 0^+ states of ^{168}Er.

All these states, indeed, are composed almost solely of pairing correlated $q\bar{q}$ components. The high energy ph configurations are practically absent. Pairing, however, acts coherently only into one RPA 0^+ state. This collects most of the two-nucleon transfer strength $\mathcal{S}(p, t)$. The coupling with the other phonons, however, depletes the (p, t) strength of the collective RPA 0^+

state, yielding a fragmented spectrum reasonably close to the experimental data (second panel of Fig. 1).

The QPM misses some detailed features of the spectrum. These discrepancies get manifest in the third panel of Fig. 1 showing that both RPA and QPM strengths reach saturation too early with respect to experiments.

It is remarkable that the calculation yields very little strength above 3 MeV, in very good agreement with experiments. This reflects the fact that the levels above 3 MeV correspond predominantly to two-phonon excitations and are, therefore, poorly populated in transfer reactions. It is also worth of notice that the QPM strength sums up to a value ~ 0.25, which approaches closely the experimental integrated strength.

A similar picture is obtained also for the nuclei of the actinide region. As shown in Fig. 2, the QPM energies and normalized spectroscopic fac-

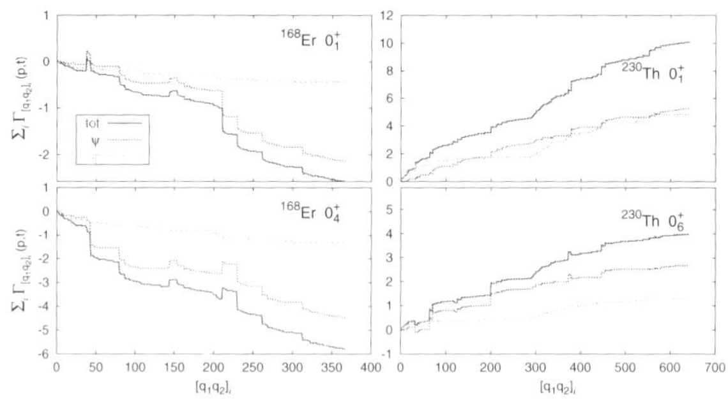

Fig. 3. The contribution to the running sum of the (p, t) transfer amplitudes coming from the forward ψ (dash line) and backward ϕ (dot-dashed line) RPA amplitudes.

tors $\mathcal{S}(p, t)$ are in fair agreement with experiments. The calculation yields correctly one strong peak in the actinides close in magnitude to the corresponding experimental value. The other peaks are small consistently with experiments. We have analyzed also the RPA spectroscopic amplitudes $\Gamma_\imath$ for some typical cases to better understand the role of pairing. The contribution of the "small" backward RPA amplitudes ϕ to $S(p, t)$ are almost negligible in ^{168}Er (Fig. 3) suggesting that the pairing properties of the excited 0^+ states in ^{168}Er might not arise from the fluctuations of the ground state pairing field.

The ϕ amplitudes contribute considerably to the spectroscopic factors in the actinides, indicating that pairing fluctuations are quite pronounced in their ground state. We may, therefore, make the reasonable guess that the 0^+ states in these nuclei describe pairing vibrations arising from ground state fluctuations. Indeed, the phonon coupling has a very modest damping effect, since the (p,t) strength remains concentrated predominantly into the lowest 0^+.

A closer look shows that the two-phonon content is more pronounced in the 0^+ states of the actinides. On the other hand, the octupole phonons have a relatively modest impact, contrary to expectations. The actinides, indeed, are nuclei of strong octupole collectivity. Indeed, the lowest unperturbed octupole RPA phonons fall at low-energy, consistently with experiments, but the enforcement of the Pauli principle on the octupole two-phonon components, entering the 0^+ states, has a repulsive effect and spoils the octupole coherence. As a result, these components spread among several QPM 0^+ states at high energy.

The $E2$ collectivity in the actinides is considerably stronger than in ^{168}Er. The lowest 0^+ excited state, in particular, is basically a one-phonon state and collects an appreciable E2 strength, in agreement with the available data. In ^{230}Th, in fact, the measured E2 strength is $B^{(exp)}(E2;0^+_1 \rightarrow 2^+_0) = 1.10$ W.u,[16] close to the computed value $B^{(QPM)}(E2;0^+_1 \rightarrow 2^+_0) = 1.71$ (W.u.).

The monopole transition strength collected by the first 0^+ is estimated to be $10^3\,\rho^2_{exp}(E0;0^+_\beta \rightarrow 0^+_g) = 50(20)$,[15] few times larger than the QPM value and orders of magnitude larger than in ^{168}Er. Still, neither the $E2$ or the $E0$ transitions are sufficiently strong to qualify the 0^+ of the actinides explored here as beta vibrational levels.[17]

3. Conclusion

Our analysis suggests that the appearance of so many low-lying 0^+ states in deformed nuclei is a manifestation of pairing collectivity. Indeed, almost all low-lying 0^+ levels are described as one-phonon states. These came out to be built out of pairing correlated configurations giving rise to a pairing vibrational mode,[18] strongly populated through (p,t) reactions.

According to thee QPM calculation, none of the QPM 0^+ states is quadrupole collective. This prediction needs to be tested by more reliable measurements of the $E2$ transition strengths.

On the whole, the present QPM calculation provides a fairly satisfactory description of the existing properties of the 0^+ states in deformed nuclei. In order to assess its complete reliability, however, it is desirable to complete the characterization of these states by systematic measurements of their $E2$ and $E0$ decay strengths.

Acknowledgments

This work is partly supported by the Ministero dell' Istruzione, Università e Ricerca (MIUR).

References

1. S. R. Lesher *et al*, Phys. Rev. C **66**, 051305(R) (2002).
2. H.-F. Wirth *et al*, Phys. Rev. C **69**, 044310 (2004).
3. D. Bucurescu *et al*, Phys. Rev. C **73**, 064309 (2006).
4. D. A. Meyer *et al.*, J. Phys. G: Nucl. Part. Phys. **31**, S1399 (2005).
5. D. A. Meyer *et al.*, Phys. Lett. **B 638**, 44 (2006).
6. D. A. Meyer *et al.*, Phys. Rev. C **74**, 044309 (2006).
7. N. V. Zamfir, Jing-ye Zhang, and R. F. Casten, Phys. Rev. C **66**, 057303 (2002)
8. Y. Sun, A. Aprahamian, J. Zhang, C. Lee, Phys. Rev. C **68**, 061301(R) (2003).
9. N. Lo Iudice, A. V. Sushkov, and N. Yu. Shirikova, Phys. Rev. C **70**, 064316 (2004)
10. N. Lo Iudice, A. V. Sushkov, and N. Yu. Shirikova, Phys. Rev. C **72**, 034303 (2005)
11. V. G. Soloviev, *Theory of Atomic Nuclei: Quasiparticles and Phonons* (Institute of Physics Publishing, Bristol, 1992).
12. T. Härtlein, M. Heinebrodt, D. Schwalm, C. Fahlander, Eur. Phys. J. A **2**, 253 (1998)
13. W. F. Davidson, D. D. Warner, R. F. Casten, K. Schreckenbaçh, H. G. Börner, J. Simić, M. Stojanović, M. Bogdanović, S. Koički, W. Gelletly, G. B. Orr, and M. L. Stelts, J. Phys. G **7**, 455 (1981).
14. H. Lehmann, J. Jolie, F. Corminboeuf, H. G. Börner, C. Doll, M. Jentschell, R. F. Casten, N. V. Zamfir, Phys. Rev. C **57**, 569 (1998)
15. J. L. Wood, E. F. Zganjar, C. De Coester, K. Heyde, Nucl. Phys. A **651**,323 (1999).
16. Y. A. Akovali, Nuclear Data Sheets **66**, 505 (1992).
17. P. E. Garrett, J. Phys. G **27**, R1 (2001).
18. D.R. Bes and R. A. Broglia, Nucl. Phys. **80**, 289 (1966).

NEW MICROSCOPIC APPROACH TO MULTIPHONON NUCLEAR SPECTRA

F. ANDREOZZI[1], F. KNAPP[2], N. LO IUDICE[1], A. PORRINO[1], J. KVASIL[2]

[1] *Dipartimento di Scienze Fisiche, Universitá di Napoli "Federico II" and Istituto Nazionale di Fisica Nucleare, Monte S Angelo, Via Cintia I-80126 Napoli, Italy*

[2] *Institute of Particle and Nuclear Physics, Charles University, V.Holešovičkách 2, CZ-18000 Praha 8, Czech Republic*

The method outlined here consists in generating a multiphonon basis by constructing and solving iteratively a set of equations of motion within a subspace spanned by states which are tensor products of n Tamm Dancoff phonons. In such a basis, the Hamiltonian takes on a simple form and can be easily diagonalized. For illustrative purposes, the method is applied to ^{16}O.

Keywords: Equations of motion; multiphonon TDA basis; ^{16}O

1. Introduction

The growing experimental evidence of multiphonon collective modes at low[1,2] as well as high[3,4] energies has triggered a series of theoretical investigations of phenomenological and microscopic nature. The interacting-boson model (IBM)[5] has been adopted with success for a systematic study of the low-energy multiphonon modes. More detailed and specific investigations have been carried out within microscopic schemes, which extend in various ways the random-phase-approximation (RPA). Most of these extensions are based on or inspired by the Fermion-Boson mapping technique.[6] The IBM itself is to be considered a phenomenological realization of such a mapping.[5]

Among the microscopic approaches inspired by the Fermion-Boson mapping, we mention the nuclear field theory,[7] especially suitable for characterizing the anharmonicities of the vibrational spectra and the spreading widths of the giant resonances, and the quasiparticle-phonon model (QPM).[8] In the QPM, a Hamiltonian of generalized separable form is expressed in terms of RPA quasi-boson operators and then diagonalized in a severely truncated space which includes a selected set of two and three RPA phonons. The method has been extensively adopted to describe both low and high energy multiphonon excitations, like the mixed-symmetry states[9] and the double giant dipole resonance.[10] Other microscopic methods have tried to go beyond the quasi boson approximation but with more limited success.[11,12]

We report here on an iterative equation of motion method which we have proposed[13] in order to construct a basis of multiphonon states built out of phonons generated in the Tamm-Dancoff approximation (TDA). The basis is then used to solve exactly the nuclear eigenvalue problem in the space spanned by such a correlated basis using a general Hamiltonian.

While a complete and detailed description of the method can be found elsewhere,[13] we will outline here the the method and show how it cab implemented, with no approximations, in the case of ^{16}O, a nucleus of highly complex shell structure, which represents a severe test for our multiphonon approach.

2. Brief outline of the method

Our goal is to generate a set of multiphonon states $\mid n;\alpha\rangle$ which diagonalize the Hamiltonian H within each separate subspace spanned by states with n phonons, so that

$$< n;\beta|H|n;\alpha >= E_\alpha^{(n)})\delta_{\alpha\beta}. \tag{1}$$

Under this request we obtain

$$< n;\beta|\,[H,b_{ph}^\dagger]\,|n-1;\alpha >= (E_\beta^{(n)} - E_\alpha^{(n-1)}) < n;\beta|b_{ph}^\dagger|n-1;\alpha >, \tag{2}$$

where $b_{ph}^\dagger = a_p^\dagger a_h$ is a bilinear form in the operators $a_p^\dagger$ and a_h which create respectively a particle (p) and a hole (h) with respect to the unperturbed ground state (ph vacuum). The above equation follows from the property of the ph operator $b_{ph}^\dagger$, which couples only states differing by one phonon.

We then write the Hamiltonian in second quantized form and expand the commutator $[H,a_p^\dagger a_h]$ on the left-hand side of the equation. After a linearization procedure, we obtain for the n-phonon subspace the eigenvalue equation

$$\sum_{\gamma p'h'} A_{\alpha\gamma}^{(n)}(ph\ ;\ p'h')\ X_{\gamma\beta}^{(n)}(p'h') = E_\beta^{(n)}\ X_{\alpha\beta}^{(n)}(ph), \tag{3}$$

where

$$X_{\alpha\beta}^{(n)}(ph)\ \equiv < n;\beta|b_{ph}^\dagger|n-1;\alpha > \tag{4}$$

are the vector amplitudes and

$$
\begin{aligned}
A^{(n)}_{\alpha\gamma}(ph;p'h') = \delta_{hh'}\delta_{pp'}\delta^{(n-1)}_{\alpha\gamma}\left[(\epsilon_p - \epsilon_h) + E^{(n-1)}_{\alpha}\right] \\
+ \sum_{h_1} V_{p'h_1h'p}\, \rho^{(n-1)}_{\alpha\gamma}(h_1h) \\
- \sum_{p_1} V_{p'hh'p_1}\rho^{(n-1)}_{\alpha\gamma}(pp_1) \\
+ \frac{1}{2}\delta_{hh'} \sum_{p_1p_2} V_{p'p_1pp_2}\rho^{(n-1)}_{\alpha\gamma}(p_1p_2) \\
- \frac{1}{2}\delta_{pp'} \sum_{h_1h_2} V_{hh_1h'h_2}\{\rho^{(n-1)}_{\alpha\gamma}(h_1h_2)\}.
\end{aligned} \tag{5}
$$

The symbols ϵ_p (ϵ_h) are single particle (hole) energies, V_{ijkl} the matrix elements of the two-body potential, and

$$
\rho^{(n)}_{\alpha\gamma}(kl) =< n;\ \gamma|a^\dagger_k a_l|n;\ \alpha > \tag{6}
$$

defines the density matrix with $a^\dagger_k a_l$ written in normal order with respect to the ph vacuum. This is a crucial quantity and is seen to weight the particle-hole, particle-particle and hole-hole interaction.

For $n = 1$, the density matrices appearing in the Eq. (5) take the values

$$
\rho^{(0)}_{\alpha\gamma}(pp') = \delta_{\alpha 0}\delta_{\gamma 0} < 0 \mid a^\dagger_p a_{p'} \mid 0 >= 0
$$

$$
\rho^{(0)}_{\alpha\gamma}(hh') = \delta_{\alpha 0}\delta_{\gamma 0} < 0 \mid a^\dagger_h a_{h'} \mid 0 >= \delta_{hh'}
$$

so that, Eqs. (3) and (5) yield the standard TDA equations. Our method is, therefore, nothing but the extension of TDA to multiphonon spaces.

It is easy to infer from the expression (4) of the vector amplitudes that the eigenvectors generated from solving such a system of equations are linear combinations of N_r states $b^\dagger_{ph}|n-1;\ \alpha >$. These are linearly dependent and, therefore, form an overcomplete set. In order to extract a linear independent basis, we expand $|n;\ \beta >$ in terms of the redundant N_r states

$$
|n;\ \beta >= \sum_{\alpha\, ph} C^{(n)}_{\alpha\beta}(ph)\, b^\dagger_{ph}\ |n-1;\ \alpha > . \tag{7}
$$

Upon insertion in Eqs. (4) and (5), we get

$$
X = \mathcal{D}C \tag{8}
$$

$$
A\mathcal{D}C = E\mathcal{D}C, \tag{9}
$$

where $\mathcal{D}$ is the overlap or metric matrix

$$
d^{(n-1)}_{\alpha\beta}(ph;p'h') = \langle n-1;\beta \mid b_{p'h'}b^\dagger_{ph} \mid n-1;\alpha\rangle. \tag{10}
$$

Eq. (9) defines an eigenvalue equation of general form.

In order to solve this equation we have first to evaluate the matrices A and D. Our equation of motion method provides recursive formulas, which make the computation of these quantities quite straightforward.

We have then to face the non trivial task of eliminating the redundancy of the basis vectors $b^{\dagger}_{ph} \mid n-1;\alpha\rangle$, which renders the matrix $\mathcal{D}$ singular. The Choleski decomposition method[13] provides a fast and efficient prescription for extracting a basis of linear independent states $b^{\dagger}_{ph} \; |n-1;\, \alpha>$ spanning the physical subspace of the correct dimensions $N_n < N_r$.

We can then solve the generalized eigenvalue problem (9) in each subspace and generate iteratively the multiphonon basis $\{\mid n;\alpha_n\rangle\}$. This is highly correlated and reduces the Hamiltonian into diagonal blocks mutually coupled by off-diagonal terms of simple structure, which can be computed recursively. Once this is done H can be diagonalized yielding exact eigenvalues and eigenvectors. These are given by

$$\mid \Psi_\nu\rangle = \sum_{n,\alpha_n} C^{(\nu)}_{\alpha_n} \mid n;\alpha_n\rangle. \tag{11}$$

Their phonon structure allows to use simple recursive relations also for computing the transition amplitudes of one and two-body operators.

3. A numerical illustrative application of the method for ^{16}O

For illustrative purposes, we have applied the method to ^{16}O, whose low-lying excitations are known to have a highly complex *ph* structure since the pioneering work of Brown and Green.[16]

The low-energy positive parity spectrum of this nucleus was studied in a shell model calculation which included up to $4p - 4h$ and $4\hbar\omega$ configurations.[17] The same spectrum was studied very recently within a no-core shell model and an algebraic symplectic shell model[18] up to $6\hbar\omega$.

For our illustrative purposes, we have included all *ph* configurations up to $3\hbar\omega$, which limits our phonon space up to $n = 3$. We used a modified harmonic oscillator one-body Nilsson Hamiltonian plus a bare G-matrix deduced from the Bonn-A potential.[19] We have adopted the method of Palumbo[20] to separate the intrinsic from the center of mass motion.

Being the space confined to $3-\hbar\omega$, the ground state contains correlations up to 2-phonons only. These account for about 20% of the state, while the remaining 80% pertains to the *ph* vacuum. These values are reasonably close to the estimates of the consistent no-core symplectic shell model. This

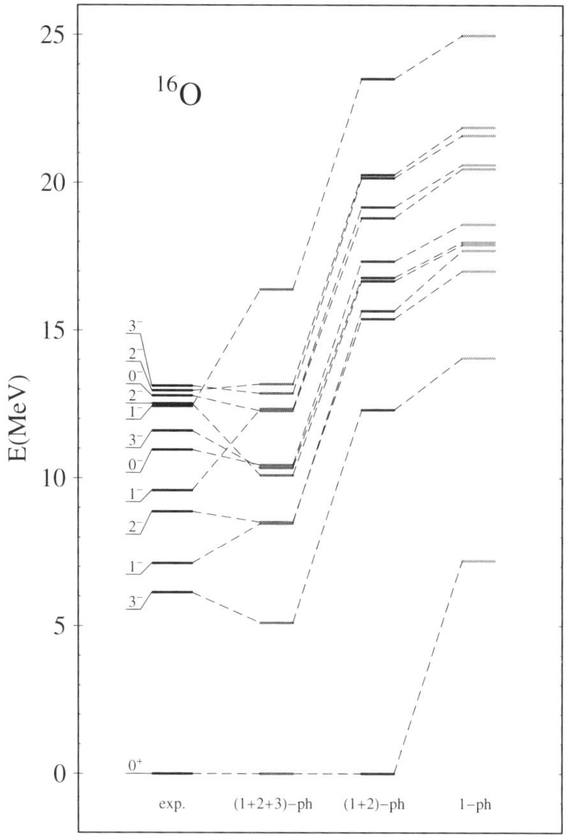

Fig. 1. Negative parity spectrum of ^{16}O.

yielded about 60% for the $0p - 0h$, 20% for $2p - 2h$ and 20% for the other more complex configurations,[18] excluded from our restricted space.

As shown in Fig. 1, the effect of the multiphonon configurations on the low-lying negative parity spectrum is very important. The coupling with the two phonons pushes down the one-phonon levels. The energy shifts, however, are much smaller than the ones induced by the positive parity two-phonons on the ground state, so that the gap between excited and ground state levels increases with respect to the one phonon case. The three phonon configurations are much more effective and bring all the states down in energy within the range of the corresponding experimental levels.[21]

We have investigated also the anharmonicities induced by the multi-phonon configurations on some selected giant resonances. To this purpose, we have computed the strength function

$$\mathcal{S}(\lambda,\omega) = \sum_{\nu} B_{\nu}(\lambda)\,\delta(\omega-\omega_{\nu}) \approx \sum_{\nu} B_{\nu}(\lambda)\,\rho_{\Delta}(\omega-\omega_{\nu}), \qquad (12)$$

where ω is the energy variable, ω_{ν} the energy of the transition of multipolarity λ from the ground to the ν_{th} excited state $\Psi_{\lambda}^{(\nu)}$ of spin $J=\lambda$, given by Eq. (11), and $\rho_{\Delta}(\omega-\omega_{\nu})$ is a Lorentzian of width Δ,[22] which replaces the δ function as a weight of the reduced transition probability $B^{(\nu)}(\lambda)$.

The electric $(E\lambda)$ transitions operator have the standard shell model form. For the isoscalar giant dipole resonance (squeezed dipole mode) we have used the operator

$$\mathcal{M}_{IS}(\lambda=1\mu) = \frac{1}{2}\sum_{i=1}^{A} r_i^3 Y_{1\mu}(\hat{r_i}). \qquad (13)$$

It is important to notice the absence of the corrective term generally included in order to eliminate the spurious contribution due to the center of mass excitation. Such a term is not necessary in our approach which guarantees a complete separation of the center of mass from the intrinsic motion.

While the isovector $E1$ response $S_{IV}(E1,\omega)$ is little sensitive to the number of phonons, the other responses undergo dramatic changes as the number of phonons increases.

The most dramatic effects are observed in the isoscalar $E1$ giant resonance. The strength gets spread over a much larger energy range as we increase the number of phonons (Fig. 2). The high energy peak is almost entirely due to $2p-2h$ as well as $3p-3h$ configurations.

4. Conclusions

The method we have proposed is not only exact but also of easy implementation for a Hamiltonian of general form. Moreover, it generates at once the whole nuclear spectrum. It is therefore suitable for studying the low-lying spectroscopic properties as well as the high energy giant resonances.

Its exact numerical implementation in ^{16}O was confined to a space which included up to three-phonons and 3 $\hbar\omega$, sufficient for our illustrative pur-

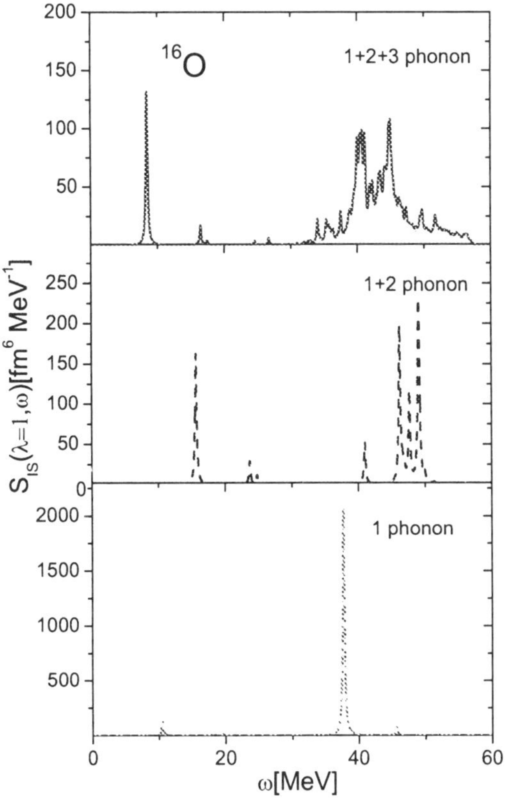

Fig. 2. Isoscalar $E1$ strength distributions in ^{16}O .

poses, but too restricted to describe exhaustively and faithfully all spectroscopic properties of this complex nucleus. In order to extend the method to spaces with a larger number of phonons we need to truncate the dimensions of the phonon subspaces. This should be done with little detriment to the accuracy of the solutions. We can rely in fact on the correlated nature of the phonon states.

To this purpose we are developing a new formulation which should render the truncation procedure efficient and, at the same time, accurate.

Acknowledgments

Work supported in part by the Italian Ministero della Istruzione Universitá e Ricerca (MIUR) and by the Czech Ministry of Education contract No. VZ MSM 0021620859.

References

1. For a review see M. Kneissl, H. H. Pitz and A. Zilges, *Prog. Part. Nucl. Phys.* **37** (1996) 439.
2. For a review see M. Kneissl, N. Pietralla and A. Zilges, *J. Phys. G: Nucl. Part. Phys.* **32** (2006) R217.
3. For a review N. Frascaria, *Nucl. Phys.* **A482** (1988) 245c.
4. For a review see T. Auman, P. F. Bortignon, H. Hemling, *Annu. Rev. Nucl. Part. Sci.* **48** (1998) 351.
5. For a review see A. Arima and F. Iachello, *Adv. Nucl. Phys.* **13** (1984) 139.
6. For a review see A. Klein and E. R. Marshalek, *Rev. Mod. Phys.* **63** (1991) 375 .
7. P. F. Bortignon, R. A. Broglia, D. R. Bes, and R. Liotta, *Phys. Rep.* **30** (1977) 305.
8. V. G. Soloviev, *Theory of Atomic Nuclei: Quasiparticles and Phonons* Institute of Physics, Bristol, 1992.
9. N. Lo Iudice and Ch. Stoyanov, *Phys. Rev. C* **62** (2000) 047302; *Phys. Rev. C* **65** (2002) 064304.
10. V. Yu. Ponomarev, P. F. Bortignon, R. A. Broglia, and V. V. Voronov, *Phys. Rev. Lett.* (2000) **85** 1400.
11. C. Pomar, J. Blomqvist, R. J. Liotta, and A. Insolia, *Nucl. Phys.* **A515** (1990) 381.
12. M. Grinberg, R. Piepenbring, K. V. Protasov, B. Silvestre-Brac, *Nucl. Phys.* **A597** (1996) 355.
13. F. Andreozzi, N. Lo Iudice, and A. Porrino, F. Knapp, J. Kvasil, Phys. Rev. C **75** (2007) 044312.
14. R. J. Liotta and C. Pomar, *Nucl. Phys.* (1982) **A382** 1.
15. K. V. Protasov, B. Silvestre-Brac, R. Piepenbring, and M. Grinberg, *Phys. Rev. C* **53** (1996) 1646.
16. G. E. Brown, A. M. Green, *Nucl. Phys.* **75** (1966) 401.
17. W. C. Haxton and C. J. Johnson, *Phys. Rev. Lett.* **65** (1990) 1325.
18. T. Dytrych, K. D. Sviratcheva, C. Bahri, J. P. Draayer, J.P. Vary, *Phys. Rev. Lett.* **98**, 162503 (2007).
19. R. Machleidt, *Adv. Nucl. Phys.* **19** (1989) 189.
20. F. Palumbo, *Nuc. Phys.* **99** (1967) 100.
21. D.R. Tilley, H.R. Weller and C.M. Cheves *Nucl. Phys.* **A564** 1 (1993).
22. See for instance J. Kvasil, N. Lo Iudice, V.O. Nesterenko, and M. Kopal, Phys. Rev. C **58**(1998) 209.

VARIATIONAL EQUATION FOR QUANTUM NUMBER PROJECTION AT FINITE TEMPERATURE

KOSAI TANABE

Department of Physics, Saitama University,
Sakura, Saitama 338-8570, Japan

HITOSHI NAKADA

Department of Physics, Chiba University,
Inage, Chiba 263-8522, Japan

To describe phase transitions in a finite system at finite temperature, we develop a formalism of the variation-after-projection (VAP) of quantum numbers based on the thermofield dynamics (TFD). We derive a new Bardeen-Cooper-Schrieffer (BCS)-type equation by variating the free energy with approximate entropy without violating Peierls inequality. The solution to the new BCS equation describes the S-shape in the specific heat curve and the superfluid-to-normal phase transition caused by the temperature effect. It simulates the exact quantum Monte Carlo results well.

Keywords: Quantum number projection; Number-projected BCS equation at finite temperature; S-shape in specific heat; Superfluid-to-normal phase transition.

1. Introduction

In describing the evolution of nuclear structure at finite temperature, the meanfield approximation has been demonstrated to be useful in the quantum number-constrained Hartree-Fock-Bogoliubov (CHFB) approximation at finite temperature[1-3] and the Badeen-Cooper-Schrieffer (BCS) approximation.[4] Several results in recent experiment[5] coincide with the CHFB predictions,[1-3] for instance, the critical temperature of the superfluid-to-normal phase transition ($T_c \sim 0.5\,\mathrm{MeV}$) and the S-shape in the specific heat curve,[1] and the behaviour of nuclear level densities at finite temperature.[3] The latter CHFB formalism is based on the grand canonical ensemble. It introduces not only constraints of definite particle numbers and angular momentum, but also projections of definite space parity and particle-number parity for quasiparticles [i.e. projection-after-variation (PAV)]. It may be worth mentioning that a theoretical prescription is introduced in Ref. 3 to avoid a notorious divergence occuring in the level density at the limit of zero temperature when the conventional Fowler-Darwin method is applied.

Atomic nucleus is a finite system whose physical states are specified in reference to a set of conserved quantum numbers, and the signature of phase transitions is usually washed out by the effect of quantum fluctuations. To provide a microscopic description beyond the meanfield approximation, we develop a formalism by introducing quantum number projections in the free energy as a basis for the variation-after-projection (VAP) scheme, since the projection plays dual roles in taking account of fluctuations as symmetry-restoring correlations as well as in guaranteeing correct quantum numbers.[6] With particle number projection, we can develop a formalism at finite temperature on the canonical ensemble.[7]

In Sec. 2, we present a general framework of the HFB approximation at finite temeprature with quantum number projections by applying the thermofield dynamics (TFD).[8,9] In Sec. 3, in order to make the VAP scheme practicable, we approximate the projected form of the entropy without violating Peierls inequality.[6] In Sec. 4, we derive a new BCS-type equation from variation of the particle number-projected free energy with the approximate entropy, and discuss the numerical results.[7] In Sec. 5, the paper is concluded.

2. Application of Thermofield Dynamics to Quantum Number-Projected Free Energy

We express the thermal average of an operator $\hat{O}$ on the canonical ensemble as $\langle\hat{O}\rangle = \mathrm{Tr}(\hat{w}_0\hat{O})$, where the trial statistical operator $\hat{w}_0$ is expressed in terms of quasiparticle operators $\{\alpha_\mu, \alpha_\mu^\dagger\}$ as

$$\hat{w}_0 = \frac{e^{-\hat{H}_0/T}}{\mathrm{Tr}(e^{-\hat{H}_0/T})}\,; \quad \hat{H}_0 = \sum_\mu E_\mu \alpha_\mu^\dagger \alpha_\mu\,. \tag{1}$$

In the thermofield dynamics (TFD),[8,9] the quasiparticle operator space is enlarged by introducing the tilded operators, i.e. $\{\alpha_\mu, \tilde{\alpha}_\mu, \alpha_\mu^\dagger, \tilde{\alpha}_\mu^\dagger\}$, which are related to a new set of temperature-dependent operators $\{\beta_\mu, \tilde{\beta}_\mu, \beta_\mu^\dagger, \tilde{\beta}_\mu^\dagger\}$ by the other Bogoliubov transformation

$$\alpha_\mu = \bar{g}_\mu \beta_\mu + g_\mu \tilde{\beta}_\mu^\dagger\,, \quad \tilde{\alpha}_\mu = \bar{g}_\mu \tilde{\beta}_\mu - g_\mu \beta_\mu^\dagger\,. \tag{2}$$

We combine this transformation with the generalized Bogoliubov transformation (GBT) connecting the original single-particle operators $\{c_i, c_i^\dagger\}$ to quasiparticle operators, i.e.

$$\begin{pmatrix} c_k \\ c_k^\dagger \end{pmatrix} = \sum_\mu W_{k\mu} \begin{pmatrix} \alpha_\mu \\ \alpha_\mu^\dagger \end{pmatrix}; \quad W_{k\mu} \equiv \begin{pmatrix} U_{k\mu} & V_{k\mu}^* \\ V_{k\mu} & U_{k\mu}^* \end{pmatrix}, \tag{3}$$

to compose an extended form of the GBT as[10]

$$\begin{pmatrix} c \\ \tilde{c} \\ c^\dagger \\ \tilde{c}^\dagger \end{pmatrix}_k = \sum_\mu \bar{W}_{k\mu} \begin{pmatrix} \beta \\ \tilde{\beta} \\ \beta^\dagger \\ \tilde{\beta}^\dagger \end{pmatrix}_\mu ; \quad \bar{W}_{k\mu} = \begin{pmatrix} \bar{U} & \bar{V}^* \\ \bar{V} & \bar{U}^* \end{pmatrix}_{k\mu}, \tag{4}$$

where

$$\bar{U}_{k\mu} = \begin{pmatrix} U\bar{g} & V^* g \\ -V^* g & U\bar{g} \end{pmatrix}_{k\mu}, \quad \bar{V}_{k\mu} = \begin{pmatrix} V\bar{g} & U^* g \\ -U^* g & V\bar{g} \end{pmatrix}_{k\mu} \tag{5}$$

and

$$g_\mu \equiv f_\mu^{1/2}, \quad \bar{g}_\mu \equiv (1 - f_\mu)^{1/2}, \quad f_\mu \equiv \langle \alpha_\mu^\dagger \alpha_\mu \rangle = \frac{1}{e^{E_\mu/T} + 1}. \tag{6}$$

The temperature-dependent vacuum $|0_T\rangle$ is annihilated by β_μ and $\tilde{\beta}_\mu$, i.e. $\beta_\mu|0_T\rangle = \tilde{\beta}_\mu|0_T\rangle = 0$, so that the ensemble average of any operator $\hat{O}$ is expressed in terms of the vacuum expectation value as

$$\langle \hat{O} \rangle = \mathrm{Tr}(\hat{w}_0 \hat{O}) = \langle 0_T | \hat{O} | 0_T \rangle. \tag{7}$$

The tilded operators are interpreted to represent the effect of heat reservoir,[8,9] and disappear remaining their effect on the expectation value in the r.h.s. of Eq. (7).

Now we consider the product of projection operators as $\hat{P} = \hat{P}_J \hat{P}_Z \hat{P}_N$. Each operator is given as follows. The operator projecting the state of angular momentum J is given by[6]

$$\hat{P}_J = \hat{P}_J^\dagger \equiv \sum_{M=-J}^{J} \hat{P}_{MM}^J = \frac{2J+1}{4\pi^2} \int_0^{2\pi} \sin^2 \frac{\omega}{2} d\omega \int_0^{\pi} \sin\theta d\theta \int_0^{2\pi} d\phi \, \chi_J(\omega) \hat{R}(\Phi), \tag{8}$$

where

$$\hat{R}(\Phi) = e^{-i\alpha \hat{J}_z} e^{-i\beta \hat{J}_y} e^{-i\gamma \hat{J}_z} = e^{-i\omega(n_x \hat{J}_x + n_y \hat{J}_y + n_z \hat{J}_z)}, \tag{9}$$

$$\chi_J(\omega) = \sum_{M=-J}^{J} D_{MM}^J(\Phi) = \frac{\sin[(J+1/2)\omega]}{\sin(\omega/2)}. \tag{10}$$

Projection operator of neutron(proton) number $n = N(Z)$ is given by

$$\hat{P}_n = \hat{P}_n^\dagger = \frac{1}{2\pi} \int_0^{2\pi} e^{-i\varphi(\hat{N}-n)} d\varphi. \tag{11}$$

We consider also projection operator of quasiparticle number-parity (q) as

$$\hat{P}_q = \frac{1}{2}\left[1 + \eta \exp\left(i\pi \sum_{\mu} \alpha_\mu^\dagger \alpha_\mu\right)\right], \tag{12}$$

where $\eta = +1$ ($\eta = -1$) for the projection of even (odd) number-parity.[2] Similarly, we can introduce the space-parity projection operator.[2]

Thus, any projected ensemble average can be calculated by the vacuum expectation value as

$$\langle \hat{O} \rangle_{\rm P} \equiv {\rm Tr}(\hat{w}_{\rm P}\hat{O}) = \frac{{\rm Tr}(\hat{P}e^{-\beta\hat{H}_0}\hat{P}\hat{O}\hat{P})}{{\rm Tr}(e^{-\beta\hat{H}_0}\hat{P})} = \frac{\langle 0_T|\hat{P}\hat{O}\hat{P}|0_T\rangle}{\langle 0_T|\hat{P}|0_T\rangle}, \tag{13}$$

where the statistical operator including the projection operator $\hat{P}$ is defined by

$$\hat{w}_{\rm P} \equiv \frac{\hat{P}e^{-\beta\hat{H}_0}\hat{P}}{{\rm Tr}(e^{-\beta\hat{H}_0}\hat{P})}. \tag{14}$$

As shown in Ref. 6, any ensemble average including projection operators (our free energy $F_{\rm P}$ is such an example) can be reduced to a sum of the products of basic quantities by the generalized Wick theorem. This reduction technique is a direct extension of the general formalism for the quantum number projection at zero temperature.[11]

3. Approximation of Entropy and Peierls Inequality

The exact free energy $F^{\rm exact}$, which is generated by a given total Hamiltonian $\hat{H}$, is not greater than the approximate one with projection operator, $F_{\rm P}$. Variational principle is based on such a Peierls inequality, i.e.

$$F^{\rm exact} = -T\ln\left[{\rm Tr}(e^{-\hat{H}/T}\hat{P})\right] \le F_{\rm P} \equiv E_{\rm P} - TS_{\rm P} \tag{15}$$

with

$$E_{\rm P} = {\rm Tr}(\hat{w}_{\rm P}\hat{H}) = \frac{{\rm Tr}(\hat{w}_0\hat{H}\hat{P})}{{\rm Tr}(\hat{w}_0\hat{P})} = \frac{\langle 0_T|\hat{H}\hat{P}|0_T\rangle}{\langle 0_T|\hat{P}|0_T\rangle}, \tag{16a}$$

$$S_{\rm P} = -{\rm Tr}(\hat{w}_{\rm P}\ln\hat{w}_{\rm P}) = -{\rm Tr}\left[\frac{e^{-\hat{H}_0/T}\hat{P}}{{\rm Tr}(e^{-\hat{H}_0/T}\hat{P})}\ln\left(\frac{e^{-\hat{H}_0/T}\hat{P}}{{\rm Tr}(e^{-\hat{H}_0/T}\hat{P})}\right)\right]. \tag{16b}$$

Since direct treatment of the entropy in the form of Eq. (16b) is difficult, some sort of approximation is necessary for the practice of calculation. As presented in detail in Ref. 6, we can prove an inequality

$$S'_{\rm P} \equiv \frac{1}{T}\frac{{\rm Tr}(e^{-\hat{H}_0/T}\hat{H}_0\hat{P})}{{\rm Tr}(e^{-\hat{H}_0/T}\hat{P})} + \ln{\rm Tr}(e^{-\hat{H}_0/T}\hat{P}) \le S_{\rm P}. \tag{17}$$

Hence, we confirm that the approximate free energy with the approximate entropy $S'_{\rm P}$ does not violate Peierls inequality, i.e. $F'_{\rm P} \equiv E_{\rm P} - TS'_{\rm P} \geq F_{\rm P} \geq F^{\rm exact}$. Therefore, such an approximate free energy, which is explicitly given by

$$F'_{\rm P} = \frac{\langle 0_T|(\hat{H} - \hat{H}_0)\hat{P}|0_T\rangle}{\langle 0_T|\hat{P}|0_T\rangle} - T\ln\langle 0_T|\hat{P}|0_T\rangle + T\sum_{\mu({\rm all})}\ln(1-f_\mu), \qquad (18)$$

can be regarded as the basis for the variational principle. Variational parameters are $\{E_\mu\}$ appearing in $\hat{w}_0$ together with the Bogoliubov transformation coefficitents $\{U_{k\mu}, V_{k\mu}, U^*_{k\mu}, V^*_{k\mu}\}$.

4. Particle Number-Projected BCS-Type Equation and Superfluid-to-Normal Phase Transition

In order to perform numerical analysis based on the proposed formalism, we adopt a simple BCS model, whose Hamiltonian with pairing interaction is given by

$$H = \sum_k (t_k - \mu)N_k - g\,B^\dagger B\,; \quad B = \sum_{k>0} c_{\bar{k}}c_k\,, \qquad (19)$$

where the parameter t_k is chosen to be $t_k = t_{\bar{k}} = -\Lambda + (k-1)d$ with the level distance $d = 2\Lambda/(\Omega-1)$ $(k = 1, 2, \cdots, \Omega)$, and the model space is cut off by $|t_k| \leq \Lambda = 10$. We consider only the projection of particle number for the half-filled case, i.e. $n = \Omega$. The HFB scheme is coverted to the BCS approximation by putting $U_{k\mu} = U_{\bar{k}\bar{\mu}} = u_k\delta_{k\mu}$ and $V_{k\bar{\mu}} = -V_{\bar{k}\mu} = v_k\delta_{k\mu}$. Further details of the theoretical framework are presented in Refs. 6 and 7. Variating the free energy in Eq. (18) with respect to $\{u_k, v_k\}$ and $\{f_\mu\}$, we derive a new BCS-like VAP equation with particle number projection,[7] i.e.

$$2u_kv_k\tilde{h}_k - (u_k^2 - v_k^2)\tilde{\Delta}_k = 0\,; \qquad u_k^2 + v_k^2 = 1\,. \qquad (20)$$

The quantity $\tilde{\Delta}_k$ plays a similar role to the gap parameter in the usual BCS theory. We performed numerical analysis of this equation for three cases of particle number, $n = 10, 26$ and 56. We express the gap parameter calculated in the meanfield approximation on the grand canonical ensemble without projection by $\Delta_{\rm G}(= g\langle B\rangle_{\rm G})$. The coupling g is adjusted so as to give $\Delta_{\rm G} = 1$ at $T = 0$. We express the maximum (minimum) value of $\tilde{\Delta}_k$ by $\tilde{\Delta}_{\rm max}$ ($\tilde{\Delta}_{\rm min}$). Furthermore, we introduce an alternative definition of the gap parameter, i.e.

$$\Delta^{\rm av} \equiv g\sqrt{\langle B^\dagger B\rangle - \sum_{k>0}\langle N_k\rangle\,\langle N_{\bar{k}}\rangle}. \qquad (21)$$

We define three kinds of gap parameters depending on the method of calculating the expectation values $\langle \ldots \rangle$ in Eq. (21). We denote Δ_C^{av} for the ensemble average with particle number projection (i.e. the VAP calculation on the canonical ensemble), $\Delta_{C'}^{av}$ with the particle number projection-after-variation (PAV), and Δ_q^{av} with the number-parity projection-after-variation (PAV).

In Fig. 1, we compare various gap parameters as functions of T.[7] We recognize that the superfluid-to-normal phase transition exists even in the VAP case, i.e. Δ_C^{av} (thick solid lines), and the critical temperature rapidly decreses with increasing n, while the critical temperature is not much dependent on n in the meanfield result, i.e. Δ_G (dashed lines). For $n = 26$, we see that both the PAV gap $\Delta_{C'}^{av}$ (dot-dashed line) and the gap with number-parity projection Δ_q^{av} (thin solid line) are close to the meanfield value Δ_G (dashed lines). For $n = 26$, we see that the VAP gap value is located between two dotted lines representing $\tilde{\Delta}_{max}$ and $\tilde{\Delta}_{min}$.

In Fig. 2, we plot the magnitudes of the paring energy as functions of T for $n = 26$ calculated in several different ways.[7] Any sharp phase transition as predicted by the projected statistics is not seen in the result of Monte Carlo simulation (diamonds) which is regarded as exact solution. The VAP value Δ_C^{av} is the best to simulate the exact solution so far.

In Fig. 3, we plot the specific heat calculated as functions of T.[7] The observed S-shape in the heat capacity[5] is well simulated by any of the theoretical curves in low temperature side of the phase transition, whose critical temperature becomes lower in the meanfield and the PAV results.

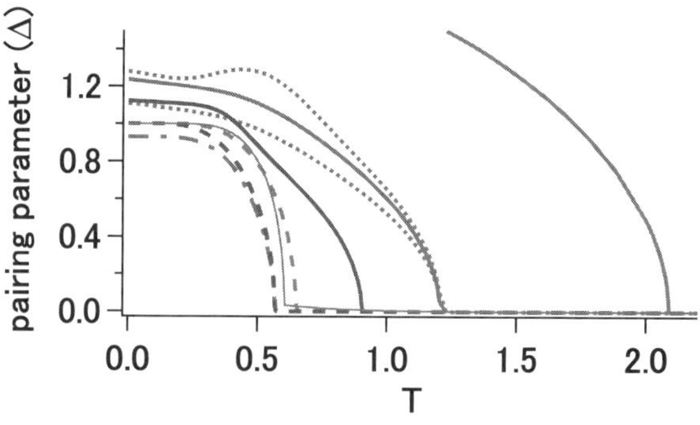

Fig. 1. Comparison among various pairing parameters. Colours distinguish n: green for $n = 10$, blue for $n = 26$ and red for $n = 56$. In each colour the thick solid line corresponds to Δ_C^{av} and the dashed line to the meanfield result Δ_G. For $n = 26$, $\tilde{\Delta}_{min}$ and $\tilde{\Delta}_{max}$ (dotted lines), $\Delta_{C'}^{av}$ (dot-dashed line) and Δ_q^{av} (thin solid line) are also given.

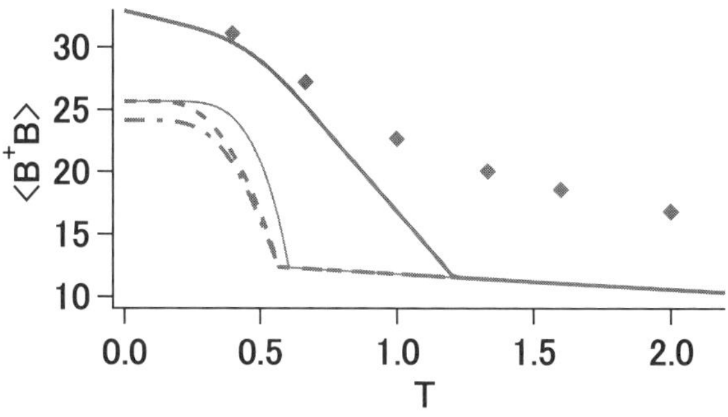

Fig. 2. Pairing energy $\langle B^{\dagger}B\rangle$ for $n = 26$. Diamonds represents the exact Monte Carlo results. We adopt the same conventions for lines as in Fig. 1.

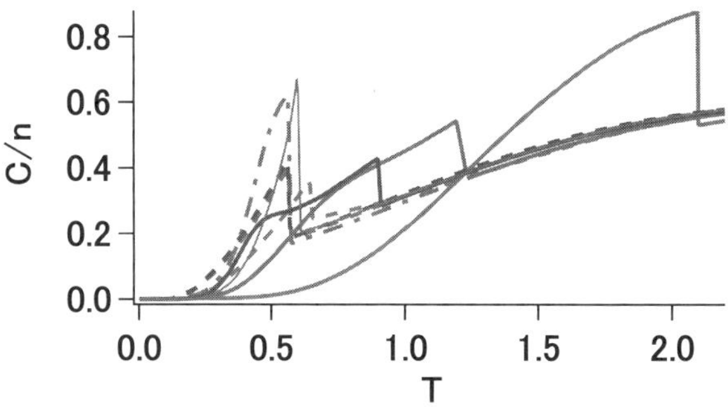

Fig. 3. Behaviour of the specific heat C/n. We adopt the same conventions for lines as in Fig. 1.

5. Conclusion

On the basis of the quantum number-projected free energy with an approximate entropy, we have developed the HFB and BCS formalism at finite temperature. Especially the BCS-type equation is derived by applying the variational principle to the free energy on the canonical ensemble. The particle number projection takes into account the effect of symmetry restoring fluctuations. The results of numerical analysis adopting a simple pairing model show that the critical temperature of the superfluid-to-normal phase transition depends on the particle number and is shifted towards higher temperature compared with that obtained in the meanfield approximation on the grand canonical ensemble. The BCS calculation with particle number projection in the VAP scheme well reproduces the experimental S-shape in the specific heat curve, but not good enough to reproduce the exact bahaviour of the pairing gap and the smeared superfluid-to-normal phase transition.

The latter difficulty must be mainly due to the BCS approximation, in which different single-particle levels are not mixed, and the collapse of pairing gaps takes place suddenly at a common critical temperature. In our simple model represented by the Hamiltonian in Eq. (19), the pairing gap takes a common value at $T = 0$ independent of single-particle levels in the BCS approximation without number projection, and its level dependent split is still small even with number projection at finite temeprature. As the result, pairing collapse takes place suddenly at a critical temperature. In case of the HFB approximation, we may expect that its larger number of variational parameters works to attain lower minimum of free energy, where the pairing energy, $-g\langle B^\dagger B\rangle$, remains finite even at higher temeprature.

The proposed formalism of the quantum number projection at finite temperature will be useful for a wide class of phase transitions in finite physical systems.

References

1. K. Tanabe and K. Sugawara-Tanabe, *Phys. Lett.* **97B**, 337 (1980).
2. K. Tanabe, K. Sugawara-Tanabe and H.-J. Mang, *Nucl. Phys. A* **357**, 20 (1981).
3. K. Sugawara-Tanabe, K. Tanabe and H.-J. Mang, *Nucl. Phys. A* **357**, 45 (1981).
4. M. Sano and S. Yamasaki, *Prog. Theor. Phys.* **29**, 397 (1963).
5. A. Schiller *et al*, *Phys. Rev. C* **63**, 021306(R) (2001).
6. K. Tanabe and H. Nakada, *Phys. Rev. C* **71**, 024314 (2005).
7. H. Nakada and K. Tanabe, *Phys. Rev. C* **74**, 061301(R) (2005).
8. Y. Takahashi and H. Umezawa, *Collect. Phenom.* **2**, 55 (1975).
9. H. Umezawa, H. Matsumoto and H. Tachiki, *Thermo Field Dynamics and Condensed States* (North-Holland, Amsterdam, 1982).
10. K. Tanabe and K. Sugawara-Tanabe, *Phys. Lett. B* **247**, 202 (1990).
11. K. Tanabe, K. Enami and N. Yoshinaga, *Phys. Rev. C* **59**, 2494 (1999).

THERMAL PAIRING IN NUCLEI

NGUYEN DINH DANG

1) Heavy-ion nuclear physics laboratory, Nishina Center for Accelerator-Based Science, RIKEN, 2-1 Hirosawa, Wako, 351-0198 Saitama, Japan
2) Institute for Nuclear Science and Technique, Hanoi, Vietnam

Particle-number projection is applied to the modified BCS (MBCS) theory. The resulting approach, called particle-number-projected MBCS theory, taking into account the effects due to fluctuations of particle and quasiparticle numbers at finite temperature, is tested within the exactly solvable multilevel model for pairing as well as the realistic ^{120}Sn nucleus. The results confirm that quasiparticle-number fluctuations indeed smooth out the sharp superfluid-normal phase transition in finite nuclei.

1. Introduction

The BCS and Hartree-Fock-Bogolyubov (HFB) theories use a simple ground-state wave function consisting of a product of Cooper pairs acting on the particle vacuum. Although the pairing problem can be exactly solved,[1–3] the simplicity of the BCS and the HFB theories still makes them a preferable choice in the study of finite systems such as atomic nuclei, whose realistic configuration space quite often prevents the feasibility of the exact solutions. However, when the BCS and HFB theories are applied to small systems at finite temperature such as hot nuclei modifications are required because of the large quantal and statistical (thermal) fluctuations, which are ignored in these theories.

Quantal fluctuations (QF) exist within the BCS (HFB) theory even at zero temperature $T = 0$ because of the violation of particle number in the BCS ground-state wave function. To eliminate this deficiency various methods of particle-number projection (PNP) have been proposed, which project out that component of the BCS wave function which corresponds to the right number of particles. The variation after projection (VAP) proposed by Lipkin and Nogami is quite popular as it is computationally simple albeit approximate. The LN equations have non-trivial solutions at any $G \neq 0$, which are quite close to the exact ones when tested in schematic models.[4–6]

Statistical fluctuations (SF) in the pairing field have been studied within several approaches based on the Landau's theory of phase transitions.[7] The results of these studies show that the pairing gap does not collapse at the critical temperature T_c as has been predicted by the BCS theory, but decreases monotonously as the temperature increases, and remains finite even at rather high T. The recently proposed modified BCS (MBCS) theory[8,9] and its generation, the modified HFB (MHFB) theory,[10] have taken into account the effects due to quasiparticle-number fluctuations (QNF) based on a microscopic foundation. This is realized through a secondary Bogolyubov transformation from quasiparticle operators to modified quasiparticle ones that allows to include the QNF, which is ignored in the conventional BCS and HFB theories at finite temperature, in the generalized single-particle matrix.[7] It has been shown within the MBCS theory[8,9] that it is the QNF that smoothes out the sharp SN phase transition and lead to the non-vanishing thermal pairing in finite systems.

Since the effects of QF are still neglected within the MBCS (MHFB) theory, to give a conclusive answer to the question for SN phase transition in finite systems it is necessary to carry out a PNP in combination with the MBCS (MHFB) theory. In the present paper we will first modify the LN approach by using the secondary Bogolyubov transformation to include the effect of QNF. We also present a second way of combining PNP with the MBCS theory by applying the PAV to the total energy of the system obtained within the MBCS theory to extract the effective thermal gap. These two ways of combining the PNP with the MBCS theory will be called as the modified Lipkin-Nogami (MLN) method,[4] and PAV-MBCS theory, respectively. The results of numerical calculations are carried out and analyzed within the exactly solvable equidistant multi-level model called the Richardson model for pairing as well as for the neutron spectrum obtained within the Woods-Saxon potential for ^{120}Sn.

2. Outline of MBCS theory

The MBCS theory includes the quasiparticle-number fluctuations by using the secondary Bogolyubov transformation from quasiparticle operators, $\alpha^{\dagger}_{jm}$ and α_{jm}, to the modified quasiparticle ones, $\bar{\alpha}^{\dagger}_{jm}$ and $\bar{\alpha}_{jm}$, where the indices j and m denote the angular-momentum quantum numbers of single-particle orbitals, while the sign $\tilde{}$ stands for the time-reversal operation, e.g. $a_{j\tilde{m}} = -\mathcal{T}a_{j-m} \equiv (-)^{j-m}a_{j-m}$. By applying successively the original Bogolyubov transformation from particle operators to quasiparticle ones and the secondary Bogolyubov transformation, one obtains a combined transformation between particle and modified quasiparticle operators as

$$a^{\dagger}_{jm} = \bar{u}_j\bar{\alpha}^{\dagger}_{jm} + \bar{v}_j\bar{\alpha}_{j\widetilde{m}} \ , \qquad a_{j\widetilde{m}} = \bar{u}_j\bar{\alpha}_{j\widetilde{m}} - \bar{v}_j\bar{\alpha}^{\dagger}_{jm} \ , \tag{1}$$

where coefficients $\bar{u}_i$ and $\bar{v}_i$ of the combined transformation (1) are given as

$$\bar{u}_j = u_j\sqrt{1-n_j} + v_j\sqrt{n_j} \ , \qquad \bar{v}_j = v_j\sqrt{1-n_j} - u_j\sqrt{n_j} \ . \tag{2}$$

Using Eq. (1), one rewrites the pairing Hamiltonian in the modified-quasiparticle representation. Because of the formal Eq. (2) the result has the same form as that of the quasiparticle representation for the pairing Hamiltonian, but with the modified-quasiparticle operators $\bar{\alpha}^{\dagger}_i$, $\bar{\alpha}_i$ replacing the quasiparticle ones $\alpha^{\dagger}_i$, α_i, and coefficients $\bar{u}_i$, $\bar{v}_i$ replacing u_i, v_i. The rest of the derivation is similar to that for the BCS equation. The final result yields the MBCS equation in the form

$$\bar{\Delta} = G\sum_j \Omega_j\bar{\tau}_j \ , \qquad N = 2\sum_j \Omega_j\bar{\rho}_i \ , \qquad \Omega_j = j + 1/2 \ , \tag{3}$$

where the modified single-particle density matrix $\bar{\rho}_j$ and modified particle-pairing tensor $\bar{\tau}_j$ are different from the conventional single-particle density matrix ρ_j and particle-pairing tensor τ_j by the terms containing the QNF on j-th orbitals, $\delta\mathcal{N}_j$, namely

$$\bar{\rho}_j = \rho_j - 2u_jv_j\delta\mathcal{N}_j \ , \qquad \bar{\tau}_j = \tau_j - (u_j^2 - v_j^2)\delta\mathcal{N}_j \tag{4}$$

with

$$\rho_j = v_j^2 + (1-2v_j^2)n_j \ , \qquad \tau_j = u_jv_j(1-2n_j) \ , \qquad \delta\mathcal{N}_j = \sqrt{n_j(1-n_j)} \ . \tag{5}$$

The MBCS internal energy is given as

$$\mathcal{E}_{\rm MBCS} = \sum_j \Omega_j\left[2\epsilon_j\bar{v}_j^2 - G\bar{v}_j^4\right] - \frac{\bar{\Delta}^2}{G} \ . \tag{6}$$

The gap and number equations (3) clearly show how the QNF is included within the MBCS theory. This leads to the appearance of the thermal component $\delta\Delta = -G\sum_j\Omega_j(u_j^2 - v_j^2)\delta\mathcal{N}_j$, in addition to the quantal one, $\Delta = G\sum_j\Omega_j\tau_j$, so that $\bar{\Delta} = \Delta + \delta\Delta$. The thermal component $\delta\Delta$ is generated only by QNF in a similar way as that of the phase-fluctuation model, where a gradual projection into a state with exact particle number makes the phases become more and more uncorrelated with increasing temperature. The main feature of each state in the grand canonical ensemble as a coherent superposition of states with different particle numbers is gradually lost so that the BCS-phase ordered regime gradually transforms into a new phase-disordered pseudogap regime. Therefore, we also call the thermal component $\delta\Delta$ of the MBCS gap $\bar{\Delta}$ as the pseudogap.

3. Modified Lipkin-Nogami (MLN) method

The MLN method consists of two self-consistent steps. In the first step, the LN method is applied to remove the particle-number fluctuations inherent in the BCS theory. This leads to a renormalization of the single-particle and quasiparticle energies as

$$\widetilde{\epsilon}_j = \epsilon_j + (4\lambda_2 - G)v_j^2 \ , \qquad \widetilde{E}_j = \sqrt{(\widetilde{\epsilon}_j - \lambda)^2 + \bar{\Delta}^2} \ , \tag{7}$$

where

$$\lambda = \lambda_1 + 2\lambda_2(N+1) \ , \qquad v_j^2 = \frac{1}{2}\left[1 - \frac{\widetilde{\epsilon}_j - \lambda}{\widetilde{E}_j}\right] \ , \qquad u_j^2 = 1 - v_j^2 \ . \tag{8}$$

In the next step one determines the modified pairing gap $\bar{\Delta}$ and λ from the same MBCS equations (3), where coefficients u_j and v_j from Eq. (8) are used to determine $\bar{u}_j$ and $\bar{v}_j$ in Eq. (2). The coefficient λ_2 is given as

$$\lambda_2 = \frac{G}{4}\frac{\sum_j \Omega_j(1-\bar{\rho}_j)\bar{\tau}_j \sum_{j'} \Omega_{j'}\bar{\rho}_{j'}\bar{\tau}_{j'} - \sum_j \Omega_j(1-\bar{\rho}_j)^2\bar{\rho}_j^2}{\left[\sum_j \Omega_j\bar{\rho}_j(1-\bar{\rho}_j)\right]^2 - \sum_j \Omega_j(1-\bar{\rho}_j)^2\bar{\rho}_j^2} \ . \tag{9}$$

The set of Eqs. (2), (7), (8), and (9) is solved self-consistently, and forms the MLNBCS equations. The total energy is given as

$$\mathcal{E}_{\rm MLN} = \sum_j \Omega_j\left[\epsilon_j - G\bar{\rho}_j\right]\bar{\rho}_j - \frac{\bar{\Delta}^2}{G} - \lambda_2\Delta N^2 \ , \tag{10}$$

where the particle-number fluctuation ΔN^2 is calculated as $\Delta N^2 = 4\sum_j \Omega_j(\bar{u}_j\bar{v}_j)^2$.

4. PAV-MBCS theory

The PNP energy is realized by applying the PNP operator

$$P^N = \frac{1}{2\pi}\int d\phi e^{-i\phi(\hat{N}-N)} \ , \tag{11}$$

on the Hamiltonian whose expectation (average) value in the ground-state (grand canonical ensemble) corresponds to the energy under consideration.

The PNP pairing energy is given as

$$\mathcal{E}_{\rm pair}^{\rm N} = -G\frac{\int d\phi e^{i\phi N}\left\{\left[\sum_j \Omega_j \frac{\bar{u}_j\bar{v}_j}{\bar{D}_j(\phi)}\right]^2 + \sum_j \Omega_j \frac{\bar{v}_j^4}{\bar{D}_j(\phi)}\right\}\det(e^{i\phi})\left[\det\bar{\mathcal{C}}(\phi)\right]^{-1/2}}{\int d\phi e^{i\phi N}\left[\det\bar{\mathcal{C}}(\phi)\right]^{-1/2}} \tag{12}$$

where $\bar{\mathcal{C}}(\phi) = e^{2i\phi}/\bar{D}(\phi)$, and $D(\phi) = 1+\bar{\rho}(e^{2i\phi}-1)$. For comparison with the pairing gaps determined within the MBCS theory and MLN method we define an effective gap from the PNP pairing energy as $\Delta_{PNP} = \sqrt{-G\mathcal{E}_{\rm pair}^{\rm N}}$. Notice that the term $-G\sum_j \Omega_j \bar{v}_j^4$ is included in the definition of the effective gap Δ_{PNP}.

5. Numerical analysis

The calculations were carried out within the Richardson model and for ^{120}Sn. The Richardson model consists of Ω doubly-folded equidistant levels, which interact via a pairing force with a constant parameter G. The single-particle energies take the values $\epsilon_j = j\epsilon$ with index j running over all Ω levels. The model is called half-filled when the number Ω of levels is equal to the number N of particles. In general, the number Ω_h of hole levels is not necessary to be the same as the number Ω_p of particle levels, i.e. $\Omega \neq N$. The level distance $\epsilon = 1$ MeV will be used in the present paper. The exact solutions of this model can be found using a number of different methods .[1–3] To use the exact solutions at finite temperature one needs to average them over a statistical ensemble. As the number N of particles in the system is fixed, the canonical ensemble is used here. The only shortcoming of such extension is that, for large N, the exact solutions weighed up to high temperature are impracticable. At the same time, for small N, the small configuration space for the ph-symmetric cases ($\Omega = N$) significantly reduces the limiting temperature up to which the MBCS theory is valid. The reason comes from the symmetry of the QPF profile as a function of single-particle energies with respect to the Fermi level. This profile becomes asymmetric at rather low temperature when $\Omega = N$ is small. It has been demonstrated in[9] that, for $N \leq 14$, its is sufficient to enlarge the space by one more level, $\Omega = N + 1$ to restore the symmetry of the QPF profile up to high temperatures. In the present the predictions within the PNP-MBCS approaches will be compared with the exact solutions for $\Omega = 11$ and $N = 10$. As for the realistic nucleus ^{120}Sn the single-particle energies obtained within the Woods-Saxon potential will be used.

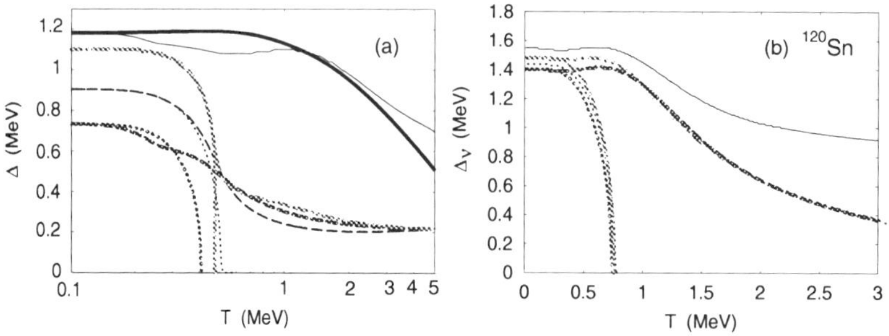

Fig. 1. Pairing gaps as functions of temperature obtained within the Richardson model for ($\Omega = 11$, $N = 10$) (a), and for neutrons in ^{120}Sn (b). The BCS solutions obtained without and including the self-energy term $-Gv_j^2$ in the single-particle energies are shown with the thin and thick dotted lines, respectively. The corresponding MBCS solutions are shown with the thin and thick dashed lines, respectively. The results obtained within the LN and MLN methods are denoted by the dot-dashed and double-dot – dashed lines, respectively. The thin solid line shows the PAV+MBCS effective gap $\Delta_{\rm PNP}$ while the thick solid line in (a) stands for the exact result.

Shown in Fig. 1 are pairing gaps as functions of T. For the Richardson model it is seen in Fig. 1 (a) that, as the particle number is small ($N =$ 10), the effect of self-energy term $-Gv_j^2$ in the single-particle energies is rather strong. In the region $T \leq T_{\rm cr}$ the BCS and MBCS gaps obtained including this term are significantly smaller that the values predicted by the BCS and MBCS theories when this term is omitted. The inclusion of the self-energy term also reduces the value of $T_{\rm cr}$ within the BCS theory. At $T \geq 0.7$ MeV the MBCS gap obtained including the self-energy term becomes slightly larger that that obtained ignoring this term. The PNP applied by using the LN method increases the gap at $T = 0$ by nearly 47%. The value of $T_{\rm cr}$ also raises closer to that obtained within the BCS theory ignoring the self-energy term. However, at $T > T_{\rm cr}$ the predictions by the MBCS theory that includes the self-energy term, and by the MLN method are nearly the same, just demonstrating that the quantal fluctuations due to particle-number violation vanish at high temperature. It is also seen that in this region both approaches significantly underestimate the exact result. The situation is largely improved within the PAV-MBCS theory. At $T \leq$ 0.2 MeV and 1 MeV $\leq T \leq 2$ MeV the values of the pairing gap predicted by the PAV-MBCS theory almost coincides with the exact one. At 0.2 MeV $\leq T \leq 1$ MeV the PAV-MBCS prediction is slightly lower than the exact result, while at $T > 2$ MeV the discrepancy increases with the PAV-MBCS overestimating the exact result.

The effect of PNP at low T becomes much weaker in ^{120}Sn, where the contribution of the self-energy term is negligible due to the large number of particles. The MBCS and MLN results are close to each other even at low T, while at high T they coalesce. The increasing discrepancy between the PAV-MBCS and MBCS results with increasing T is mainly caused by the $-G\sum_j \Omega_j \bar{v}_j^4$-term, which enters in the definition of the effective gap $G_{\rm PNP}$.

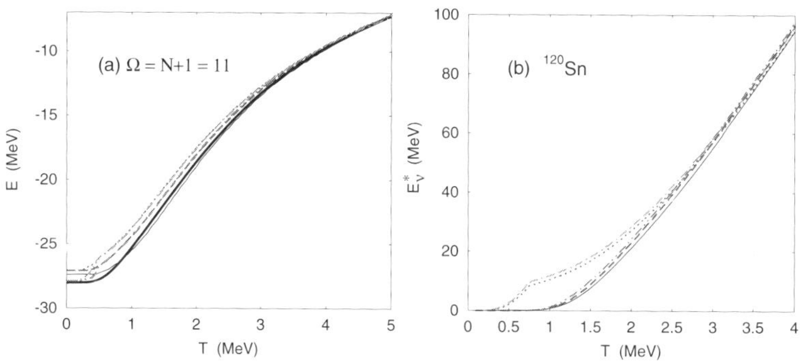

Fig. 2. Total energy (a) and excitation energy (b) as functions of temperature obtained within the Richardson model for ($\Omega = 11$, $N = 10$) (a), and for neutrons in ^{120}Sn (b). Notations are as in Fig. 1

The total energies $\mathcal{E}$ obtained within the approaches under consideration within the Richardson model are plotted as functions of T in Fig. 2 (a). For ^{120}Sn, as the absolute value of the total energy at $T = 0$ is large, we plot in panel (b) of the same figure the excitation energies $E_\nu^* = \mathcal{E}(T) - \mathcal{E}(0)$ for neutrons. It is seen that the QNF indeed smoothes out signature of the sharp SN phase transition from the total and excitation energies even after PNP is taken into account. The effect of PNP is noticeable only at very low temperature, and improves greatly the agreement with the exact result [Fig. 2 (b)]. Increasing the particle number reduces the difference between the MLN, MBCS, and PAV-MBCS, which becomes negligible in ^{120}Sn [Fig. 2 (b)]. As high T all approaches predict nearly the same energy.

The heat capacities calculated as $C = \partial\mathcal{E}/\partial T$ are shown in Fig. 3 as functions of T. It is seen that PNP within the PAV-MBCS smooths out the sharp peak at $T_{\rm cr}$, which is the signature of the SN phase transition, improving greatly the agreement between the PAV-MBCS prediction and the exact result in the schematic case (a).

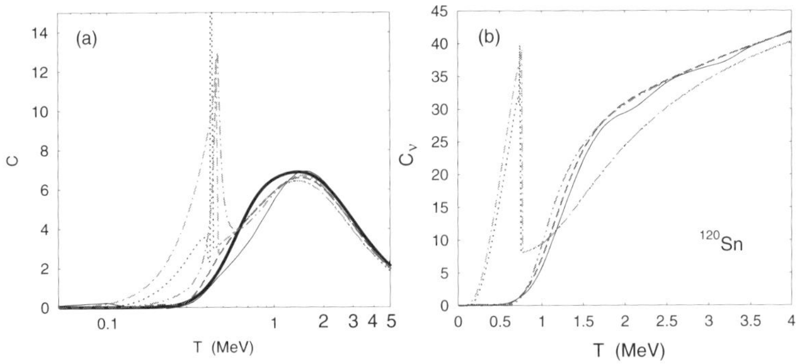

Fig. 3. Heat capacities as functions of temperature within the Richardson model for ($\Omega = 11$, $N = 10$) (a), and for neutrons in ^{120}Sn (b). Notations are as in Fig. 1.

6. Conclusions

This work performs combines the PNP with the MBCS theory. This allows to take into account both quantal and thermal fluctuation effects in the study of SN phase transitions in finite systems such as nuclei. The results in both exactly solvable model and realistic nucleus confirm the previous conclusion by the MBCS theory that thermal fluctuations of quasiparticle number smooths out the sharp SN phase transition in hot nuclei. Therefore, in order to obtain adequate conclusions regarding the superfluid properties of finite systems at finite temperature the approaches based on BCS and HFB theories need to take these thermal fluctuations into account.

References

1. R.W. Richardson, Phys. Lett. **3**, 277 (1963), Phys. Lett. **5**, 82 (1963), Phys. Lett. **14**, 325 (1965).
2. F. Pan, J.P. Draayer, W.E. Ormand, Phys. Lett. B **422**, 1 (1998).
3. A. Volya, B.A. Brown, and V. Zelevinsky, Phys. Lett. B 509 (2001) 37.
4. H.C. Pradhan, Y. Nogami, and J. Law, Nucl. Phys. A **201**, 357 (1973)
5. N. D. Dang, Eur. Phys. J. A **16**, 181 (2003).
6. N. Quang Hung and N. Dinh Dang, submitted to Phys. Rev. C.
7. L.G. Moretto, Nucl. Phys. **A182**, 641 (1972); A.L. Goodman, Phys. Rev. C **29**, 1887 (1984); N. Dinh Dang and N. Zuy Thang, J. Phys. **G 14**, 1471 (1988); N.D. Dang, P. Ring, and R. Rossignoli, Phys. Rev. C **47**, 606 (1993).
8. N. Dinh Dang and V. Zelevinsky, Phys. Rev. C **64**, 064319 (2001); N. Dinh Dang and A. Arima, Phys. Rev. C **67**, 014304 (2003).
9. N. Dinh Dang, Nucl. Phys. A **784**, 147 (2007).
10. N. Dinh Dang and A. Arima, Phys. Rev. C **68**, 014318 (2003).

COMPOSITE BOSONS AND QUASIPARTICLES IN A NUMBER CONSERVING APPROACH

F. PALUMBO

INFN-Laboratori Nazionali di Frascati,
Frascati, P.O. Box 13, 00044, Italy
E-mail: Fabrizio.Palumbo@lnf.infn.it

I recently proposed a method of bosonization valid for systems of an even number of fermions whose partition function is dominated at low energy by bosonic composites. This method respects all symmetries, in particular fermion number conservation. I extend it to treat odd systems and excitations involving unpaired fermions.

Keywords: Composite bosons and quasiparticles in superconducting systems

1. Introduction

There are many ways to introduce composite bosons as elementaty degrees of freedom in fermion systems, but most of them violate fermion number conservation, which is an important symmetry in atomic nuclei and other finite systems.[1] I recently developed a method to bosonize even systems which respects all symmetries, including fermion number conservation.[2] This method was applied to relativistic field theories at zero[3] and nonvanishing[4] fermion number. I present here an application of this technique to nonrelativistic systems. In Section 2 I make an outline of the formalism, giving in Section 3 its physical interpretation. In Section 4 I report the effective action of composite bosons and finally in Section 5 the effective action of composite bosons plus unpaired fermions.

2. Outline of the approach

Consider the partition function of a fermion system

$$Z = \mathrm{Tr}^F \left\{ \exp \left[-\frac{1}{T} (\hat{H}_F - \mu_F \, \hat{n}_F) \right] \right\} \tag{1}$$

where Tr^F is the trace over the Fock space, T the temperature, μ_F the chemical potential and $\hat{n}_F$ the fermion number operator. A sector of n_F fermions can be selected by the constraint

$$T \frac{\partial}{\partial \mu_F} \ln Z = n_F . \tag{2}$$

A functional form of Z can be found by performing the trace over coherent states

$$|c\rangle = \exp\left(\sum_m \hat{c}_m^\dagger \, c_m\right) |0\rangle \, , \tag{3}$$

where m are the fermion quantum numbers, $\hat{c}_m^\dagger$ their canonical creation operators and c_m Grassmann variables. Coherent states satisfy the basic or defining equations

$$\hat{c}_m |c\rangle = c_m |c\rangle \tag{4}$$

where $\hat{c}_m$ are canonical destruction operators. In terms of these states I can write the identity in the fermion Fock space in the form

$$\mathcal{I} = \int \prod_m dc_m^* dc_m \, \langle c|c\rangle^{-1} |c\rangle\langle c| \, . \tag{5}$$

Using the above resolution of the identity the partition function can be written

$$Z = \mathrm{Tr}^F \left\{ \mathcal{I} \exp\left[-\tau (\hat{H}_F - \mu_F \, \hat{n}_F) \right] \right\}^{L_0} \tag{6}$$

where I introduced Euclidean time spacing τ and number of temporal sites L_0 by setting

$$T = \frac{1}{\tau L_0} \, . \tag{7}$$

The trace can be evaluated exactly at the relevant order in τ with the result

$$Z = \int [dc^* dc] \exp\left[-S_F(c^*, c) \right] \, . \tag{8}$$

S_F is the fermion action in Euclidean time

$$S_F = \tau \sum_t \left\{ c_t^* \nabla_t c_{t-1} + H_F(c_t^*, c_{t-1}) - \mu_F \, c_t^* c_{t-1} \right\} \, , \tag{9}$$

with the discrete time derivative

$$\nabla_t \, f = \frac{1}{\tau} \left(f_{t+1} - f_t \right) \, . \tag{10}$$

In a superconducting system one has to perform a Bogoliubov transformation which introduces quasiparticles creation-annihilation operators

$$\begin{aligned}\hat{\alpha} &= R^{\frac{1}{2}}(\hat{c} - \mathcal{B}\,\hat{c}^{\dagger})\\ \hat{\alpha}^{\dagger} &= (c^{\dagger} - \hat{c}\,\mathcal{B}^{\dagger})R^{\frac{1}{2}}\,,\end{aligned} \tag{11}$$

where

$$R = \left(1 + \mathcal{B}\,\mathcal{B}^{\dagger}\right)^{-1}\,. \tag{12}$$

They satisfy canonical commutation relations for an arbitrary matrix $\mathcal{B}$. To write down the Bogoliubov transformation of coherent states I introduce some definitions. First I define the creation operator of a bosonic composite with quantum number K

$$\hat{B}^{\dagger}_{K} = \frac{1}{2\sqrt{\Omega}} \sum_{m_1,m_2} \hat{c}^{\dagger}_{m_1} \left(B^{\dagger}_{K}\right)_{m_1,m_2} \hat{c}^{\dagger}_{m_2}\,. \tag{13}$$

B_K is the structure function of the composite and Ω its index of nilpotency (assumed to be independent of K), defined as the largest integer such that

$$\left(\hat{B}_K\right)^{\Omega} \neq 0. \tag{14}$$

Next I expand the matrix $\mathcal{B}$ on the matrices B_K

$$\mathcal{B} = \frac{1}{\sqrt{\Omega}} \sum_K b_K B^{\dagger}_K \equiv \frac{1}{\sqrt{\Omega}}\, b \cdot B^{\dagger}\,. \tag{15}$$

Lastly I define new coherent states

$$|\alpha, b\rangle = \exp\left(-\sum_m \alpha_m\, \hat{\alpha}^{\dagger}_m\right) |b\rangle \equiv \exp(-\alpha \cdot \hat{\alpha}^{\dagger})|b\rangle \tag{16}$$

where

$$|b\rangle = \exp(b \cdot \hat{B}^{\dagger})|0\rangle \tag{17}$$

is the quasiparticle vacuum.The Bogoliubov transformation changes the coherent states $|c\rangle$ according to

$$\mathcal{S}|c\rangle = \langle b|b\rangle^{-\frac{1}{2}}|c, b\rangle \tag{18}$$

because[4]

$$\langle c, b|c, b\rangle = \exp(c^{*}c)\langle b|b\rangle\,. \tag{19}$$

I can now write a new expression of the identity in the fermion Fock space in terms of the transformed coherent states

$$\mathcal{I}(b^{*}, b) = \int [dc^{*}dc]\langle c, b|c, b\rangle^{-1}|c, b\rangle\langle c, b| \tag{20}$$

and use it to get a new expression of the partition function

$$Z = \mathrm{Tr}^F \left\{ \mathcal{I}(b^*, b) \exp\left[-\tau(\hat{H}_F - \mu_F \hat{n}_F) \right] \right\}^{L_0} . \tag{21}$$

Evaluation of the trace yields an action of the system strictly equivalent to the original one, but whose individual terms do not respect several symmetries, in particular fermion number conservation.

There are many cases in which all symmetries can be enforced term by term proceeding in a slightly different way. I introduce at each Euclidean time an independent Bogoliubov transformation $\mathcal{S}_t$. I do this by letting the expansion coefficients b_K to depend on time, $b_K \to b_{K,t}$ while keeping the basis matrices B_K fixed. Since nothing depends on the $\mathcal{B}$ and therefore on the expansion coefficients $b_{K,t}$ I can integrate over the latter ones in the partition function with an arbitrary measure $d\mu(b^*, b)$

$$Z = \int d\mu(b^*, b)\, \mathrm{Tr}^F \left\{ \prod_t \mathcal{I}(b^*, b_t) \exp\left[-\tau(\hat{H}_F - \mu_F \hat{n}_F) \right] \right\} . \tag{22}$$

Requiring the variables b^*, b to transform in the proper way all symmetries are restored.

Performing the trace, which can be done exactly[4] at the relevant order in τ, I get a new functional form of the transfer matrix

$$Z = \int d\mu(b^*, b) \prod_{m,t} \left[d\alpha^*_{m,t} d\alpha_{m,t} \right] \exp\left[-S(b^*, b, \alpha^*, \alpha, B^\dagger, B) \right] . \tag{23}$$

3. Physical interpretation

The last expression of the partition function contains, in addition to the Grassmann variables α^*, α, the bosonic variables b^*, b. Is it possible to associate these variables to physical bosons?

To answer this question let us analyze the nature of the states $|\alpha, b\rangle$. They are constructed in terms of quasiparticles and composite boson creation operators $\hat{\alpha}^\dagger, \hat{B}^\dagger_K$. I call these states coherent because they are obtained by a unitary transformation from truly coherent states of fermions and they share with coherent states of elementary bosons the property of a fixed phase relation among components with different number of composites. But the basic property of coherent states cannot be fulfilled

$$\hat{B}_K |b\rangle \neq b_K |b\rangle . \tag{24}$$

This is a consequence of the composites commutation relations, which are not canonical

$$\left[\hat{B}_J, \hat{B}_K^\dagger\right] = \frac{1}{2\,\Omega}\,\mathrm{Tr}\,(B_J B_K^\dagger) - \frac{1}{\Omega}\,\hat{c}^\dagger B_K^\dagger B_J\,\hat{c}\,. \tag{25}$$

In states with a number of composites $n << \Omega$, the above equations can approximate the canonical ones by an appropriate normalization of the structure functions provided they are sufficiently smooth. Indeed in such a case the last term is of order n/Ω. But in states with $n \sim \Omega$ this is impossible even with an absolute freedom about the form of the structure functions (which are instead determined by the dynamics). The best we can do[2] is to satisfy them for states with $n+k$ composites, for fixed $n \sim \Omega$ and $|k| << \Omega$. The above considerations can be made more precise, also taking quasiparticles into account, by showing that under the above conditions and for a number of quasiparticles much smaller than Ω

$$\langle(\hat{\alpha}_{m_1})^{r_1}...(\hat{\alpha}_{m_n})^{r_n}(\hat{b}_{K_1})^{r_{K_1}}...(\hat{b}_{K_t})^{r_{K_t}}|\int\prod_{K,t}\left[\frac{db^*_{K,t}db_{K,t}}{2\pi i}\right]\mathcal{I}(b^*,b)|$$
$$(\hat{\alpha}^\dagger_{m_1})^{s_1}...(\hat{\alpha}^\dagger_{m_n})^{s_n}(\hat{b}_{K_1})^{s_{K_1}}...(\hat{b}_{K_t})^{s_{K_t}}\rangle \sim \delta_{r_1,s_1}...\delta_{r_t,s_t}\,. \tag{26}$$

Only for such states can the holomorphic variables b^*_K, b_K be interpreted as the fields of bosons with quantum numbers "K" in the presence of fermions with quantum numbers "m" associated to the Grassmann variables α_m. Restricting myself to such states I can assume the measure

$$d\mu(b^*,b) = \prod_{K,t}\left[\frac{db^*_{K,t}db_{K,t}}{2\pi i}\right]\,. \tag{27}$$

The property

$$\hat{\alpha}|b\rangle = 0 \tag{28}$$

ensures that there is no double counting: There are no quasiparticles in the bosonic composites.

What to do with states which do not satisfy the above requirements? We should integrate them out (which would require a different choice of the measure) leaving only states which have a physical interpretation. But in some cases we can reasonably assume that we can ignore them, in the spirit of a variational calculation.

4. The effective action of composite bosons

In a system of fermions whose low energy excitations are dominated by fermion composites I can restrict the trace to these composites neglecting the quasiparticles. The restricted partition function can be written

$$Z_C = \mathrm{Tr}^F \left\{ \mathcal{P}_C \exp\left[-\frac{1}{T}(\hat{H}_F - \mu_F \hat{n}_F) \right] \right\} \tag{29}$$

where $\mathcal{P}_C$ is a projection operator in the subspace of the composites for which I assume the approximate form

$$\mathcal{P}_C = \int \frac{db^*\, db}{2\pi i} \langle b|b\rangle^{-1} |b\rangle\langle b| \,. \tag{30}$$

To proceed further I write the fermion Hamiltonian in the form

$$H_F = \mu_F \hat{n}_F + \hat{c}^\dagger h_0 \hat{c} - \sum_K g_K \frac{1}{2} \hat{c}^\dagger F_K^\dagger \hat{c}^\dagger \frac{1}{2} \hat{c} F_K \hat{c}. \tag{31}$$

The one-body term includes the single-particle energy with matrix e, the fermion chemical potential μ_F and any single-particle interaction with external fields included in the matrix $\mathcal{M}$

$$h_0 = e - \mu_F + \mathcal{M}. \tag{32}$$

The matrices F_K are the form factors of the potential, normalized according to

$$\mathrm{tr}(F_{K_1}^\dagger F_{K_2}) = 2\,\Omega\, \delta_{K_1 K_2}. \tag{33}$$

Now the trace in the partition function can be evaluated [2] yielding its functional form

$$Z_C = \int \left[\frac{db^* db}{2\pi i} \right] \exp\left(-S_C(b^*, b, B^\dagger, B) \right) \tag{34}$$

in which

$$\begin{aligned} S_C(b^*, b, B^\dagger, B) &= \tau \sum_t \frac{1}{2} \mathrm{tr} \left\{ \frac{1}{\tau} \ln \left[1\!\!1 + \tau\, R\text{"}\, \mathcal{B}^\dagger \nabla_t \mathcal{B} \right] \right. \\ &+ 2\, R\text{"}\, \mathcal{B}^\dagger h\, \mathcal{B} - \sum_K g_K \left[(R\text{"} - 1)\, F_K^\dagger F_K \right. \\ &\left.\left. + (R\text{"}\, \mathcal{B}^\dagger F_K^\dagger) \frac{1}{2} \mathrm{tr}(R\text{"}\, F_K \mathcal{B}) - R\text{"}\, \mathcal{B}^\dagger F_K^\dagger R\text{"}\, F_K \mathcal{B} \right] \right\} . \end{aligned} \tag{35}$$

In this expression "tr" is the trace over fermion quantum numbers,

$$h = h_0 - \sum_K g_K F_K^\dagger F_K \tag{36}$$

and the variables b^*, b must be understood at times $t, t-1$ respectively, unless otherwise specified. So, for instance

$$R\text{"} \rightarrow \left(1 + \mathcal{B}_t^\dagger \mathcal{B}_{t-1} \right)^{-1} . \tag{37}$$

The latter function should be distinguished from the function R introduced in Eq.(12)

$$R = \left(1 + \mathcal{B}\,\mathcal{B}^\dagger\right)^{-1} \rightarrow \left(1 + \mathcal{B}_{t-1}\mathcal{B}_t^\dagger\right)^{-1} \equiv R_{t,t-1}\,. \tag{38}$$

The derivation of the above functional form of the partition function is based only on the physical assumption of boson dominance and the approximation adopted for $\mathcal{P}_C$. *The action* S_C *is a functional of the structure matrices* B_K *which can be determined by a variational calculation.*

In many-body physics it is often preferred the Hamiltonian formalism. The Hamiltonian of the effective bosons, H_B, cannot be read directly from the effective action, because S_C does not have the form of an action of elementary bosons. Indeed it contains anomalous time derivative terms, anomalous couplings of the chemical potential and nonpolynomial interactions, which are all features of compositeness. Therefore it has been necessary to devise an appropriate procedure to derive H_B, which is given[2] in terms of *canonical* boson operators $\hat{b}^\dagger, \hat{b}$, (not to be confused with the composite operators $\hat{B}^\dagger, \hat{B}$), so that

$$Z_C = \mathrm{Tr}^B \left[-\frac{1}{T}(\hat{H}_B - \mu_B\,\hat{n}_B)\right]. \tag{39}$$

Tr^B is the trace on the boson Fock space, μ_B is the boson chemical potential and $\hat{n}_B$ the boson number operator.

$\hat{H}_B$ has a closed form but, for a practical use, it is necessary to perform an expansion in inverse powers of the index of nilpotency Ω.

5. The effective action of composite bosons plus quasiparticles

Finally I consider the case in which one needs to retain, in addition to composite bosons, also quasiparticles. To find the total effective action I first observe that the new coherent states can be rewritten

$$|\alpha, b\rangle = \exp\left[-\frac{1}{2}\alpha\,(R^*)^{\frac{1}{2}}\mathcal{B}^\dagger\,R^{\frac{1}{2}}\alpha\right]|\alpha\,, b\rangle\rangle \tag{40}$$

where I introduced the definition

$$|\alpha\,, b\rangle\rangle = \exp\left(\hat{c}^\dagger\,R^{-\frac{1}{2}}\alpha\right)|b\rangle\,. \tag{41}$$

Notice that in the latter states there appear the original fermion operators, not the quasiparticle ones.

According to Eqs. (22,23) the action is given in terms of the ratios

$$\langle \alpha_1 b_1 | \alpha_2 b_2 \rangle^{-1} \langle \alpha_1 b_1 | \exp(-\tau \hat{H}) | \alpha_2 b_2 \rangle \,. \tag{42}$$

The matrix element of any operator $\mathcal{O}$ factorizes according to

$$\begin{aligned} \langle \alpha_1, b_1 | \mathcal{O} | \alpha_2, b_2 \rangle = \exp\left(-\frac{1}{2} \alpha_1^* R_{11}^{\frac{1}{2}} \mathcal{B}_1 (\mathrm{R}_{11}^T)^{\frac{1}{2}} \alpha_1^* - \frac{1}{2} \alpha_2 (R_{22}^*)^{\frac{1}{2}} \mathcal{B}_2^\dagger R_{22}^{\frac{1}{2}} \alpha_2 \right) \\ \times \langle\langle \alpha_1, b_1 | \mathcal{O} | \alpha_2, b_2 \rangle\rangle \,. \end{aligned} \tag{43}$$

The first factor cancels out in the ratios (42) and the second factor is given by

$$\langle\langle \alpha_1, b_1 | \exp(-\tau \hat{H}_F) | \alpha_2, b_2 \rangle\rangle = \langle b_1 | \exp(-\tau \hat{H}_F) | b_2 \rangle \exp(-\int - \tau H_{qp}) \,. \tag{44}$$

The first matrix element yields the composite boson term S_C of the action reported in the previous Section, while the second one is the quasiparticle contribution. To shorten its expression I define the Grassmann variables

$$\begin{aligned} w^* &= R^T \left[\left(R^T \right)^{-\frac{1}{2}} \alpha^* - \mathcal{B}^\dagger R^{-\frac{1}{2}} \alpha \right] \\ w &= R \left[R^{-\frac{1}{2}} \alpha + \mathcal{B} \left(R^T \right)^{-\frac{1}{2}} \alpha^* \right] . \end{aligned} \tag{45}$$

In terms of these variables

$$\begin{aligned} \int &= -w^* w - \frac{1}{2} w^* \mathcal{B} w^* - \frac{1}{2} w \mathcal{B}^\dagger w \\ H_{qp} &= w^* \left(h + \sum_K g_K F_K^\dagger R^T F_K \right) w - \frac{1}{4} \sum_K g_K \ \mathrm{tr} \ \left[\left(\mathcal{B}^\dagger R F_K^\dagger \right) \right. \\ &\quad \times \left. w F_K w + (F_K R \mathcal{B}) w^* F_K w^* \right] - \sum_K g_K \frac{1}{2} w^* F_K^\dagger w^* \frac{1}{2} w F_K w \,. \end{aligned} \tag{46}$$

Now I restrict myself to the case of a number of quasiparticles much smaller than Ω. Then assuming $\nabla_t b_K \sim \nabla_t b_K^* \sim 1$ the noncanonical temporal terms in the action are of order $\Omega^{-\frac{1}{2}}$

$$\frac{1}{2} w_t^* (\mathcal{B}_{t-1} - \mathcal{B}_t) w_t^* + \frac{1}{2} w_{t-1} (\mathcal{B}_t^\dagger - \mathcal{B}_{t-1}^\dagger) w_{t-1} = O(\Omega^{-\frac{1}{2}}) \tag{47}$$

because $\mathcal{B} = \Omega^{-\frac{1}{2}} b \cdot B^\dagger$ and $b_2 - b_1 = O(1)$. To this approximation the temporal terms in the action become $\alpha_1^*(\alpha_1 - \alpha_2)$, which are canonical, and therefore the total Hamiltonian can be written in operator form

$$\hat{H} = \hat{H}_B + \hat{H}_{qp} \tag{48}$$

where

$$\hat{H}_{qp} = : \Big\{ -\hat{w}^\dagger (h + \sum_K g_K F_K^\dagger R^T F_K)\hat{w} - \frac{1}{4}\sum_K g_K \,\mathrm{tr}\, \Big[\Big(\mathcal{B}^\dagger R F_K^\dagger\Big)$$

$$\times \; \hat{w} F_K \hat{w} + (F_K \, R\mathcal{B}) \; \hat{w}^\dagger F_K^\dagger \hat{w}^\dagger\Big] - \sum_K g_K \,\frac{1}{2} \hat{w}^\dagger F_K^\dagger \hat{w}^\dagger \,\frac{1}{2} \hat{w} F_K \hat{w} \Big\} : \quad (49)$$

In this expression the operators $\hat{w}^\dagger, \hat{w}$ must be regarded as functions of the boson and quasiparticle operators $\hat{b}^\dagger, \hat{b}, \hat{\alpha}^\dagger, \hat{\alpha}$ and the colons mean normal order with respect to the latters.

I do not have the space to discuss the properties of the Hamiltonian I have presented. I only make the following observations

i) it respects term by term all fermion symmetries, in particular fermion number conservation, as it can be seen remembering that the bosonic operators $\hat{b}^\dagger$ have fermion number 2

ii) it contains quadratic quasiparticles terms of the type $\hat{\alpha}^\dagger \hat{\alpha}^\dagger$ and $\hat{\alpha}\hat{\alpha}$, similar to the "dangerous" terms of Bogoliubov. These terms however do not break fermion number conservation because they are accompanied by the operators $\hat{b}, \hat{b}^\dagger$ respectively. Using the variational equations satisfied by the structure functions of the composites, Eq.(81) of Ref. 2, one can see that the coefficients of these terms vanish and the term involving the operator $\hat{\alpha}^\dagger \hat{\alpha}$ takes the usual form. Therefore the difference with respect to the standard Bogoliubov transformation is that the quartic quasiparticle term involves interactions whith composite bosons.

6. Acknowledgments

This work was partly supported by PRIN 2006021029 "Complex problems in statistical mechanics and field theory" of Ministero dell'Universita' e della Ricerca Scientifica.

References

1. See for instance J. Dukelski and G. Sierra, Phys. Rev. B61 (2000-II) 12302; F. Braun and J. von Delft, Phys. Rev. Lett. 81 (1998) 47121; K. Tanaka and F. Marsiglio, Phys. Rev. 60B (1999-I) 3508
2. Eur. Phys. J. B56 (2007) 335
3. S. Caracciolo, V. Laliena and F. Palumbo, JHEP02 (2007) 034, [arXiv: hep-lat/0611012 v1]
4. F.Palumbo, arXiv:hep-lat/0702001

5. F. Iachello and A. Arima, The Interacting Boson Model, Cambridge University Press, Cambridge, 1987
6. J.M.Blatt, Theory of Superconductivity, Academic Press New York and london, 1964

KINETIC EQUATION FOR NUCLEAR RESPONSE WITH PAIRING

V. I. ABROSIMOV

Institute for Nuclear Research,
Kiev, 03028, Ukraine
E-mail: abrosim@kinr.kiev.ua

D. M. BRINK

Oxford University,
Oxford, OX1 3NP, U.K.
E-mail: thph0032@herald.ox.ac.uk

A. DELLAFIORE* and F. MATERA†

Istituto Nazionale di Fisica Nucleare, Sezione di Firenze
and Dipartimento di Fisica, Universitá degli Studi di Firenze,
via Sansone 1, I 50019 Sesto Fiorentino (Firenze), Italy
**E-mail: della@fi.infn.it* †*E-mail: matera@fi.infn.it*

The solutions of the Wigner-transformed time-dependent Hartree–Fock–Bogoliubov equations are studied in the constant-Δ approximation, in spite of the fact that this approximation is known to violate both local and global particle-number conservation. As a consequence of this symmetry breaking, the longitudinal response function given by this approximation contains spurious contributions. A simple prescription for restoring both broken symmetries and removing the spurious strength is proposed. It is found that the semiclassical analogue of the quantum single-particle spectrum, has an excitation gap of 2Δ, in agreement with the quantum result. The effects of pairing correlations on the density response functions of three one-dimensional systems of different size are shown.

Keywords: Pairing; Vlasov equation.

1. Introduction

The problem of extending the Vlasov equation to systems in which pairing correlations play an important role has been tackled some time ago by Di Toro and Kolomietz[1] in a nuclear physics context and, more recently, by Urban and Schuck[2] for trapped fermion droplets. These last authors derived the TDHFB equations for the Wigner transform of the normal density matrix ϱ and of the pair correlation function κ (plus their time-reversal conjugates) and used them to study the dynamics of a spin-saturated trapped Fermi gas. In the time-dependent theory one obtains a system of four coupled differential equations for ϱ, κ, and their conjugates[2] and, if one wants an analytical solution, some approximation must be introduced. Here we try to find a solution of the equations of motion derived by Urban and Schuck in the approximation in which the pairing field $\Delta(\mathbf{r}, \mathbf{p}, t)$ is treated as a constant. It is well known that such an approximation violates both particle-conservation and gauge invariance (see e.g. sect. 8-5 of Ref. 3 and Ref. 4), nonetheless we study it because of its simplicity, with the aim of correcting the final results for its shortcomings.

2. Basic equations

We assume that our system is saturated both in spin and isospin space and do not distinguish between neutrons and protons, so we can use directly the equations of motion of Urban and Schuck.

First we recall the equations of motion derived in[2] for the Wigner-trasformed density matrices $\varrho = \varrho(\mathbf{r}, \mathbf{p}, t)$ and $\kappa = \kappa(\mathbf{r}, \mathbf{p}, t)$, with the warning that the sign of κ that we are using agrees with that of[1] , hence it is opposite to that of[2] . Moreover we find convenient to use the odd and even combinations of the normal density introduced in[2] :

$$\rho_{ev} = \frac{1}{2}[\rho(\mathbf{r}, \mathbf{p}, t) + \rho(\mathbf{r}, -\mathbf{p}, t)] \,, \tag{1}$$

$$\rho_{od} = \frac{1}{2}[\rho(\mathbf{r}, \mathbf{p}, t) - \rho(\mathbf{r}, -\mathbf{p}, t)] \,. \tag{2}$$

Thus, the equations of motion given by Eqs.(15a...d) of Ref. 2 read

$$i\hbar\partial_t \rho_{ev} = i\hbar\{h, \rho_{od}\} - 2i\mathrm{Im}[\Delta^*(\mathbf{r}, \mathbf{p}, t)\kappa] \tag{3}$$

$$i\hbar\partial_t \rho_{od} = i\hbar\{h, \rho_{ev}\} + i\hbar\mathrm{Re}\{\Delta^*(\mathbf{r}, \mathbf{p}, t), \kappa\} \tag{4}$$

$$i\hbar\partial_t \kappa = 2(h-\mu)\kappa - \Delta(\mathbf{r}, \mathbf{p}, t)(2\rho_{ev} - 1) + i\hbar\{\Delta(\mathbf{r}, \mathbf{p}, t), \rho_{od}\} \,. \tag{5}$$

Here h is the Wigner-transformed Hartree–Fock hamiltonian $h(\mathbf{r}, \mathbf{p}, t)$, while $\Delta(\mathbf{r}, \mathbf{p}, t)$ is the Wigner-transformed pairing field. Since the time-dependent part of κ is complex, $\kappa = \kappa_r + i\kappa_i$, the last equation gives two independent equations for the real and imaginary parts of κ.

Moreover, from the supplementary normalization condition (Ref. 5, p. 252)

$$\mathcal{R}^2 = \mathcal{R} \tag{6}$$

satisfied by the generalized density matrix $\mathcal{R}$, the two following independent equations are obtained:

$$\rho_{od}\kappa + i\frac{\hbar}{2}\{\rho_{ev}, \kappa\} = 0\,, \tag{7}$$

$$\rho_{ev}(\rho_{ev} - 1) + \rho_{od}^2 + \kappa\kappa^* = 0\,. \tag{8}$$

We shall use the equations of motion (3–5), together with these equations, as our starting point.

3. Constant-Δ approximation and static limit

In a fully self-consistent approach, the pairing field $\Delta(\mathbf{r}, \mathbf{p}, t)$ is related to $\kappa(\mathbf{r}, \mathbf{p}, t)$, however here we introduce an approximation and replace the pairing field of the HFB theory with the phenomenological pairing gap of nuclei, hence in all our equations we put

$$\Delta(\mathbf{r}, \mathbf{p}, t) \approx \Delta_0(\mathbf{r}, \mathbf{p}) \approx \Delta = const.\,, \tag{9}$$

with $\Delta \approx 1$MeV.

In the constant-Δ approximation the equilibrium solutions of the semi-classical equations (3–5) and (7,8) are[6]

$$\rho_0(\epsilon) = \frac{1}{2}\Big(1 - \frac{\epsilon - \mu}{E(\epsilon)}\Big)\,, \tag{10}$$

$$\kappa_0(\epsilon) = -\frac{\Delta}{2E(\epsilon)}\,, \tag{11}$$

with the quasiparticle energy

$$E(\epsilon) = \sqrt{\Delta^2 + (\epsilon - \mu)^2}\,, \tag{12}$$

and

$$\epsilon = h_0(\mathbf{r}, \mathbf{p}) = \frac{\mathbf{p}^2}{2m} + V_0(\mathbf{r}) \tag{13}$$

the particle energy in the equilibrium mean field.

The chemical potential μ is determined by the condition

$$A = \frac{4}{(2\pi\hbar)^3}\int d\mathbf{r}d\mathbf{p}\rho_0(h_0(\mathbf{r}, \mathbf{p}))\,, \tag{14}$$

where A is the number of particles in the system. This integral should remain constant also when ρ is time-dependent.

4. Dynamics

Always in constant-Δ approximation, the time-dependent equations (3–5) become instead

$$i\hbar\partial_t\rho_{ev} = i\hbar\{h, \rho_{od}\} - 2i\Delta\mathrm{Im}(\kappa) \tag{15}$$

$$i\hbar\partial_t\rho_{od} = i\hbar\{h, \rho_{ev}\} \tag{16}$$

$$i\hbar\partial_t\kappa = 2(h-\mu)\kappa - \Delta(2\rho_{ev} - 1)\,. \tag{17}$$

This is the simplified set of equations that we want to study here. The sum of the first two equations gives an equation that is similar to the Vlasov equation of normal systems, only with the extra term $-2i\Delta\mathrm{Im}(\kappa)$. This extra term couples the equation of motion of ρ with that of κ, thus, instead of a single differential equation (Vlasov equation), now we have a system of two coupled differential equations (for ρ and κ_i).

A good starting point for the solution of the normal Vlasov equation is to assume that the average mean field keeps its static value. This approximation, that was termed "zero-order" in,[7] corresponds to the quantum single-particle approximation and is a preliminary step that allows us to include collective effects in a second stage. More explicitly, the hamiltonian h in Eqs. (15–17) is replaced by $h = h_0 + \delta h$, where h_0 is the equilibrium hamiltonian and $\delta h = \beta\delta(t)V^{ext}(\mathbf{r})$ is a weak external driving field acting at time $t = 0$.

In this case Eqs. (15–17) can be easily solved in linear approximation (see Ref. 8 for details), giving an explicit expression for the new eigenfrequencies of the system and for the normal-density fluctuations

$$\delta\varrho(\mathbf{r}, t) = \int d\mathbf{p}\delta\rho(\mathbf{r}, \mathbf{p}, t)\,. \tag{18}$$

However, as anticipated, there is a problem because this solution is seen to violate both the continuity equation

$$\partial_t\delta\varrho(\mathbf{r}, t) = -\nabla\cdot\mathbf{j}(\mathbf{r}, t)\,, \tag{19}$$

and the energy-weighted sum rule.

The current density $\mathbf{j}(\mathbf{r}, t)$ in the last equation is defined as

$$\mathbf{j}(\mathbf{r}, t) = \int d\mathbf{p}\frac{\mathbf{p}}{m}\delta\rho(\mathbf{r}, \mathbf{p}, t)\,, \tag{20}$$

hence, while $\varrho(\mathbf{r}, t)$ is determined by the even part of $\rho(\mathbf{r}, \mathbf{p}, t)$, the current density is determined by ρ_{od}. Our simplified set of equations (15–17) is actually incompatible with Eq. (19) and this is a consequence of our constant-Δ approximation, since the problem does not arise for the initial equations (3–5) if the self-consistency relation between $\Delta(\mathbf{r}, \mathbf{p}, t)$ and $\kappa(\mathbf{r}, p, t)$ is maintained.

In Ref. 8 we have shown that both problems with the continuity equation and with the energy-weighted sum rule can be avoided by re-defining the density fluctuations through Eq. (19). Although this prescription might seem somewhat arbitrary, it has the merit of respecting both the continuity equation and the energy-weighted sum rule, moreover it does not change the eigenfrequencies of the normal modes of the system.

5. One-dimensional systems

In order to give a more quantitative description of the effects of pairing correlations on the density response of a many-body system, we show here the result of calculations[8] for the response function of finite one-dimensional systems of fermions with given density but different sizes $L = L_1, L_2, L_3$. Starting with L_1, the size is increased by a factor 10 each time, so that $L_2 = 10L_1$ and $L_3 = 10L_2$. The external field $V^{ext}(x)$ is supposed to be of the form $V^{ext}(x) = x^2$.

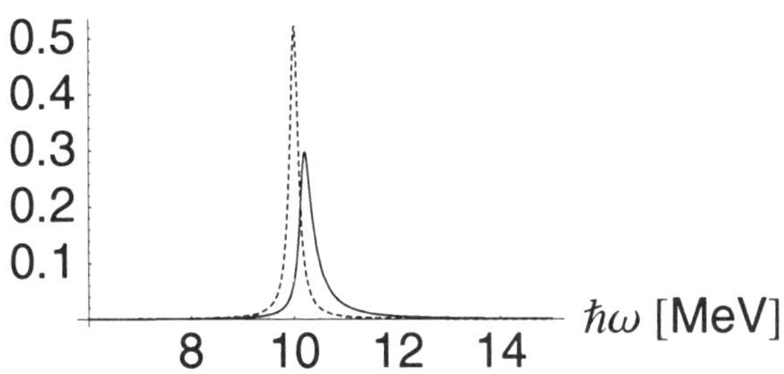

Fig. 1. Uncorrelated (dashed) and correlated (solid) strength functions (divided by L^4), as a function of excitation energy $\hbar\omega$ expressed in MeV, the units for the vertical axes are MeV^{-1}. There are other similar peaks around $\hbar\omega = 20, 30 \ldots$ MeV, but their strength decreases rapidly with increasing excitation energy.

Figure 1 shows the main peak of the response function in the case $L = L_1$ This size is chosen so that the fundamental frequency associated with the motion of the particles in the uncorrelated system, ω_0, gives $\hbar\omega_0 = 10$ MeV, hence $\hbar\omega_0 >> \Delta$. In this case pairing correlations have a rather small effect: essentially they shift slightly the uncorrelated peak towards higher excitation energy, in agreement with[1] .

Figure 2 instead shows the response in the case $L = L_2 = 10\,L_1$. In this case $\hbar\omega_0$ has the same value as the pairing parameter Δ. The pairing gap 2Δ at low excitation is clearly visible.

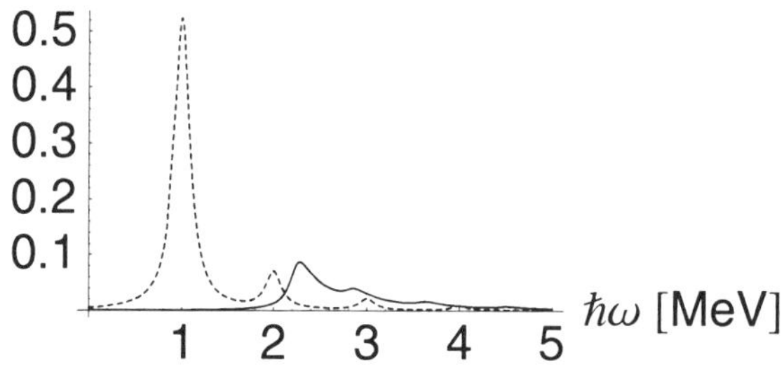

Fig. 2. Same as Fig. 1. The main peak of the uncorrelated strength function (dashed) at $\hbar\omega = 1$ MeV is pushed to higher energy by the pairing correlations and a gap 2Δ is created at low energy.

Finally Fig. 3 shows the response in the case $L = L_3 = 10\,L_2$. In this case $\hbar\omega_0$ is much smaller than Δ and we can clearly see that the correlated response is dominated by a peak around 2Δ. The small tail extending into the gap region is due to an unphysical smoothing parameter introduced for numerical reasons.

It can be easily checked that the correlated zero-order response function obtained here gives exactly the same energy-weighted sum rule as the uncorrelated one.

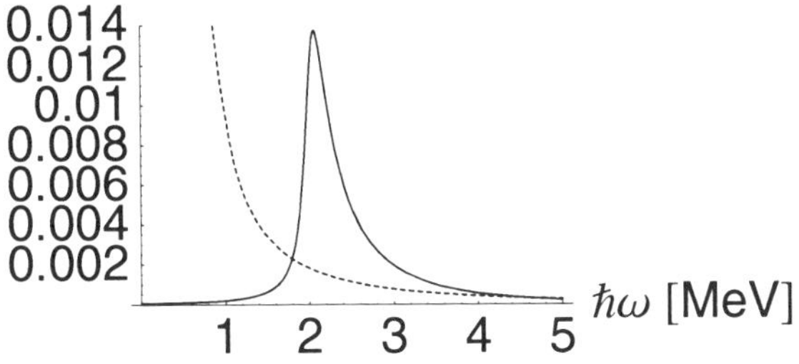

Fig. 3. Same as Fig. 2, but with $\hbar\omega_0 = .1$ MeV. All the low-enegy peaks of the uncorrelated strength function (dashed, practically only the tail of the main peak at $\hbar\omega = .1$ MeV is visible) are pushed to higher energy by the pairing correlations and a gap 2Δ is created at low energy.

6. Conclusions

The solutions of the semiclassical time-dependent Hartree–Fock–Bogoliubov equations have been studied in a simplified model in which the pairing field $\Delta(\mathbf{r}, \mathbf{p}, t)$ is treated as a constant phenomenological parameter. Such an approximation is known to violate both local and global particle-number conservation.

Both symmetries can be restored by introducing a new density fluctuation that is related to the current density by the continuity equation. This prescription changes the strength associated with the various eigenmodes of the density fluctuations, but not the eigenfrequencies of the system. The energy-weighted sum rule calculated according to this prescription has exactly the same value as for normal, uncorrelated systems.

The qualitative features of the response depend essentially on the ratio between two characteristic frequencies of the system: $\Delta/\hbar$ and ω_0. For nuclei, the parameter Δ is fixed by experiment and depends weakly on the size of the system, while the sigle-particle frequency ω_0 strongly depends on the system size. In one dimension, we must distinguish between small and large systems: in small systems, for which $\omega_0 >> \Delta/\hbar$, the effect of pairing correlations is rather small and gives essentially an increase of the value of the (zero-order) system eigenfrequencies, while in larger systems, for which $\omega_0 << \Delta/\hbar$, the zero-order system response is dominated by a single peak at a frequency $\approx 2\Delta/\hbar$. In all cases, the zero-order response (analogous to the quantum single-particle response) displays an excitation energy gap for $0 < \hbar\omega < 2\Delta$. The effects of this gap become important as the system becomes larger.

These considerations will change in three-dimensional sytems because in that case the frequency ω_0 does not depend solely on the system size (for example, in spherical systems there is also a dependence on the value of the particle angular momentum), thus in three dimensions we expect to find features typical both of small and large one-dimensional systems. Work for extending the present method to spherical systems is in progress.

References

1. M. Di Toro and V.M. Kolomietz, *Zeit. Phys. A–Atomic Nuclei* **328**, 285 (1987).
2. M. Urban and P. Schuck, *Phys. Rev. A* **73**, 013621 (2006) .
3. J.R. Schrieffer, *Theory of superconductivity*, (W.A. Benjamin, Inc., New York, 1964).
4. R. Combescot, M. Yu. Kagan, S. Stringari, *Phys. Rev. A* **74**, 042717 (2006).
5. P. Ring and P. Schuck, *The Nuclear Many–Body Problem*, (Springer, New York, 1980).
6. R. Bengtsson and P. Schuck, *Phys. Lett.* **89B**, 321 (1980).
7. D.M. Brink, A. Dellafiore, M. Di Toro, *Nucl. Phys.* **A456**, 205 (1986).
8. V. I. Abrosimov, D. M. Brink, A. Dellafiore and F. Matera, *arXiv:0704.0152*.

SELF-CONSISTENT QUASIPARTICLE RPA FOR MULTI-LEVEL PAIRING MODEL

N. QUANG HUNG*† AND N. DINH DANG

Heavy-Ion Nuclear Physics Laboratory, RIKEN Nishina Center
2-1 Hirosawa, Wako City, 351-0198 Saitama, Japan
Contact e-mail: nqhung@riken.jp

Particle-number projection within the Lipkin-Nogami (LN) method is applied to the self-consistent quasiparticle random-phase approximation (SCQRPA), which is tested in an exactly solvable multi-level pairing model. The SCQRPA equations are numerically solved to find the energies of the ground and excited states at various numbers Ω of doubly degenerate equidistant levels. The comparison between results given by different approximations such as the RPA, SCRPA, QRPA, LNQRPA, SCQRPA and LNSCQRPA is carried out.

1. Introduction

The random-phase approximation (RPA), which includes correlations in the ground state, provides a simple theory of excited states of the nucleus. However, the RPA breaks down at a certain value $G_{\rm cr}$ of interaction parameter G, where it yields imaginary eigenvalues. The reason is that the RPA equations, linear with respect to the X and Y amplitudes of the RPA excitation operator, are derived based on the quasi-boson approximation (QBA). The latter neglects the Pauli principle between fermion pairs and its validity is getting poor with increasing the interaction parameter G. The collapse of the RPA at the critical value $G_{\rm cr}$ of G invalidates the use of the QBA. The RPA therefore needs to be extended to correct this deficiency.

One of methods to restore the Pauli principle is to renormalize the conventional RPA to include the non-zero values of the commutator between the fermion-pair operators in the correlated ground state. These so-called ground-state correlations beyond RPA are neglected within the QBA. The

*RIKEN Asia Program Associate (APA)

†[On leave of absence from the] Institute of Physics and Electronics, Vietnam Academy of Science and Technology, Hanoi, Vietnam

interaction in this way is renormalized and the collapse of RPA is avoided. The resulting theory is called the renormalized RPA (RRPA).[1,2] However, the test of the RRPA carried out within several exactly solvable models showed that the RRPA results are still far from the exact solutions.[2,3]

Recently, a significant development in improving the RPA has been carried out within the self-consistent RPA (SCRPA).[3,4] Based on the same concept of renormalizing the particle-particle (*pp*) RPA, the SCRPA made a step forward by including the screening factors, which are the expectation values of the products of two pairing operators in the correlated ground state. The SCRPA has been applied to the exactly solvable multi-level pairing model, where the energies of the ground state and first excited state in the system with $N+2$ particles relative to the energy of the ground-state level in the N-particle system are calculated and compared with the exact results. It has been found that the agreement with the exact solutions is good only in the weak coupling region, where the pairing-interaction parameter G is smaller than the critical values $G_{\rm cr}$. In the strong coupling region ($G >> G_{\rm cr}$), the agreement between the SCRPA and exact results becomes poor.[3] In this region a quasiparticle representation should be used in place of the *pp* one, as has been pointed out in Ref.[5] As a matter of fact, an extended version of the SCRPA in the superfluid region has been proposed and is called the self-consistent quasiparticle RPA (SCQRPA), which was applied for the first time to the seniority model in Ref.[6] and a two-level pairing model in Ref.[7] However, the SCQRPA also collapses at $G = G_{\rm cr}$. It is therefore highly desirable to develop a SCQRPA that works at all values of G and also in more realistic cases, e.g. multi-level models. The aim of the present work is to construct such an approach. Obviously, the collapse of the SCQRPA at $G = G_{\rm cr}$, which is the same as that of the non-trivial solution for the pairing gap within the Bardeen-Cooper-Schrieffer theory (BCS), can be removed by performing the particle-number projection (PNP). The Lipkin-Nogami method,[8,9] which is an approximated PNP before variation, will be used in such extension of the SCQRPA in the present paper because of its simplicity. This approach shall be applied to a multi-level pairing model, the so-called Richardson model,[3,4] which is an exactly solvable model extensively employed in literature to test approximations of many-body problems.

2. FORMALISM

2.1. *Model Hamiltonian*

The Richardson model was described in detail in Refs.[3,4] It consists of Ω two-fold equidistant levels interacting via a pairing force with a constant parameter G. The model Hamiltonian is given as

$$H = \sum_{j=1}^{\Omega} (\epsilon_j - \lambda) N_j - G \sum_{j,j'=1}^{\Omega} P_j^+ P_{j'}, \tag{1}$$

where $N_j = a_j^+ a_j + a_{-j}^+ a_{-j}$ is the particle-number operator and $P_j^+ = a_j^+ a_{-j}^+$, $P_j = (P_j^+)^+$ are pairing operators. These operators fulfill the following exact commutation relations

$$[P_j, P_{j'}^+] = \delta_{jj'}(1 - N_j), \tag{2}$$

$$[N_j, P_{j'}^+] = 2\delta_{jj'} P_{j'}^+, \quad [N_j, P_{j'}] = -2\delta_{jj'} P_{j'}. \tag{3}$$

By using the Bogolyubov transformation from particle operators a_j^+ and a_j to quasiparticle ones α_j^+ and α_j, the pairing Hamiltonian (1) is transformed into the quasiparlicle Hamiltonian as[10]

$$\begin{aligned} H = a &+ \sum_j b_j \mathcal{N}_j + \sum_j c_j \mathcal{A}_j^+ \mathcal{A}_j \\ &+ \sum_{jj'} d_{jj'} \mathcal{A}_j^+ \mathcal{A}_{j'} + \sum_{jj'} g_j(j')(\mathcal{A}_{j'}^+ \mathcal{N}_j + \mathcal{N}_j \mathcal{A}_{j'}) \\ &+ \sum_{jj'} h_{jj'} (\mathcal{A}_j^+ \mathcal{A}_{j'}^+ + \mathcal{A}_{j'} \mathcal{A}_j) + \sum_{jj'} q_{jj'} \mathcal{N}_j \mathcal{N}_{j'}, \end{aligned} \tag{4}$$

where $\mathcal{N}_j = \alpha_j^+ \alpha_j + \alpha_{-j}^+ \alpha_{-j}$ is the quasiparticle-number operator and $\mathcal{A}_j^+ = \alpha_j^+ \alpha_{-j}^+$, $\mathcal{A}_j = (\mathcal{A}_j^+)^+$ are a pair of time-conjugated quasiparticle operators. The coefficients $a, b_j, c_j, d_{jj'}, g_j(j'), h_{jj'}, q_{jj'}$ in Eq. (4) are given as functions of the Bogolyubov transformation coefficients u_j and v_j (see, e.g. Ref.[10]). The single particle energies are defined as $\epsilon_j = j\epsilon$ ($j = 1 \to \Omega$), with ϵ=1 MeV being the level distance. The chemical potential λ and the coefficients u_j and v_j are determined by solving the gap equations discussed in the next section.

2.2. *Renormalized Lipkin-Nogami gap equations*

The main drawback of the BCS is that its wave function is not an eigenstate of the particle-number operator $\hat{N}$. The BCS, therefore, suffers from an inaccuracy caused by the particle-number fluctuations. The collapse of the BCS at a critical value $G_{\rm cr}$ of the pairing parameter G, below which it has only a trivial solution with zero pairing gap, is intimately related to the particle-number fluctuations within BCS.[9] This defect is cured by projecting out the component of the wave-function that corresponds to the right number of particles. The Lipkin-Nogami (LN) method is an approximated particle-number projection (PNP), which has been shown to be simple and yet efficient in many realistic calculations. This method, discussed in detail in Refs.,[8,9] is an PNP before variation based on the BCS wave function, therefore the Pauli principle between the quasiparticle-pair operators is still neglected within the original version of this method. In the present work, to restore the Pauli principle we propose a renormalization of the LN method, which we refer to as the renormalized LN (RLN) method. The RLN includes the quasiparticle correlations in the correlated ground state $|\bar{0}\rangle$, and the RLN equations are obtained by carrying out the variational calculation to minimize Hamiltonian $\tilde{H} \equiv H' - \lambda\hat{N} - \lambda_2\hat{N}^2$. The RLN equations obtained in this way have the form

$$\tilde{\Delta} = G\sum_j \tilde{\tau}_j, \qquad N = 2\sum_j \tilde{\rho}_j, \tag{5}$$

$$\tilde{\epsilon}_j = \epsilon_j + (4\lambda_2 - G)\tilde{v}_j^2 , \quad \lambda = \lambda_1 + 2\lambda_2(N+1), \tag{6}$$

where

$$\tilde{\tau}_j = \tilde{u}_j\tilde{v}_jD_j , \quad \tilde{\rho}_j = \tilde{v}_j^2D_j + \frac{1}{2}(1 - D_j) , \tag{7}$$

$$\tilde{u}_j^2 = \frac{1}{2}\left(1 + \frac{\tilde{\epsilon}_j - \lambda}{\tilde{E}_j}\right) , \quad \tilde{v}_j^2 = \frac{1}{2}\left(1 - \frac{\tilde{\epsilon}_j - \lambda}{\tilde{E}_j}\right) , \quad \tilde{E}_j = \sqrt{(\tilde{\epsilon}_j - \lambda)^2 + \tilde{\Delta}^2} . \tag{8}$$

The coefficient λ_2 has the following form[11]

$$\lambda_2 = \frac{G}{4}\frac{\sum_j(1-\tilde{\rho}_j)\tilde{\tau}_j\sum_{j'}\tilde{\rho}_{j'}\tilde{\tau}_{j'} - \sum_j(1-\tilde{\rho}_j)^2\tilde{\rho}_j^2}{\left[\sum_j\tilde{\rho}_j(1-\tilde{\rho}_j)\right]^2 - \sum_j(1-\tilde{\rho}_j)^2\tilde{\rho}_j^2}, \tag{9}$$

which becomes the expression given in the original paper[9] of the LN method when $D_j = 1$. The RLN ground-state energy is given as

$$E_{\rm g.s.}^{\rm RLN} = 2\sum_j(\epsilon_j - \lambda)\tilde{\rho}_j - \frac{\tilde{\Delta}^2}{G} - G\sum_j\tilde{\rho}_j^2 - 4\lambda_2\Delta N^2 , \tag{10}$$

where the expression for the particle-number fluctuation ΔN^2 in terms of $\tilde{u}_j$, $\tilde{v}_j$ and $n_j \equiv (1 - D_j)/2$ has been derived in Ref.[10] The RLN equations return to the BCS ones in the limit case when $\lambda_2 = 0$ and $D_j = 1$.

2.3. *SCQRPA equations*

The derivation of the SCQRPA equations is based on the renormalized QRPA (RQRPA) operators as

$$\bar{Q}^{+}_{\nu} = \sum_{j} \left(\frac{\mathcal{X}^{\nu}_{j}}{\sqrt{D_j}} \mathcal{A}^{+}_{j} - \frac{\mathcal{Y}^{\nu}_{j}}{\sqrt{D_j}} \mathcal{A}_{j} \right), \quad \bar{Q}_{\nu} = (\bar{Q}^{+}_{\nu})^{+}, \tag{11}$$

and the renormalized quasiboson approximation (RQBA)

$$\langle \bar{0} | \left[\bar{Q}_{\nu}, \bar{Q}^{+}_{\nu'} \right] | \bar{0} \rangle = \delta_{\nu\nu'}. \tag{12}$$

The SCQRPA equations are presented in the matrix form as follow

$$\begin{pmatrix} A & B \\ B & A \end{pmatrix} \begin{pmatrix} \mathcal{X}^{\nu}_{j} \\ \mathcal{Y}^{\nu}_{j} \end{pmatrix} = \omega_{\nu} \begin{pmatrix} \mathcal{X}^{\nu}_{j} \\ -\mathcal{Y}^{\nu}_{j} \end{pmatrix}, \tag{13}$$

where the SCQRPA sub-matrices are given as

$$\begin{aligned} A_{jj'} = 2 \Bigg[& b_j + 2q_{jj'} + 2\sum_{j''} q_{jj''}(1 - D_{j''}) \\ & - \frac{1}{D_j} \left(\sum_{j''} d_{jj''} \langle \mathcal{A}^{+}_{j''} \mathcal{A}_j \rangle - 2 \sum_{j''} h_{jj''} \langle \mathcal{A}_{j''} \mathcal{A}_j \rangle \right) \Bigg] \delta_{jj'} \\ & + d_{jj'} \sqrt{D_j D_{j'}} + 8q_{jj'} \frac{\langle \mathcal{A}^{+}_{j} \mathcal{A}_{j'} \rangle}{\sqrt{D_j D_{j'}}}, \end{aligned} \tag{14}$$

$$\begin{aligned} B_{jj'} = -2 \Bigg[& h_{jj'} + \frac{1}{D_j} \left(\sum_{j''} d_{jj''} \langle \mathcal{A}_{j''} \mathcal{A}_j \rangle + 2 \sum_{j''} h_{jj''} \langle \mathcal{A}^{+}_{j''} \mathcal{A}_j \rangle \right) \Bigg] \delta_{jj'} \\ & + 2h_{jj'} \sqrt{D_j D_{j'}} + 8q_{jj'} \frac{\langle \mathcal{A}_{j} \mathcal{A}_{j'} \rangle}{\sqrt{D_j D_{j'}}}, \end{aligned} \tag{15}$$

with $\langle \mathcal{A}^{+}_{j} \mathcal{A}_{j'} \rangle = \sqrt{D_j D_{j'}} \sum_{\nu} \mathcal{Y}^{\nu}_{j} \mathcal{Y}^{\nu}_{j'}$, $\langle \mathcal{A}_{j} \mathcal{A}_{j'} \rangle = \sqrt{D_j D_{j'}} \sum_{\nu} \mathcal{X}^{\nu}_{j} \mathcal{Y}^{\nu}_{j'}$ and $D_j = \langle \bar{0} | \alpha^{+}_{j} \alpha_j | \bar{0} \rangle = [1 + \sum_{\nu} (\mathcal{Y}^{\nu}_{j})^2]^{-1}$. The SCQRPA ground state energy is obtained by calculating the expectation value of the Hamiltonian (4) over the quasiparticle vacuum $|\bar{0}\rangle$. The Lipkin-Nogami SCQRPA (LNSCQRPA) equations have the same form as that of the SCQRPA ones given in Eqs. (14) and (15), but the chemical potential and coefficients of the Bogoliubov transformation are determined by solving the RLN gap equations (5), (6) instead of the BCS ones.

3. Analysis of numerical calculations

We carried out the calculations of the ground-state energy, $E_{\rm g.s}$, and energies of excited states, $\omega_\nu \equiv \mathcal{E}_\nu - \mathcal{E}_0$, in the quasiparticle representation using the BCS, QRPA, SCQRPA as well as their renormalized and PNP versions, namely the LN, RLN, RBCS, LNQRPA, and LNSCQRPA, at several values of particle number N. The detailed discussion is given for the case with $N = 10$.

3.1. *Ground-state energy*

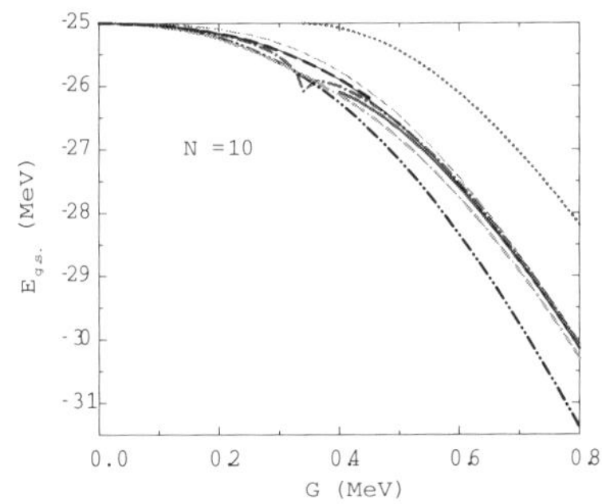

Fig. 1. Ground state energies as functions of G for $N = 10$. The exact result is represented by the thin solid line. The dotted line denotes the BCS result. The thin dashed line stands for the LN result. The thick dashed line depicts the SCRPA result. The dash-dotted line shows the *pp* RPA result at $G \leq G_{\rm cr}^{\rm BCS}$ and the QRPA one at $G > G_{\rm cr}^{\rm BCS}$. The SCQRPA, LNQRPA, and LNSCQRPA are shown by the thick solid, dash–double-dotted, and double-dash–dotted lines, respectively.

Shown in Fig. 1 are the results for the ground-state energies obtained within the BCS, LN, SCRPA, QRPA, LNQRPA, SCQRPA, and LNSCQRPA in comparison with the exact one for $N = 10$. The exact result is obtained by directly diagonalizing the Hamiltonian in the Fock space.[14] It is seen that the BCS strongly overestimates the exact solution. The LN result comes much closer to the exact one even in the vicinity of the BCS (QRPA) critical point, while the QRPA (RPA) result agrees well with the exact solution only at $G \gg G_{\rm cr}^{\rm BCS}$ ($G \ll G_{\rm cr}^{\rm BCS}$). The improvement given by the SCRPA is significant as its result nearly coincides with the exact one in the weak coupling region. However the convergence of the SCRPA solution is getting poor in the strong coupling region. As a result, only the values up to $G \leq 0.46$ are accessible. The SCQRPA result has almost the same quality as that of the QRPA except for the region near the critical point, where it is slightly lower. The LNQRPA strongly underestimates the exact solution while the LNSCQRPA, which includes the effects due to the screening factors in combination with PNP, significantly improves the overall fit. From this analysis, we can say that, among all the approximations undergoing the test to describe simultaneously the ground and excited states, the SCRPA, SCQRPA, and LNSQRPA can be selected as those which fit best the exact ground-state energy. The LN method on the BCS (thin dashed line) also fits quite well the exact one at all G but it does not allow to describe the excited states as the approaches based on the QRPA do. Although the fit offered by the LNSCQRPA in the vicinity of the critical point is somewhat poorer than those given by the SCRPA and the SCQRPA, its advantage is that it does not suffer any phase-transition point due to the violation of particle number as well as the Pauli principle.

3.2. *Energies of excited state*

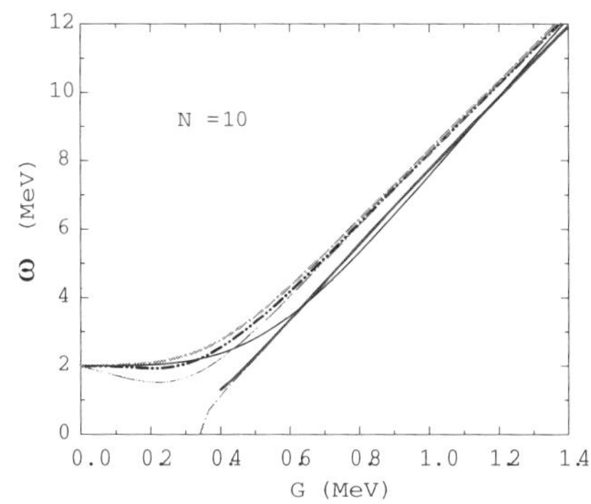

Fig. 2. The first excited state energy as functions of G at N=10. The results refer to the exact solution, $\omega_2^{\rm ex}$ (solid line), the QRPA solution, $\omega_2^{\rm QRPA}$ (dash-dotted line), the SCQRPA solution, $\omega_2^{\rm SCQRPA}$ (thick solid line), the LNQRPA solutions, $\omega_2^{\rm LNQRPA}$ (thin dash – double-dotted line) and $\omega_3^{\rm LNQRPA}$ (thick dash – double-dotted line), as well as the LNSCQRPA solutions, $\omega_2^{\rm LNSCQRPA}$ (thin double-dash – dotted line) and $\omega_3^{\rm LNSCQRPA}$ (thick double-dash – dotted line).

As discussed in Refs.,[7,12] the first solution ω_1 of the QRPA or SCQRPA equations is the energy of spurious mode, which is well separated from the physical solutions ω_ν, with $\nu \geq 2$. The first excited state energy is therefore given by ω_2. As has been discussed in Ref.,[14] the coupling in the small-G region causes only small perturbations in the single-particle levels. With increasing G the system goes to the crossover regime, where level splitting and crossing are seen. In the strong coupling regime the levels coalesces into narrow well-separated band. The approaches based on the QRPA with PNP within the LN method also splits the levels but the nature of the splitting comes from the two components featuring the addition and removal modes within the QRPA operator. One can see that within the LN(SC)QRPA each single level at $G = 0$ splits into two components in the small-G region, e.g. the pair $\omega_2^{\rm LNQRPA}$ and $\omega_3^{\rm LNQRPA}$ or $\omega_2^{\rm LNSCQRPA}$ and $\omega_3^{\rm LNSCQRPA}$ in Fig. 2. Here $\omega_2^{\rm LN(SC)QRPA}$ and $\omega_3^{\rm LN(SC)QRPA}$ can be identified by the dominations of the addition and removal modes, respectively. It is clear to see in Fig. 2 that, in the weak coupling region, the level $\omega_3^{\rm LN(SC)QRPA}$, which is generated mainly by the addition mode, fits well the exact result, while the agreement between $\omega_2^{\rm EXACT}$ and $\omega_2^{\rm QRPA}$ as well as $\omega_2^{\rm SCQRPA}$ is good only in the strong coupling region. At large values of G, predictions by all approximations and the exact solution coalesce into one band, whose width vanishes in the limit $G \to \infty$.

References

1. K. Hara, Prog. Theor. Phys. **32**, 88 (1964); K. Ikeda, T. Udagawa, and H. Yamamura, *ibid.* **33**, 22 (1965); D. J. Rowe, Phys. Rev. **175**, 1283 (1968); P. Schuck and S. Ethofer, Nucl. Phys. A **212**, 269 (1973).
2. F. Catara, N. D. Dang, and M. Sambataro, Nucl. Phys. A **579**, 1 (1994)
3. J. Dukelsky and P. Schuck, Phys. Rev. Lett. **B464**, 164 (1999); J. G. Hirsch, A. Mariano, J. Dukelsky, and P. Schuck, Ann. Phys. (NY) **296**, 187 (2002).
4. N. D. Dang, Phys. Rev. C **71**, 024302 (2005).
5. N.D. Dang and K. Tanabe, Phys. Rev. C **74**, 034326 (2006).
6. J. Dukelsky and P. Schuck, Phys. Lett. **B387**, 233 (1996)
7. A. Rabhi, R. Bennaceur, G. Chanfray, and P. Schuck, Phys. Rev. **C66**, 064315(2002).
8. H. J. Lipkin, Ann. Phys. (NY) **9** 272 (1960); Y. Nogami and I. J. Zucker, Nucl. Phys. **60** 203 (1964); Y. Nogami, Phys. Lett. **15** 4 (1965); J. F. Goodfellow and Y. Nogami, Can. J. Phys. **44** 1321 (1966).
9. H.C. Pradhan, Y. Nogami, and J. Law, Nucl. Phys. **A201**, 357 (1973)
10. N. D. Dang, Z. Phys. A **335**, 253 (1990).
11. M. V. Stoitsov, J. Dobaczewski, W. Nazarewicz, S. Pittel and D. J. Dean, Phys. Rev. C **68**, 054312 (2003)
12. N. D. Dang, Eur. Phys. A **16**, 181 (2003).
13. A. Volya, B. A. Brown, V. Zelevinsky, Phys. Lett. **B 509**, 37 (2001)
14. E. A. Yuzbashyan, A. A. Baytin, B. L. Altshuler, Phys. Rev. **B 68**, 214509 (2003).

SECTION V

SPECIAL TOPICS

THE LIQUID-VAPOR PHASE DIAGRAM OF INFINITE UNCHARGED NUCLEAR MATTER

L. G. MORETTO, J. B. ELLIOTT, L. PHAIR

Nuclear Science Division, Lawrence Berkeley National Laboratory, Berkeley, CA, USA
**E-mail: lgmoretto@lbl.gov*

We have extracted the critical parameters of infinite uncharged nuclear matter from experimental nuclear fragmentation data. To do this, three obstacles had to be overcome: finite size effects, the Coulomb interaction, and an appropriate physical picture (not particles contained in a box) that reflects particle emission into a vacuum.

Keywords: liquid-vapor phase transition, multifragmentation, finite size effects

1. Introduction

After many decades of theoretical and experimental studies, we have recently published a quantitative, credible liquid-vapor phase diagram containing the coexistence line up to the critical temperature.[1] This diagram has not been obtained, as could have been expected, through the study of caloric curves[2,3] or anomalous heat capacities.[4,5] Rather, it was generated from the fitting of the charge distributions in multifragmentation by means of a modified Fisher's formula[1,6] giving the cluster composition of a vapor:

$$n_A(T) = q_0 A^{-\tau} \exp\left[\frac{\Delta\mu A}{T} - \frac{c_0 \epsilon A^{\sigma}}{T}\right], \tag{1}$$

where: q_0 is a normalization constant;[6] τ is the critical exponent giving rise to a power law at criticality; A is the cluster number; $\Delta\mu$ is the difference of chemical potentials between the liquid and the vapor; c_0 is the surface energy coefficient; T is the temperature; $\epsilon = (T_c - T)/T_c$ is the distance from the critical temperature T_c; σ is another critical exponent (expected to be approximately 2/3, if one interprets the second term in the exponent as the surface energy of a cluster of mass A divided by the temperature). For $\Delta\mu = 0$ the liquid and the vapor are in equilibrium and Eq. (1) is the equivalent of the coexistence line. In fact, one can immediately obtain from Eq. (1) the usual p, T and ρ, T phase diagrams by recalling that in Fisher's model, the clusterization is assumed to exhaust all the non-idealities of the gas which then becomes an ideal gas of clusters. Consequently, the total pressure and density can be calculated

$$p(T) = \sum_A p_A(T) = T \sum_A n_A(T), \tag{2}$$

$$\rho = \sum_A A n_A(T), \tag{3}$$

as well as the corresponding scaled quantities p/p_c and ρ/ρ_c.

In order to make use of the Fisher description of the cluster concentrations there were three major obstacles to overcome: finite size effects, the problem of the long-range Coulomb interaction, and the (un)physical picture of particles in a box.

2. Finite size effects and the complement correction

Finite size effects are essential to the study of nuclei and other mesoscopic systems for two opposite but complementary reasons. In the nuclear case finite size effects dominate the picture macroscopically and spectroscopically at all excitation energies. The challenge is to reduce the vast amount of specific knowledge of each "drop" (nucleus) to a general characterization of uncharged, symmetric infinite nuclear matter. This reduction has been successfully achieved for cold nuclei ($T = 0$) in determining the saturation binding energy and density by means of the liquid drop model.

Here we present a general method (complement method) to deal with finite size effects in phase transitions.[7] In the case of liquid-vapor phase coexistence, a dilute nearly ideal vapor phase is in equilibrium with a dense liquid-like phase. The interesting case of finiteness is realized when the liquid phase is a finite drop. The vapor pressure of a drop can be calculated by correcting the molar vaporization enthalpy to include the surface energy of the drop.[8] The complement method (the residual drop which remains after a cluster has been emitted) further quantifies finite size effects and generalizes Fisher's theory to describe the cluster concentrations in equilibrium with extremely small systems. The complement approach consists of evaluating the change in free energy occuring when a particle or cluster is moved from one (finite) phase to another. In the case of a liquid drop in equilibrium with its vapor, this is done by extracting a vapor particle of any given size from the drop and evaluating the energy and entropy changes associated with both the vapor particle and the residual liquid drop (the complement).

This detailed accounting can be easily generalized to incorporate other energy terms present in the nuclear case, such as symmetry energy, Coulomb energy and even angular momentum.[9,10] In order to demonstrate the power of this method, we apply it to the Ising model, where a great deal of work exists on the subject of finiteness, in particular on the dependence of critical quantities on the lattice size.[11,12]

We concentrate first on the canonical lattice gas representation of the Ising model: a fixed number of up spins represents the occupied lattice sites (i.e., matter in the form of monomers, dimers, large drops etc.), while down spins are empty space. In a finite lattice, we can fix the mean density of occupied sites in such a way that, below the coexistence temperature T_0, there is a large cluster or drop of a certain size in equilibrium with its vapor, populated mostly by monomers. Two dimensional Ising calculations were performed for fixed magnetization (equivalently, fixed mean occupation density). Periodic boundary conditions were used to minimize the global finite size effects of the container in contrast with the droplet-vapor situation realized within the container where the finite drop size effect remains in force. At the lowest temperatures the up spins congregate into one liquid drop in a vacuum. At higher temperatures, the volume is filled with a vapor made from clusters of up spins. Clusters in the vapor were identified via the Coniglio-Klein algorithm[13] to insure that they were physical (i.e., cluster distributions return Ising critical exponents and not percolation exponents). The largest cluster was formed via a geometric clustering algorithm (*all* spin up nearest neighbors are bonded together) and identified as the liquid drop. The largest cluster is identified geometrically and not via the Coniglio-Klein prescription because we do not want to include clusters associated with the skin thickness of the liquid into our description of the vapor.[14]

Rather than consider the vapor globally, we examine the vapor in detail. Fisher's droplet model[6] describes the cluster concentration and can be written as

$$\begin{aligned} n_A(T) &= g(A) \exp\left(-\frac{c_0 A^\sigma}{T}\right) \\ &= q_0 A^{-\tau} \exp\left(c_0 A^\sigma \left[\frac{1}{T_c} - \frac{1}{T}\right]\right) \end{aligned} \tag{4}$$

where q_0 is the normalization, $c_0 A^\sigma$ is the surface energy of the cluster and $g(A)$ is the degeneracy of the clusters of size A (the number of ways the cluster A can be realized through different surface configurations) and is approximately $A^{-\tau}\exp(c_0 A^\sigma/T_c)$. The resulting surface entropy $S(A)$ is given by

$$S(A) \approx \ln g(A) \approx -\tau \ln A + \frac{c_0 A^\sigma}{T_c}. \quad (5)$$

Eq. (5) is a remarkably felicitous asymptotic expansion. The presence of a leading term in S proportional to A^σ permits the vanishing of the cluster free energy at a $T = T_c$ independent of cluster size. This expression, valid for a vapor in equilibrium with the infinite liquid, must be generalized for equilibrium with a finite liquid. This can be done by making a preliminary observation: in Fisher's expression the abundance of each cluster in the vapor follows a Poisson distribution by construction. This is because the resulting partial vapors are non-interacting with themselves and with each other. In the Ising/lattice gas model this same situation prevails to a surprising degree as can be seen in ref.[15] The Poisson nature of the multiplicity distributions allows us to introduce the concept of the complement.

Consider a vapor in equilibrium with a drop of its liquid. The system may be a physical system or the Ising realization of it. For each cluster of the vapor we can make the mental exercise of extracting it from the liquid, determining the change in entropy and energy of the drop and cluster system, and then putting it back in the liquid (the equilibrium condition), as if all other clusters of the vapor did not exist. Fisher's expression can now be written for a drop of size A_0 in equilibrium with its vapor as follows[7]

$$n_A(T) = q_0 \left[\frac{A\,(A_0 - A)}{A_0}\right]^{-\tau} \exp\left\{[A^\sigma + (A_0 - A)^\sigma - A_0^\sigma]\left(\frac{c_0}{T_c} - \frac{c_0}{T}\right)\right\}. \quad (6)$$

In other words, we treat the "complement" $(A_0 - A)$ in the same fashion as a cluster. The resulting expression reduces to Eq. (4) when A_0 tends to infinity. Notice that Eq. (6) contains the same T_c as that of the infinite system.

We are now prepared to compare the Ising yields of the vapor concentrations to our modified Fisher's droplet model. An example is given in Fig. 1. The yields in the upper panel show the linear Boltzmann dependence on $1/T$ with different slopes for each fragment. In the lower panel the ordinate is scaled by the Fisher power law and the abscissa is scaled to reflect the total surface energy over the temperature. The scaling works rather well.

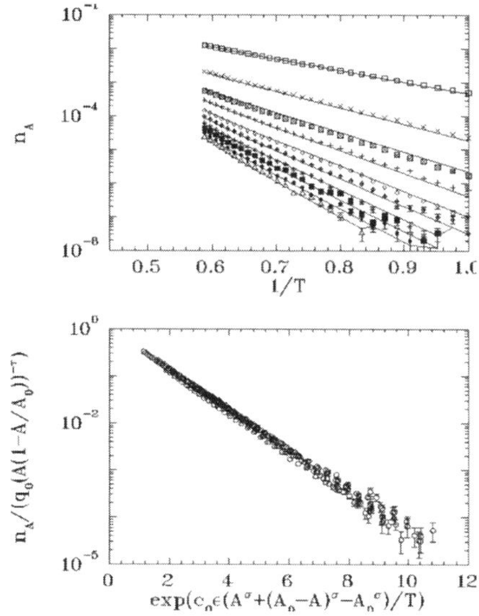

Fig. 1. Upper panel: the yields of clusters of size one (open squares) up to clusters of size ten (open triangles) as a function of temperature from a two dimensional lattice gas calculation (lattice of 40 by 40) at constant density $\rho = 0.2$ (corresponding to a cold liquid drop of $A_0 = 80$) as a function of $1/T$. The lines are fits to the yields using the modified Fisher droplet model described in Eq. (6). Lower panel: the same yields scaled by the power law on the ordinate and the surface energy over temperature on the abscissa.

From the Fisher droplet model fits to the yields comes an estimate of the critical temperature of 2.32 ± 0.02. This is to be compared to the known analytical solution of $T_c = 2/\ln(1+\sqrt{2}) = 2.269...$ Without the complement correction to the Fisher droplet model, a fit gives results of similar quality but a critical temperature of 2.07 ± 0.05, clearly in worse agreement with the known theoretical value.

The line represented by the scaled behavior shown in the lower panel of Fig. 1 contains all the information regarding coexistence between the finite liquid and vapor phases in the lattice gas model. Using that information, the coexistence pressure and density can be calculated (see the solid line in Fig. 2). The agreement between the Fisher description and the Ising calculation (open symbols) is quite good. Furthermore, one can take the finite value for A_0 used in the modified Fisher droplet model of Eq. (6), and let it go to infinity to recover the standard Fisher description for coexistence with an infinite liquid. This is shown by the long dashed curve in Fig. 2. The agreement with the known analytic solution of Onsager[16] (dotted curve) is

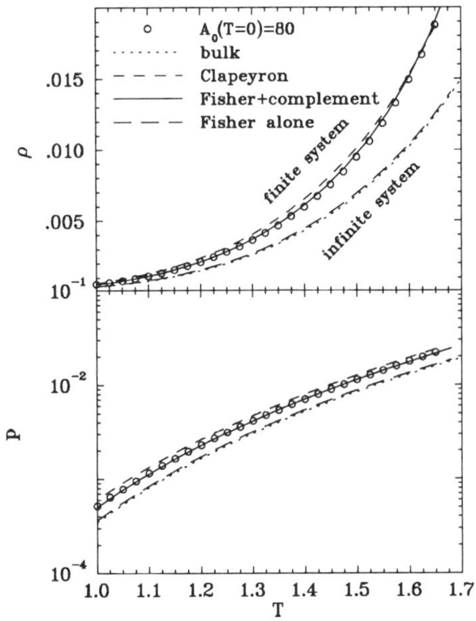

Fig. 2. The density ρ and pressure p of a vapor in coexistence with a finite drop A_0 whose size at $T = 0$ is 80 (symbols) from 2d Ising calculations at fixed magnetization. The dotted lines are the analytic bulk solutions known for the Ising model in two dimensions.[16] The solid line comes from fits to the cluster yields using the modified Fisher formula to include the complement (see Eq. (6)). The long dashed curve comes from the normal Fisher description (see Eq. (4)) of the cluster yields.

remarkable. This shows that the complement method provides a practical way of moving from finite to infinite systems with the Fisher droplet model.

3. Coulomb correction

In ref.,[9] we explored the effects of the Coulomb interaction upon the nuclear liquid phase transition. Because large nuclei are Coulomb metastable, phases, phase coexistence, and phase transitions cannot be defined with any generality and the analogy to liquid vapor is ill-posed for these heavy systems. This realization puts in jeopardy most of the work on nuclear liquid-vapor equilibrium in literature. However, we have shown that the Coulomb effect can be divided out in the decay rates and thus obtain the coexistence phase diagram for the corresponding uncharged system.

4. Physical picture

We do not believe that there is nowadays an experimentally achievable environment where nuclear liquid and vapor coexist in equilibrium. Rather, we make a time-honored assumption which we do not justify other than through the clarification it brings to the experimental picture: we assume (just as in compound nuclear decay) that after prompt emission in the initial phase of the collision, the resulting system relaxes in shape and density and thermalizes on a time scale shorter than its thermal decay. At this point, the excited nucleus emits particles in vacuum, according to standard statistical decay rate theory. In this picture, there is no surrounding vapor and no confining box; there is no need for either. By studying the outward flux of the first fragments emitted, we can study the nature of the vapor even when it is absent (the virtual vapor) because of the equivalence of the evaporation and condensation fluxes of a liquid in equilibrium with its saturated vapor. Quantitatively, the concentration $n_A(T)$ of any species A in the vapor is related to the corresponding decay rate $R_A(T)$ (or to the decay width Γ_A) from the nucleus by matching the evaporation and condensation fluxes

$$R_A(T) = \frac{\Gamma_A(T)}{\hbar} = n_A(T) \left\langle v_A(T) \sigma_{\mathrm{inv}}(v_A) \right\rangle , \qquad (7)$$

where $v_A(T)$ is the velocity of the species A (of order $(T/A)^{1/2}$) crossing the nuclear interface represented by the inverse cross section σ_{inv}. The temperature T_0 of the equilibrated, excited nucleus when the first fragment is emitted can be estimated by the thermometric equation of a Fermi gas and the calorimetrically measured excitation energy E^* such that $T_0 = \sqrt{E^*/a}$, allowing for a weak dependence of a on T and remembering that the system is most likely still in the Fermi strong degeneracy regime (where the temperature is much less than the Fermi energy: $T << \epsilon_F$).

This is the fundamental and simple connection between Eq. (7), the (compound nucleus) decay rate, and Eq. (6), the modified Fisher droplet model. In the latter, one immediately recognizes in the exponential the canonical expansion of the standard compound nucleus decay rate, namely, the Boltzmann factor. The surface part of the barrier is isolated from all other components, e.g., Coulomb,[9] symmetry, and finite size.[7] Thus, the vapor phase in equilibrium can be completely characterized in terms of the decay rate.

The physical picture described above is valid instantaneously, but not globally. The result of a global or successive evaporation in vacuum leads to abundances of various species of emitted fragments that arise from a continuum of systems at different temperatures.[17] This leads to complications in various thermometers: kinetic energy, isotope ratios, etc. One way to avoid this complication is to consider only fragments that are emitted very rarely so that, if they are not emitted first, they are effectively not emitted at all. In other words, we consider only fragments that by virtue of their large surface energy have a high emission barrier.

5. Infinite nuclear matter

The above solutions for the problems posed by finite size effects and the long-range Coulomb interaction along with an appropriate physical picture, allow us to attempt a Fisher description of yields measured in different nuclear reactions.

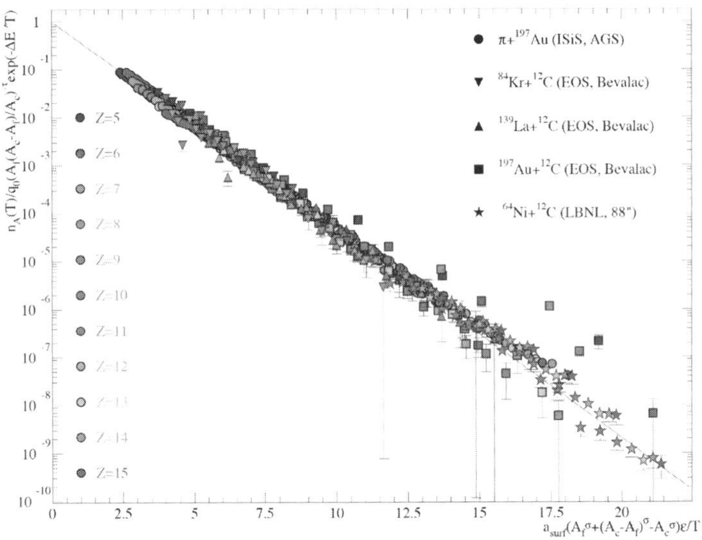

Fig. 3. The scaled yields from several different reactions, ranging from compound nucleus reactions to AGS energies, are plotted as a function of total surface energy over temperature.

Preliminary results are shown in Fig. 3 for the indicated reactions. When the yields are scaled by the Coulomb, rotational, symmetry energies, etc. (all terms that are not surface energy), and plotted as a function of the Boltzmann factor associated with surface energy, the different reactions show a similar scaling. This scaled line contains all the information of coexistence between the two phases for chargeless, symmetric, and infinite nuclear matter. The data can be fit to obtain an estimate of T_c and such a procedure yields $T_c = 18.5 \pm 1.8$ MeV. Using the ideal gas approximations (Eqs. (2) and (3)), estimates can be obtained for the critical density ρ_c and critical pressure p_c. Such calculations give $\rho_c = 0.077 \pm 0.018$ fm^{-3} and $p_c = 0.41 \pm 0.18$ MeV/fm^3.

With the results described herein, we are poised to make a definitive, experimentally based measure of the phase diagram of infinite, neutral nuclear matter. Future work will explore the sensitivity of these results to model assumptions.

References

1. J. B. Elliott *et al.*, Phys. Rev. Lett. **88**, 042701 (2002).
2. J. Pochodzalla *et al.*, Phys. Rev. Lett. **75** , 1040 (1995).
3. Chernomoretz *et al.*, Phys. Rev. C **64**, 044605 (2001).
4. F. Gulminelli *et al.*, Phys. Rev. Lett. **82**, 1402 (1999).
5. M. Dagostino *et al.*, Phys. Lett B **473**, 219 (2000).
6. M. E. Fisher, Rep. Prog. Phys. **30**, 615 (1969).
7. L. G. Moretto *et al.*, Phys. Rev. Lett. **94**, 202701 (2005).
8. L. Rayleigh, Philos. Mag. **34**, 94 (1917).
9. L. G. Moretto *et al.*, Phys. Rev. C **68**, 061602 (2003).
10. J. B. Elliott *et al.*, to be published.
11. M. E. Fisher and A. E. Ferdinand, Phys. Rev. Lett **19**, 169 (1967).
12. D. P. Landau, Phys. Rev. B **13**, 2997 (1976).
13. A. Coniglio and W. Klein, J. Phys. A **13**, 2775 (1980).
14. J. B. Elliott *et al.*, Phys. Rev. C **71**, 024607 (2005).
15. C. M. Mader *et al.*, Phys. Rev. C **68**, 064601 (2003).
16. L. Onsager, Phys. Rev. **65**, 117 (1944).
17. L. G. Moretto *et al.*, Phys. Rev. C **72**, 064605 (2005).

BREMSSTRAHLUNG ACCOMPANIED α DECAY OF ^{210}Po

H. BOIE[1], H. SCHEIT[1], U. D. JENTSCHURA[1], F. KÖCK[1], M. LAUER[1],
A. I. MILSTEIN[2], I. S. TEREKHOV[2] and D. SCHWALM[1,*]

[1] *Max-Planck-Institut für Kernphysik, D-69117 Heidelberg, Germany*
[2] *Budker Institute of Nuclear Physics, 630090 Novosibirsk, Russia*

We report on a high-statistics measurement of bremsstrahlung emitted in the α decay of ^{210}Po. The measured differential emission probabilities, which we could follow up to γ-energies of $\sim$ 500 keV, allow for the first time for a serious test of various model calculations of bremsstrahlung accompanied α decay.[1] Taking into account the interference between the electric dipole and quadrupole amplitudes, which we calculated within the framework of a refined quasi-classical approximation[2] and which modifies the angular correlation between the α particle and the emitted photon , we find good agreement of the measured γ-emission probabilities with those calculated in our quasi-classical model as well as with the fully quantum mechanical prediction of Ref. 3.

Keywords: α decay, bremsstrahlung, quasi-classical approach

1. Introduction

When discussing the α decay of a nucleus the "classical" picture one usually has in mind is that of an α particle tunnelling through the Coulomb barrier and then being accelerated from the classical turning point to its final energy. Thus Bremsstrahlung photons will be emitted during the process. But while the emission of soft photons during the Coulomb scattering of charged particles can be well described by classical electrodynamics, in α decay the implication of a trajectory partly located in a classically forbidden region immediately provokes the question: Do α particles emit photons during tunneling?

Several remarks seem to be in place with regard to this question: As the wavelengths of the photons are much larger than the extend of the tunneling barrier and even much larger than the main classical acceleration region,

*Present address: The Weizmann Institute of Science, 76100 Rehovot, Israel.
E-mail: schwalm@mpi-hd.mpg.de

it is in principle not possible to identify where the photon was emitted. This is clearly born out in the quantum mechanical perturbation approach. Here Fermi's Golden Rule provides a well defined way how to calculate the emission probability connected with the decay of the initial state (the mother nucleus) into the final state consisting of the daughter nucleus, an α particle and a photon, and the calculation of the transition matrix element involves, of course, the integration over the full coordinate space. On the other hand, within the quasi-classical approximation, which is well justified as the Sommerfeld parameter η of the α particle is large compared to unity ($\eta = 22$ for the α emitter ^{210}Po), different space regions can be connected to different time intervals. It is therefore tempting to split the transition matrix element into contributions from classically allowed and classically forbidden regions. However, such an interpretation can only have a restricted meaning because it is possible to rewrite the Bremsstrahlung matrix element in different forms using operator identities. As a result, the integrand for the matrix element, as well as the relative contributions of the regions of integration, will be different though the final outcome will be the same. Nevertheless, the issue of the tunneling during the emission process was widely discussed [3–10]. These authors used different theoretical approaches leading, not surprisingly, to partly conflicting results as to the relative contribution of the tunneling, but — more seriously — also with regard to the total γ emission probabilities.

The interest in the bremsstrahlung accompanied α decay was actually stirred up in 1994 when a first attempt to observe this rare decay mode was published [11]. But this and later experimental attempts to measure

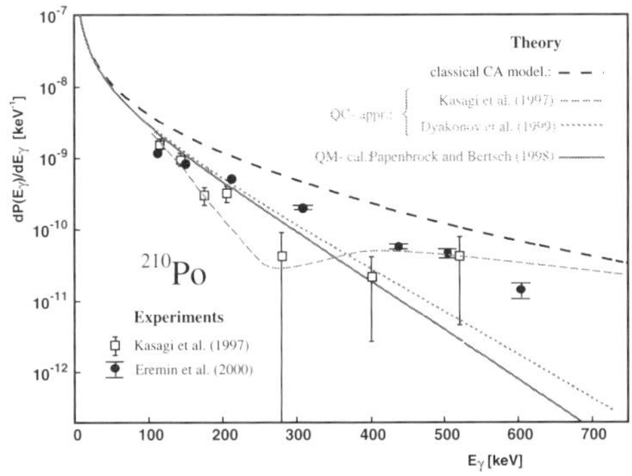

Fig. 1. Experimental and theoretical situation of the bremsstrahlung accompanied α decay of ^{210}Po before the present experiment was performed.

these elusive decays produced inconsistent results and did not reach the sensitivity to allow for a serious test of the various theoretical predictions. The unsatisfactory situation accounted prior to our investigations for the best studied case ^{210}Po is displayed in Fig. 1: Kasagi *et al.* [4] were the first to measure the bremsstrahlung accompanied α decay of ^{210}Po and to calculate the emission probability in a quasi-classical approximation. While the kind of interference pattern implied by their results was neither observed in the quantum mechanical (QM) calculation of Papenbrock and Bertsch [3] nor in the quasi-classical (QC) approach of Dyakonov [6], all calculations are certainly not in conflict with the data given the large error bars. The picture was getting confused when Eremin *et al.* [12] presented their data on ^{210}Po, which is not only in contradiction with the previous data but also with all theoretical predictions, the exception being the classical Coulomb acceleration (CA) model (see e.g. Ref. 6) expected to represent an absolute upper limit for the emission probabilities. In an attempt to clarify the situation we have therefore reinvestigated the bremsstrahlung accompanied α decay of ^{210}Po experimentally [1] as well as theoretically [2].

2. Experimental Procedure

The main experimental challenge is the very low emission probability for bremsstrahlung photons. Even with a 0.1 MBq strong ^{210}Po source, the emission rate is only about one photon per day in the energy range of 300-400 keV. Thus these measurements require a very sensitive experimental setup and stable running conditions for many months.

The principle of our measurement is explained in Fig. 2: Two extended ^{210}Po sources of 0.1 MBq each are placed in a common vacuum chamber and are viewed by two segmented 5×5 cm^2 Silicon detectors, each placed about 30 mm above the source to measure the energy of the α particles. Directly below the centre of the vacuum chamber an efficient high-purity Germanium triple cluster detector of the MINIBALL design [14] was placed to record the emitted bremsstrahlung photons in coincidence with the α particles (see Ref. 1 for a more detailed description of the experimental setup). Energy and momentum conservation can then be employed to distinguish the rare bremsstrahlung events from random coincidences between the α's and unavoidable background photons, and from true α-γ coincidences involving the 803 keV branch. Plotting the measured α energies versus the energy of the photons, the bremstrahlung events should show up on a diagonal line approximately given by $E_\alpha \approx E_{\alpha,0} - 206/210 E_\gamma$.

The measured α-γ coincidence matrix is shown in Fig. 3. The data were

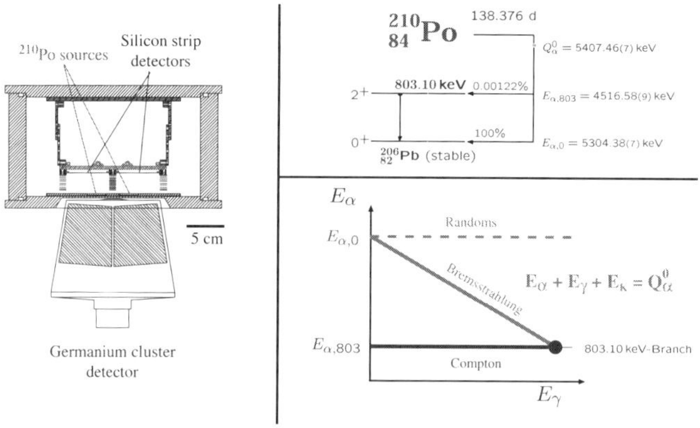

Fig. 2. Principle of the present experiment. Left panel: Cross-section of the experimental setup. Upper right panel: Properties of the ^{210}Po source (from Ref. 13). Lower right panel: Plotting the measured α energies vs. the energies of the photons, the coincident bremsstrahlung events will be located on a diagonal line given by energy and momentum conservation.

collected in 270 days, during which 4.31×10^{11} α particles were recorded. Due to the good energy and time resolution of the Germanium cluster detector and the excellent α linehape, which could be maintained during the long data taking period, the bremstrahlung photons can be well distinguished for γ energies $E_\gamma \geq 90$ keV from α-X-ray coincidences, which are caused by the knockout of K electrons from the Pb nuclei by the escaping α particles [16], and from the random band of α-γ coincidences.

To determine the differential bremsstrahlung emission probabilities dP/dE_γ, coincident α-γ events with γ energies within 20 keV up to 100 keV

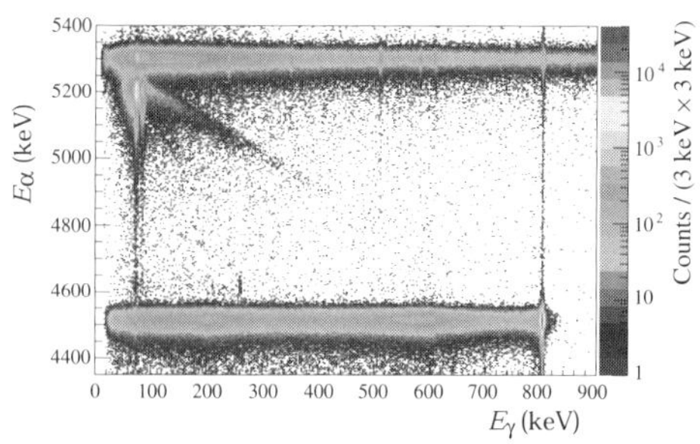

Fig. 3. Two-dimensional spectrum of α- *vs.* γ-energies, detected coincident within a time window of ± 50 ns. The bremsstrahlung events can be clearly seen on the diagonal line starting at $E_\alpha = E^0_\alpha = 5304\ keV$.

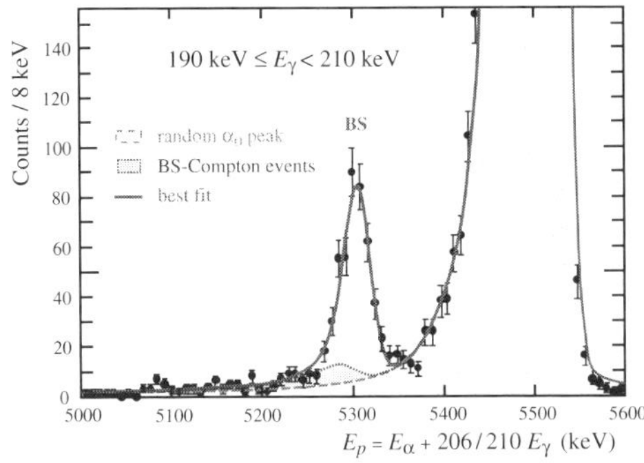

Fig. 4. Projected data for a γ energy bin 190 keV $\leq E_\gamma \leq$ 210 keV. The peak (BS) centred around 5304 keV corresponds to the full-energy peak of photons from the bremsstrahlung accompanied α decay.

broad gates were projected along the diagonal line by plotting them as a function of $E_p = E_\alpha + (206/210)E_\gamma$. In the projected spectra the bremsstrahlung events are expected to show up in a sharp peak around $E_p = E_\alpha^0 = 5304$ keV independent of the width and position of the γ energy gate. As an example, the projected energy spectrum for the 190 keV $\leq E_\gamma < 210$ keV bin is shown in Fig. 4: Riding on the low energy tail of the random α-γ line the bremsstrahlung events are clearly born out. The solid curve corresponds to the result of a least-squares fit of the data, where only the intensity and central energy of the bremsstrahlung peak was varied. The line-shape and intensity of the random α-γ peak was determined from the corresponding projected energy spectrum of the random α-γ matrix, scaled by the ratio of the widths of the respective time windows. The shaded area reflects the Compton distribution caused by bremsstrahlung events with higher energies; its intensity relative to the full energy events and its shape was determined from detailed Monte-Carlo simulations [15].

From the intensity of the bremsstrahlung line the solid angle integrated differential emission probability $dP(E_\gamma)/dE_\gamma$ can be determined if the probability to detect a bremsstrahlung photon of energy E_γ is known. This needs two ingredients: Firstly, the photons are detected preferentially at backward angles with respect to the α particle direction, hence the α-γ correlation has to be taken into account. Secondly, the extended source and the close source-detector geometry required a detailed simulation of the experimental setup.

Usually bremsstrahlung is assumed to be pure $E1$ radiation as the wavelength of the emitted radiation is much larger than the dimension of the

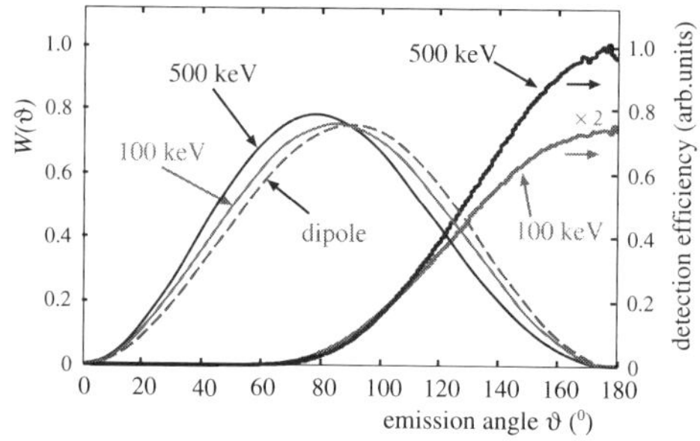

Fig. 5. Calculated α-γ correlation functions $W(\vartheta)$ for Bremsstrahlung photons of 100 and 500 keV in comparison with a pure dipole correlation. The increasing deviation from the pure $E1$ characteristic is due to the $E2$ interference. Also displayed are the ϑ dependent efficiencies (histograms) of our setup for photon energies of 100 and 500 keV.

radiating system and higher order multipole contributions are suppressed. However, in the present case the effective dipole charge amounts only to $Z_{E1}^{\text{eff}} = \mu(z/m - Z/M) = 0.40$, where z, m and Z, M denote the charge and mass of the α and the daughter nucleus, respectively, and μ the reduced mass, while the effective quadrupole charge $Z_{E2}^{\text{eff}} = \mu^2(z/m^2 + Z/M^2) = 1.95$ is almost a factor of 5 larger. Nevertheless, the $E2$ radiation still contributes even at $E_\gamma \sim 500$ keV only less than 1.5% to the total angle integrated emission probability [2], but the interference between the $E1$ and $E2$ amplitudes does result in a sizable contribution to the α-γ angular correlation. Including only the interference term the angular correlation can be expressed by $dP(\vartheta)/d\Omega \propto \sin^2\vartheta(1 + 2\chi(E_\gamma)\cos\vartheta)$, where ϑ denotes the angle between the direction of the α particle and the photon. The quantity $\chi(E_\gamma)$ is proportional to the ratio of the quadrupole to the dipole matrix element. In leading order in $1/\eta_\alpha$ we find [2] that $\chi(E_\gamma)$ increases with E_γ to take e.g. values of $+0.09$ at 100 keV up to $+0.22$ at 500 keV. The influence of the $E2$ amplitude on the angular correlation is illustrated in Fig. 5.

To determine the γ detection efficiency of our setup as a function of ϑ and E_γ, detailed simulations [15] were performed based on GEANT4 [17,18]; they were checked against source measurements and in particular against the well known 803 keV branch. The simulated acceptance at E_γ= 100 keV and 500 keV is plotted in Fig. 5.

3. Results and Discussion

The resulting solid-angle integrated and efficiency corrected differential emission probabilities $dP(E_\gamma)/dE_\gamma$ are displayed in Fig. 6 by the solid points. The 1σ errors shown comprise the statistical and systematic uncertainties and are smaller than the point size for γ energies below 250 keV; they are dominated by systematic uncertainties at small and by the statistical error at high γ energies. Note, that external bremsstrahlung contributions, which stem from the slowing down of the α particles in the Si detector material, are several orders of magnitude smaller than the measured probabilities. Also shown are the earlier results obtained by Kasagi *et al.* [4]; within their errors they are consistent with the present high statistic data. However, the previous data of Eremin *et al.* [12] (see Fig. 1) are inconsistent with the present findings.

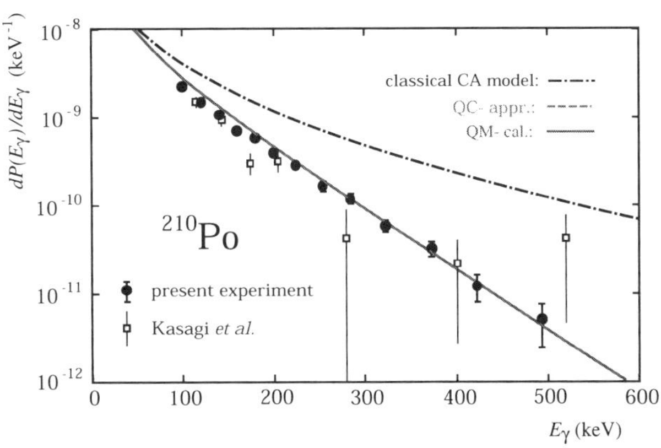

Fig. 6. Measured differential Bremsstrahlung emission probability in the α decay of ^{210}Po (solid points: present experiment, open symbols: Kasagi *et al.* [4]) in comparison with theoretical predictions. Our quasi-classical (QC) calculation [2] agrees with the quantum mechanical (QM) calculation of Papenbrock and Bertsch [3] to an extend that the results are indistinguishable on the scale of the figure.

Our high-statistic data also allow for a detailed comparison with theoretical calculations. Performing the QM calculation of Papenbrock and Bertsch [3] with the proper Q_α^0 value for ^{210}Po we obtain for the differential emission probabilities the solid curve shown in Fig. 6, which is in good agreement with our experimental data. Moreover, we revisited the QC theory of Dyakonov [6] to enlarge its range of applicability to γ energies reached in the present investigation. We find the result of our improved QC approach [2] to agree with the QM prediction to better than 2% even at γ energies as high as 500 keV. The overall agreement observed between the two theoretical approaches and between experiment and theory makes us believe that the process of the bremstrahlung accompanied α decay is now basically understood. A more detailed comparison between theory and experiment, however, does reveal some small systematic deviations of up to 2σ ($\sim 20\%$) at γ energies $E_\gamma \leq 200$ keV, the origin of which is still under investigation.

Acknowledgements

U.D.J. acknowledges support from the Deutsche Forschungsgemeinschaft (Heisenberg program), and D.S. support by a Joseph Meyerhoff Visiting Professorship granted by the Weizmann Institute of Science. A.I.M. and I.S.T. gratefully acknowledge the Max-Planck-Institute for Nuclear Physics, Heidelberg, for warm hospitality and support. The work was also supported by RFBR Grant No. 06-02-04018.

References

1. H. Boie, H. Scheit, F. Köck, M. Lauer, U.D. Jentschura, A.I. Milstein, I.S. Terekhov, D. Schwalm, *Phys. Rev. Lett.*, accepted.
2. U.D. Jentschura, A.I. Milstein, I.S. Terekhov, H. Boie, H. Scheit, D. Schwalm, arXiv:nucl-th/0606005.
3. T. Papenbrock and G. F. Bertsch, *Phys. Rev. Lett.* **80**, 4141 (1998).
4. J. Kasagi, H. Yamazaki, N. Kasajima, T. Ohtsuki, and H. Yuki, *Phys. Rev. Lett.* **79**, 371 (1997), *Phys. Rev. Lett.* **85**, 3062 (2000).
5. M. I. Dyakonov and I. V. Gornyi, *Phys. Rev. Lett.* **76**, 3542 (1996).
6. M. I. Dyakonov, *Phys. Rev. C* **60**, 037602 (1999).
7. N. Takigawa, Y. Nozawa, K. Hagino, A. Ono, and D. M. Brink, *Phys. Rev. C* **59**, R593 (1999).
8. E. V. Tkalya, *Phys. Rev. C* **60**, 054612 (1999).
9. C. A. Bertulani, D. T. de Paula, and V. G. Zelevinsky, *Phys. Rev. C* **60**, 031602(R) (1999).
10. S. P. Maydanyuk and V. S. Olkhovsky, *Prog. Theor. Phys.* **109**, 203 (2003), and *Eur. Phys. J. A* **28**, 283 (2006).
11. A. D'Arrigo, N. Eremin, G. Fazio, G. Giardina, M. G. Glotova, T. V. Klochko, M. Sacchi, and A. Taccone, *Phys. Lett. B* **332**, 25 (1994).
12. N. V. Eremin, G. Fazio, and G. Giardina, *Phys. Rev. Lett.* **85**, 3061 (2000).
13. R.B. Firetone, *Table of Isotopes* (J. Wiley&Sons, New York, 1996).
14. J. Eberth et al., *Prog. Part. Nucl. Phys.* **46**, 389 (2001).
15. H. Boie, Dissertation, Univ. Heidelberg, in preparation.
16. H.J. Fischbeck, M.S. Freedman, *Phys. Rev. Lett.* **34**, 173 (1975).
17. S. Agostinelli et al., *Nucl.Instr.Meth. A* **506**,250 (2003)
18. H. Boie, g4miniball - the MINIBALL simulation package, http://www.mpi-hd.mpg.de/mbwiki/wiki/MiniballSimulation

THE DISSOCIATION OF ^{8}B IN THE COULOMB FIELD AND THE VALIDITY OF THE CD METHOD *

MOSHE GAI

Laboratory for Nuclear Science at Avery Point, University of Connecticut, 1084 Shennecossett Rd, Groton, CT 06340-6097.
and
Department of Physics, WNSL Rm 102, Yale University, PO Box 208124, 272 Whitney Avenue, New Haven, CT 06520-8124.

e-mail: moshe.gai@yale.edu, URL: http://www.phys.uconn.edu

The GSI1, GSI2 (as well as the RIKEN2 and the corrected GSI2) measurements of the Coulomb Dissociation (CD) of ^8B are in good agreement with the most recent Direct Capture (DC) $^7Be(p,\gamma)^8B$ reaction measurement performed at Weizmann and in agreement with the Seattle result. Yet it was claimed that the CD and DC results are sufficiently different and need to be reconciled. We show that these statements arise from a misunderstanding (as well as misrepresentation) of CD experiments. We recall a similar strong statement questioning the validity of the CD method due to an invoked large E2 component that was also shown to arise from a misunderstanding of the CD method. In spite of the good agreement between DC and CD data the slope of the astrophysical cross section factor (S_{17}) can not be extracted with high accuracy due to a discrepancy between the recent DC data as well as a discrepancy of the three reports of the GSI CD data. The slope is directly related to the d-wave component that dominates at higher energies and must be subtracted from measured data to extrapolate to zero energy. Hence the uncertainty of the measured slope leads to an additional uncertainty of the extrapolated zero energy cross section factor, $S_{17}(0)$. This uncertainty must be alleviated by future experiments to allow a precise determination of $S_{17}(0)$, a goal that so far has not be achieved in spite of strong statement(s) that appeared in the literature.

Keywords: Coulomb Dissociation, Direct Capture, Astrophysical Cross Section Factor, Solar Neutrinos.

*Work Supported by USDOE grant No. DE-FG02-94ER40870

1. Introduction

The Coulomb Dissociation (CD) method was developed in the pioneering work of Baur, Bertulani and Rebel[1] and has been applied to the case of the CD of ^8B[2–5] from which the cross section of the $^7Be(p,\gamma)^8B$ reaction was extracted. This cross section is essential for calculating the ^8B solar neutrino flux. The CD data were analyzed with a remarkable success using only first order Coulomb interaction that includes only E1 contribution. An early attempt (even before the RIKEN data were published) to refute this analysis by introducing a non-negligible E2 contribution[6] was shown[7] to arise from a neglect of the angular acceptance of the RIKEN1 detector and a misunderstanding of the CD method. Indeed the CD of ^8B turned out to be a testing ground of the very method of CD. Later claims by the MSU group for evidence[8] of non-negligible E2 contribution in **inclusive measurement** of an asymmetry, were disputed in a recent **exclusive measurement** of a similar asymmetry by the GSI2 collaboration.[5]

In contrast, Esbensen, Bertsch and Snover[9] recently claimed that higher order terms and an E2 contribution are an important correction to the RIKEN2 data.[3] It is claimed that "S_{17} values extracted from CD data have a significant steeper slope as a function of E_{rel}, the relative energy of the proton and the ^7Be fragment, than the direct result". However they find a substantial correction only to the RIKEN2 CD data and claim that this correction(s) yield a slope of the RIKEN2 data in better agreement with Direct Capture (DC) data. In addition it is stated[9] that "the zero-energy extrapolated $S_{17}(0)$ values inferred from CD measurements are, on the average 10% lower than the mean of modern direct measurements". The statements on significant disagreement between CD and DC data are based on the re-analyses of CD data by the Seattle group.[10] In this paper we demonstrate that an agreement exists between CD and DC data and the statements of the Seattle group[10] are based on misunderstanding (as well as misrepresentation) of CD data.

In spite of the general agreement between CD and DC data, still the the slope of astrophysical cross section factor measured between 300 - 1,500 keV can not be extracted with high accuracy. This hampers our ability to determine the d-wave contribution that dominates the cross section of the $^7Be(p,\gamma)^8B$ reaction at higher energies and must be subtracted for extrapolating the s-wave to zero energy. Lack of accurate knowledge of the d-wave contribution to data (even if measured with high accuracy), precludes accurate extrapolation to zero energies. We show that this leads to additional uncertainty of the extrapolated $S_{17}(0)$. We doubt the strong statement that $S_{17}(0)$ was measured with high accuracy (see for example[10]).

2. The Slope of S_{17} Above 300 keV

Early on it was recognized that s-wave capture alone yields an s-factor with a negative slope. This is due to the Coulomb distortion of the s-wave at very low distances. The observation of a positive slope of S_{17} measured at energies above 300 keV was recognized as due to the d-wave contribution. It was also recognized that the d-wave contribution is very large at measured energies and in fact it dominates around 1.0 MeV. The d-wave contribution must be subtracted to allow an accurate extrapolation of the s-wave to zero energy (where the d-wave contribution is very small, of the order of 6%). The (large) contribution of the d-wave at energies above 300 keV leads to a linear dependence of S_{17} on energy (with a positive slope). An accurate extrapolation of S_{17} must rely on an accurate knowledge of the d-wave contribution or the slope at energies above 300 keV.

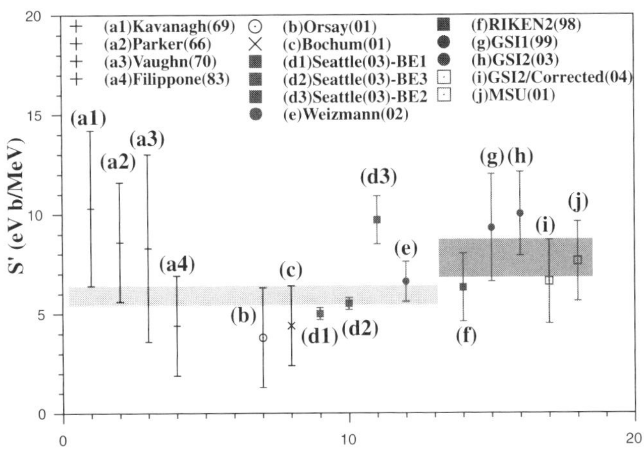

Fig. 1. The measured slopes (S' = dS/dE) of world data measured between 300 and 1500 keV, as discussed in the text. The range of "average values" is indicated and discussed in the text.

In Fig. 1 we show the slope parameter (S' = dS/dE) extracted from both DC and CD data in the energy range of 300 - 1500 keV. We refer the reader to[11] for detailes on data used to extract the slope shown in Fig. 1. We conclude from Fig. 1 that the slope parameter can not be extracted from DC data[10,13–19] with high accuracy as claimed. The DC data are not sufficiently consistent to support this strong statement;[10] for example there is not a single data point measured by the Bochum group[14] that agrees with that measured by the Seattle group,[10] where we observe that some of the individual data points disagree by as much as five sigma. The disagreement of the three slopes measured by the Seattle group and the disagreement with the Weizmann slope are most disturbing. In the same time the dispersion among slopes measured in CD is also of concern. However, it is clear that the over all agreement between CD and DC data (1.7 sigma) is better than the agreement among specific DC data. We do not support the strong claim of substantial disagreement between slopes measured in DC and CD.[10]

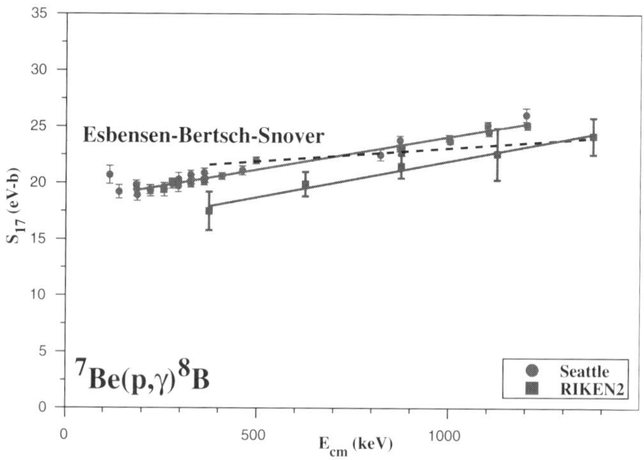

Fig. 2. Extracted S_{17} from the RIKEN2 CD data[3] using first order electric dipole interaction as shown in,[5] compared to the DC capture data published by the Seattle group[10] and the so called reconciled slope calculated by EBS.[9] The shown RIKEN2 data include systematic uncertainties (equal or slightly smaller) as published.[3]

The lack of evidence for substantial difference between CD and DC results leads to doubt on the very need to reconcile these data.[12] Furthermore, in Fig. 2 we show the slope obtained by EBS after their attempt to reconcile the slope of CD with the slope of DC data. Clearly the original slope of the RIKEN2 data obtained using only first order E1 interactions is in considerably better agreement with DC data than the so called reconciled slope.

3. $S_{17}(0)$ Extracted From CD Data

In Fig. 20 of the Seattle paper[10] they show extracted $S_{17}(0)$ from CD using the extrapolation procedure of Descouvemont and Baye,[20] and based on this analysis it is stated[9] that "the zero-energy extrapolated $S_{17}(0)$ values inferred from CD measurements are, on the average 10% lower than the mean of modern direct measurements". The extracted $S_{17}(0)$ shown in Fig. 20[10] are only from data measured at energies below 425 keV and the majority of CD data points that were measured above 425 keV were excluded in Fig. 20.[10]

This arbitrary exclusion of (CD) data above 425 keV has no physical justification (especially in view of the fact that the contribution of the 632 keV resonance is negligible in CD). For example as shown by Descouvemont[21] the theoretical error increases to approximately 5% at 500 keV and in fact it is slightly decreased up to approximately 1.0 MeV, and there is no theoretical justification for including data up to 450 keV but excluding data between 500 keV and 1.0 MeV.

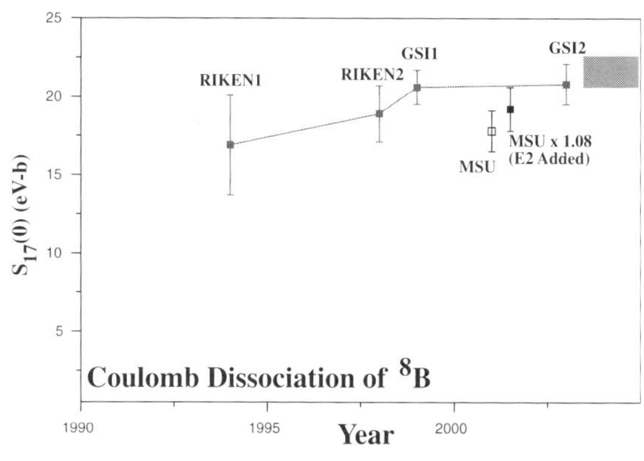

Fig. 3. Measured $S_{17}(0)$ as originally published by the authors who performed the CD experiments. These analyses include all measured data points[2–5,8] using the extrapolation procedure of Descouvemont and Baye.[20] We also plot the MSU data as published as well as with the E2 correction ($\approx 8\%$)[8] added back to the quoted $S_{17}(0)$, as discussed in the text. The range of $S_{17}(0)$ results from the measurements of DC by the Seattle[10] and Weizmann groups[15] is indicated.

Thus when excluding the CD data above 425 keV, the Seattle group excluded the data that were measured with the best accuracy and with smallest systematical uncertainty. If in fact one insists on such an analysis of

CD data, one must estimate the systematic uncertainty due to this selection of data. This has not been done in the Seattle re-analyses of CD data.[10]

Instead we rely here on the original analyses of the authors that published the CD data. In Fig. 3 we show the $S_{17}(0)$ factors extracted by the original authors who performed the CD experiments. These results include all measured data points up to 1.5 MeV, and are analyzed with the same extrapolation procedure of Descouvemont and Baye.[20]

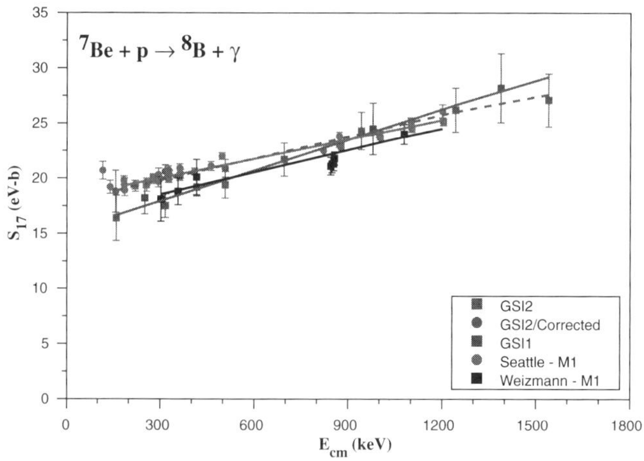

Fig. 4. A comparison of the most recent DC data with the GSI1 and GSI2 results.

We note that the (four) CD results are consistent within the quoted error bars, but they show a systematic trend of an increased $S_{17}(0)$ (to approximately 20.7 eV-b), while the error bars are reduced. We obtain a $1/\sigma$ weighted average of $S_{17}(0) = 20.0 \pm 0.7$ with $\chi^2 = 0.5$, which is in excellent agreement with the measurement of the Weizmann group[15] and in agreement with the measurement of the Seattle group.[10]

4. Extrapolating $S_{17}(0)$ From World Data

The current situation with our knowledge of S_{17} and the extrapolated $S_{17}(0)$ is still not satisfactory. The main culprit are major disagreements among DC data. It is clear for example that the systematic disagreements between the Orsay-Bochum[13,14] and the Weizmann-Seattle[10,15] results must be resolved before these data are included in a so called "world average". In Fig. 4 we compare the most recent Seattle-Weizmann data (with M1 contribution subtracted) with the GSI1 and GSI2 (as well as corrected GSI2) results. While the data appear in agreement we still observe a systematic disagreement between all measured slopes. The DC data of the Seattle and the Weizmann groups have different slopes as do the GSI1, GSI2 and corrected GSI2 data. The slope above 300 keV is directly related to the d-wave contribution that dominates at measured laboratory energies, but must be subtracted to extrapolate to solar burning energies. This disagreement does not allow for an accurate (better than 5% accuracy) extrapolation of $S_{17}(0)$ and must be resolved by future experiments. A reasonable systematic error of +0.0 -3.0 eV-b due to extrapolation seems to be required by current data.

References

1. G. Baur, C.A. Bertulani, and H. Rebel; Nucl. Phys. **A458**(1986)188.
2. T. Motobayashi *et al.*; Phys. Rev. Lett. **73**(1994)2680.
3. T. Kikuchi *et al.*; Phys. Lett. **B391**(1997)261, ibid E. Phys. J. **A3**(1998)213.
4. N. Iwasa *et al.*; Phys. Rev. Lett. **83**(1999)2910.
5. F. Schumann *et al.*; Phys. Rev. Lett. **90**(2003)232501, ibid Phys. Rev. **C73**(2006)015806.
6. K. Langanke and T.D. Shoppa; Phys. Rev. C52(1995)1709.
7. M. Gai and C.A. Bertulani; Phys. Rev. **C52**(1995)1706.
8. B.S. Davids *et al.*; Phy. Rev. **63**(2001)065806.
9. H. Esbensen, G.F. Bertsch, and K. Snover; Phys. Rev. Lett. **94**(2005)042502.
10. A.R. Junghans *et al.*; Phys. Rev. **C68**(2003)065803.
11. M. Gai; Phys. Rev. **C74**(2006)025810.
12. M. Gai; Phys. Rev. Lett. **96**(2006)159201.
13. F. Hammache *et al.*; Phys. Rev. Lett. **86**(2001)3985.
14. F. Strieder *et al.*; Nucl. Phys. **A696**(2001)219.
15. L.T. Baby Phys. Rev. **C67**(2003)065805, ER **C69**(2004)019902(E).
16. B.W. Filippone *et al.*; Phys. Rev. **C28**(1983)2222.
17. F.J. Vaughn *et al.*, Phys. Rev. **C2**(1970)1657.
18. P.D. Parker, Phys. Rev. **150**(1966)851.
19. R.W. Kavanagh, T.A. Tombrello, T.A. Mosher, and D.R. Goosman, Bull. Amer. Phys. Soc., **14**(1969)1209.
20. P. Descouvemont and D. Baye; Nucl. Phys. **A567**(1994)341.
21. P. Descouvemont; Phys. Rev. **C70**(2004)065802.

NUCLEAR MANY-BODY PHYSICS WHERE STRUCTURE AND REACTIONS MEET*

NAUREEN AHSAN AND ALEXANDER VOLYA

Department of Physics, Florida State University, Tallahassee, FL 32306, USA

The path from understanding a simple reaction problem of scattering or tunneling to contemplating the quantum nuclear many-body system, where structure and continuum of reaction-states meet, overlap and coexist, is a complex and nontrivial one. In this presentation we discuss some of the intriguing aspects of this route.

1. Introduction

Structure and reactions have traditionally been separate subjects of nuclear physics; however, recently the need for unification of our approach to open many-body systems has become apparent. The advances in experimental techniques led to observation of exotic nuclei which exist only due to complex interplay between many-body structural effects and reaction dynamics. Open mesoscopic systems like microwave cavities, quantum dots, nuclei, and even hadrons, play an increasingly important role in science and technology. A number of different theoretical techniques and approaches have been recently proposed on the path to a unified theory[1,2,3]. In this presentation we discuss simple examples that stress the complexity of the structure-reaction borderline physics. We concentrate on the effects that the intrinsic structure of composite objects plays in quantum-mechanical scattering and tunneling processes. These processes are governed by non-perturbative physics with exponential sensitivity to various quantities. We examine two examples which, in our view, illuminate the physics of interest from two opposite sides. The first example is a model of the one-dimensional scattering of a two-body system from a potential, and the second case is a realistic example of ^{6}He-neutron scattering. Despite complexity and non-perturbative nature we obtain an exact quantum-mechanical solution to

*The collaboration with V. Zelevinsky is highly appreciated. The work was supported by the U. S. Department of Energy, grant DE-FG02-92ER40750.

the model example, while in the realistic case the problem is solved with a Continuum Shell Model approach[4,5] using an effective Hamiltonian. Both our examples highlight similar features related to the interplay of internal structure and reaction channels and emphasize the roles of symmetries and unitarity.

2. Reactions with a composite object: a model study

As our model example we consider a one-dimensional scattering problem where a projectile is a composite object made up of two particles bound by a potential $v(x_1 - x_2)$ which depends only on the relative distance between the particles. Here the particle coordinates are x_1 and x_2, and the masses are m_1 and m_2, respectively. This composite system interacts with an external scattering potential V. We assume that the external potential acts only on the second particle so that it depends only on the coordinate x_2. Below we use the usual center-of-mass coordinate R and relative coordinate $r = x_1 - x_2$ and assume M and μ to be the total and reduced masses of the system, respectively. In these coordinates the total Hamiltonian is

$$H = -\frac{1}{2M}\frac{\partial^2}{\partial R^2} + h + V\,, \quad \text{where} \quad h = -\frac{1}{2\mu}\frac{\partial^2}{\partial r^2} + v(r)$$

is the intrinsic Hamiltonian of the system.

The channels $|n\rangle$ are the eigenstates of the intrinsic Hamiltonian: $h|n\rangle = \epsilon_n|n\rangle$, so that the asymptotic forms for the incoming-plus-reflected and transmitted waves are

$$|\Psi_-\rangle = e^{iK_0R}|0\rangle + \sum_{n=0}^{\infty} C_{-,n}e^{-iK_nR}|n\rangle, \text{ and } \quad |\Psi_+\rangle = \sum_{n=0}^{\infty} C_{+,n}e^{iK_nR}|n\rangle. \tag{1}$$

Here $|\Psi_-\rangle$ includes the incoming wave in $n = 0$ channel, the ground state of the intrinsic potential; K_n is the channel momentum $K_n(E) = \sqrt{2M(E-\epsilon_n)}$ at a given total energy E. The sums in (1) implicitly contain both open and closed channels depending on whether the corresponding $K_n(E)$ is real or purely imaginary. In the expression we use the principal value of the square root so that the reflected waves in closed channels exponentially fall off. The conservation of the center-of-mass flux in the open channels leads to the unitarity relation

$$\sum_{n\,\mathrm{open}} (R_n + T_n) = 1, \text{ where } R_n = \frac{K_n}{K_0}\,|C_{-,n}|^2 \text{ and } T_n = \frac{K_n}{K_0}\,|C_{+,n}|^2 \tag{2}$$

are, respectively, the reflection and transmission probabilities in the n-th channel.

We model the external potential with a delta-peak, and only the second particle is assumed to interact with it: $V(x_2) = \alpha\delta(x_2)$. We implement a usual treatment of a delta-potential by separating left and right regions denoted by $-$ and $+$ subscripts, respectively. The principal complication in this problem comes from the boundary condition on x_2 being incompatible with the center-of-mass coordinates. The boundary conditions projected onto the m-th quantum state result in a system of linear equations:

$$\sum_n [C_{+,n}\langle m|D(k_n)|n\rangle - C_{-,n}\langle m|D(-k_n)|n\rangle] = \langle m|D(k_0)|0\rangle, \qquad (3)$$

$$\sum_n [C_{+,n}\langle m|Q(k_n) - 2m_2\alpha D(k_n)|n\rangle - C_{-,n}\langle m|Q(-k_n)|n\rangle] = \langle m|Q(k_0)|0\rangle.$$

Here for simplicity of notations we use an intrinsic momentum shift operator $D(k) = e^{ikr}$ and the operator $Q(k) = i[\rho k D(k) - D(k)p]$ where $p = -i\partial/\partial r$ and $\rho = m_2/m_1$.

We show our results for a case where the binding potential is given by that of a harmonic oscillator. The expectation value of the momentum shift operator can be expressed analytically with the Associated Laguerre Polynomials. We choose ω as our energy scale leaving relative kinetic energy $\mathcal{E} = E/\omega - 1/2$ and relative energy scale of delta-peak $\Delta = M\alpha^2/\omega$ as energy parameters.

The problem expressed by equations (3) is that of an infinite set of linear equations which had to be solved by truncating the space and including only a finite number of intrinsic states. The momentum-shifts for highly virtual channels occur along the imaginary axis; therefore the momentum-shift operator matrix elements are exponentially divergent for these channels, and so are the corresponding coefficients $C_{\pm,n}$. This shows a mathematically complex behavior near the barrier where the boundary conditions are satisfied by cancellations of exponentially divergent terms. Physically, however, highly-virtual excitations decay fast leading to a regular behavior away from the delta peak. Thus, this well-formulated problem of quantum mechanics appears to be quite challenging mathematically. The proof of validity of the truncation mentioned above is an important issue addressed in Ref.[6]. The flux conservation (2) provides an additional test of consistency and convergence.

The transmission probabilities for the two lowest channels resulting from the scattering of an oscillator-bound system off a delta function are shown in Fig.1. The figure demonstrates some of the generic features inherent to the composite-particle scattering. The incident wave contains a composite particle in the ground state and the number of open reflection/transmission channels depends on the incident kinetic energy. For harmonic oscillator the threshold energies for channel-opening correspond to integral values of $\mathcal{E}$. At low energies, $\mathcal{E} < 1$, transmission and reflection only in the ground state are possible, and $T_0 + R_0 = 1$. Once the $\mathcal{E} = 1$ threshold is crossed transmission and reflection in the first excited oscillator state are also possible and the total flux is then shared among all four processes: $T_0 + T_1 + R_0 + R_1 = 1$. The number of open channels increases with each integral value of $\mathcal{E}$. The redistribution of probabilities at the threshold values leads to cusps in the cross sections[7,8]. A careful examination of Fig.1 reveals the appearance of such sharp points at thresholds. In addition to these, an interesting resonant-type behavior can be noted in the transmission (and reflection, not shown here) probabilities associated with peaks that do not coincide with threshold energies. This resonant behavior is related to the intrinsic structure. The case of a non-composite particle is a standard textbook example, where the reflection $R = (1 + 2[\mathcal{E}/\Delta])^{-1}$ depends only on kinetic energy relative to the delta strength. In the figure, this non-composite limit for the corresponding kinematic conditions is shown with a thick solid line that has no oscillations. Within our model this limit can be continuously reached when $\rho \to \infty$, namely, when the mass of a non-interacting particle approaches zero.

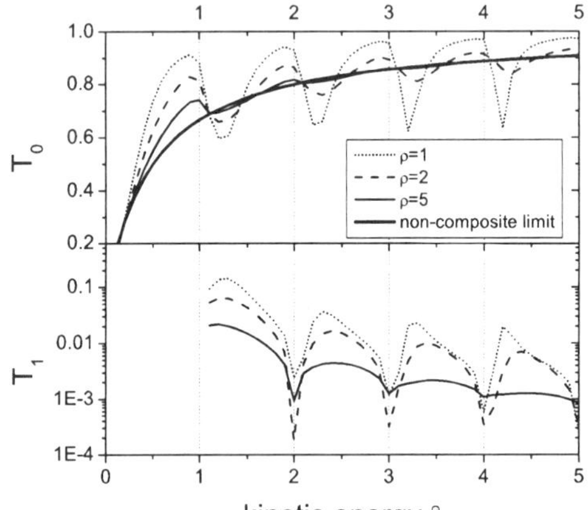

Figure 1. Transmission probabilities for the first two channels of a composite particle through the delta barrier as a function of kinetic energy $\mathcal{E}$ at $\Delta = 1$ with different ρ's.

For the mass ratio ρ of the orders of 1 or smaller the internal structure plays an increasingly important role in the dynamics. The sensitivity of R and T to the parameters of the model becomes large and there appear resonances associated with the internal structure.

3. Interplay of structure and reactions in the unified continuum shell model approach

We study our second, realistic example of ^{7}He+n scattering using a Continuum Shell Model approach which is discussed in a series of recent publications[9,4]. The projection formalism that lies in the foundation of the model dates back to Feshbach[10]. The detailed study can be found in the textbooks[11,12], and the previous development of the method was reviewed by Rotter[13]. This method is actively used in diverse areas of open quantum many-body systems from molecular and condensed matter physics[14] to multi-quark systems[15]. We review some of the important ingredients below.

3.1. *Structure*

The part of the Hilbert space related to the particle(s) in the continuum is eliminated with the projection formalism. As a result the "intrinsic" dynamics is given by the effective Hamiltonian

$$\mathcal{H}(E) = H_0 + \Delta(E) - \frac{i}{2}W(E). \tag{4}$$

Here the full Hamiltonian H_0 is restricted to the intrinsic space, and is supplemented with the Hermitian term $\Delta(E)$ that describes virtual particle excitations into the excluded space. The imaginary term $iW(E)/2$ represents irreversible decays to the continuum. These new terms in the projected Hamiltonian (4) are given in terms of the matrix elements $A_1^c(E) = \langle 1|H_0|c; E\rangle$ of the full original Hamiltonian that link the internal states $|1\rangle$ with the energy-labeled external states $|c; E\rangle$ in the following manner:

$$\Delta_{12}(E) = \text{P.v.} \int dE' \sum_c \frac{A_1^c(E')A_2^{c*}(E')}{E - E'}, \quad W_{12}(E) = 2\pi \sum_{c(\text{open})} A_1^c A_2^{c*}. \tag{5}$$

The properties of the effective Hamiltonian (4) are as follows:

1. For unbound states, above the decay thresholds, the effective Hamiltonian is non-Hermitian which reflects the loss of probability from the intrinsic space.

2. The Hamiltonian has explicit energy dependence, making the internal dynamics highly non-linear.

3. The solution for each individual nucleus is coupled to all the daughter systems via a chain of reaction channels.

4. Even with two-body forces in the full space, the many-body interactions appear in the projected effective Hamiltonian.

5. The eigenvalue problem $\mathcal{H}(E)|\alpha\rangle = \mathcal{E}|\alpha\rangle$ represents a condition for the many-body resonant Siegert states, for which the regular wave function is matched with the purely outgoing one at infinity. Below all decay thresholds the imaginary part disappears and the problem is equivalent to that of a traditional shell model. Above decay thresholds, in general, there are no real energy solutions, i.e., the stationary state boundary condition cannot be satisfied. The complex energy eigenvalues correspond to poles of the scattering matrix.

3.2. *Reactions and Unitarity*

The picture where the nuclear system is probed from "outside" is given by the transition matrix defined within the general scattering theory[11],

$$T^{ab}(E) = \sum_{12} A_1^{a*}(E) \left(\frac{1}{E - \mathcal{H}(E)} \right)_{12} A_2^b(E). \tag{6}$$

The poles of this transition matrix and the related full scattering matrix $S = 1 - 2\pi i T$ are the eigenvalues of the effective Hamiltonian. The reaction theory is fully consistent with the structure description in Sec. 3.1. However, many-body complexity, numerous poles, overlapping resonances and energy dependence can make the observable cross-section quite different from a collection of individual resonance peaks.

The transition matrix (6) with the dimensionality equal to the number of open channels can be written as $T = \mathbf{A}^\dagger \mathcal{G} \mathbf{A}$, where the full effective Green's function $\mathcal{G}(E) = 1/(E - \mathcal{H})$ includes the loss of probability into all decay channels. The non-Hermitian part of Eq.(4) is factorized as $W = 2\pi \mathbf{A}\mathbf{A}^\dagger$, where $\mathbf{A}$ represents a channel matrix (a set of column-vectors A_1^c for all the channels c). As shown in Refs.[16,17], iteration of the Dyson equation using the definitions $\mathcal{H} = H - iW/2$ and $G = (E - H)^{-1}$ leads to the following transition and scattering matrices

$$T = \frac{R}{1 + i\pi R}, \qquad S = \frac{1 - i\pi R}{1 + i\pi R}. \tag{7}$$

The matrix $R = \mathbf{A}^\dagger G \mathbf{A}$ is analogous to the R-matrix of the standard reaction theory; it is based on the Hermitian part of the Hamiltonian $H =$

$H_0 + \Delta$. Thus, the factorized nature of the intrinsic Hamiltonian and the appearance of the same effective operator in the scattering matrix are important consequences of unitarity.

3.3. *The He example*

In Fig.2 we consider a realistic example of ^{6}He-neutron scattering. The parameters of the model are given by the intrinsic shell model Hamiltonian from Ref. [5] within the p-shell valence space. The continuum reaction physics is modelled by the Woods-Saxon Hamiltonian. The model is discussed in-depth in Ref.[5], where the entire chain of He isotopes is solved in a coupled manner. For simplification of this discussion we consider states in ^{6}He to be bound, and concentrate on the role of the internal structure in scattering with a neutron. Energies are quoted here relative to the alpha particle ground state.

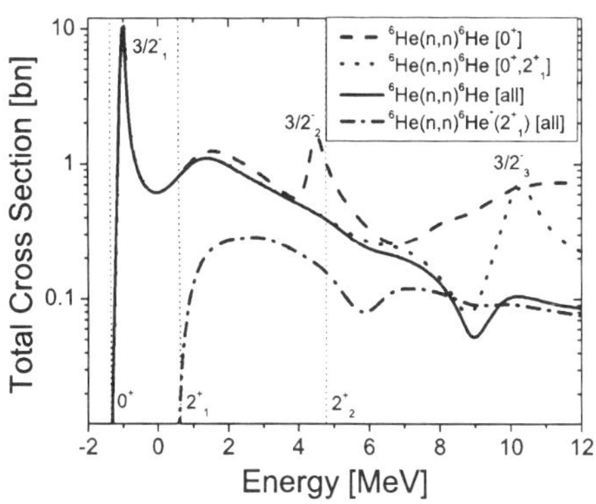

Figure 2. Cross-section for the neutron scattering of ^{6}He in 0^+ ground state. The solid curve is the total elastic cross-section, while the dashed and dotted curves correspond to the cases when only 0^+ channel (final state of ^{6}He) and both 0^+ and 2_1^+ channels are included, respectively. The dash-dot curve shows inelastic cross-section with ^{6}He in the 2_1^+ final state. Thresholds for the lowest three channels, 0^+, 2_1^+ and 2_2^+, are marked with vertical grid lines. Locations for three $3/2^-$ resonances in ^{7}He are indicated.

1. At low energies the peak in the cross-section corresponds to a narrow resonance, E_r=-1.02 MeV, and a width of 91 keV which corresponds well to a spectroscopic factor C^2S= 0.498.

2. The threshold to a second decay-channel (decay to a $2^+, 1.89MeV$ excited state in ^{6}He) is at 0.515 MeV. At this energy there is a cusp in the cross-section; however, for a p-wave neutron the curve is smooth unlike that for an s-wave.

3. The resonance corresponding to the second $3/2^-$ state (5.510 MeV excitation energy) in ^{7}He appears at 4.494 MeV only when all other decay channels are ignored (dashed line in Fig.2). This state has a large width to decay into 2^+ final state in ^{6}He, the C^2S =1.03 which makes resonance peak impossible to observe.

4. Conclusions

In this work we target the issue of internal degrees of freedom in scattering processes. We use two models, which are very different in their nature and nicely show different aspects of the physics of interest. The exactly solved simple one-dimensional scattering shows an unusual resonant behavior associated with the composite nature of the incident system. An application of the Continuum Shell Model was demonstrated and discussed within a realistic example. Interplay between channels, unitarity, and distribution of flux are common to both examples and lead to generic near-threshold behavior. We emphasize the importance of future theoretical developments toward a unified description of structure and reactions.

References

1. N. Michel et al., Phys. Rev. C **67** (2003) 054311; Phys. Rev. C **70** (2004) 064313.
2. K. Bennaceur et al., Nucl. Phys. A 671 (2000) 203.
3. D. Eppel and A. Lindner, Nucl. Phys. **A 240** (1975) 437.
4. A. Volya and V. Zelevinsky, Phys. Rev. Lett. **94** (2005) 052501.
5. A. Volya and V. Zelevinsky, Phys. Rev. C **74** (2007) 064314.
6. N. Ahsan, Role of internal degrees freedom of a composite object in reactions, 2007, Prospectus for Ph.D., Department of Physics, Florida State University.
7. L. D. Landau and E. M. Lifshitz, Quantum mechanics. Non-relativistic theory. (Pergamon Press, New York, 1981).
8. A. I. Baz, I. B. Zeldovich and A. M. Perelomov, Scattering, reactions and decay in nonrelativistic quantum mechanics (Nauka, Moscow, 1971).
9. A. Volya and V. Zelevinsky, Phys. Rev. C **67** (2003) 054322.
10. H. Feshbach, Ann. Phys. 5 (1958) 357; Ann. Phys. 19 (1962) 287.
11. C. Mahaux and H. A. Weidenmüller, Shell-model approach to nuclear reactions (North-Holland Pub. Co., Amsterdam, London, 1969).
12. H. Feshbach, Theoretical nuclear physics : nuclear reactions (Wiley, New York, 1991).
13. I. Rotter, Rep. Prog. Phys. **54** (1991) 635.
14. A. Volya and V. Zelevinsky, J. Opt. B **5** (2003) S450.
15. N. Auerbach, V. Zelevinsky and A. Volya, Phys. Lett. **B 590** (2004) 45.
16. L. Durand, Phys. Rev. D **14** (1976) 3174.
17. V. V. Sokolov and V. G. Zelevinsky, Nucl. Phys. **A 504** (1989) 562.

RECENT DEVELOPMENTS IN THE SPECTRAL FLUCTUATIONS OF NUCLEI, HADRONS AND OTHER QUANTUM SYSTEMS

J. M. G. GÓMEZ, L. MUÑOZ AND J. RETAMOSA
Departamento de Física Atómica, Molecular y Nuclear, Universidad Complutense de Madrid, E-28040 Madrid, Spain

R. A. MOLINA AND A. RELAÑO
Instituto de Estructura de la Materia, CSIC, E-28006 Madrid, Spain

E. FALEIRO
Departamento de Física Aplicada, E. U. I. T. Industrial, Universidad Politécnica de Madrid, E-28012 Madrid, Spain

A survey of chaotic dynamics in atomic nuclei is presented, using on the one hand standard statistics of quantum chaos studies, and on the other a new approach based on time series analysis methods. We emphasize the energy and isospin dependence of nuclear chaoticity, based on shell-model energy spectra fluctuations in Ca, Sc and Ti isotopes, which are analyzed using standard statistics such as the nearest level spacing distribution $P(s)$ and the Dyson-Mehta Δ_3 statistic. We also present a recent study of the low-lying baryon spectrum up to 2.2 GeV which has shown that experimental data exhibit a $P(s)$ distribution close to GOE and, on the contrary, quark models predictions are more similar to the Poisson distribution. Finally, we discuss quantum chaos using a new approach based on the analogy between the sequence of energy levels and a discrete time series. Considering the energy spectrum fluctuations of quantum systems as a discrete time series, we suggest the following conjecture: *The energy spectra of chaotic quantum systems are characterized by* $1/f$ *noise.* Moreover, we show that the spectra of integrable quantum systems exhibit $1/f^2$ noise.

1. Introduction

The understanding of quantum chaos has greatly advanced during the last two decades. It is well known that there is a clear relationship between the energy level fluctuation properties of a quantum system and the large time scale behavior of its classical analogue. The spectral fluctuations of a quantum system whose classical analogue is fully integrable are well described by

Poisson statistics, i. e. the successive energy levels are not correlated [1]. On the contrary the fluctuation properties of generic quantum systems, which are fully chaotic, coincide with those of random matrix theory (RMT) [2]. A review of later developments can be found in [3,4].

However, real and complex quantum systems are usually not fully ergodic or integrable, and many questions on their chaotic and regular motion are still open. In this context, the atomic nucleus can be considered as a small laboratory where the principles obtained in schematic systems may be tested. The information on regular and chaotic nuclear motion available from experimental data is rather limited, because the analysis of energy levels requires the knowledge of sufficiently large pure sequences, i.e. consecutive level samples all with the same quantum numbers (J, π, T) in a given nucleus. The situation is quite clear above the one-nucleon emission threshold, where a large number of neutron and proton $J^\pi = 1/2^+$ resonances are identified. The agreement between this Nuclear Data Ensemble (NDE) [5] and the GOE predictions is excellent. In the low energy domain, however, it is rather difficult (if not impossible) to get large enough pure sequences, and for this reason the conclusions are less clear. [6,7].

In order to get a deeper understanding of what happens in the low energy region we can use the shell model with configuration mixing. Although, most nuclei exhibit chaotic motion, we will show that the low energy spectra of Ca isotopes exhibit fluctuation properties closer to regular than to chaotic motion. Moreover, comparison of level fluctuations in different $pf-$isotopes shows a clear isospin dependence of nuclear chaoticity.

Another interesting application of spectral fluctuations studies has been recently reported [8]. The experimental low-lying baryon spectrum up to 2.2 GeV exhibits a nearest neighbor spacing distribution $P(s)$ close to GOE, while quark models predictions are more similar to the Poisson distribution. This result shows that the prediction of these quark models about the problem of *missing* resonances and about the statistical properties of baryon spectra are not reliable.

Finally, we present a different approach to quantum chaos based on traditional methods of time series analysis. The essential feature of chaotic energy spectra in quantum systems is the existence of level repulsion and correlations. To study these correlations, we can consider the energy spectrum as a discrete signal, and the sequence of energy levels as a time series. As we shall see, examination of the power spectrum of energy level fluctuations reveals very accurate power laws for completely regular or completely chaotic Hamiltonian quantum systems. It turns out that chaotic systems have $1/f$ noise, in contrast to the brown noise of regular systems [9,10,11]

2. Atomic nuclei in the pf-shell

To study the T dependence of nuclear chaos, we have performed a detailed comparative analysis of spectral properties of the A = 46, 48, 50 and 52 Ca and Sc isotopes and ^{46}Ti [12]. The $T = T_z$ states are considered in all the cases. We follow the shell-model procedure to obtain the low energy levels for a given nucleus. The valence space is made of the $f_{7/2}$, $p_{3/2}$, $f_{5/2}$ and $p_{1/2}$ shells, and the the single-particle energies are taken from the experiment. The residual two-body interaction is a monopole improved version of the Kuo-Brown interaction called KB3 [13]. The construction and subsequent diagonalization of the JT matrices was carried out using the code NATHAN [14]. The dimensions of the shell-model $J^\pi T$ matrices involved are very large, up to 36287.

The first neighbor spacing distribution $P(s)$ has been studied including all the levels up to 5 MeV and 10 MeV above the yrast line, and without any cutoff. The unfolding is performed for each $J^\pi, T = T_z$ pure sequence separately and then the unfolded spacings are gathered into a single set for each nucleus to get better statistics. The values of the Brody parameter ω are displayed in Fig. 1 separated in three sub-panels according to the energy cutoff. Up to 5 MeV above the yrast line, Ca isotopes show spectral fluctuations intermediate between those of regular and chaotic systems, except ^{52}Ca which essentially is a regular system. On the contrary, all the Sc isotopes and ^{46}Ti are very close to GOE fluctuations. For a given A, the big differences between Sc and Ca isotopes must be due to the residual two-body interaction, because the single particle energies are the same in both cases. It was argued [15] that the neutron-neutron interaction is much weaker that the neutron-proton interaction and thus the central field motion is less affected by the former interaction.

Another interesting feature observed in Fig. 1 is that the ω parameter for the Ca isotopes shows a strong fall from A = 48 to A = 52, where $\omega = 0.25$. This astonishing result means that the two-body interaction is almost unable to perturb the single-particle motion in the low energy levels of ^{52}Ca. The ω value for ^{46}Ti has been included in Fig. 1 to show that, for all the energies, the Brody parameter already reaches its maximum for ^{46}Sc. Therefore, according to short range level correlation behavior, replacing a single neutron by a proton in Ca isotopes causes a transition from a quasi-regular to a chaotic regime. It is remarkable that this transition takes place

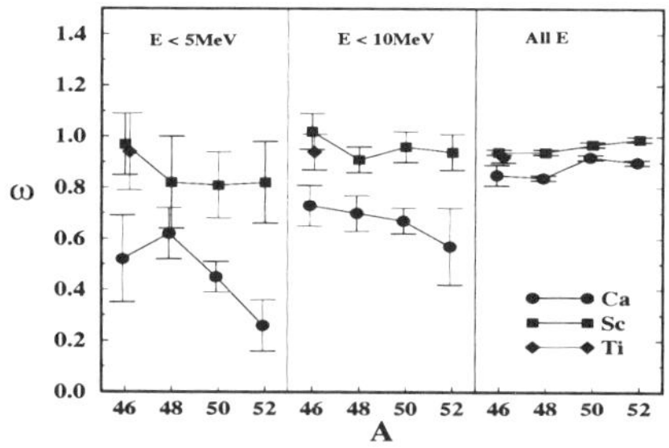

Figure 1. Brody parameter ω for the $A = 46$, 48, 50 and 52 Ca and Sc isotopes and ^{46}Ti, using all the energy levels up to 5 MeV and 10 MeV over the yrast line, and the full spectrum.

abruptly at all excitation energies in all the isotopes. A second replacement of a neutron by a proton does not seem to produce appreciable effects.

In order to confirm the previous results, we have computed the $\Delta_3(L)$ statistic for some $J^\pi, T = T_z$ sequences. To obtain the $\Delta_3(L)$ value for each L, we take its average value over many overlapping intervals of L unfolded spacings along the whole spectrum. Therefore, the results given below concern the full spectrum and not only the low energy region. Fig. 2 shows $\Delta_3(L)$ values for $L \leq 50$, using the $J^\pi = 0^+$, $T = T_z$ levels of ^{46}Ca, ^{46}Sc and ^{46}Ti. Of the three nuclei, only ^{46}Ti follows the GOE line, at least until $L = 50$. For ^{46}Sc the Δ_3 is close to GOE predictions up to a certain separation value, $L_{sep} \simeq 30$ where it upbends from the GOE curve. In ^{46}Ca the upbending starts at a smaller value $L_{sep} \simeq 10$. The upbending from the GOE curve and a linear growth of this statistic reveals a departure from the chaotic regime. The Δ_3 behavior clearly shows a strong isospin dependence in the $A = 46$ nuclei, with chaoticity increasing as T decreases. This happens not only from Ca to Sc, but also from Sc to Ti. The same phenomenon is observed for other J values.

As another example, we compare in Fig. 3 the Δ_3 values for the $J^\pi = 0^+$, $T = T_z$ spectra of ^{52}Ca and ^{52}Sc. Here we see again that ^{52}Sc is clearly more chaotic than ^{52}Ca. Notice that in this case $L_{sep} > 50$ for ^{52}Sc, and comparing with Fig. 2 it seems to be more chaotic than ^{46}Sc. The main reason may be that there are more proton-neutron interactions in ^{52}Sc.

Summarizing the analysis of the spectral fluctuations, there exists a clear excitation energy and isospin dependence in the chaoticity degree of nuclear motion. It increases from Ca to Sc and from Sc to Ti. It is observed

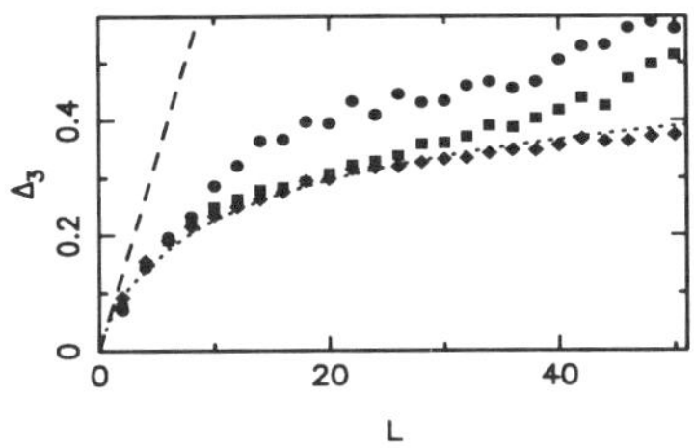

Figure 2. Average Δ_3 for all the $J^\pi = 0^+, T = T_z$ levels of ^{46}Ca (dots), ^{46}Sc (squares) and ^{46}Ti (diamonds). The dotted and dashed curves represent the GOE and Poisson Δ_3 values, respectively.

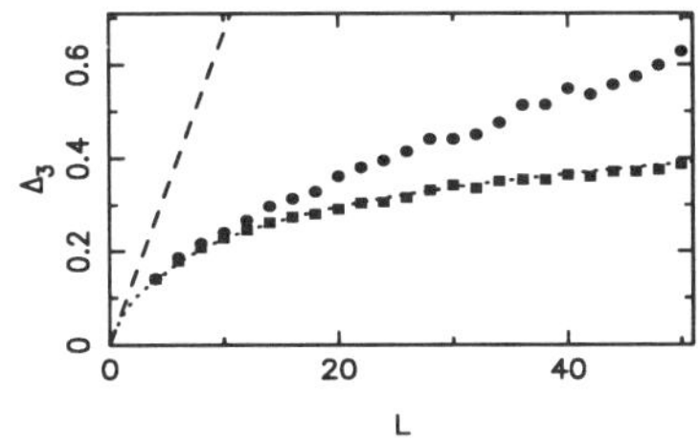

Figure 3. Same as Fig. 2 for all the $J^\pi = 5^+, T = T_z$ levels of ^{52}Ca and ^{52}Sc.

not only in the ground state region, but along the whole spectrum. When the full spectrum is taken into account, the $P(s)$ distribution is not very sensitive to the isospin dependence, but the effect is clearly seen in the Δ_3 statistic.

3. Spectral statistics of baryons

It is well known that the number of baryons predicted by quark models [16,17] is larger than what is observed experimentally. This fact raises the problem of *missing* resonances, which has lead to an important experimental effort to search for these missing states [18,19]. Baryons are aggregates of partons, and consequently their low-lying mass spectrum consists on all the possible bound states and resonances that stem from an interacting many-body quantum system. Recently, spectral fluctuation analysis techniques have been applied to the problem of baryon *missing* resonances [8]. In order to perform this analysis the spectra were separated into sets of levels having the same spin, isospin, parity and strangeness.

The nearest neighbor spacing distribution $P(s)$ has been calculated and compared for quark models and experimental spectra. Results are shown in Figs. 4 and 5. It is clearly seen that the experimental spectrum is close to GOE predictions. By contrast, the theoretical quark model results are close to the Poisson distribution. In order to perform a finer analysis of the

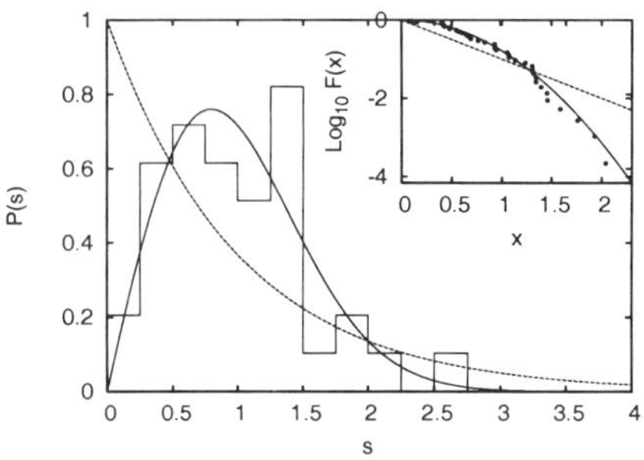

Figure 4. Experimental $P(s)$ distribution for the baryon mass spectrum (histogram) compared to the GOE result(solid line) and to the Poisson distribution (dashed line). Inset shows the hole function $F(s)$ in logarithmic scale.

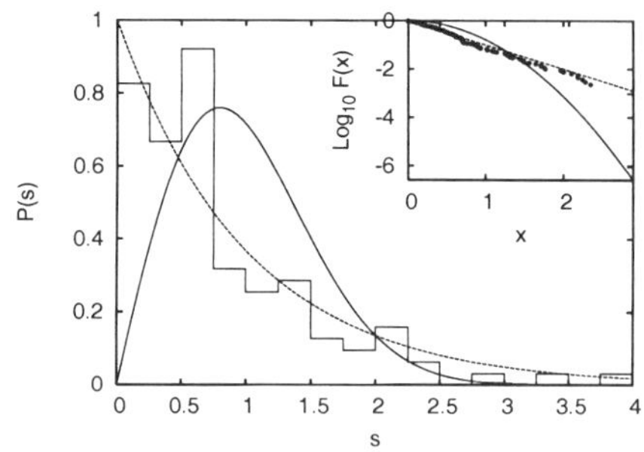

Figure 5. Same as Fig. 4 for the theoretical spectrum predicted by the quark model reported on reference 17.

tail region one can compare the behavior of the hole function

$$F(s) = 1 - \int_0^s P(x)dx. \tag{1}$$

The results are displayed in the same figures and it is clearly seen that the experimental $F(s)$ fits very well the GOE curve, while the theoretical $F(s)$ follows the Poisson one. If quark model spectra were correct we should expect a good agreement with the real experimental spectral fluctuations. If the observed experimental spectra is incomplete, the effect of the missing levels on the $P(s)$ distribution would be a clear deviation towards the Poisson limit. Thus, the experimental $P(s)$ distribution should be much closer to Poisson than the theoretical one. But, as Figs. 4 and 5 show, the situation is just the opposite.

In summary, the statistical analysis of spectral fluctuations clearly shows that present quark models are not able to give reliable predictions on missing resonances in the experimental baryon mass spectrum.

4. Time series approach to quantum chaos

As we have seen in the previous sections, generally two suitable statistics are used to study fluctuation properties of quantum energy spectra. The $P(s)$

distribution gives information on the short range correlations among the energy levels. The $\Delta_3(L)$ statistic makes it possible to study correlations of length L.

However, a new approach to study spectral fluctuations in quantum systems has been recently presented. [9] The basic idea is to consider the energy spectrum as a time series where energy plays the role of time. Using this formal analogy spectral fluctuations can be studied by means of techniques borrowed from time series analysis. Since the fluctuation properties of chaotic and regular quantum systems have universal character we can investigate whether the emerging properties from this analogy are related to some of the universal features that appear in many complex systems.

We can characterize the spectral fluctuations by the statistic δ_n [21] defined by

$$\delta_n = \sum_{i=1}^{n} (s_i - \langle s \rangle), \quad n = 1, 2, \dots N-1, \tag{2}$$

where $s_i = \epsilon_{i+1} - \epsilon_i$ is the spacing between two consecutive unfolded energy levels, and N is the total number of levels. Since $< s >= 1$, the δ_n function represents the deviation of the unfolded excitation energy from its mean value n.

One of the simplest methods to analyze correlations in a time series is the study of its power spectrum, which provides information on the correlations at all time scales. The power spectrum $S(k)$ of a discrete and finite series δ_n is given by $S(k) = \left|\hat{\delta}_k\right|^2$, where $\hat{\delta}_k$ is the Fourier transform of δ_n,

$$\hat{\delta}_k = \frac{1}{N} \sum_n \delta_n \exp\left(\frac{2\pi i k n}{N}\right). \tag{3}$$

The first systems studied by this method were the spectra of atomic nuclei calculated using the shell model with configuration mixing. Fig. 6 shows the results for a typical stable *sd* shell nucleus, ^{24}Mg, and for a very exotic nucleus, ^{34}Na, in the *sd* proton and *pf* neutron shells. Clearly, the power spectrum of δ_n follows closely a power law. We may assume the simple functional form $\langle S(k) \rangle \sim 1/k^{\alpha}$.

A least squares fit to the data of Fig. 6 gives $\alpha = 1.11 \pm 0.03$ for ^{34}Na, and $\alpha = 1.06 \pm 0.05$ for ^{24}Mg. These results raised the question of whether there is a general relationship between quantum chaos and the power spectrum of the δ_n fluctuations of the system. A similar numerical calculation performed for the three classical random matrix ensembles GOE, GUE and

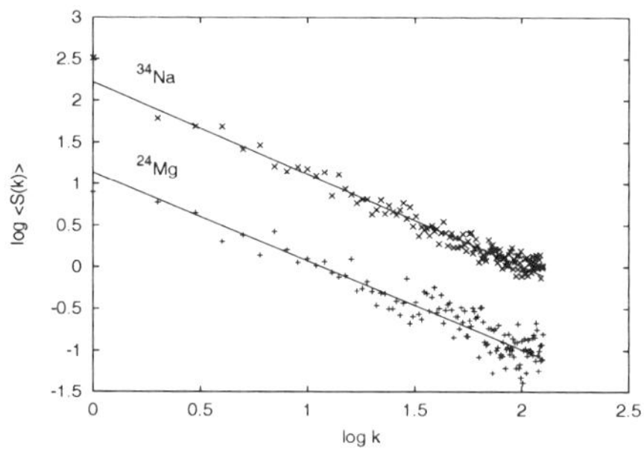

Figure 6. Average power spectrum of the δ_n function for ^{24}Mg and ^{34}Na, using 25 sets of 256 levels from the high level density region. The plots are displaced to avoid overlapping.

GSE showed that the power spectrum $\langle S(k)\rangle$ follows quite accurately a power law of type $1/k$ for all of them. Therefore, the spectral fluctuations of chaotic quantum systems described by the δ_n function exhibit $1/f$ noise [9]. As is well known, the existence of $1/f$ noise is a remarkable and very ubiquitous property of many complex systems in nature and in social sciences.

Finding the origin of the $1/f$ noise in the time fluctuations of these systems is an important open problem. In the case of quantum systems, an exact and complete proof of the $1/f$ noise behavior in the power spectrum of the δ_n statistic seems to be extremely difficult. However, it can be theoretically studied in semiclassical systems or random matrix ensembles, where the mathematical tractability of these systems may help to understand the origin of the $1/f$ noise in quantum systems.

We have recently shown that the power spectrum of δ_n for fully chaotic or integrable systems can be written in terms of the ensemble form factor $K(\tau)$ as follows, [10]

$$\langle S(k)\rangle_\beta = \frac{N^2}{4\pi^2}\left\{\frac{K\left(\frac{k}{N}\right)-1}{k^2}+\frac{K\left(1-\frac{k}{N}\right)-1}{(N-k)^2}\right\} + \frac{1}{4\sin^2\left(\frac{\pi k}{N}\right)} + \Delta, \quad k=1,2,\cdots,N-1, \quad N\gg 1. \tag{4}$$

where β is the repulsion parameter of RMT ensembles and takes the values

$\beta = 1$ for GOE, $\beta = 2$ for GUE, and $\beta = 4$ for GSE [21]. Here $\Delta = 0$ for integrable systems and $\Delta = -1/12$ for chaotic systems. This equation, together with the appropriate values of $K(\tau)$, gives explicit expressions of $\langle S(k) \rangle$ for specific ensembles or systems.

When $k \ll N$ the first term of Eq. (4) becomes dominant and we can write $\langle S(k) \rangle_\beta = \dfrac{N}{2\beta\pi^2 k}$ for chaotic systems and $\langle S(k) \rangle_\beta = \dfrac{N^2}{4\pi^2 k^2}$ for integrable ones. These expressions show that, for small frequencies, the excitation energy fluctuations exhibit $1/f$ noise in chaotic systems and $1/f^2$ noise in integrable systems. As we shall see below, these power laws are also approximately valid through almost the whole frequency domain, due to partial cancellation of higher order terms. Only near $k = N/2$ the effect of these terms becomes appreciable.

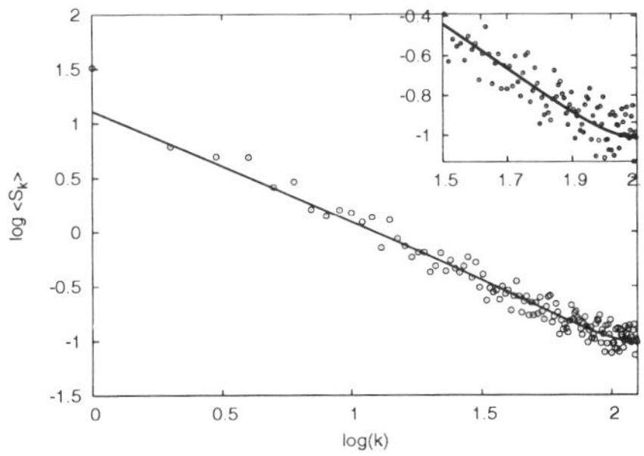

Figure 7. Numerical average power spectrum of the δ_q function for ^{34}Na, calculated using 25 sets of 256 consecutive levels from the high level density region, compared to the parameter free theoretical values (solid line) for GOE.

To test all these theoretical expressions we have compared their predictions to numerical results obtained for different RMT ensembles, shell model spectra of atomic nuclei and chaotic and regular quantum billiards. In all the cases the agreement between Eq. 4 and numerical calculations of the power spectrum $\langle S(k) \rangle$ is excellent. As an example Fig. 7 displays the theoretical values of $\langle S(k) \rangle$ compared with the shell model results for ^{34}Na. As it can be seen Eq. 4 gives a $1/k$ behavior up to quite high values of the frequency counter k, and the agreement with the power spectrum of ^{34}Na is excellent.

5. Conclusions

In contrast with the broadly accepted idea that the atomic nucleus is a chaotic system we have shown the existence of an energy and isospin dependence of the degree of chaoticity. In some nuclei, like in Ca isotopes, the spectral fluctuations of the low-energy spectrum are closer to the Poisson limit than to GOE. This result is strongly related to the weakness of the $T = 1$ part of the residual nucleon-nucleon interaction, which is not able to disturb the nucleon mean-field motion. By contrast, when one of the neutrons in Ca isotopes is replaced by a proton, the $T = 0$ part of the residual nucleon-nucleon interaction strongly disturbs the mean field and Sc isotopes become fully chaotic, even at low energies.

The statistical analysis of the baryon mass spectrum shows that the nearest neighbor $P(s)$ distribution predicted by quark models is closer to the Poisson distribution, while the experimental result is close to GOE. This is opposite to what one should expect if the experimental spectrum is plagued with the missing levels predicted by quark models. Thus, it can be concluded that these models are not able to give reliable predictions on missing resonances in the experimental baryon mass spectrum.

Finally, we have presented a new approach to the study of spectral fluctuations in quantum systems. The δ_n function can be considered as a time series, where the level order index n plays the role of a discrete time. The power spectrum $\langle S(k)\rangle$ of δ_n has been studied for representative energy spectra of regular and chaotic quantum systems. It has been shown that a power law of type $\langle S(k)\rangle \sim 1/k^{\alpha}$ arises both in chaotic and regular systems. For Poisson spectra, we get $\alpha = 2$, as expected for independent random variables. For spectra of atomic nuclei at higher energies, in regions of high level density, and for the GOE, GUE and GSE ensembles, we obtain $\alpha = 1$.

These results suggest the conjecture that *chaotic quantum systems are characterized by $1/f$ noise in the energy spectrum fluctuations.* This property is not a mere statistic to measure the chaoticity of the system. It provides an intrinsic characterization of quantum chaotic systems without any reference to the properties of RMT ensembles. As is well known $1/f$ noise is quite ubiquitous. It characterizes sunspot activity, the flow of the Nile river, music, and chronic illness [26]. And we believe that it characterizes quantum chaos as well.

This work is supported in part by Spanish Government grants FIS2006-12783-C03-02 and FTN2000-0963-C02.

References

1. M. V. Berry and M. Tabor, Proc. R. Soc. London A **356**, 375 (1977).
2. O. Bohigas, M. J. Giannoni and C. Schmit, Phys. Rev. Lett. **52**, 1 (1984).
3. T. Guhr, A. Müller-Groeling and H. A. Weidenmüller, Phys. Rep. **299**, 189 (1998).
4. H. J. Stöckmann, *Quantum Chaos*, (Cambridge University Press, Cambridge, 1999).
5. R. U. Haq, A. Pandey and O. Bohigas, Phys. Rev. Lett. **48**, 1086 (1982).
6. J. F. Shriner jr., G. E. Mitchell and T. von Egidy, Z. Phys. A**338**, 309 (1991).
7. J. D: Garret, J. R. German, L. Courtney and J. M. Espino, in Future Directions in Nuclear Physics, eds. J. Dudek and B. Haas (A. I. P., N.Y., 1992) p. 345.
8. C. Fernández-Ramírez and A. Relaño, Phys. Rev. Lett. **98**, 062001 (2007).
9. A. Relaño, J.M.G. Gómez, R. A. Molina, J. Retamosa, and E. Faleiro, Phys. Rev. Lett. **89**, 244102 (2002).
10. E. Faleiro, J.M.G. Gómez, R. A. Molina, L. Muñoz, A. Relaño, and J. Retamosa, Phys. Rev. Lett. **93**, 244101 (2004).
11. J. M. G. Gómez, A. Relaño, J. Retamosa, E. Faleiro, L. Salasnich, and M. Vranicar, and M. Robnik, Phys. Rev. Lett. **94**, 084101 (2005)
12. R. A. Molina, J. M. G. Gómez and J. Retamosa, Phys. Rev. C**63**, 014311 (2000).
13. A. Poves and A. Zuker, Phys. Rep. 70,4 (1981).
14. E. Caurier, G. Martínez-Pinedo, F. Nowacki, A. Poves, J. Retamosa, and A. P. Zuker, Phys. Rev. C **59**, 2033 (1999).
15. J. M. G. Gómez, V. R. Manfredi L. Salasnich and E. Caurier, Phys. Rev. C**58**,2108 (1998).
16. A. J. Hey and R. L. Kelly, Phys. Rep. **96**, 71 (1983).
17. S. Capstick and W. Roberts, Prog. Nucl. Phys **45** S241 (2000).
18. W.-M. Yao *et al.* J. Phys. G **33**, 1 (2006).
19. W. Crede *et al.* Phys. Rev. Lett **94**, 012004 (2005).
20. J. M. G. Gómez, R. A. Molina, A. Relaño and J. Retamosa, Phys. Rev. E, in press, (2002).
21. M. L. Mehta, *Random Matrices*, (Academic Press, 1991)
22. O. Bohigas, P. Leboeuf and M.J. Sánchez, Physica D **131**, 186 (1999).
23. N. P. Greis and H. S. Greenside, Phys. Rev A **44**, 2324 (1991).
24. A. Relaño, J. M. G. Gómez and E. Faleiro, to be published.
25. H. G. Schuster, *Deterministic Chaos: an introduction* (Weinheim VCH, 1995).
26. B. B. Mandelbrot, *Multifractals and 1/f noise* (Springer, New York, 1999).

LIST OF PARTICIPANTS

V. Abrosimov	Institute for Nuclear Research 47 Prospect Nauki, 03028 Kiev, UKRAINE abrosim@kinr.kiev.ua
F. Andreozzi	Dipartimento di Scienze Fisiche Università di Napoli Federico II Complesso Universitario di Monte S. Angelo Via Cintia, 80126 Napoli, ITALY andreozzi@na.infn.it
G. de Angelis	Laboratori Nazionali di Legnaro Viale dell'Università 2, 35020 Legnaro (Padova), ITALY deangelis@lnl.infn.it
S. Antalic	Comenius University Mlynska dolina, Bratislava 84248, SLOVAKIA s.antalic@gsi.de
J. Äystö	Department of Physics, P.O. Box 35 (YFL) FI-40014 University of Jyvaskyla, FINLAND juha.aysto@phys.jyu.fi
F. Azaiez	Institut de Physique Nucléaire CNRS-IN2P3 – Université Paris-Sud F-91406 Orsay Cedex, FRANCE azaiez@ipno.in2p3.fr
C. Baktash	Oak Ridge National Laboratory, Physics Division BLDG 6000, Oak Ridge, TN 37831-6371, USA baktashc@ornl.gov
D. Balabanski	Institute for Nuclear Research and Nuclear Energy Bulgarian Academy of Sciences 72 Blvd. Tzarigradsko chaussee Sofia 1784 , BULGARIA dimiter.balabanski@unicam.it
N. Benczer-Koller	Department of Physics and Astronomy Rutgers University, New Brunswick, NJ 08903, USA nkoller@physics.rutgers.edu

P. G. Bizzeti

Dipartimento di Fisica, Università di Firenze
Via G. Sansone 1, 50019 Sesto Fiorentino, ITALY
bizzeti@fi.infn.it

A. M. Bizzeti-Sona

Dipartimento di Fisica, Università di Firenze
Via G. Sansone 1, 50019 Sesto Fiorentino, ITALY
Annamaria.Sona@fi.infn.it

A. Brondi

Dipartimento di Scienze Fisiche
Università di Napoli Federico II
Complesso Universitario di Monte S.Angelo
Via Cintia, 80126 Napoli, ITALY
brondi@na.infn.it

B. A. Brown

NSCL, Michigan State University
East Lansing, MI 48824-1321, USA
brown@nscl.msu.edu

X. Cai

Shanghai Institute of Applied Physics
Chinese Academy of Sciences, P.O. Box 800-204
Shanghai 201800, CHINA
caixz@sinap.ac.cn

L. Coraggio

Istituto Nazionale di Fisica Nucleare, Sezione di Napoli
Complesso Universitario di Monte S. Angelo
Via Cintia, 80126 Napoli, ITALY
coraggio@na.infn.it

A. Covello

Dipartimento di Scienze Fisiche
Università di Napoli Federico II
Complesso Universitario di Monte S. Angelo,
Via Cintia, 80126 Napoli, ITALY
covello@na.infn.it

N. D. Dang

RIKEN
2-1 Hirosawa, Wako, 351-0198 Saitama, JAPAN
dang@rikaxp.riken.go.jp

A. Dellafiore

Istituto Nazionale di Fisica Nucleare, Sezione di Firenze
Via G.Sansone 1, 50019 Sesto Fiorentino, ITALY
della@fi.infn.it

S. Dimitrova
Institute for Nuclear Research and Nuclear Energy
72 Blvd. Tzarigradsko chaussee
Sofia 1784, BULGARIA
sevdim@inrne.bas.bg

J. P. Draayer
Southeastern Universities Research Association/ SURA
1201 New York Avenue NW, Suite 430,
Washington DC, 20005, USA
draayer@sura.org

T. Dytrych
Department of Physics and Astronomy
Lousiana State University
Baton Rouge, Lousiana, 70803-4001, USA
tdytrych@phys.lsu.edu

P. Egelhof
GSI Darmstadt
Planckstrasse 1, 64291 Darmstadt, GERMANY
p.egelhof@gsi.de

A. Faessler
University of Tübingen, Institute for Theoretical Physics
Auf der Morgenstelle 14
D-72076 Tübingen, GERMANY
amand.faessler@uni-tuebingen.de

C. Fahlander
Division of Cosmic and Subatomic Physics
Lund University, Box 118
Lund, SE-22100, SWEDEN
claes.fahlander@nuclear.lu.se

D. Fang
Shanghai Institute of Applied Physics
Chinese Academy of Sciences, P. O. Box 800-204
Shanghai 201800 CHINA
dqfang@sinap.ac.cn

L. S. Ferreira
Centro de Fisica Int. Fundamentais, IST
Av. Rovisco Pais, Lisboa, 1049-001 PORTUGAL
flidia@ist.utl.pt

B. Fornal
Institute of Nuclear Physics PAN
ul. Radzikowskiego 152, 31-342 Kraków, POLAND
bogdan.fornal@ifj.edu.pl

M. Gai
University of Connecticut
1084 Shennecossett Rd., Groton, CT 06340, USA
moshe.gai@yale.edu

A. Gargano — Istituto Nazionale di Fisica Nucleare, Sezione di Napoli
Complesso Universitario di Monte S. Angelo
Via Cintia, 80126 Napoli, ITALY
gargano@na.infn.it

J. Ginocchio — Los Alamos National Laboratory, MS B283
Los Alamos, New Mexico 87545, USA
gino@lanl.gov

J. M. Gómez — Departamento de Fisica Atómica, Molecular y Nuclear
Facultad de Ciencias Fisicas, Universidad Complutense
28040 Madrid, SPAIN
gomezk@nuc1.fis.ucm.es

P. M. Gore — 629 Country Club Lane
Nashville, TN 37205-4692, USA
p.m.gore.phys@earthlink.net

P. Guazzoni — Dipartimento di Fisica, Università di Milano,
Via Celoria 16, 20133 Milano, ITALY
paolo.guazzoni@mi.infn.it

G. Hagen — ORNL
Physics Division, Bldg. 6025, M.S. 6373
Oak Ridge, TN 37831-6373, USA
hageng@ornl.gov

I. Hamamoto — Matematisk Fysik, LTH, University of Lund
P.O.Box 118, Lund S-22100, SWEDEN
ikuko@matfys.lth.se

J. H. Hamilton — Vanderbilt University
6301 Stevenson Center, Nashville, TN 37235, USA
j.h.hamilton@vanderbilt.edu

A. Heusler — Max-Planck-Institut für Kernphysik
Gustav-Kirchhoff-Strasse 7/1
Heidelberg, 69120 GERMANY
A.Heusler@mpi-hd.mpg.de

M. Horoi — Central Michigan University, Department of Physics,
Mount Pleasant, Michigan 48859 , USA
horoi@phy.cmich.edu

N. Q. Hung	Heavy Ion Nuclear Physics Laboratory Nishina-center for Accelerator-based Science, RIKEN 2-1 Hirosawa, Wako city, 351-0198 Saitama, JAPAN nqhung@riken.jp
N. Itaco	Dipartimento di Scienze Fisiche Università di Napoli Federico II Complesso Universitario di Monte S. Angelo Via Cintia, 80126 Napoli, ITALY itaco@na.infn.it
E. F. Jones	629 Country Club Lane Nashville, TN 37205-4692, USA e.f.jones.phys@earthlink.net elizabethfj@earthlink.net
F. Knapp	Institute of Particle and Nuclear Physics Charles University, V Holesovickach 2 180 00 Praha 8, CZECH REPUBLIC knapp@ipnp.troja.mff.cuni.cz
T. T. S. Kuo	Physics Department, State University of New York at Stony Brook, Stony Brook, NY 11794-3800, USA kuo@tonic.physics.sunysb.edu
J. Kvasil	Institute of Particle and Nuclear Physics Charles University, V Holesovickach 2 180 00 Praha 8, CZECH REPUBLIC kvasil@ipnp.troja.mff.cuni.cz
G. La Rana	Dipartimento di Scienze Fisiche Università di Napoli Federico II Complesso Universitario di Monte S.Angelo Via Cintia, 80126 Napoli, ITALY larana@na.infn.it
S. Leoni	Dipartimento di Fisica, Università di Milano, Via Celoria 16, 20133 Milano, ITALY silvia.leoni@mi.infn.it
N. Lo Iudice	Dipartimento di Scienze Fisiche Università di Napoli Federico II Complesso Universitario di Monte S. Angelo Via Cintia, 80126 Napoli, ITALY loiudice@na.infn.it

S. Lunardi — Dipartimento di Fisica, Università di Padova
Via Marzolo 8, 35131 Padova, ITALY
lunardi@pd.infn.it

Y. Ma — Shanghai Institute of Applied Physics
Chinese Academy of Sciences, P. O. Box 800-204
Shanghai 201800 CHINA
ygma@sinap.ac.cn

H. Mach — Uppsala University, Angströmlaboratoriet
Lägerhyddsv. 1, Uppsala, SE-751 21, SWEDEN
Henryk.Mach@tsl.uu.se

L. Moretto — Lawrence Berkeley National Laboratory
BLDG88, Berkeley, CA 94720, USA
lgmoretto@lbl.gov

R. Moro — Dipartimento di Scienze Fisiche
Università di Napoli Federico II
Complesso Universitario di Monte S.Angelo
Via Cintia, 80126 Napoli, ITALY
moro@na.infn.it

P. von Neumann-Cosel — Institut für Kernphysik, TU Darmstadt,
Schlossgartenstrasse 9, 64289 Darmstadt, GERMANY
vnc@ikp.tu-darmstadt.de

R. Orlandi — Laboratori Nazionali di Legnaro
Viale dell'Universitá 2,
35020 Legnaro (Padova), ITALY
orlandi@lnl.infn.it

F. Palumbo — Laboratori Nazionali di Frascati
00044 Frascati, ITALY
Fabrizio.Palumbo@lnf.infn.it

N. Pietralla — Institut für Kernphysik, TU Darmstadt
Schlossgartenstrasse 9, 64289 Darmstadt, GERMANY
pietralla@ikp.tu-darmstadt.de

G. Pisent — Dipartimento di Fisica, Università di Padova
Via Marzolo 8, 35131 Padova, ITALY
pisent@pd.infn.it

G. Prete	Laboratori Nazionali di Legnaro Viale dell'Universitá 2 35020 Legnaro (Padova), ITALY prete@lnl.infn.it
A. A. Raduta	University of Bucharest and IFIN-HH 407 Atomistilor Str. Bucharest-Magurele, 077125 ROMANIA raduta@ifin.nipne.ro
C. M. Raduta	IFIN-HH 407 Atomistilor Str. Bucharest-Magurele, 077125 ROMANIA raduta@rocketmail.com
A. V. Ramayya	Vanderbilt University, Physics Department Box 1807 Station B, Nashville, TN 37235, USA a.v.ramayya@vanderbilt.edu
R. Schiavilla	Jefferson Lab Theory Group, Jefferson Laboratory 12000 Jefferson Avenue Newport News, VA 23606, USA schiavil@jlab.org
D. Schwalm	Max Planck Institut für Kernphysik Saupferch-Eckweg 1, D 69117 Heidelberg, GERMANY schwalm@mpi-hd.mpg.de
R. Schwengner	Institut für Strahlenphysik Forschungszentrum Rossendorf, PF 510119 01314 Dresden, GERMANY R.Schwengner@fz-rossendorf.de
A. Schwenk	TRIUMF, 4004 Wesbrook Mall Vancouver, B.C. V6T 2A3, CANADA schwenk@triumf.ca
A. Shotter	TRIUMF, 4004 Wesbrook Mall Vancouver, B.C. V6T 2A3, CANADA ashotter@triumf.ca
I. Sick	Dept. für Physik und Astronomie Klingelbergstrasse 82, CH4056 Basel, SWITZERLAND ingo.sick@unibas.ch

P. Sona — Dipartimento di Fisica, Università di Firenze
Via G.Sansone 1, 50019 Sesto Fiorentino, ITALY
sona@fi.infn.it

Ch. Stoyanov — Institute for Nuclear Research and Nuclear Energy
72 Blvd. Tzarigradsko chaussee
Sofia 1784, BULGARIA
stoyanov@inrne.bas.bg

K. Sugawara-Tanabe — Otsuma Women's University
Karakida, Tama, Tokyo 206-8540, JAPAN
tanabe@otsuma.ac.jp

A. Sushkov — Joint Institute for Nuclear Research
Bogoliubov Laboratory of Theoretical Physics
141980 Dubna, RUSSIA
sushkov@theor.jinr.ru

I. Talmi — Department of Physics
The Weizmann Institute of Science
Rehovot 76100, ISRAEL
igal.talmi@weizmann.ac.il

K. Tanabe — Department of Physics, Saitama University
Shimo-Okubo 255, Sakura-ku
Saitama 338-8570, JAPAN
kosai@muj.biglobe.ne.jp

M. Thoennessen — NSCL
1 Cyclotron, East Lansing, MI 48824, USA
thoennessen@nscl.msu.edu

M. Tomaselli — GSI
Planckstrasse 1, D64291 Darmstadt, GERMANY
m.tomaselli@gsi.de

H. Ueno — RIKEN
2-1 Hirosawa, Wako City 351-0198, Saitama, JAPAN
ueno@riken.jp

N. Van Giai — Institut de Physique Nucléaire
Université Paris-Sud, F-91406 Orsay Cedex, FRANCE
nguyen@ipno.in2p3.fr

J. P. Vary	Department of Physics and Astronomy Iowa State University, Ames, IA 50011, USA jvary@iastate.edu
P. Vesely	Institute of Particle and Nuclear Physics Charles University, V Holesovickach 2 180 00 Praha 8, CZECH REPUBLIC vesely@ipnp.troja.mff.cuni.cz
A. Volya	Florida State University 208 Keen building, Tallahassee, FL 32306, USA volya@phy.fsu.edu
V. Voronov	Bogoliubov Laboratory of Theoretical Physics Joint Institute for Nuclear Research 141980 Dubna, Moscow Region, RUSSIA voronov@thsun1.jinr.ru
J. Wambach	Institut für Kernphysik, TU Darmstadt, Schlossgartenstrasse 9, 64289 Darmstadt, GERMANY j.wambach@gsi.de
L. Zetta	Dipartimento di Fisica, Università di Milano Via Celoria 16, 20133 Milano, ITALY luisa.zetta@mi.infn.it
A. Zilges	Universität zu Köln Zülpicher Str. 77, 50937 Köln, GERMANY zilges@ikp.uni-koeln.de

AUTHOR INDEX

Abrosimov, V.I. 495
Ahn, T. 325
Ahsan, N. 539
Andersson, L.-L. 65
Andreozzi, F. 419, 461
Andrighetto, A. 111
Antonucci, C. 111
Arndt, O. 263
Atanasova, L. 65
Äystö, J. 29
Azaiez, F. 39

Bahri, C. 183, 191
Balabansky, D. L. 65
Barbui, M. 111
Baumann, T. 23
Bayman, B.F. 315
Bazin, D. 23
Becker, F. 65
Bednarczyk, P. 65
Beghini, S. 283
Benczer-Koller, N. 49
Benouaret, N. 355
Bentley, M.A. 65
Beyer, R. 355
Biasetto, L. 111
Bisoffi, G. 111
Bizzeti, P.G. 363
Bizzeti-Sona, A.M. 363
Blazhev, A. 65, 263
Boelaert, N. 263
Boie, H. 523
Borge, M.J.G. 263
Boutami, R. 263
Bracco, A. 307
Bradley, H. 263
Brandau, C. 65
Brandolini, F. 307
Braun, N. 263
Brink, D.M. 495
Broda, R. 283
Brown, B.A. 243, 263
Brown, G.E. 153
Brown, J. 23, 65
Burda, O. 325

Caceres, L. 65
Cai, X.Z. 81
Carpenter, M.P. 283, 307
Carturan, S. 111
Celona, L. 111
Cervellera, F. 111
Cevolani, S. 111
Che, X.L. 387
Chen, J.G. 81
Chiara, C.J. 307
Chines, F. 111
Cinausero, M. 111
Cole, J.D. 387
Colombo, P. 111
Comunian, M. 111
Coraggio, L. 253, 273
Corradi, L. 283
Covello, A. 253, 263, 273, 315
Crowell, H.L. 395
Cuttone. G. 111

Dainelli, A. 111
Dang, N.D. 477, 503
Daniel, A.V. 57
Dean, D.J. 173
de Angelis, G. 73, 283, 307
Dellafiore, A. 495
Della Vedova, F. 283, 307
DeYoung, P.A. 23
Di Bernardo, P. 111

Dimitrov, V. 387
Dlouhy, Z. 263
Donangelo, R. 387
Dönau, F. 355
Doornenbal, P. 65
Draayer, J.P. 183, 191
Dytrych, T. 183, 191

Egelhof, P. 97
Elliott, J.B. 513
Entem, D.R. 273
Erhard, M. 355

Faessler, A. 199
Faestermann, T. 293, 315, 371
Fagotti, E. 111
Fahlander, C. 65
Faleiro, E. 547
Fang D.Q. 81
Farnea, E. 283, 307
Finck, J.E. 23
Fioretto, E. 283
Fornal, B. 283
Fraile, L.M. 263
Frank, N. 23
Fransen, C. 263
Frauendorf, S. 387
Fritzsche, S. 89
Fynbo, H.O.U. 263

Gade, A. 23
Gadea, A. 283, 307
Gai, M. 531
Gargano, A. 253, 263, 273, 315
Garnsworthy, A.B. 65
Gelberg, A. 387
Gerl, J. 65
Giacchini, M. 111
Ginocchio, J.N. 301
Gómez, J.M.G. 547
Goodin, C. 57, 387, 395
Gore, P.M. 387, 395
Górska, M. 65
Gramegna, F. 111
Graw, G. 293, 315, 371
Grębosz, J. 65
Greene, J. 307
Grosse, E. 355
Guazzoni, P. 315, 371
Gueorguiev, V.G. 163
Guiot, B. 283
Guo, W. 81

Hagemann, G.B. 379
Hagen, G. 173
Hamamoto, I. 379
Hamilton, J.H. 57, 387, 395
Hellström, M. 65
Hertenberger, R. 293, 315, 371
Heusler, A. 293
Hinke, Ch. 263
Hinnefeld, J. 23
Hjorth-Jensen, M. 173
Hoff, P. 263
Hoischen, R. 65
Holt, J.D. 153
Holt, J.W. 153
Honma, M. 283
Hosoi, M. 81
Howes, R. 23
Hung, Q.N. 503
Hwang, J.K. 57, 387, 395

Itaco, N. 253, 273
Izumikawa, T. 81

Janssens, R.V.F. 283
Jaskóla, M. 315, 371
Jentschura, U.D. 523
Jiang, Z. 387
Johansson, E.K. 65
Joinet, A. 263
Jokinen, A. 263
Jolie, J. 263, 293, 387
Jones, E.F. 387, 395
Jungclaus, A. 65
Junghans, A.R. 355

Kanungo, R. 81
Kleinig, W. 437
Klug, J. 355
Knapp, F. 419, 461
Köck, F. 523
Kojouharov, I. 65
Korgul, A. 263
Kosev, K. 355
Köster, U. 263
Kratz, K.-L. 263
Królas, W. 283
Kröll, T. 263
Krücken, R. 293
Kühl, T. 89
Kuo, T.T.S. 153, 273
Kurcewicz, W. 263
Kurz, N. 65
Kvasil, J. 419, 437, 461

Lauer, M. 523
Lauritsen, T. 283
Lecouey, J.-L. 23
Lee, I.Y. 57
Lenzi, S. 307
Leoni, S. 307, 403
Li, K. 57, 387, 395
Lister, C.J. 307
Lo Iudice, N. 419, 453, 461
Lollo, M. 111
Long, W.H. 221
Lunardi, S. 283, 307
Luo, Y.X. 57, 387, 395
Luther, B. 23

Ma, Y. G. 81
Ma, W.C. 81
Mach, H. 263
Machleidt, R. 153, 273
Maggioni, G. 111
Maier, K.H. 293
Mantica, P.F. 283
Manzolaro, M. 111
Mărginean, N. 283, 307
Maris, P. 163
Mason, P. 283
Matera, F. 495
Meneghetti, G. 111
Meng, J. 221
Mengoni, D. 307
Messina, G.E. 111
Milstein, A.I. 523
Molina, R.A 547
Montagnoli, G. 283
Moretto, L.G. 513
Mücher, D. 293
Muñoz, L. 547

Nair, C. 355
Nakada, H. 469
Nakajima, S. 81
Nankov, N. 355
Napoli, D.R. 283, 307
Nara Singh, B.S. 307
Navarro-Pérez, R. 263
Navrátil, P. 163
Nesterenko, V.O. 437
Nogga, A. 163
Nyberg, J. 263

Ohnishi, T. 81
Ohtsubo, T. 81
Orlandi, R. 307
Ormand, W.E. 163
Otsuka, T. 283
Ozawa, A. 81

Palmieri, A. 111
Palumbo, F. 485
Papenbrock, T. 173
Pawlat, T. 283
Pechenaya, O.L. 307
Peters, W.A. 23
Petrovich, C. 111
Phair, L. 513
Pietralla, N. 293, 325
Pietri, S. 65
Piga, L. 111

Pisent, A. 111
Podolyák, Zs. 65
Pollarolo, G. 283
Ponomarev, V. Yu. 371
Porrino, A. 419, 461
Prete, G 111
Prokopowicz, W. 65

Raduta, A.A. 209
Raduta, C.M. 209
Rainovski, G. 325
Ramayya, A.V. 57, 387, 395
Rasmussen, J.O. 57, 387, 395
Re, M. 111
Recchia, F. 307
Regan, P.H. 65
Reillo, E.-M. 263
Reinhard, P.-G. 437
Relaño, A. 547
Ren, Z.Z. 81
Retamosa, J. 547
Reviol, W. 307
Richter, W. 243
Riess, F. 293
Rizzi, V. 111
Rizzo, D. 111
Ruchowska, E. 263
Rudolf, D. 65
Rusev, G. 355

Sagawa, H. 221
Sahin, F. 307
Sarantites, D.G. 307
Savran, D. 345
Scarlassara, F. 283
Schaffner, H. 65
Scheit, H. 23, 523
Schiavilla, R. 135
Schiller, A. 23
Schilling, K.D. 355
Schwalm, D. 523
Schwerdtfeger, W. 263
Seweryniak, D. 283, 307
Seweryukhin, A.P. 445
Schwengner, R. 355
Shen, W.Q. 81
Shirikov, A. 163
Shirikova, N. Yu. 453
Shotter, A. 3
Sick, I. 143
Simpson, G.S. 263
Singh, B. 263
Stanjou, M. 263
Steer, S.J. 65
Stefanescu, I. 387
Stefanini, A.M. 283
Stone, J.R. 57, 387
Stone, N.J. 57, 387
Stoyanov, Ch. 427
Stoyer, M. 57, 387
Suda, T. 81
Sugawara, K. 81
Sugawara-Tanabe, K. 411
Sun, Z.Y. 81
Sushkov, A.V. 453
Suzuki, T. 81
Sviratcheva, K.D. 183, 191
Szilner, S. 283

Takisawa, K. 81
Talmi, I. 231
Tanabe, K. 411, 469
Tanaka, K. 81
Tanihata, I. 81
Tarpanov, D. 427
Tengblad, O. 263
Ter-Akopian, G.M. 57
Terekhov, I.S. 523
Thirolf, P.G. 263
Thoennessen, M. 23
Tian. W.D. 81
Tomaselli, M. 89
Tonev, D. 307
Tonezzer, M. 111
Trotta, M. 283

Ueno, H. 13
Ugryumov, V. 263
Ur, C. 283, 307
Ursescu, D. 89

Valiente-Dobón, J.J. 283, 307
Van Giai, N. 221, 445
Van Isacker, P. 387
Vary, J.P. 163, 183, 191
Vesely, P. 437
Volya, A. 539
von Brentano, P. 293, 387
von Neumann-Cosel, P. 335
Voronov, V.V. 445

Wadsworth, R. 307
Wagner, A. 355
Walters, W.B. 263
Wambach, J. 123
Wang, K. 81
Werner, V. 293
Wiedemann, K.T. 307
Wirth, H.-F. 293, 315, 371
Wood, J.L. 387
Wollersheim, H.-J. 65
Wrzesiński, J. 283

Yamaguchi, T. 81
Yan, T.Z. 81

Zafiropulos, D. 111
Zanonato, D. 111
Zetta, L. 315, 371
Zhang, J.Y. 387
Zhong, C. 81
Zhu, S. 283, 307
Zhu, S.J. 57, 387, 395
Zilges, A. 345